Marine Biodiversity, Climatic Variability and Global Change

D0222730

Biodiversity loss in terrestrial environments associated with human activities has been appreciated as a major issue for some years now. What is less well documented is the effect of such activities, including climate change, on marine biodiversity. This pioneering book is the first to address this important but neglected topic, which is likely to be the key challenge for marine scientists in the near future.

Using a multidisciplinary and a holistic approach, the book reveals how climatic variability controls biodiversity at timescales ranging from synoptic meteorological events to millions of years and at spatial scales ranging from local sites to the whole ocean. It shows how global change, including anthropogenic climate change, ocean acidification and more direct human influences such as exploitation, pollution and eutrophication may alter biodiversity, ecosystem functioning and regulating and provisioning services. The author proposes a theory termed the 'macroecological theory on the arrangement of life', which explains how biodiversity is organized and how it responds to climatic variability and anthropogenic climate change.

The book concludes with recommendations for further research and theoretical development to identify oceanic areas in need of observation and gaps in current scientific knowledge. Many references and comparisons with the terrestrial realm are included in all chapters to better understand the universality of the relationships between biodiversity, climate and the environment. The book will serve as a textbook for all students and researchers of marine science and environmental change, but will also be accessible to the more general reader.

Grégory Beaugrand is a research scientist at the National Centre for Scientific Research (CNRS), currently based at the Laboratory of Oceanography and Geosciences in Wimereux, France. He is a research associate at the Sir Alister Hardy Foundation for Ocean Science (Plymouth, UK) and a former member of the French National Committee for Scientific Research (Earth System Science).

Earthscan Oceans

Marine Biodiversity, Climatic Variability and Global Change
Grégory Beaugrand

Governing Marine Protected Areas
Resilience through Diversity
Peter J.S. Jones

Marine Policy
An Introduction to Governance and International Law of the Oceans
Mark Zacharias

The Great Barrier Reef
An Environmental History
Ben Daley

Marine Biodiversity Conservation
A Practical Approach
Keith Hiscock

For further details, please visit the series page on the Routledge website:
www.routledge.com/books/series/ECOCE

'This is no ordinary textbook. The author presents clear, historically informed material on a huge range of relevant topics in climate science, physical and biological oceanography, ecology (terrestrial as well as marine) and of course biodiversity. There are excellent introductions to existing classifications of observed spatial and temporal variation in biodiversity and to the competing hypotheses to account for them. We urgently need unifying theories of marine biodiversity and he sets out a candidate theory together with a call for the global monitoring needed to test such theories and to track and predict changes. I warmly recommend this book not only as the first major volume on marine biodiversity, but as an outstanding introduction to marine ecology and climate change. It is well written and provocative, but in the best sense of clarifying issues and presenting differing views.'

—*Keith Brander, Johan Hjort Professor Emeritus, Danish Institute of Aquatic Resources, Technical University of Denmark.*

'With *Marine Biodiversity, Climatic Variability and Global Change*, Grégory Beaugrand joins the select club of scholars who have written a major textbook, i.e. a work that develops an original idea in great depth, based on a broad knowledge of the primary literature. Several textbooks have decisively influenced the development of scientific disciplines. Examples are, among others, *Perspectives in Ecological Theory* (Margalef), *Dynamics of Marine Ecosystems* (Mann & Lazier), and *Ecological Geography of the Sea* (Longhurst). It is likely that, in years and decades to come, graduate students and researchers will refer to Beaugrand's textbook as the seminal work in the field of marine biodiversity, as they refer to our own *Ecologie Numérique* and *Numerical Ecology* when they ground their studies in numerical ecology.'

—*Louis Legendre, Sorbonne Universités UPMC Paris 6, Villefranche Oceanography Laboratory, France.*

'One of the most important, appealing, and enduring "big ideas" in oceanography has been that global patterns of oceanic biology productivity and composition are driven by and predictable from patterns of ocean climate and seasonality. Data and theory on this topic have produced hundreds of good journal papers and book chapters, and several remarkable books.

Grégory Beaugrand extends the tradition with his new book *Marine Biodiversity, Climatic Variability and Global Change*. The book begins with an overview comparison of biodiversity between differing taxonomic levels and between terrestrial, marine pelagic, and marine benthic environments. It then reviews (Part 1) "natural" spatial and temporal patterns of climate and biodiversity, and how these have changed over geologic time scales. This is followed (Part 2) by an important and timely forecast of how (very rapid) anthropogenic alterations of climate and ocean chemistry are likely to alter marine biogeographic patterns, and (Part 3) by a potentially unifying theory, and needs for long-term monitoring of the physical and biological ocean.

The book is wide-ranging, and will earn its place in institutional and personal libraries in several ways. For example, Part 1 is a good cross-disciplinary introduction to topics such as the global heat engine, modes of physical climate variability, biogeography, and biodiversity theory. Part 2 provides more detailed review and synthesis of how various marine populations and communities have responded to recent climate variability (and to other modes of anthropogenic environmental change), and how changing biodiversity might affect ecosystem resilience and human well-being. Part 3 (the shortest but my favourite) makes the case for a new theoretical paradigm based on the ecological niche, and predicts how climate change will alter projections of species niches onto the earth's surface and within the annual seasonal cycle.'

—*Dave Mackas, Senior Researcher, Institute of Ocean Sciences, Canada.*

'This is a very special book on a number of fronts that summarises for the reader a vast array of information on marine biodiversity and the world we live in, reinforced by a large bibliography. Grégory Beaugrand takes us on a journey through time and space documenting in a systematic way the regulation, laws and theories behind the patterns of marine biodiversity seen today and in the geological past. He rounds off the book by proposing a new macroecological theory to explain patterns of biodiversity and their interaction with climate and environmental change.

Beautifully illustrated with drawings and graphs of an exceptional quality, and liberally with superb coloured maps and plates, the book has an immediate appeal. Both the text and illustrations are well balanced reflecting the care and thought that has gone into the composition as almost all are drawn, redrawn or simplified by the author. Perhaps its greatest recommendation is the book's readability. Even complex mathematical and statistical relationships are presented in a way that is understandable and accessible to a wide readership while at the same time giving the necessary foundation for an expert in the topic. If anyone is in any doubt about the huge impact humankind is having on the oceans and world at present they should read this book. Referencing Beaugrand's first sentence – Aristotle would have approved – as this book is truly interdisciplinary, the work of a talented polymath.'

—*Philip (Chris) Reid, Professor of Oceanography, Marine Institute,*
University of Plymouth, UK; Senior Research Fellow, Sir Alister Hardy
Foundation for Ocean Science, UK; and Honorary Fellow,
Marine Biological Association of the UK.

Marine Biodiversity, Climatic Variability and Global Change

Grégory Beaugrand

First published 2015
by Routledge
2 Park Square, Milton Park, Abingdon, Oxon OX14 4RN

and by Routledge
711 Third Avenue, New York, NY 10017

Routledge is an imprint of the Taylor & Francis Group, an informa business

British Library Cataloguing-in-Publication Data
A catalogue record for this book is available from the British Library

Library of Congress Cataloging-in-Publication Data
  Beaugrand, Grégory.
  Marine biodiversity, climatic variability and global change/Grégory
  Beaugrand.
  pages cm. – (Earthscan oceans)
  Includes bibliographical references and index.
  1. Marine biodiversity. 2. Marine ecology. 3. Climatic changes –
  Environmental aspects. 4. Nature – Effect of human beings on. I. Title.
  QH91.8.B6B427 2014
  577.6 – dc23
  2014019873

ISBN: 978-1-84407-678-9 (hbk)
ISBN: 978-0-415-51703-4 (pbk)
ISBN: 978-0-203-12748-3 (ebk)

Typeset in Goudy and Gill Sans
by Florence Production Ltd, Stoodleigh, Devon, UK

Printed and bound in Great Britain by
TJ International Ltd, Padstow, Cornwall

This book is dedicated to my family, and especially my grandparents Andrée and Alfred Gatoux, Christiane and Georges Beaugrand, my parents Christian and Mireille Beaugrand, my sister Karine Leroy, Mathéo and Timéa Leroy, Dominique Beaugrand and Stéphane Leroy. I would like to thank Elisabeth Mauger, Jean-Luc and Micheline Mascot, and Gérard and Valérie Mascot.

To Geneviève and my children Caroline, Sarah and Maxime.

Contents

Acknowledgements   xii

1 **Introduction**   1

   1.1 Biodiversity from the land to the ocean 3
   1.2 Classification and census of marine biodiversity 3
   1.3 Organisation of the book 14

PART I
Natural environmental variability and marine
biodiversity   17

2 **Large-scale hydro-climatic variability**   19

   2.1 The composition and structure of the atmosphere 19
   2.2 The earth radiation budget 21
   2.3 The thermal engine 23
   2.4 Main climatic regions 31
   2.5 The oceanic hydrosphere 33
   2.6 Hydro-climatic variability 42

3 **Large-scale biogeographic patterns**   61

   3.1 Biogeography: from the terrestrial to the marine realm 61
   3.2 Primary compartments of the marine ecosphere 66
   3.3 Ecogeographic patterns 79

4 **Large-scale biodiversity patterns**   84

   4.1 The search for a primary cause 84
   4.2 Neutral and null models or theories 85
   4.3 The area hypothesis 87
   4.4 History 89
   4.5 Hypotheses based on fundamental processes 90
   4.6 The climatic influence 98
   4.7 Temperature 107

4.8 Environmental hypotheses 109
4.9 Evolutionary rate 113
4.10 Biotic interactions 113

5 Marine biodiversity through time   120

5.1 Palaeoclimatic changes 120
5.2 Natural causes of extinction 142
5.3 Natural contemporaneous changes 150
5.4 Is climate the primary factor? 173

6 Temperature and marine biodiversity   174

6.1 Temperature from the origin of the universe to early life 174
6.2 Basics and first principles 175
6.3 Effects of temperature at the physiological level 181
6.4 Influence of temperature at the species level 198
6.5 Thermal influence at the community level 205

PART 2
Marine biodiversity changes in the Anthropocene   **207**

7 Biodiversity and anthropogenic climate change   209

7.1 Human alteration of the greenhouse effect and the radiative
        budget of the planet 212
7.2 Increase in global air and sea surface temperature 215
7.3 Species responses to anthropogenic climate change 219
7.4 Community/ecosystem response to climate change 244
7.5 Anthropogenic climate change and natural hydro-climatic variability 257

8 Marine biodiversity and ocean acidification   259

8.1 Introduction 259
8.2 Anthropogenic acidification 261
8.3 Effects of acidification on biodiversity 264
8.4 Limitations of past studies on ocean acidification 270
8.5 Conclusions 282

9 Biodiversity and direct anthropogenic effects   283

9.1 Exploitation of marine biodiversity 284
9.2 Pollution 302
9.3 Nutrient enrichment and eutrophication 317
9.4 Oxygen depletion 323
9.5 Introduction and invasion of exotic species 329
9.6 UV-B radiation 333
9.7 Tourism 337

9.8 Extinction 338
9.9 Interactive effects 340

10 Marine biodiversity, ecosystem functioning, services and
    human well-being   346

    10.1 Biodiversity and ecosystem functioning 346
    10.2 Biodiversity changes and ecosystem goods and services 366
    10.3 Potential effects of changes in marine biodiversity for global biogeochemistry 384
    10.4 Potential feedbacks 390

PART 3
Theorising and scenarising biodiversity   393

11 Theorising and scenarising biodiversity   395

    11.1 Introduction 395
    11.2 The concept of the ecological niche 396
    11.3 Rationale of the METAL theory 405
    11.4 The METAL theory 408
    11.5 Strength and assumptions of the METAL theory 419
    11.6 Limits to predictions in the context of global change 421
    11.7 Scenarising biodiversity 428

12 Conclusions   436

    12.1 A macroscopic approach 436
    12.2 Global monitoring 436
    12.3 Towards a unifying ecological theory 437

    References   439
    Index   466

The colour plate section can be found between pages 244 and 245

Glossary – terms highlighted in bold in the text are available online at:
http://sm-wimereux.univ-lille1.fr/beaugrand/.

Acknowledgements

I wish to express my grateful thanks to Richard Kirby, Stephane Gasparini, Christophe Luczak, Sylvie Malardel, Philip C. Reid, Thierry Comtet, Martin Edwards, Ulf Riebesell, Justin Ries, Will Howard, Mark Vermeij, Anthony Barnosky, Gudrun Marteinsdottir, Ken Whelan, Ruth Callaway, Stan Proboszcz, Helen Fox, Rebecca Lewison, Lucie Courcot, Lis Lindal Jørgensen, Gloria Manney, John Sibbick, Michael Benton, Jordi Bascompte, Les Watling, John C. Briggs, Mark Spalding, Chris Yesson and Vincent Bouchet for providing figures and materials. I also would like to thank my former Ph.D. and Master's students Pierre Helaouët, Isabelle Rombouts, Sylvain Lenoir, Eric Goberville, Aurélie Chaalali, Gabriel Reygondeau and Virginie Raybaud for the interesting discussions we had together. I also thank very much the Conseil Régional Nord-Pas-de-Calais for funding the colour plates used in this book as part of the project BIODIMAR.

I dedicate this book to my former Ph.D. director Dr Frédéric Ibañez who was a man of great talent, heart and insight.

Dr Frédéric Ibañez (1944–2013) in his office in Villefranche-sur-Mer (Photo: Isabelle Palazzoli)

Chapter 1

Introduction

Since Aristotle, science (and biology in particular) has become fractionated into many disciplines, which have been further subdivided into specialisms as formalisms and concepts have diverged progressively. As a result, communication between scientists has become more and more difficult. Although science specialisation reflects our increasing scientific knowledge, fractionation due to specialisation has become a serious issue now that our planet is under global pressure. Indeed, the direct (e.g. exploitation, destruction) and the indirect effects (e.g. anthropogenic climate change) of human activities on our planet, and their global effects on biodiversity and ecosystems, require the integration of many different branches of science.

Many scientists have provided evidence for the strength of having a multidisciplinary and holistic view to resolve global problems. One of the best examples is perhaps given by the discovery of the theory of continental drift by the German meteorologist Alfred Lothar Wegener (1). His theory resulted from a synthesis of many different scientific disciplines (e.g. geology, biogeography, palaeoclimatology). The scientist stressed in the preface of his book that it was essential to develop a global view of the planet (1), so as to examine what Alexander von Humboldt called the Unity of Nature. Indeed, the earth should be viewed as a system composed of functional units that exchange energy and matter by physical, chemical and biological processes. Functional units are intricately coupled.

Eduard Suess introduced the word biosphere and Wladimir Ivanovich Vernadsky developed the concept. The biosphere interacts with the geosphere (all other non-living functional units), creating a unique planetary ecosphere, and subsystems of significant importance for mankind such as water quality, soil fertility and fisheries. Lovelock and Margulis (2) proposed that organisms interact with their abiotic environment on earth, creating a self-regulating, complex system that contributes to maintain suitable environmental conditions for life. Even if the theory remains controversial (3, 4), many authors have stressed the need to develop an integrative science devoted to the study of the response of the biosphere to global change. In this book, I will try to adopt such a global approach and integrate as many scientific fields as necessary to better understand how biodiversity is organised in space and time and how it is influenced by climatic and environmental variability, as well as global change.

Our planet has entered a new era, the Anthropocene (5), where human activities are interfering with the natural functioning of the planetary ecosphere. Interference occurs in the atmosphere through an increase in greenhouse gas concentrations such as carbon dioxide and nitrous oxide, in the hydrosphere (oceans, seas, estuaries, lakes, ponds, rivers) through pollution, eutrophication and acidification, and in the biosphere through **habitat** fragmentation and destruction, species introduction and over-exploitation. As a result, biodiversity has been altered at an unprecedented level (6). Biodiversity, a term coined by Walter G. Rosen in 1985 and popularised by Edward O. Wilson in 1988, has been defined in the Convention on Biological Diversity in Rio de Janeiro in 1992 as 'the variability among living organisms from all sources including, inter alia, terrestrial, marine and other aquatic ecosystems and the ecological complexes of which they are part; this includes diversity within

Table 1.1 Phylum composition of the marine benthic and pelagic realms in comparison to the terrestrial domain.

Phylum	Marine domain		Terrestrial domain (including freshwater)
	Benthic domain	Pelagic realm	
Acanthocephala	P	P	P
Annelida	1	1	1
Arthropoda	1	1	1
Brachiopoda	1	0	0
Bryozoa	1	0	1
Chaetognatha	1	1	0
Chordata	1	1	1
Cnidaria	1	1	1
Ctenophora	1	1	0
Cycliophora	P	0	0
Echinodermata	1	1	0
Echiura	1	0	0
Entoprocta/Kamptozoa	1	0	1
Gastrotricha	1	1	1
Gnathostomulida	1	0	0
Hemichordata	1	0	0
Kinorhyncha	1	0	0
Loricifera	1	0	0
Mesozoa/Dicyemida	P	P	0
Mollusca	1	1	1
Nematoda	1	0	1
Nematomorpha	1	0	1
Nemertea	1	1	1
Onychophora	0	0	1
Orthonectida	P	0	0
Phoronida	1	0	0
Placozoa	1	0	0
Platyhelminthes	1	1	1
Pogonophora	1	0	0
Porifera	1	0	1
Priapula	1	0	0
Rotifera	1	1	1
Sipuncula	1	0	0
Tardigrada	1	0	1
Totals	33	14	17
Endemic	13		1

Note: The presence is indicated by a 1, the absence by a 0, and parasitic phyla are indicated by a 'P'.
Source: Modified from Herring (12).

species, between species and of ecosystems'. Biodiversity is currently studied at four main levels of organisation: the genetic, molecular, organismal and ecological levels.

1.1 Biodiversity from the land to the ocean

About 1.8 million species have been described in both the terrestrial and the marine realms (7). Because of a lack of a central catalogue and the existence of a probable important fraction of synonymous names, the number of currently valid species is around 1.2 million (8). Assessments of total biodiversity range between 3 million and 100 million species (9), with many species undoubtedly waiting to be described or even to be discovered (10). For example, insects are among groups in which new species are reported every year (11). Mora and colleagues (8) estimated that the global number of eukaryotic species may be ~8.7 ± 1.3 million, of which 2.2 ± 0.2 million may be marine. Their results suggest that 86% of terrestrial species and 91% of the marine species remain unknown to science.

At the phylum level, marine metazoan biodiversity is by far greater than terrestrial biodiversity (Table 1.1). On the total number of metazoan phyla (a total of 34), 33 are present in the sea. Table 1.1 shows that huge differences exist between the pelagic and the benthic realms, however. All phyla are present in the benthos, whereas only 14 are detected in the pelagic realm. Only 17 occur in the terrestrial realm, with only one endemic.

At the species level, marine biodiversity does not seem high, despite the fact that the oceans cover ~71% of our planet. Estimates from May (7) suggest that only 15% of all inventoried animal and plant species alive are marine. Benton (13) provided evidence that in contrast to the marine realm, the diversification process is more rapid on land. In the oceans, examination of well-known marine fossils suggests that the diversification shows rapid radiation followed by a plateau, which suggest that equilibria may exist (Figure 1.1). In contrast, in the terrestrial realm, the study of fossil records reveals exponential diversification from the Silurian to the present.

Widdicombe and Somerfield (14) also stressed other important differences between the marine and the terrestrial realms. Marine biodiversity is ancient. Chemoautotrophic prokaryotes and photosynthetic cyanobacteria appeared 3 billion years ago and eukaryotic cells 2 billion years ago. Marine **biomass** per unit area is much smaller than terrestrial biomass. The terrestrial domain is mainly two-dimensional (the third dimension being between 0 and 100 m), whereas the marine domain is three-dimensional (mean depth of 3,500 m). Only a fraction of the marine domain can therefore perform photosynthesis. In addition, marine primary producers are mobile in contrast to their terrestrial counterparts.

1.2 Classification and census of marine biodiversity

1.2.1 Classification

Biodiversity has been subdivided many times, and research on life classification continues. Carl Linnaeus classified biodiversity into domains and kingdoms, with those divided into phyla, classes, orders and families up to the species level. Different classification schemes have been proposed. The first classification proposed by Carl Linnaeus divided life into two groups: Vegetabilia and Animalia (15). More recently, biodiversity was divided into five kingdoms (16): (1) Monera (prokaryotic organisms); (2) Protista (unicellular eukaryotic organisms such as dinoflagellates, coccolithophores and foraminifers); (3) Fungi; (4) Plantae

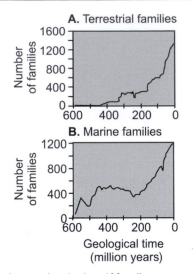

Figure 1.1 Pattern of life diversification for the last 600 million years.

Source: Redrawn from Benton (*13*).

(Metaphyta or multicellular eukaryotic plants); and (5) Animalia (Metazoa or multicellular eukaryotic animals). In 1977, a phylogenic classification scheme based on ribosomal RNA sequences divided biodiversity into three kingdoms: (1) Eubacteria; (2) Archaebacteria; and (3) Eukaryota (*17*). The most recent division of life is due to Cavalier-Smith (*18, 19*). The molecular biologist distinguished two empires (Prokaryota and Eukaryota) and six kingdoms: (1) Bacteria (Eubacteria and Archaebacteria), the only member of Prokaryota; (2) Protozoa; (3) Animalia; (4) Fungi; (5) Plantae; and (6) Chromista (e.g. diatoms, haptophytes and dinoflagellates). The kingdom Bacteria is subject to the International Code of Bacteriological Nomenclature, the two 'zoological' kingdoms, Protozoa and Animalia to the International Code of Zoological Nomenclature, and the three 'botanical' kingdoms (Plantae, Fungi, Chromista) to the International Code of Botanical Nomenclature (*18*).

1.2.2 Census of biodiversity

Table 1.2 shows the catalogued and estimated biodiversity of the different kingdoms in both the terrestrial and the marine realms. At the species level, all kingdoms are more diverse in the terrestrial realm.

Because scientists have inventoried living organisms for 250 years, many species have been described many times and many different names have frequently been attributed to the same species. Accounting for the problem of synonyms, Mora and colleagues (*8*) estimated that there are ~1.2 million described species and that ~70% of these species are animals, and especially insects. They estimated that 193,756 marine species have been labelled. This estimate is smaller than others that ranged between 250,000 (*20*) and 274,000 (*21*). Mora and colleagues assessed that about 2 million species remain to be described in the marine environment.

Bouchet (*22*) synthesised the number of marine species known per taxonomic group and estimated ~230,000 described species (Table 1.3). Many new species are discovered each

Table 1.2 Described and estimated species richness of different kingdoms for terrestrial and marine systems.

Species	Land			Ocean		
	Catalogued	Estimated	± SE	Catalogued	Estimated	±SE
Prokaryotes						
Archea	502	455	160	1	1	0
Bacteria	10,358	9,680	3,470	652	1,320	436
Total	10,860	10,100	3,630	653	1,320	436
Eukaryotes						
Animalia	953,434	7,770,000	958,000	171,082	2,150,000	145,000
Chromista	13,033	27,500	30,500	4,859	7,400	9,640
Fungi	43,271	611,000	297,000	1,097	5,320	11,100
Plantae	215,644	298,000	8,200	8,600	16,600	9,130
Protozoa	8,118	36,400	6,690	8,118	36,400	6,690
Total	1,233,500	8,740,000	1,300,000	193,756	2,210,000	182,000
Grand total	1,244,360	8,750,000	1,300,000	194,409	2,210,000	182,000

Source: From Mora and colleagues (8).

year. A well-known example of this is the coelacanth (*Latimeria chalumnae*), which was discovered off the eastern coast of South Africa in 1938. Before then, coelacanths were thought to have become extinct around 80 million years ago. We now know that a small population lives off the Comoros Islands between Mozambique on the African continent and Madagascar, with other populations in North Sulawesi (*Latimeria menadoensis*) and in the Saint Lucia Marine Protected Area (South Africa).

Collecting samples originating from some seamounts off Tasmania, Koslow and colleagues (23) found a high marine invertebrate biodiversity (262 species). Of the invertebrate species they inventoried, between 24 and 43% were new to science. Since 2000, as part the programme **Census of Marine Life**, researchers have discovered more than 5,300 likely new species, of which at least 110 have been subsequently confirmed. The deep sea has been poorly sampled, and it is likely that many new species will be discovered in the decades to come (24).

Bouchet (22) estimated that 1,635 marine species were discovered in 2002–2003 (Figure 1.2).

An estimated 439 Crustacea and 354 Mollusca were discovered. Note that a new species of Cetacea was also described. Bouchet stressed that it is, however, likely that between 10 and 20% of these new species have already been described in the past, so that between 1,300 and 1,500 species were ultimately discovered.

I now present briefly the biodiversity of some key marine groups.

1.2.3 Marine viruses

Viruses are not part of the previous biological classifications because of their pseudo-living nature. They are not considered as actual living entities. Nevertheless, viruses are extremely abundant in marine systems. Observations by transmission electron microscopy suggest that

Table 1.3 Described number of marine species per taxon.

Taxon	Estimates by Groombridge and Jenkins (2000)	Estimates by Bouchet (2006)
Bacteria	4,800	4,800
Cyanophyta		1,000
Chlorophyta	7,000	2,500
Phaeophyta	1,500	1,600
Rhodophyta	4,000	6,200
Other Protoctista	23,000	
Bacillariophyta		5,000
Euglenophyta		250
Chrysophyceae		500
Sporozoa		?
Dinomastigota		4,000
Ciliophora		?
Radiolaria		550
Foraminifera		10,000
Porifera	10,000	5,500
Cnidaria	10,000	9,795
Ctenophora	90	166
Platyhelminthes	15,000	15,000
Nemertina	750	1,180–1,230
Gnathostomulida	80	97
Rhombozoa	65	82
Orthonectida	20	24
Gastrotricha	400	390–400
Rotifera	50	50
Kinorhyncha	100	130
Loricifera	10	18
Acanthocephala	600	600
Cycliophora		1
Entoprocta	4,000–5,000	5,700
Phoronida	16	10
Brachiopoda	350	550
Mollusca	75,000	52,525
Priapulida	8	8
Sipuncula	150	144
Echiura	140	176
Annelida	12,000	12,000
Tardigrada	'few'	212
Chelicerata	1,000	2,267
Crustacea	38,000	44,950
Pogonophora	120	148
Echinodermata	7,000	7,000
Chaetognatha	70	121
Hemichordata	100	106
Urochordata	2,000	4,900
Cephalochordata	23	32
Pisces	14,673	16,475
Mammalia	110	110
Fungi	500	500
Total	242,135	229,602

Source: From Bouchet (22).

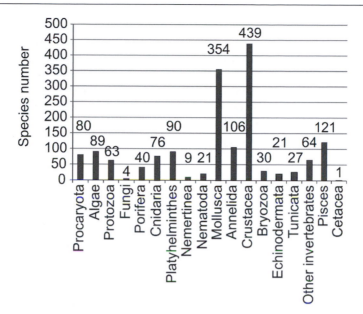

Figure 1.2 Mean annual number of new marine species discovered per taxonomic group in 2002–2003.

Source: Redrawn from Bouchet (22).

there are ~10^7 viruses ml^{-1}. Generally, virus abundance diminishes with depth and distance from the shore. Viruses are abundant when primary productivity, the number of bacteria and chlorophyll concentration are high. Virus concentration is also elevated in marine sediments when 10^8–10^9 viruses cm^{-3} are generally enumerated. Phages are the most abundant entities on our planet, with a vast reservoir estimated at 4×10^{30} in the oceans (25).

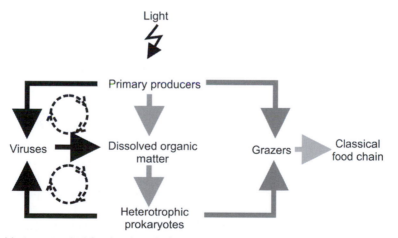

Figure 1.3 Marine microbial food web with the viral shunt in black and grazing in grey. The viral shunt produces dissolved organic matter that serves to other heterotrophic prokaryotes. In contrast, grazers fuel the carbon towards the more classical food web.

Source: Redrawn from Breitbart (26).

Assuming that each virus contains ~0.2 fg (one femto g = 10^{-15} g) of carbon and is ~100 nm long, Suttle (25) estimated ~200 million tonnes of carbon to be the biomass of marine viruses. This estimate corresponds to the carbon contained in ~75 million blue whales, assuming 10% of carbon by weight.

Marine viruses increase the efficiency of the biological carbon pump and play a key role in the trophodynamics of marine ecosystems (Figure 1.3). The lysis of bacteria by phages diverts carbon from the classical food web, a process called viral shunt. Dissolved organic matter, estimated at ~25% of the carbon produced by photosynthesis, can subsequently be used by prokaryotes.

Metagenomic analyses have revealed that environmental viral communities are diverse, with an estimated 5,000 viral genotypes for 200 L of seawater. If all biomes and provinces had endemic viruses, global viral biodiversity would be equal to ~100 million genotypes.

1.2.4 Marine prokaryotes

The number of prokaryotes may reach 10,100 species, of which ~1,320 are marine (Table 1.2). Archaea are the first prokaryotes, and many live in extreme environments; they are extremophiles. Some of them can be extreme halophiles (from the Greek *halo* for salt) or hyperthermophiles (live at temperatures ranging from 80 to 122°C). Others are, however, mesophiles, living in more moderate environments (e.g. Methanogens). As Archaea, bacteria are also prokaryotic cells lacking a membrane-enclosed nucleus. They are well known and are represented in the marine domain by cyanobacteria, the only prokaryotes to do oxygen-generating photosynthesis. Marine picoplankton, which are between 0.2 and 2 μm, include both **autotrophic** and **heterotrophic** bacteria and cyanobacteria. Autotrophic picoplankton is represented by the two genera *Prochlorococcus* and *Synechococcus* (27). These cyanobacteria contribute significantly to global primary productivity.

Synechococcus are ubiquitous, occurring from polar to tropical waters, and are generally more abundant in nutrient-rich surface waters. In contrast, *Prochlorococcus* mainly occur between 40°S and 40°N, and are generally absent from brackish or well-mixed waters. Furthermore, *Prochlorococcus* is also generally present in higher concentration deeper in the water column than *Synechococcus* (Figure 1.4).

The difference in depth between *Prochlorococcus* and *Synechococcus* is probably related to their distinct photosynthetic light-harvesting antennas. Although the light-harvesting complex of *Prochlorococcus* uses chlorophyll a_2/b_2, the antennas of *Synechococcus* contains phycobilisomes (28). This biochemical difference in the composition of **photosystems** makes *Prochlorococcus* particularly efficient in absorbing blue wavelengths of light, which probably explains its vertical distribution.

Primitive cyanobacteria invented oxygenic photosynthesis and modified atmospheric chemistry. Free molecular oxygen, a by-product of oxygenic photosynthesis started to accumulate 2.4 billion years ago (29). Prior to this period, **anoxygenic photosynthesis**, performed by other phototrophic cyanobacteria, predominated. The oxygen-producing apparatus of cyanobacteria was subsequently incorporated in heterotrophic eukaryotes that fed on cyanobacteria and retained their genome as a **plastid**. Currently, cyanobacteria are thought to contribute to half of the marine primary production, the remaining being attributed to phytoplankton.

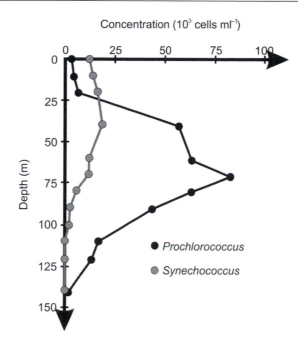

Figure 1.4 Vertical distribution of *Synechococcus* (grey circle) and *Prochlorococcus* (black circle) in the subtropical Atlantic (32°08′N, 70°02′W) during the boreal summer (10 June 1996).

Source: Redrawn from Ting and colleagues (*28*).

1.2.5 Marine eukaryotes

1.2.5.1 Plankton

Plankton form a taxonomically diverse group. The size of this group ranges from 10 nm (e.g. marine viruses) to ~2 m for the largest metazooplanktonic species (Figure 1.5). For example, the Nomura's jellyfish *Nemopilema nomurai* can reach 2 m and weigh up to 200 kg (*30*).

The picoplankton is also composed of marine eukaryotes named picoeukaryotes. These species are mainly abundant in oligotrophic regions and also contribute to global primary production (*32*). **Chromalveolata** (kingdom Chromista) such as pelagophytes, bolidophytes, chlorophytes and prasinophytes have species that belong to picoeukaryotes.

Nanoplankton are made of autotrophic (e.g. nanodiatoms, some dinoflagellates and prymnesiophycea) and heterotrophic protists between 2 and 20 μm. They are important as a source of food for many zooplankton species, including copepods (*33*). Prymnesiophytes are probably the most common nanoplanktonic species, being detected from tropical to polar systems. Some genera such as *Phaeocystis* or the coccolithophore *Emiliania* make significant contributions to the global sulphur and carbon cycles. Heterotrophic nanoplankton, mainly composed of flagellated protozoa, are major grazers of bacteria (*34*).

Phytoplankton creates ~50% of the oxygen produced annually by photosynthesis at a global scale. Plankton play a key role in the regulation of the earth's climate through its influence on the global carbon cycle, a process termed the biological carbon pump. Plankton are at the basis of the ocean's food web. The totality of the ocean's primary production is

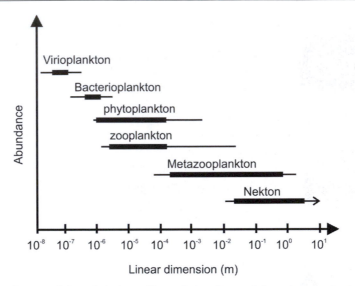

Figure 1.5 Abundance and size of plankton. Size and abundance of the nekton is also indicated for comparison.

Source: Adapted from Sieburth and colleagues (*31*).

estimated at ~48.5 Pg (1 Pg = 10^{15} g) of carbon each year (*35*). The carbon produced by phytoplankton is then transferred to the zooplankton, which are at the basis of a global production of 240 million tonnes of fish. Of this, ~80 million tonnes of fish is harvested annually by humans.

Sournia and colleagues (*36*) estimated that between 3,500 and 4,500 phytoplankton species are described in the sea, which can be compared to the 250,000 plants catalogued in the terrestrial realm. A total of 280 species of coccolithophores has been described. The phylum Chaetognatha, exclusively marine, contains 100 species, 20 of which are benthic coastal species (*37*). The author further estimated that 250 pteropod molluscs, 400 pelagic cephalopods, 2,200 marine pelagic copepods, 400 pelagic mysids (mostly neritic) and 86 euphausiids (mainly oceanic) have been described so far. Pierrot-Bults (*37*) stressed that the number of zooplankton species was by several orders of magnitude less than the number of insects.

About 7,000 species of holozooplankton distributed in 15 phyla have been described (*38*). Despite its low biodiversity, this taxonomic group plays an important role in the global carbon cycle and in the trophodynamics of marine ecosystems. Copepoda are the most diverse taxonomic group, and ~2,200 marine species have been catalogued (Figure 1.6).

One net sample from oceanic waters may catch ~220 copepod species. Although zooplankton biomass maximum is often found in the first two 200 m of the water column, biodiversity is greatest at 800–900 m (*39*).

1.2.5.2 Marine fish and other vertebrates

Based on Ocean Biogeographic Information System (OBIS) data, Mora and colleagues (*40*) reported a global number of 15,716 marine fish species publicly described. Based on species

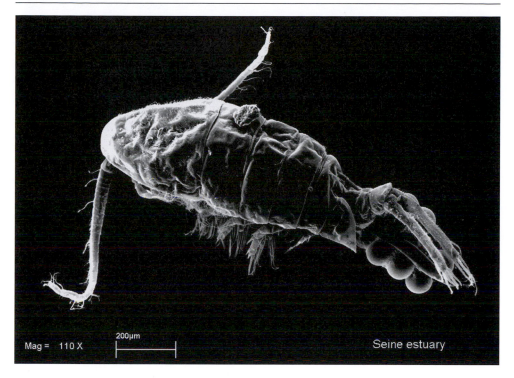

Figure 1.6 Scanning electron microscopy of the copepod *Eurytemora affinis* collected in the Seine Estuary (France).

Source: Photo courtesy of Dr Courcot, Laboratoire Océanologie et Géosciences.

accumulation curves, they estimated that the global census should be ~19,800 species and that ~4,084 marine fish species remain to be described. Other authors considered that 850 cartilagineous fishes and 11,500 bony fishes have been inventoried so far (*41*). The same authors catalogued 58 marine reptiles and 114 marine mammals.

1.2.5.3 Benthos

In contrast to shallow waters where photosynthesis remains active at the bottom and in some specific regions where chemosynthesis is possible, benthic organisms depend on organic compounds that sink from the surface to the sediments. About 98,100 benthic species have been inventoried, and some authors estimated that there may be between 1 million and 1 billion (Table 1.4). If these estimates are correct, this means that less than 1% of the present benthic biodiversity is known.

Benthic biodiversity is composed of infauna (i.e. animals living within the sediments) and epifauna (i.e. animals living on the sediments). Benthic biodiversity is also classified as a function of their size in four categories: (1) microbenthos; (2) meiobenthos; (3) macrobenthos; and (4) megabenthos.

Table 1.4 Described and estimated numbers of species in marine sedimentary habitats.

Taxonomic groups	Number of described species	Number of estimated species
Bacteria	500	10^9
Fungi	600	2,000
Protists	3,000	30,000
Meiofauna	7,000	10^8
Macrofauna	87,000	725,000
Total	98,100	10^6–10^9

Note: The number of bacteria remains poorly known.

Source: From Snelgrove (42).

1.2.5.3.1 MICROBENTHOS (<40 μM SIZE RANGE)

Biodiversity smaller than 40 μm is mainly composed of bacteria and protists (e.g. diatoms and amoeba) and remains poorly known. In the photic zone, microalgal community forms biofilms at the sediment-water interface that influence nutrients fluxes and stabilises sediments (43). The microfauna, mainly represented by benthic foraminifera and small ostracods, is sensitive to environmental disturbances and pollution, and has been used as a bioindicator (44).

1.2.5.3.2 MEIOBENTHOS (40–500 μM SIZE RANGE)

This category gathers together very diverse taxonomic groups such as nematodes, large foraminifers (Figure 1.7), ciliata and harpacticoid copepods (45).

Gnathostomulida (~80 species), Gastrotricha (160 marine and freshwater species) and Tardigrada (25 marine species) are examples of taxa that compose exclusively the **meiofauna**.

Benthic meiofauna are grazers of bacteria and microalgae, and influence primary production, nutrients turnover and other benthic processes (46). Some members only belong temporarily to this category, while others are permanent members. Some Holothuroidea (Echinodermata) of the order Apodida are permanent members of the meiofauna. Some molluscs such as Solengasters and some Prosobranchia are also members of this category. Species of the meiofauna have short generation times, from weeks to months. The ratio meiofaunal nematodes/copepods has been used as an indicator of pollution because nematodes are less sensitive than copepods (47).

1.2.5.3.3 MACROBENTHOS (0.5–20 MM SIZE RANGE)

The macrobenthos, visible to the naked eye, is composed of many taxonomic groups. For example, members of this category are Porifera, Cnideria, Annelida, Sipunculida, Echiurida, Crustacea, Pycnogonida, Mollusca, Bryozoa, Brachiopoda and Echinodermata. Macrobenthic biodiversity is sensitive to pollution, and many species are used as bioindicators (48, 49). Some species such as *Capitella capitata* occur in organically enriched areas where hypoxia is likely.

Figure 1.7 Scanning electron microscopy of the benthic foraminifer *Quinqueloculina carinatastriata*.
Source: Courtesy of Dr Bouchet, University of Lille.

1.2.5.3.4 MEGABENTHOS (>2 CM SIZE RANGE)

The megabenthos (>2 cm) also represents an important fraction of the benthos, and is recognised in bottom photographs and video images. In **abyssal** ecosystems, holothurians enable the quick removal of phytodetritus on the sediment surface. These deposit-feeders play an important role in the remineralisation of organic matter, even if community biomass is dominated by bacteria. Biodiversity estimates of the megabenthos are also difficult, especially in the deep-sea ocean. Remotely operated vehicles (ROVs) can detect new species, but their scientific description can only be undertaken when specimens are collected. Many species belonging to the megafauna continue to be discovered. For example, the Yeti crab *Kiwa hirsuta* was discovered in the hydrothermal vents located along the Pacific Antarctic Ridge (37°46S and 110°55W) at a depth of 2,228 m (Figure 1.8).

1.2.5.3.5 OTHER BENTHIC CLASSIFICATIONS

Benthic biodiversity is also classified into several groups according to their feeding mode (51). Suspension feeders are benthic organisms that feed by capturing particles or species

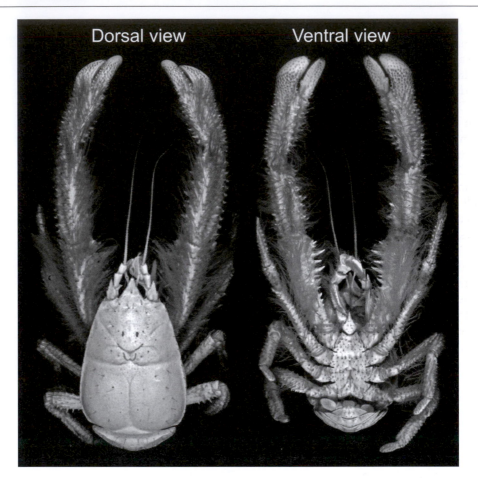

Figure 1.8 Picture of a male Yeti crab *Kiwa hirsuta*.

Source: Courtesy of the National Museum of Natural History, Paris (*50*).

present in the water. Corals feed on zooplankton, bivalves and polychaetes on phytoplankton and sponges and ectoprocts on bacteria. Deposit feeders such as sea cucumbers, polychaetes and gastropods ingest sediments and retain both organic matters and microbes. Suspension-feeders are more abundant in high-energy environments (e.g. sandy environment with large sediment grain and low organic matter) and deposit-feeders in depositional areas with large quantities of organic matter (*42*). Benthic organisms can also be separated into herbivores, carnivores and scavengers.

1.3 Organisation of the book

The first part of this book (Chapters 2–6) investigates the influence of natural hydro-climatic and environmental changes on biodiversity at all spatial and temporal scales and all organisational levels. Chapter 2 makes an overview of the atmosphere and the earth's radiation budget, describes the main climatic regions on earth, summarises global atmospheric and oceanic circulations, and proposes a short description of the oceanic hydrosphere. Hydro-climatic processes and mechanisms that can potentially affect marine biodiversity patterns are also reviewed. This chapter contains the information needed to better understand how hydro-climatic and environmental changes, as well as meteorological events, may affect marine biodiversity.

Chapter 3 investigates how the biosphere has been partitioned in both the terrestrial and the marine realms, and examines which abiotic factors control the locations of the primary ecological compartments on earth. Ecogeographic patterns observed in the marine environment and mechanisms causing these patterns are subsequently examined.

Chapter 4 investigates in great detail latitudinal biodiversity patterns and reviews the main factors, hypotheses or theories that have been proposed to explain them. Neutral and null models or theories, the area hypothesis, the effect of history, hypotheses based on fundamental processes, climate and environmental hypotheses, the effects of evolutionary rates and biotic interactions are all examined.

Chapter 5 investigates the relationships between marine biodiversity, climate and the environment at temporal scales ranging from million years to intraseasonal and synoptic scales. Palaeoclimatic changes are examined at tectonic, orbital, millennial and sub-millennial frequency bands. Global extinctions that punctuated the geological time period and their likely causes are reviewed. Natural contemporaneous changes are explored from the effects of meteorological events (e.g. cyclones, heatwaves) to seasonal, year-to-year and decadal hydro-climatic changes, including long-term and abrupt ecosystem shifts. The chapter concludes by showing why climate is a strong driver of temporal changes in marine biodiversity.

Chapter 6 shows how temperature affects marine biodiversity at the physiological, behavioural, biological and ecological levels. First, the role of this parameter prior to the establishment of the planetary ecosphere is explained. Second, some fundamental laws involving temperature are reviewed. Third, the effects of temperature at the physiological level are explored. Finally, the influence of temperature at the species and community levels is evaluated. This chapter therefore shows why temperature is so fundamental for biological and ecological processes.

The second part of the book (Chapters 7–10) explores the potential effects of global change (e.g. pollution, eutrophication, exploitation, aquaculture), including anthropogenic climate change and ocean acidification, on marine biodiversity. The impacts of marine biodiversity loss for ecosystem functioning and both provisioning and regulating services are also examined.

Chapter 7 reviews the potential influences of anthropogenic climate change on marine biodiversity. Human alteration of the greenhouse effect and its influence on the earth's radiative budget and the oceans are reviewed. The implications of current anthropogenic climate change are explored. Four types of species responses to climate change are detailed: (1) physiological adjustment; (2) adaptive evolution; (3) species niche tracking (pheno-logical, biogeographical shifts); and (4) species extinction. Ecosystem responses to anthropogenic climate change are also explored. The chapter ends by comparing anthro-pogenic climate change with natural sources of climatic variability.

Chapter 8 is about the effects of ocean acidification on marine biodiversity. The chapter explains why increasing CO_2 concentrations leads to ocean acidification. Then, the effects of ocean acidification on different taxonomic groups and ecosystems are explored and an emphasis is made on the complexity of the responses of species and ecosystems to ocean acidification, highlighting the potential drawbacks of past studies, and especially the lack of consideration of the synergistic effects of acidification and temperature.

Chapter 9 reviews the direct effects of human activities on marine biodiversity. The consequences of species exploitation for marine biodiversity are reviewed. The influences of all types of pollution are then investigated. Consequences of hypoxia due to eutrophication or climate change are also explored. Biodiversity erosion due to species invasion is studied and the alteration of the biodiversity by ultraviolet radiation and tourism are outlined. I also review marine species vulnerability to human activities and conclude the chapter by showing some examples of interactive effects among anthropogenic stressors.

Chapter 10 reviews the relationships between biodiversity, ecosystem functioning, ecosystem goods and services. In the first section, the chapter investigates the relationships between biodiversity and ecosystem functioning (stability, ecosystem productivity, functional diversity, trophic food webs) at all spatial and temporal scales, including the geological timescale, and examines the debate on the role of biodiversity for ecosystem functioning. In the second section, the relationships between biodiversity and ecosystem goods and services are reviewed. This section covers the economic quantification of biodiversity, the role of some important marine ecosystems, meta-analyses on the effects of biodiversity for goods and services, potentially harmful species and the great potential of marine biodiversity in medicine. The third section talks about the potential effects of changes in marine biodiversity for global biogeochemistry and potential feedbacks of changes in marine biodiversity for the earth's system.

The third and final part of the book (Chapter 11) is devoted to theorising and scenarising the effects of hydro-climatic and environmental changes on marine biodiversity. The MacroEcological Theory on the Arrangement of Life (METAL) is outlined. This unifying ecological theory explains how marine biodiversity is organised in space and time and how it reacts to climate and environmental changes. As the elementary macroscopic unit of this theory is based on the ecological niche, the chapter first reviews this concept in detail. In the next section, the fundamentals of the theory are presented, and the chapter shows that the theory explains the arrangement of marine life and most biological responses documented so far in the context of climate change: local changes in abundance including local extirpation, phenological and biogeographical shifts and both long-term gradual and abrupt community/ecosystem shifts. Strengths and weakness of the present version of the theory is reviewed and current or absolute limits to predictions in the context of global change explored. This chapter concludes by showing how we currently establish scenarios of changes in marine biodiversity and the limits to such projections.

I conclude this book (Chapter 12) by highlighting that we urgently need to adopt a macroscopic approach, based on joint global monitoring programmes to better understand the effects of global change on marine biodiversity. I think it is also high time to develop a unifying ecological theory to explain how marine biodiversity is organised and how it may respond to global change. Such a theory is not only needed to anticipate the responses of marine biodiversity to climatic variability and global change, but also to guide future research by defining research priorities, areas that need to be monitored and gaps in our current scientific knowledge.

All words in bold in the text are defined in a short glossary, available at: http://sm-wimereux.univ-lille1.fr/beaugrand/.

Part I

Natural environmental variability and marine biodiversity

Chapter 2

Large-scale hydro-climatic variability

In this chapter, I make an overview on climatic and hydrographic factors that are particularly relevant for the understanding of spatial and temporal patterns in marine biodiversity. I believe that climate is the most important driving force controlling marine biodiversity, and accordingly I will try to emphasise some climatic processes or parameters that are of primary importance to explain ecogeographical patterns in diversity. The objective of this chapter is also to provide a primer in applied climatology and physical oceanography to understand more deeply the potential links between hydro-climatic processes and marine biodiversity, which I will detail in the next chapters of this book. I also hope it will solve the problem of selection of hydro-climatic variables when relationships between climate, the marine environment and biodiversity are examined.

2.1 The composition and structure of the atmosphere

The atmosphere is an essential component of the **ecosphere**. Although it remains difficult to give an upper limit, the atmospheric envelope has a thickness of about 1,000 km, corresponding to the last molecules that can be retained by the earth's gravity. About 50% of the atmospheric mass is below 5,500 m. The planet has a thick atmosphere composed of 78% nitrogen, 21% oxygen and 1% other gases, predominantly argon. Among rare gases, two are of prime importance for climate: carbon dioxide, which reached 401.3 ppm (parts per million) in April 2014 in Mauna Loa; and water vapour, which varies from 0.1 ppm at the South Pole to 40,000 ppm in some equatorial regions (52). The atmosphere is composed of a number of layers that differ by their thermal and chemical property. All meteorological phenomena that influence marine biodiversity occur in the troposphere, which has a height ranging from 7 km at the poles to 17 km at the equator (Figure 2.1).

2.1.1 The troposphere

As its name suggests ('tropos' comes from the Greek and means 'change'), this layer is characterised by intense horizontal and vertical movements. The reduction in atmospheric pressure with altitude is remarkable. The atmospheric pressure is about 1,013 hPa at sea level, and decreases to 500 hPa at 5,500 m and 100 hPa at 16 km. The troposphere contains 90% of the atmospheric mass and nearly all the water vapour (53). This atmospheric layer contains a low quantity of water (~12,900 km³), representing only 0.001% of the total water available on the planet, or 0.04% of total freshwater. Both evaporation and transpiration allow water to reach the atmosphere, while both condensation and

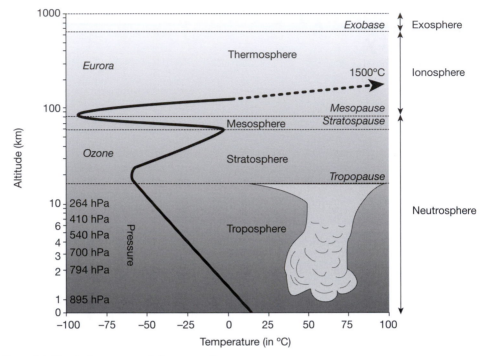

Figure 2.1 Vertical structure of the atmosphere.

precipitation enable water to return back to the earth's surface. The average residence time of water in the troposphere is rapid, about 10 days. This contrasts with the oceanic turnover, which is estimated to be ~2,500 years.

2.1.2 The stratosphere

The temperature continually decreases with altitude until the tropopause, which indicates the beginning of the stratosphere. In the stratosphere, temperature increases because of the absorption of ultraviolet radiation by the ozone layer. As its name suggests, this layer is stratified and therefore stable. A substance that reaches this layer can remain for a long time. After some major volcanic eruptions (e.g. the eruption of Mount Pinatubo on 15 June 1991), ash can be hurled into the stratosphere and remain for a few years, diminishing the amount of solar radiation that reaches the earth's surface. After the eruption of Mount Pinatubo, global tropospheric temperatures decreased by about 0.5°C. Plate 2.1 shows the ash layer in the lower part of the stratosphere well above the basis of the cumulonimbus clouds that indicate the tropopause.

2.1.3 Other atmospheric layers

Other layers have been defined, but they are less relevant for marine biodiversity. For example, both troposphere and stratosphere roughly correspond to the neutrosphere, the layer of the atmosphere where molecules are electrically neutral (Figure 2.1). At 50 km, temperatures close to −3°C mark the stratopause and the beginning of the next layer, called the mesosphere. In this layer, temperatures restart to diminish with altitude. This is the layer where temperatures are the coldest on earth. Meteors burn up in this part of the atmosphere, and temperatures of −100°C are common. This layer ends at the mesopause, above which the thermosphere begins. Temperature increases to reach values of about 1,500°C (Figure 2.1). The height of the thermosphere varies between 350 and 800 km as a function of solar activity. This layer is separated by the exobase from the exosphere composed mainly of hydrogen and helium. In this layer, these molecules can reach the velocity needed to escape the earth's gravity. The ionosphere corresponds to the part of the atmosphere where molecules are ionised by solar radiation. Auroras are normally observed in this layer. This layer encompasses the upper mesosphere, the thermosphere and parts of the exosphere at daytime. However, ionisation ceases at night-time in the mesosphere, which explains why auroras are typically observed in the thermosphere and the lower part of the exosphere.

2.2 The earth radiation budget

The primary engine of climate is related to incoming solar radiation. The total energy produced by the sun is estimated to be close to $L_0 = 3.9 \times 10^{26}$ W. To obtain the energy produced per square metre, we consider the surface of a sphere ($4\pi r^2$) of radius the distance r between the sun and the earth ($r = 1.5 \times 10^{11}$ m). Solar energy S_0 emitted in space is then calculated as follows:

$$S_0 = L_0 / 4\pi r^2 \tag{2.1}$$

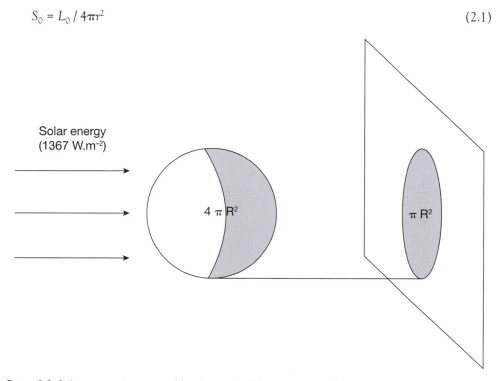

Solar energy
(1367 W.m⁻²)

$4\pi R^2$

πR^2

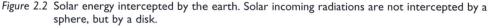

Figure 2.2 Solar energy intercepted by the earth. Solar incoming radiations are not intercepted by a sphere, but by a disk.

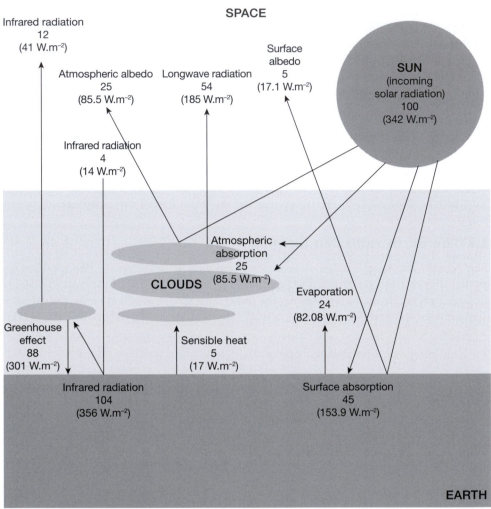

Figure 2.3 Simplified earth global energy budget. Energy fluxes are given in normalised units (total of 100 normalised units = 342 W·m^{-2}) and estimated values of the average incoming solar radiation.

Source: Redrawn from Kump and co-workers (54).

Therefore, S_0 equals 1,367 W·m^{-2}. To know which amount is intercepted by the earth, we divide S_0 by 4, which gives an average amount of 1,367/4 = 342 W·m^{-2} that arrive at the top of the atmosphere (Figure 2.2). The use of the **Wien's displacement law** enables us to calculate the wavelength of maximum emission of the sun:

$$\lambda = 2,897 / T \tag{2.2}$$

Here T is the temperature (in kelvin). As the temperature of the photosphere is $T = 5,780$ K, the wavelength of maximum emission from the sun is 0.5 μm (visible spectrum). The earth ($T = 288$ K) has a wavelength of maximum emission of 10 μm (infrared radiation).

One part of solar energy reaches the earth's surface, the other part being reflected (Figure 2.3). Therefore, a basic estimate of the solar energy that reaches the earth's surface in the form of electromagnetic radiation is:

$$E = \frac{S_0}{4} - \frac{S_0}{4}\alpha = \frac{S_0}{4}(1-\alpha) \tag{2.3}$$

The energy α reflected is called albedo. The average global albedo is close to 0.3. The global albedo can be divided into two components: (1) the atmospheric albedo α_a ($\alpha_a = 0.25$) due to clouds; and (2) the surface albedo α_s ($\alpha_s = 0.05$) due to the characteristics of the earth's surface.

A total of 25% of the globally averaged incoming solar radiation is absorbed by the atmosphere. The stratosphere (e.g. the ozone) absorbs 2% and the troposphere (e.g. aerosol and clouds) absorbs the remaining (~23%). This energy is therefore converted into heat. Of the 154 W·m⁻² (45%) that eventually arrives to the surface (30% of incoming solar radiation is reflected and 25% absorbed), about half (~82 W·m⁻²) is used in the process of evaporation (latent heat) and 5% (~17 W·m⁻²) is used as **sensible heat** (conduction and convection). An estimated 54% of the incoming shortwave radiation is returned to space by the atmosphere as long-wave radiation.

The earth's surface then emits an energy that is proportional to its temperature. This is equal to 356 W·m⁻², representing an amount of 104% of the incoming short-wave radiation. On this amount, 16% of the incoming solar radiation is returned back to space. The difference, equal to 88%, or 301 W·m⁻², represents the greenhouse effect. The incoming total short-wave radiation is therefore equal to outcoming long-wave radiation at the top of the atmosphere, according to the principle of planetary energy balance (Figure 2.3).

2.3 The thermal engine

Climate is defined by the World Meteorological Organization (WMO) as the statistical description of the mean and variability of some climatic parameters (temperature, precipitation and wind) at timescales ranging from months to thousands or millions of years (55). In simple terms, it constitutes the average weather. Climate (from the Ancient Greek *klima*, meaning inclination) is largely determined by the unequal distribution of energy around the planet. The regional imbalance of incoming and outcoming radiation triggers planetary circulation. This inequality is related to three factors.

The first factor is latitude, which governs the zenith angle of the sun at a particular point. Figure 2.4 shows that for the same amount of energy arriving at the top of the atmosphere (342 W·m⁻²), the energy per square metre that arrives at the earth's surface P (larger surface) is lower than the one striking the equator (lower surface). Because the atmosphere crossed by solar radiation is thicker towards high latitudes, the amount of solar radiation that touches the earth's surface decreases. While about 420 W·m⁻² arrives on average at the equator, less than half of this amount (175 W·m⁻²) arrives to the pole (56). Because the earth's axis tilts at 23.5°, this amount also varies seasonally.

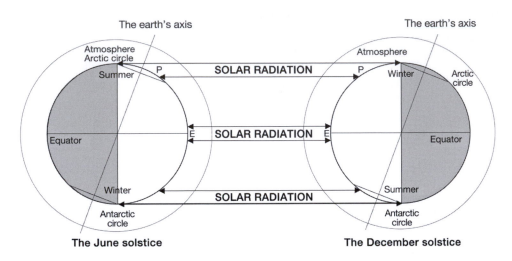

Figure 2.4 Unequal distribution of the incoming solar radiations at the June (aphelion) and December (perihelion) solstices. The amount of energy that touches the earth is smaller in high latitudes (see P) than in the equator (see E). At the December solstice, the southern hemisphere receives more energy than the northern hemisphere.

Source: Modified from Robinson and Henderson-Sellers (*61*).

The second factor is related to the transparency of the atmosphere, also called the atmospheric albedo. This depends on the amount, nature and type of clouds. For example, a stratus cloud (i.e. cloud close to ground with a horizontal layering and a uniform base) contains mainly water and reflects the incoming solar radiation. A cirrus cloud (i.e. a cloud that looks like hair-like filaments that forms at high altitude and is composed of ice crystals) allows the incoming solar radiation to reach the earth's surface but reinforces the greenhouse effect. The atmospheric albedo also depends on the atmospheric thickness, a function itself of the concentration of aerosols and water vapour.

The third factor is related to the state of the earth's surface (ice, desert, forest, water). This is called the surface albedo α_S. Some values of the surface albedo are indicated in Table 2.1.

Table 2.1 Examples of albedo values.

Surface type	Albedo (%)
Water	2–10
Sea-ice	30–45
Snow	50–90
Dry sand	25–30
Thin cloud	30–50
Thick cloud	60–90
Tundra	18–25
Boreal forest	10–20
Equatorial forest	5–10

Note: A low value means a small reflectivity and a high value indicates a greater reflectivity.

2.3.1 Average global atmospheric circulation

The unequal distribution of incoming solar energy is at the origin of the global circulation of the earth system (Figure 2.5). The figure shows an excess of energy in the tropics and a deficit over higher latitudes. An average equilibrium is reached at ~40°.

The tropics receive a high level of solar energy in comparison to higher latitudes. In a hemisphere, the unbalance is more pronounced in winter. Without any atmospheric or oceanic circulation, these zones would tend to be warm or cold indefinitely. The global atmospheric circulatory system of our planet is sometimes compared to the circulatory system of humans (57). As with the internal circulation of organisms, it enables the earth to maintain its **homeostasis**. It also allows matters (e.g. dissolved nutrients), water and chemicals to be disseminated all around the globe. The pumps, which can be a simple contractile vacuole in some freshwater protozoa or the more sophisticated heart of the mammals, are here represented by the tropical oceans, Meridional Overturning Circulation (MOC) and, at a longer timescale, the heat production in the earth's interior by radioactive decay and the movement of plate tectonics. The latter may be important on very long timescales. Kump and colleagues (54) indeed speculated that this is the cessation of the movement of plate tectonics that stops, in part, the life process on Mars.

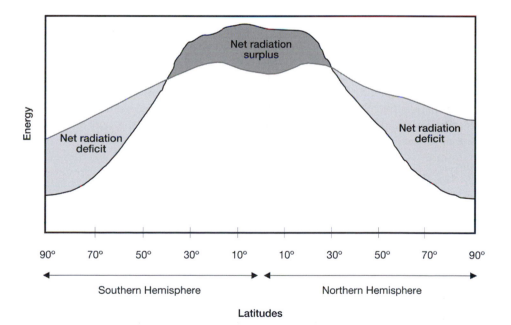

Figure 2.5 Latitudinal changes in the solar (line in black) and infrared (line in grey) radiation at the earth's surface. While there is an excess of energy in the tropics, higher latitudes have an insufficient amount of energy, which must be compensated by both oceanic and atmospheric circulations.

Source: Redrawn from Kump and co-workers (54).

2.3.1.1 Vertical motion at the equator

The excess of energy at the equator triggers vertical motion. At the equator, the high temperature at the surface triggers a change in buoyancy of the atmospheric gas at the surface (i.e. the upward force that maintains a molecule afloat in a fluid; Archimedes' Principle). This can be better understood by examining the Ideal Gas Law:

$$pV = nRT \tag{2.4}$$

Here p is the absolute pressure of the gas, V the volume of the gas, n the number of moles (i.e. a mole has 6.0221415×10^{23} atoms or molecules of a pure substance) and T the temperature (in K). The universal constant of idealised gas $R = 8.314\ \text{J·mol}^{-1}\text{·K}^{-1}$. The Ideal Gas Law can only be applied on idealised gases (i.e. gas composed of molecules with no attractive force), but it is a very convenient way to understand the phenomenon of buoyancy. The law can be applied to a mixture of idealised gas with n becoming the sum of different atoms of molecules in the system. From Equation 2.4, one can see the relationships between the volume, the temperature and the pressure. If the volume is held constant, an increase in temperature involves an increase in pressure. Equation 2.4 can be written as:

$$p = \rho R_{gas} T \tag{2.5}$$

Here ρ is the density of molecules. $\rho = m\,/\,V$ (m the mass in kg). R_{gas} is the constant of the gas. For dry air, $R_{gas} = R_a = 287\ \text{J·K}^{-1}\text{·kg}^{-1}$ (53). If the pressure remains constant, the density of molecules decreases when temperature increases. Therefore, the rise in temperature involves an increase in buoyancy and the gas rises in the atmosphere (Figure 2.6).

This process is at the origin of the convection, which occurs from the mesoscale to the megascale. For example, it allows small cumulus clouds to form in summer. It enables the air to go upwards at the meteorological equator. On a non-rotating planet, the redistribution of this energy would occur directly from the pole to the equator at the surface and in the opposite direction below the tropopause (Figure 2.7).

2.3.1.2 The Coriolis force

The rotation of the planet makes more complex the redistribution of the energy. This is in large part explained by the **Coriolis force**. The French scientist Gustave Gaspard Coriolis (1792–1843) described this effect mathematically. This force, originating from the earth's rotation, is the tendency of a moving object to veer to the right in the northern hemisphere and to the left in the southern hemisphere (Figure 2.8).

This force exerts a strong influence on the motion of large air masses, but not on small ones more controlled by the orography. In the northern hemisphere, the Coriolis force gives around high pressure systems (anticyclone) clockwise circulation and around low pressure systems (cyclone) an anticlockwise circulation. An opposite circulation is observed in the southern hemisphere. The Coriolis force cancels at the equator.

2.3.1.3 A three-cell model

Based on large-scale physical constraints and observations, a three-cell model was proposed to explain the average pattern in global atmospheric circulation (Figure 2.9).

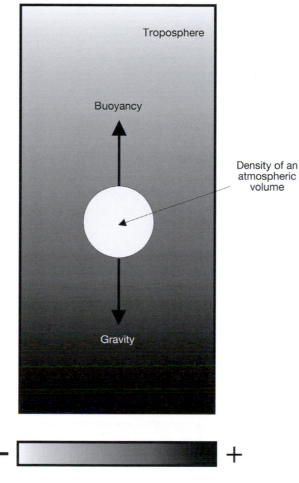

Figure 2.6 Forces influencing the buoyancy of molecules in the atmosphere.

At the equator, the excess of energy involves a change in buoyancy and the air rises. This process can be invoked to explain the ascending branch of the Hadley cell. However, the ascension of the air is also reinforced by the trade winds that converge from both southern and northern hemispheres, exerting an opposite force at the equator in near surface that forces the air to elevate. The phenomenon is called convergence. When the air arrives at the tropopause, it diverges to the north and to the south towards mid-latitudes. This cell is mainly driven by temperature. When the air arrives to about 30°N or 30°S, the air meets the flows of the Ferrel cell and a convergence occurs at the tropopause. The air descends, warms and contains less saturated water vapour. The air is therefore dry and there is not much precipitation in those regions. The phenomenon is at the origin of high-pressure cells at subtropical latitudes (see Plate 2.4).

The processes are more complex at the contact between the Ferrel and the Polar cell. In this region at surface, cold air goes equatorwards, whereas warmer air circulates

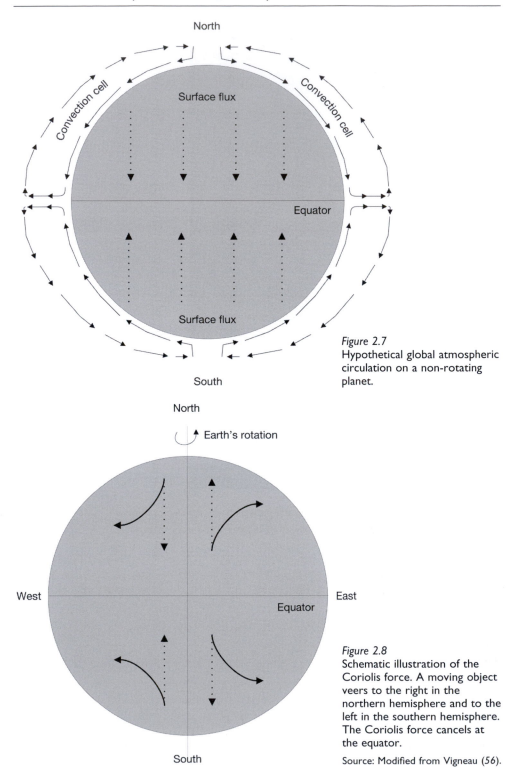

Figure 2.7
Hypothetical global atmospheric circulation on a non-rotating planet.

Figure 2.8
Schematic illustration of the Coriolis force. A moving object veers to the right in the northern hemisphere and to the left in the southern hemisphere. The Coriolis force cancels at the equator.

Source: Modified from Vigneau (56).

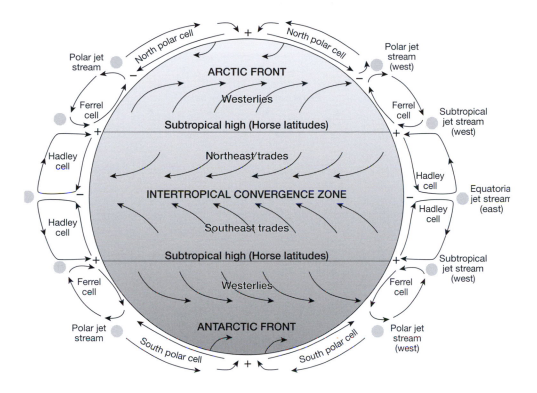

Figure 2.9 The three-cell model. Average global atmospheric circulation in the earth's atmosphere. The symbols + and − denote the processes of convergence and divergence, respectively.

Source: Modified from Vigneau (56).

polewards. Convergence occurs at the surface and triggers regions of low pressures. The Polar Front is located in this region. This structure, detected at the beginning of the twentieth century, is identified by a strong thermal contrast and has been associated with the birth and the motion of perturbations. The Polar Front, also called Temperate Discontinuity (56), is associated with the polar jet stream at mid and high altitude. This front is not permanent at an annual scale and meanders like waves, typically called Rossby or planetary waves. This front is also called circumpolar vortex. This alternation of the atmospheric masses, stronger in winter, is reflected by the alternation of low- and high-pressure systems that enable energy transfer between the pole and mid-latitudes. The location of the Polar Front has strong consequences for the distribution of the energy in the mid-latitudes (see Section 2.6.1).

2.3.2 Seasonal variability in atmospheric circulation

The model of atmospheric circulation is rarely observed and varies at all spatial and temporal scales. The three-division framework varies seasonally.

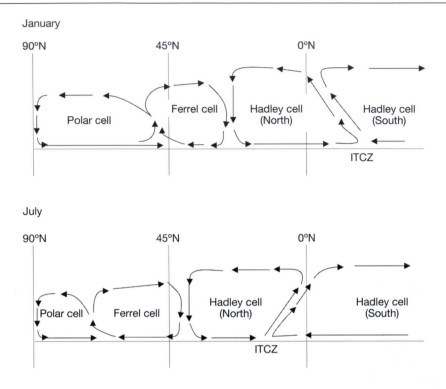

Figure 2.10 Seasonal variability in the location of the main circulation cells in the northern hemisphere. ITCZ: Intertropical Convergence Zone.

Source: Simplified from Vigneau (56).

2.3.2.1 Boreal winter

In January, the Polar cell reaches its maximum size in the northern hemisphere (Figure 2.10). In winter, the production of cold air is highest and the circumpolar vortex is reinforced. The Ferrel cell contracts and meridional exchanges of warm and cold airs are reinforced. The North Hadley cell retreats to the south and across the equator. The Intertropical Convergence Zone (ITCZ) and the associated monsoon are therefore located in the southern hemisphere. The position of ITCZ influences drastically precipitations. At the geographical equator, ITCZ is present most of the time while it varies more seasonally towards higher latitudes. Therefore, in the tropics, the season is more influenced by fluctuations in precipitations than temperatures, although precipitations also affect the thermal regime. I think that these seasonal movements in ITCZ have been too often forgotten in the explanation of the ecogeographical patterns in marine biodiversity, especially the generally observed reduction in diversity at the equator in comparison to mid-latitudes (58).

2.3.2.2 Boreal summer

In July, the production of cold air by the Polar cell is reduced in the northern hemisphere (Figure 2.10). As a result, the Polar cell contracts. The Ferrel cell expands and atmospheric meridional exchanges are less intense. The North Hadley cell moves northwards and the South Hadley cell appears in the northern hemisphere, triggering the monsoon.

2.4 Main climatic regions

Climate has been defined by the Intergovernmental Panel on Climate Change as 'the average weather', or more rigorously as 'the statistical description in terms of the mean and variability of relevant quantities over a period of time ranging from months to thousands or millions of years' (55). Among quantities of importance, there are temperature, wind and precipitation. While the first two parameters have a patent influence on ecosystems (59, 60), precipitation, by its impact on river discharge, has only a local or regional (coastal) effect on marine biodiversity. Precipitation is as important for terrestrial biodiversity as temperature is for global biodiversity. To define a climate, the WMO recommends an averaging of those parameters for a period of 30 years. These inform on the state of the climate system.

The climate system results from the interaction of some functional units of the earth system: atmosphere, hydrosphere, cryosphere and biosphere, being under the influence of its own internal dynamics and some external factors such as volcanism, solar energy and more recently anthropogenic factors. The following definition of the climatology by Hufty (52) is quite interesting. Hufty defined this discipline as the study of the exchange of energy and water between the earth's surface and the atmosphere (physical climatology or climatonomy), the study of frequency of meteorological events (statistical climatology and dynamical climatology) and phenomena that influence directly or indirectly living organisms (applied climatology or bioclimatology) (52).

2.4.1 Classification of climates

Our planet has a large number of climates, some being well defined while others representing more transitional areas between major climate regions. Climate can be characterised by a large number of parameters, which interact with each other and vary in space and time. There are, therefore, many ways to determine a classification scheme of climates. The **climatogram** is perhaps the simplest classification scheme. It is obtained by averaging both average total precipitations and temperatures for every month of the year. Robinson and Henderson-Sellers (61) distinguished empirical from genetic classifications. Empirical classifications are based on observations, whereas genetic classifications emphasise factors that control climate.

2.4.1.1 Empirical classifications of climates

Wladimir Köppen, in his classification of climates (*Handbuch der Klimatologie*), used the two parameters temperature and precipitation to determine what he called the climatic landscapes of the earth. This technique was empirical and based on some thresholds of temperature and precipitation. Köppen was neither a meteorologist nor a geographer, but instead a botanist. This simple way to classify climate continues to be used nowadays (62). Some observations attest for the relevance of this type of classification. For example, the isotherm 10°C for the warmest month is a good indicator of the polar extension of the taiga (52). Missing water limits photosynthesis of terrestrial plants. Not only did Köppen take into consideration the annual mean of precipitation and temperature, but he also considered the annual regime of both precipitation and temperature throughout the year (Plate 2.2).

The classification scheme divides the global climate into five (63):

1   Tropical (megathermal) climates (A). All months have an average temperature above 18°C. Tropical climates include tropical rainforest climate (Af), tropical monsoon climate (Am) and savannah (tropical wet and dry) climates (Aw).
2   Dry (arid and semi-arid) climates (B). These regions have low precipitation and the thresholds are fixed upon the examination of potential evapotranspiration (BS for Steppe climate and BW for desert climate).
3   Mild temperate (mesothermal) climates (C). Minimum average temperatures are between −3°C and 18°C. These climates include warm-temperate climate with dry summer (Cs), warm-temperate climate with dry winter (Cw) and warm-temperate fully humid climate (Cf).
4   Continental (microthermal) climates (D), also termed snow climates. In these climates, average minimum temperatures are always lower than −3°C. These climates include snow climate with dry summer (Ds), snow climate with dry winter (Dw) and snow fully humid climate (Df).
5   Polar climates (E). In these climates, average maximum temperatures are always strictly lower than 10°C. These climates include tundra climate (ET) where average maximum temperatures are between 0°C and 10°C and frost climate (EF) where average maximum temperatures are lower than 0°C.

2.4.1.2 Genetic classification of climates

Many other classifications have been proposed. The genetic classification of climates of Alissov (52) was based on the seasonal position of the main frontal structures and air masses (Figure 2.11).

According to the location of the main fronts and their associated air masses, Alissov distinguished nine regions. The regions were then defined over the oceans and the continents and on the eastern and western sides of the main land masses. From the Arctic (or Antarctic) Front to the pole, Alissov distinguished the Arctic zone. Then, the author proposed to call the subarctic zone the region between the Arctic Front and the Polar Front in summer. The zone where the Polar Front oscillates seasonally was termed the subtropical zone. The author named tropical zone the area between the equatorial front and the location of the polar front in summer and equatorial zone the region between the two equatorial fronts. Air masses over the ocean and land masses and in the eastern and western sides of continents are then distinguished in each primary zone. This classification is interesting because it discriminates, in a simple way, areas where the seasonality is absent from regions where it does occur. The tropical zone is the only area with no seasonality. This is also the region where diversity often peaks. Towards the equator, the seasonality tends to increase because of the seasonal change in the position of ITCZ. However, at the middle of the zone, the seasonality is reduced because precipitation occurs permanently. Although this classification scheme was too general to be applied practically, it reveals that frontal structures (mean position and fluctuations) that have a strong effect on climate probably play a key role in shaping oceanic currents and marine biodiversity.

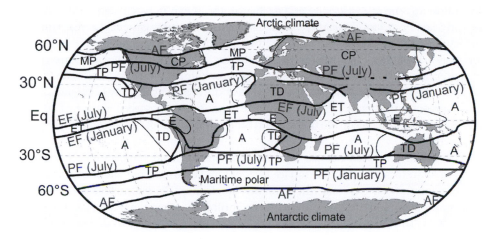

Figure 2.11 The Alissov (1954) classification of climate. AF: Arctic or Antarctic Fronts; PF: Polar Fronts; MP: Maritime Polar; CP: Continental Polar; TD: Tropical Dry; E: Equatorial; ET: Equatorial Tropical; A: Anticyclone.

Source: Redrawn from Hufty (52).

2.5 The oceanic hydrosphere

The oceanic hydrosphere is composed of 1.35 billion km^3 of water and covers 71% of the earth's surface. Water in the ocean has a mean residence time of about 2,500 years. This earth's functional unit has peculiar property due to the water it contains. Water can buffer variation in atmospheric forcing, and therefore the impact of meteorological variability is much smaller than in the terrestrial realm. The specific heat capacity (or specific heat) of water is high (4,186 $J \cdot kg^{-1} \cdot K^{-1}$). In other words, to raise the temperature of one kilogram of water, one needs to provide 4,186 joules. The relationship between heat energy and temperature can be expressed as follows:

$$Q = mc \, \Delta T \tag{2.6}$$

Where m is the mass of the substance (kg), c the heat capacity (c = 4,186 $J \cdot kg^{-1} \cdot K^{-1}$) and ΔT the difference in temperature (expressed in K). In contrast, the heat capacity of the air at 25°C is equal to 1,012 $J \cdot kg^{-1} \cdot K^{-1}$, about four times less than water. As a result, the difference between day and night sea temperatures are reduced, generally <2°C (56). The seasonality of sea temperatures is also considerably attenuated in contrast to what can be observed over the continent. Therefore, at first sight, one could argue that climate warming could have less impact on marine organisms and biodiversity. However, as we will see in the following chapters, organisms have adapted, being often more stenotherm than their terrestrial counterparts and probably more sensitive to temperature change.

The albedo of water is low – between 0 and 10%. Therefore, water absorbs nearly all incoming solar radiation. Because of the size of the oceanic hydrosphere, the low albedo and high specific heat capacity of water, the ocean has a high heat storage capacity (1.57×10^{27} J), about 1,207 times greater than the one of the atmosphere (1.3×10^{24} J). The atmosphere and the ocean interact and exchange energy in 71% of the earth's surface.

In addition to the role of storage and exchange, the ocean transports energy. Atmospheric perturbations can last about three weeks, whereas an oceanic anomaly can persist several months in surface and a few years in the ocean (64). For example, the great salinity anomaly was detectable in the North Atlantic for several years. The ocean, therefore, has a greater memory of perturbations than the atmosphere.

In contrast to the troposphere, the structure of the ocean is generally stable and is warmed by solar radiation at its top. About 90% of the radiations are absorbed in the first 100 metres of the ocean (54). The bottom of the ocean is much colder than the surface, which prevents convection.

Thermocline is often defined as a vertical zone of rapid temperature change in the water column located below the surface layer of rapid mixing. Thermocline can be seasonal or permanent (65). Seasonal thermoclines are located in extratropical regions, being shallow in spring and summer, deep at the beginning of autumn, and disappearing in winter (Plate 2.3). In the tropics, winter cooling is not strong enough to destroy the seasonal thermocline, leading to a permanent thermocline called the tropical thermocline (66). At high latitude, temperature profiles vary seasonally as a function of wind stress and ice cover. From spring to autumn, a layer separates the upper from the deeper zones of the water column, and in winter the water column is homogenous. Pickard and Emery (67) described a depth range of seasonal and tropical thermoclines from 0 to 500 m. Below 500 m and up to 1,000 m, the seasonal thermocline is known as the permanent or oceanic thermocline (i.e. transition zone from warm to cold waters of great oceanic depth).

2.5.1 The global oceanic circulation assists the atmosphere to attenuate the deficit of energy towards the poles

2.5.1.1 Mean surface currents

One of the main climatic functions of the global oceanic circulation is to assist the atmospheric motion to balance the deficit in energy observed polewards. When the oceanic circulation is virtually cancelled in ocean-atmosphere general circulation models (AO-GCM), the temperature difference between the equator and the poles increases (68). Without oceanic circulation, the difference between the equator and the poles would be as great as several hundred degrees Celsius. Currents transport heat from the equator to the poles while the atmospheric circulation transports both sensible (transfer by conduction) and latent (transfer through evaporation and condensation) heat. The main surface oceanic circulation is primarily driven by the atmospheric circulation (Plate 2.4).

The white regions on the map (Plate 2.4A) indicate regions where currents are low (<0.1 m·s⁻¹). The location of those regions coincides well with the location of atmospheric pressure highs (Plate 2.4). On the western side of the ocean and between 0 and 45°, strong surface currents (e.g. the Gulf Stream and the Kuroshio Current) carry out heat from the equator to mid-latitudes. These currents release heat from the ocean to the atmosphere in the form of both sensible and latent heat. On the eastern side of the ocean, the surface currents are weaker and equatorwards. They tend to cool the oceanic regions they cross, and as a result a clear east-west asymmetry takes place. This asymmetry is clearly evident when mean sea surface heights are examined (Plate 2.5).

The sea surface height is greater on the western than the eastern margins of the oceans. For example, a difference greater than 60 cm is observed between Florida and the northern part of France. Similar sea surface height differences are observed between the Warm Pool and the Chilean coast and between the Brazilian and the South African coasts. We will see in subsequent chapters that this asymmetry is also detected in global patterns of biodiversity (69).

2.5.1.2 Heat transportation

The equatorial zone is the region that provides the greatest amount of heat in the form of sensible heat and through oceanic currents (Figure 2.12). Here, the role of the ocean is not to stock the energy, but to transport it.

The tropical regions export energy mainly as latent heat. Regions between 30° and 50° receive the surplus of energy mostly by oceanic currents such as the Gulf Stream and the Kuroshio Current. The latter regions represent a transitional area between regions that export and those that import energy. Between 50°and 70°, the deficit of heat is compensated by the import of latent and sensible heat, with again a significant amount of energy transported by oceanic currents. Regions above 70° receive energy mainly as sensible heat. It should be noted that sensible and latent heat are not independent on the oceanic circulation because air masses exchange permanently latent and sensible heat with the ocean during the atmospheric transport, so that the contribution of the ocean is more important than what can be interpreted directly from Figure 2.12.

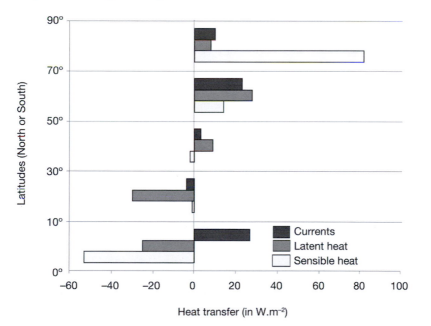

Figure 2.12 Estimated heat transfer from the equator to the poles. Negative values indicate an exportation of heat, while positive values mean an energy input. Atmospheric transports are indicated as sensible and latent heat. Oceanic transport is indicated by currents.

Source: Data from Hufty (52).

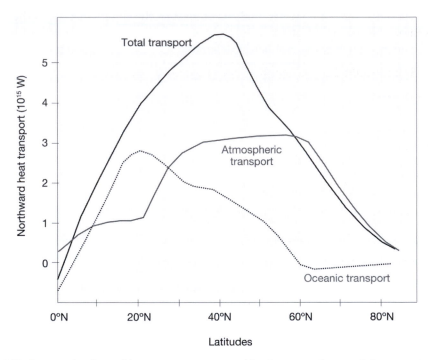

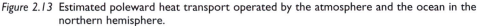

Figure 2.13 Estimated poleward heat transport operated by the atmosphere and the ocean in the
northern hemisphere.

Source: Redrawn from Merle (*68*).

Current estimates tend to indicate that the ocean and the atmosphere play an equal
role, the function of the ocean being more important in lower latitudes than in higher
latitudes (Figure 2.13). Above 50°, the role of the ocean is to absorb the excess of energy
during summer and to release it in winter. Its role of heat reservoir is therefore more important
than heat transport in these latitudes.

Furthermore, at an annual scale, the excess of heat of the summer hemisphere feeds
the other hemisphere (68). Jacques Merle stressed in his book (68) that the role of the oceans
is not identical. At the equator, the Pacific Ocean transports 1 petawatt (10^{15} W) of heat
northwards while transporting 2 petawatts southwards. The Indian Ocean moves 0.5
petawatts of heat southwards. The Atlantic Ocean (three times smaller than the Pacific
Ocean) enables the transfer of 1 petawatt of heat northwards. For comparison, 1 petawatt
represents the energy of 1 million nuclear power plants of 1,000 megawatts each.

2.5.2 The global oceanic circulation: mechanisms, role and consequences

As we have seen in this chapter, oceanic circulation is the result of the unequal distribution
of the energy at the earth's surface due to the inclination of our planet. However, some
mechanisms have to be taken into account to fully explain the circulation of the oceans.
I am detailing some important mechanisms involved in the oceanic circulation, as we will

see that they are also important to fully explain large-scale ecogeographical patterns in biodiversity.

2.5.2.1 Relationships between surface currents and temperature

As we have seen, the surface oceanic circulation plays a prominent role in the redistribution of heat throughout the planet. Salted waters are transported to the poles where convection takes place. The surface oceanic circulation is animated by movements that are closely tied to atmospheric circulations (Plate 2.6). When zonal anomalies of sea surface temperature are calculated, the effects of oceanic circulation on both sea surface and air temperatures become apparent. Surface currents (e.g. California, Humboldt, Canary and Benguela Currents) coming from the poles cool the oceanic surface and the lower part of the atmosphere on the eastern margins of the low-latitude oceanic provinces. These currents are driven by the prevailing atmospheric circulation (Plate 2.6). The easterlies along the equator cool both the oceans and the atmosphere and the westerlies bring warm waters along the western side of the continents towards high latitudes (latitudes >45°). For example, the North Atlantic Current is mainly driven by the westerlies.

2.5.2.2 Relationships between oceanic and atmospheric circulations

We have seen that atmospheric circulation is related to the location of main centres of action (Plate 2.4). A deep examination of the relationships between oceanic and atmospheric circulation indicates that there are some differences. Mechanisms at the origin of these differences are detailed here. As a fluid in movement, the ocean is also influenced by the Coriolis effect. This effect is at the origin of the phenomenon of convergence observed at the centre of the subtropical gyre. This convergence can be explained by the Ekman transport. The Norwegian Fridtjof Nansen observed that the direction of a water mass was not exactly the same as the atmospheric circulation, about 45° to the direction of the atmospheric circulation. The Swedish physical oceanographer Vagn Walfrid Ekman elaborated a theory to explain this observation (Figure 2.14).

The Coriolis force deflects the movement of a fluid to the right in the northern hemisphere and to the left in the southern hemisphere. Therefore, the resulting force in surface is at 45° to the right of the wind. The force subsequently becomes a stress for underlying water and the same process applies. However, as part of the kinetic energy is transformed into heat because of friction, the resulting force decreases from the surface to deeper regions down to a point where this effect cancels. The resulting spiralling effect has been termed the Ekman spiral. The net movement of the ocean is perpendicular to the wind direction and has been called the Ekman transport.

2.5.2.3 Convergence

In the subtropical gyre of the northern hemisphere, the Ekman transport implicates a convergence of water at the centre of the gyre (Figure 2.15). In regions where convergence does occur, the water accumulates. This accumulation can be at the origin of a downwelling. Kump and co-workers (54) stated that this elevation can be as high as 50 cm at the centre of the gyre. As sea-surface elevation increases in these areas, a downslope force due to gravity,

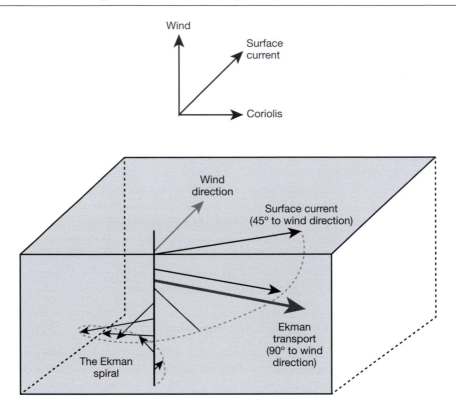

Figure 2.14 The action of the Coriolis force and wind on surface current and the Ekman spiral in the northern hemisphere. The speed and the direction of the current change as a function of depth.

Source: Modified from Kump and co-workers (54).

called a pressure gradient force, takes place, but is then counteracted by the Ekman transport at the edge of the gyre. This process creates a geostrophic current that reinforces the rotation of the gyre.

2.5.2.4 Vorticity

From the examination of the global distribution of surface currents (Plate 2.4), one can see that the subtropical gyres are not symmetrical. For example, the slow and large circulation of the eastern part of the North Atlantic subtropical gyre contrasts with the narrow and more intense surface current detected in the western part of the gyre (the Florida Current and the Gulf Steam). In the North Pacific, the same pattern is observed with the Kuroshio Current (Plates 2.4 and 2.6). This asymmetry can also be explained by the earth's rotation. The process we need to invoke here is called vorticity. Kump and co-workers (54) defined the vorticity as the tendency for a fluid to rotate. The vorticity is positive when the rotation of the fluid is anticlockwise and negative when it is clockwise. The planetary (or earth) vorticity is the rotation of a fluid related to the earth's rotation. At the poles, the planetary vorticity is maximal, while it vanishes at the equator. The vorticity of a water mass at a

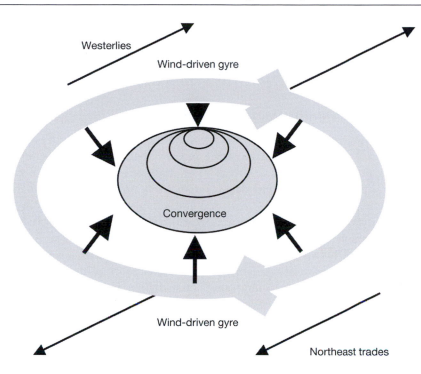

Figure 2.15 The phenomenon of convergence associated with a subtropical gyre in the northern hemisphere. This convergence increases the sea level at the centre of the gyre.

Source: Modified from Kump and co-workers (54).

given point also depends on the atmospheric circulation (cyclonic or anticyclonic). This is called relative vorticity. The absolute vorticity V_A is the sum of the planetary vorticity V_P and relative vorticity V_R:

$$V_A = V_P + V_R \tag{2.7}$$

An important property is the conservation of the absolute vorticity. When a fluid travels, it conserves its initial absolute vorticity. On the eastern side of the North Atlantic subarctic gyre, because the direction of the currents is equatorwards, the positive vorticity related to the earth's rotation decreases. Because the absolute vorticity is conserved, this has to be accompanied by a decrease in negative vorticity, which explains why currents are lower and larger in this area. Meteorologists also distinguish the shear vorticity that occurs along an air column characterized by different wind speed. Shear vorticity can also occur by friction along a coastline, which we neglect here, as its effect is weak and similar on both sides of the gyre.

On the contrary, along the western side of the gyre, the planetary vorticity increases. The negative relative vorticity must increase, which explains why currents are stronger and thinner in these regions. The role of oceanic currents for the biodiversity is important. Currents enable the biodiversity to be redistributed (e.g. advection of plankton species).

Currents also influence marine organisms through their effects on the thermal regime (Plate 2.6). A clear difference between the diversity of calanoid copepods in the eastern and western sides of the North Atlantic and the South Atlantic subtropical gyre was observed by Rombouts and colleagues (69).

2.5.2.5 The global Meridional Overturning Circulation

Surface circulation is mostly driven by wind stress and the effect of the earth's rotation (i.e. the Coriolis force). The oceanic surface can be cooled or heated by atmospheric forcing in the first few hundred metres. Below, it tends to be stratified and very stable. Both surface circulation and the deep oceanic circulation interact in a non-linear way and are part of the global Meridional Overturning Circulation (MOC), also called by Wallace Broecker the giant conveyor belt. MOC is influenced by wind, tides and the thermohaline ocean circulation. Thermohaline and MOC are not synonymous (70, 71). The thermohaline circulation is part of MOC. The former is concerned with the density mechanisms that influence the MOC, and the latter refers to the meridional south-to-north surface flux and its return at depth in the opposite direction. Thermohaline and wind-driven circulation interact in a complex way and cannot be disentangled by oceanic measurements. A change in the characteristics of the MOC could lead to major changes in the regional climate of some areas of the world, and therefore could strongly affect the marine biodiversity. It is therefore useful to detail some characteristics of MOC (Plate 2.7).

The MOC is strongly influenced by wind stress, which pushes warm water polewards in the North Atlantic. For example, the North Atlantic Current reaches the Norwegian Sea because of the westerlies, themselves constrained by the location of major land masses and the relief. As salty subtropical waters travel to the north, they start to cool and become denser. Density (or buoyancy) is a function of the temperature and the salt concentration. Kump and colleagues (54) recall that maximum density of freshwater occurs at 4°C, equal to 2°C for a water with a salinity of 10, and reaches the freezing point when the salinity of water is equal to 24.6. The freezing point itself then decreases when salt concentration continues to increase. These salty dense waters sink currently in some regions of the Greenland Sea, the Norwegian Sea and the Labrador Sea, where they form the North Atlantic Deep Water (NADW), and in the Weddell and the Ross Seas, where they form the Antarctic Bottom Water (AABW) (Plate 2.7).

Ganachaud and Wunsch (72) estimated a volume of NADW of 15 ± 2Sv (1 Sverdrup = 10^6 m^3·s^{-1}). Deep-water outflow was measured to 2.9 Sv across the sills between Greenland and Iceland, 2.5 Sv between Iceland and the Faeroe Islands and 0.2 Sv between these islands and Scotland (73). Rahmstorf (71) indicated that tracer data suggest an equal amount of deep-water formation in Antarctica (15 Sv) although available measurements indicate again an outflow of only 5 Sv. The author explained these discrepancies by the process of entrainment, which is the increase in the transport volume induced by the mixing of dense water.

However, although the increase in density is an important condition, convection is equally essential to maintain the density gradient that enables the seawater to sink. Convection enables heat to penetrate deeper in the water column, and thereby to reduce the density needed for the phenomenon to reiterate. The convection that triggers the mixing of the water column is primarily related to wind stress and secondarily to tidal motions.

Toggweiler and Samuels (74) even hypothesised that the combined effect of the westerlies and the Coriolis force in the Southern Ocean involve the establishment of a strong northward current. Because of the Drake Passage between South America and the Antarctic Peninsula, the only possible return of water to compensate the northward flow is through the southward deep-water current that is established in the Nordic seas and constrained in the Southern Ocean by the submarine ridges at the Drake Passage. A lot remains to be done to better understand the global MOC. However, models clearly indicate that the role of the wind and the thermohaline circulation is important. When one of these mechanisms is turned off in models, MOC is reduced.

Wunsch (70) stressed that the oceanic circulation should be viewed as mechanically driven, capable of storing, importing and exporting large amounts of heat and freshwater. He added that the nature of wind field is important, as it determines the surface flow and the turbulence that controls deep stratification and convection. Modelling and palaeoclimatological studies also showed the sensitivity of the thermohaline circulation to salinity. This parameter is at the origin of the non-linearity of the system with two possibilities of stability. Rahmstorf (71) explained the role of the salinity, showing that the parameter is involved in a positive feedback. When salinity is increased in the region of convective sinking, this reinforces MOC. If the amount of freshwater is increased, because precipitations are greater than evaporation (or sea-ice melting), freshwater accumulates and the salinity decreases. This involves a reduction in MOC. Similar non-linearities have been detected in the convective mixing process (71).

To close the flow of the MOC, the amount of deep-water formation has to be balanced by upwelling of deep waters. It is presently thought to occur mainly in the Antarctic Circumpolar Current region. This would occur to replace the northward flow triggered by the westerlies in the Southern Ocean. This would occur to the east of the Drake Passage (74).

2.5.2.6 Interaction between the oceanic circulation and climate

The oceanic circulation influences the climate system in a multitude of ways. The Atlantic component of the MOC transports about 1 PW (one peta = 10^{15}) of heat in the North Atlantic (see Plate 2.6). Rahmstorf (71) suggested that this amount of heat should increase the temperature of the Atlantic by 5°C in comparison to regions of similar latitudes in the North Pacific. This estimate corresponds to what we found in Plate 2.6. Seager and colleagues (75) showed that planetary waves constrained by topographical features such as mountains (e.g. the Rockies) were also responsible for the milder climate in Europe.

When the thermohaline circulation is turned off in general circulation models by increasing freshwater forcing, this reduces by 10°C the average temperature over the Nordic seas because of the positive feedback involved by sea-ice (71). In the extratropical North Atlantic, the decrease in temperature ranges from 9°C south of Iceland to 0°C at 40°N. In seas surrounding the French coast, the reduction in temperature is more modest (about 2°C). When the thermohaline circulation is switched off, the cross-equatorial transport is reduced, triggering a warming of the southern hemisphere and a cooling of the northern hemisphere. This leads to a modification in the seasonal range of the ITCZ. The author stressed that this result is consistent with palaeoclimatic and modelled data.

2.6 Hydro-climatic variability

The Intergovernmental Panel on Climate Change (55) defined climatic variability as 'variation in the mean state and other statistics of the climate on all spatial and temporal scales beyond that of individual weather events'. They further indicate that the cause of this variability can be related to the natural inherent variability of the climate system or anthropogenic external forcing. Here, I detail some natural sources of climatic variability commonly identified in the scientific literature and extend the term climatic variability to hydro-climatic variability.

2.6.1 Sources of hydro-climatic variability

The global atmospheric circulation exhibits some recurrent spatial and temporal patterns of variability. They are often localised by their centre of action. The spatial pattern of these centres of high and low pressure is, on average, well defined, resulting in a coherent pattern of weather over a large part of the earth (typically the scale of an oceanic basin). As a result, the regional climate of an area can vary in phase or out of phase with another remote place. Such an effect has been termed by climatologists as a teleconnection. These teleconnections are more clearly identified in the boreal winter in the northern hemisphere because the circulation is more dynamically active. Some indices are described here as they are commonly used by scientists who try to understand if a biological or an ecological system can be influenced by climate. Such approaches have led to interesting discoveries, especially in the 1990s. I describe here some commonly employed indices and discuss their strengths and weaknesses.

2.6.1.1 Solar variation

The incoming solar radiation is not constant through time. Therefore, the sun, the intensity of which controls climate, is a potentially important source of climatic variability. Recent estimates based on measurements provided by the Earth Radiation Budget Satellite (ERBS) launched in 1984 showed that the solar constant (1,368 $W \cdot m^{-2}$) fluctuates in the range of 4 $W \cdot m^{-2}$, representing a variation of 0.29% (76). These fluctuations are synchronous with dark colder cells located on the sun's photosphere and called sunspots. Sunspots were rapidly detected after the invention of the telescope in 1609. Galileo Galilei initiated the study of sunspots and the Zurich observatory started daily observations in 1749. Sunspots are related to the strength of the magnetic activity at the surface of the sun that can stop the process of convection. They are often grouped. The temperature of a sunspot can be as low as 3,700°C, while the average temperature of the photosphere is 5,500°C. A sunspot can last from a few hours to several months and covers, on average, an area of about 37,000 km^2. They are associated with strong magnetic field. Their number shows a cyclical fluctuation of about 11 years: the higher the number of sunspots, the greater the solar activity. The existence of a solar cycle was first suggested by Samuel Heinrich Schwabe in 1843 and then confirmed by Rudolf Wolf, who compiled data on sunspots back to 1745. The solar cycle, or Schwabe's cycle, has a main cyclicity of 10.66 years, ranging from 9 to 14 years (Figure 2.16).

More atypical periods have also been detected. For example, only a few sunspots were observed during a period known as the minimum of Maunder (1645–1715) and few sunspots were also observed during the Dalton minimum (1795–1820). Indirect evidence

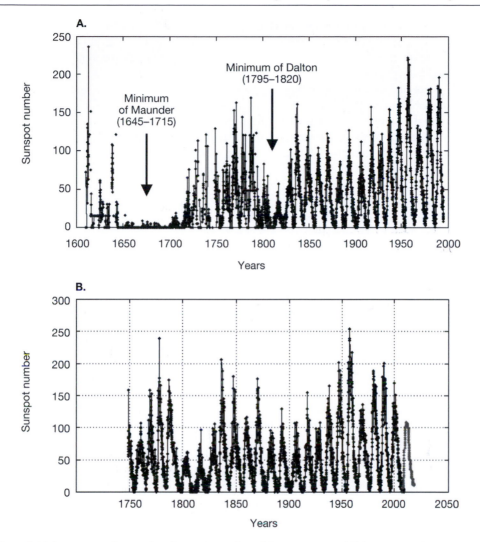

Figure 2.16 Long-term changes in solar activity inferred from sunspots. (A) Long-term changes (1610–1995) in the group sunspot number (Rg). Such a time series is more self-consistent and less noisy than the Wolf sunspot number. (B) Long-term changes from January 1749 to October 2009 in the international sunspot number. The last minimum was reached in December 2008. The grey line represents a preliminary forecast (2007–2018) of cycle 24.

Source: (A) Data and methods from Hoyt and Schatten (77). (B) Data from the World Data Center for the Sunspot Index, Royal Observatory of Belgium.

also revealed the minimum of Sporer (1450–1540). On the contrary, the maximum of the solar cycle happened in 1770, 1845 and 1940. Some studies have shown that the sun, as other stars, can spend 25% of its time with very few sunspots. Other cycles have been suggested. The solar-magnetic cycle is a cycle of 22 years, also called the Hale cycle. Another cycle has been named the Gleissberg sunspot envelope-modulation cycle. This is a periodic fluctuation of about 80–90 years in the length of the sunspots cycle.

Scientists have not found any relationships between the state of the weather and the number of sunspots. However, climatologists found some interesting links between cold periods and observed minimum number of sunspots. For example, the minimum of Maunder coincided with the coldest period of the Little Ice Age. Solar irradiance dropped by 0.24%, which decreased global temperature by 0.5°C. Half of the warming observed since the minimum of Maunder has been attributed to solar variability (78). Some climatologists have associated the sunspot cycle with thunderstorm frequency and precipitation or drought. Others suggest potential links with sea surface temperatures (79). Some bioclimatologists and ecologists have sometimes found intriguing relationships between the number of sunspots and the state of some living systems, but time series used to detect these relationships were often short. Between 1820 and 1842, year-to-year changes in the number of sunspots and the number of lynx (log-transformed) in Canada covaried positively (80). The examination of the relationships between the two variables may have led to the conclusion that the sunspot cycle influences the number of lynx. However, afterwards, the cycles of the two variables started to be asynchronous. When the correlation between variables that exhibit cycles is examined, it is often difficult to conclude because data are not statistically independent. In the marine domain, a relationship between the number of sunspots and long-term ecosystem changes has been suggested in the Black Sea (81) and in the Atlantic Iberian Peninsula for sardine *Sardina pilchardus* (82), but mechanisms by which this link may occur are poorly understood. During the period 1948–2012, correlations between annual sea surface temperatures and yearly sunspots were weak and explained a few percentage of variance when they were significant (Plate 2.8).

The link between solar variability and biodiversity should be considered with caution because the intermediate mechanisms by which solar variability may affect marine biodiversity are not yet identified and remain debated in the literature (83). Other ways exist to measure solar activity. Palaeoclimatologists measure the concentration of radioactive elements ^{14}C (radioactive carbon) in tree cellulose or ^{10}Be (radioactive beryllium) from ice cores. Radioactive carbon forms at high altitude in the atmosphere by cosmic bombardment on nitrogen. Its concentration, therefore, depends on solar activity. About 100 kg of ^{14}C would be produced annually through this process (83). However, the formation of ^{14}C would also depend on changes in the earth's magnetic field.

Radioactive beryllium is also a function of the solar and cosmic ray variability. But complexity arises from the fact that wind field affects its deposition in ice. More information here would go far above the scope of this book, but this way to measure solar activity is important when one wants to assess long-term covariation between climatic variability and biodiversity.

2.6.1.2 Global indices of temperature

Indices of global temperature anomalies have been frequently used to examine the possible consequences of global increase in temperature on ecosystems and their biodiversity. The global average surface temperature has augmented over the twentieth century by ~0.6°C. This warming mainly occurred during the periods 1910–1945 and 1976–2000. In the northern hemisphere, the 1990s was the warmest decade since the start of the instrumental record (1861) and 1998 the warmest year (55). This warming is suspected to be responsible for a plethora of physical changes such as the decrease in Arctic sea-ice (84), snow cover and ice extent (85), as well as global ocean heat content (86, 87).

This index has often been associated with major change in marine biodiversity over the North Atlantic basin (60).

The correlations between this global index of temperature (land and sea global temperature anomalies) and annual SST are highly positive, with the exception of the north-eastern part of the Pacific Ocean, the Subarctic Gyre and some regions of the Southern Ocean (Plate 2.9).

2.6.1.3 The North Atlantic Oscillation and the Arctic Oscillation

The North Atlantic Oscillation (NAO) is the most prominent source of atmospheric variability in the northern hemisphere, identified in all months of the years. This oscillation represents a basin-scale alternation of atmospheric masses over the North Atlantic between the subtropical and subarctic regions (88). Other authors prefer to refer more to the Arctic Oscillation (AO) or the Northern Annular Mode (NAM).

The AO (or NAM) index is derived from performing a PCA on Mean Sea Level Pressures (MSLP) in winter in regions above 20°N in the northern hemisphere. The conventional index of NAO activity is based on the pressure difference between the Azores high- (Lisbon or Gibraltar) and the Icelandic low-pressure centre (Stykkisholmur or Reykjavik) (89). Although they are based on different regions, the NAO and the AO are highly correlated ($r = 0.76$; $p_{ACF} < 0.01$; df = 61; 1950–2012). The correlations between annual sea surface temperatures and the two indices give similar patterns of correlations (Plate 2.10).

Quadrelli and Wallace (90) showed that the AO and the NAO are characterised by similar spatial patterns and temporal changes. Their modes are linked to the configuration of extratropical storm tracks and jet streams.

The effects of the NAO are stronger on both annual SSTs of the subarctic and subtropical gyres and less important in regions between the two gyres, including the North Sea (Plate 2.11). The opposite is true for the AO.

The NAO is the dominant mode of atmospheric behaviour in the North Atlantic, accounting for 32% of the variance explained when a PCA is performed on Sea Level Pressure for all months in the North Atlantic sector (91) and up to 36% if only winter months (December–March) are considered (92). The NAO, and especially the Icelandic low-pressure centre, may be associated with tropopause height and the strength of the stratospheric vortex (93). Ambaum and Hoskins (93) suggest that when the NAO is positive, the Icelandic low-pressure centre reinforces and is associated with a decrease in the height of the tropopause (Figure 2.17).

The upward-propagating Rossby waves move more equatorwards, which reinforce the stratospheric vortex, implicating an increase of the tropopause and an associated diminution in the pressure over the North Pole. Ambaum and Hoskins (93) found that the NAO led to a reinforced strastospheric vortex four days later. The NAO is more evident during the boreal winter because the atmosphere is more dynamically active as the contrast between the Polar cell and the Hadley cell in the northern hemisphere intensifies. A high positive value indicates reinforced low pressure around Iceland and high pressure over the Azores, whereas a significant negative value corresponds to a weakened pressure gradient. The phenomenon reflects a meridional gradient in sea level pressure and therefore affects the strength of the zonal circulation, which in turn influences transport of heat and moisture in the North Atlantic and adjacent continents (Plate 2.12).

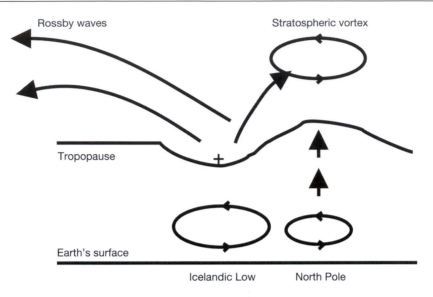

Figure 2.17 Schematics of the potential link between the NAO, the tropopause and the stratospheric vortex. The NAO is here in its positive phase.

Source: Redrawn from Ambaum and Hoskins (*93*).

When the NAO is in its positive phase, both meridional (south to north) and zonal (west to east) circulation (i.e. the westerlies) clearly reinforce between 45°N and 65°N all over in the North Atlantic, whereas decreasing zonal circulation occurs over the Mediterranean Sea and in the subtropical gyre (Plate 2.12). Easterly trade winds reinforce over the tropical Atlantic during a positive phase of the NAO because of the increase in the Azores High and its associated stronger clockwise flow. The meridional component of the wind is correlated positively to the NAO over the path of the Gulf Stream, but mainly north of the Oceanic Polar Front between 45°N and 65°N and negatively in the Labrador Sea. This change in the atmospheric circulation brings more warm moist maritime air during winter, but therefore more precipitation over the northern part of Europe (Plate 2.12). It influences the position of the North Atlantic storm tracks, shifted to the north when the NAO is positive. During a positive phase of the NAO, winters are therefore anomalously warm and wet over the northern part of Europe, colder and drier than usually in the Mediterranean Sea, parts of the Middle East and colder and drier in the eastern part of North America and the western part of Greenland. Opposite effects occur during a negative phase of the NAO. It is important to bear in mind that the correlations between SST and the winter NAO index are not high, explaining at best about 30% of the total variance in SST. Therefore, the NAO is far from explaining a great amount of SST variability in most parts of the North Atlantic section.

Fluctuations in the state of the NAO influence a large range of physical phenomena. As seen previously (Plate 2.12), temperature and precipitation in Scandinavia are highly positively correlated to the state of the NAO. This influence of the NAO substantially affects the Scandinavian energy sector (94). The topography and the climate make Norway suitable for hydroelectric power generation, which represents 99% of the electricity produced in this country. During the prolonged positive phase of the NAO in the 1980s and the 1990s, the increase in precipitation was at the origin of an excess in energy. This led this country to sell contracts with Sweden and to export electricity. In the beginning of the 1990s, the NAO was positive, and hydroelectricity worked well. However, the NAO came to a very strong negative phase in 1996. The reservoirs, filled at about 95% of their capacity in 1995, were only filled at 65% in 1996. The cold and the absence of significant precipitation prevented Norway from producing enough hydroelectricity to export to Sweden. In that year, Norway had to buy electricity at a high cost from Denmark.

The NAO may be responsible for about half of the warming experienced by the Eurasian land mass during the period 1968–1997 (95). The NAO has also been associated with Eurasian snow cover and Arctic sea-ice (96, 97). A see-saw in sea-ice extent is clearly detected in the Labrador Sea and Greenland Sea, the sea-ice extending further south in the Labrador Sea when the NAO is positive. Norwegian glaciers advanced in the 1990s. This behaviour, opposite to what has been generally found in recent years all over the world, may be a response to the positive phase of the North Atlantic Oscillation. However, recent findings indicate that glaciers have started to retreat since 2000, probably because of a reduction of winter accumulation and a greater melting in summer (55).

During a positive phase of the NAO, the Subpolar Gyre expands eastwards, implicating a reduction in salinity towards the centre of the gyre. The NAO has been related to the formation of intermediate waters known as mode waters. Such waters form when deep convection is favoured by the atmospheric circulation in winter. Hurrell and Dickson (96) stressed that the intermediate water allows the climate signal to propagate at depth. Places where such waters form are the Greenland Sea, the Labrador Sea and the Sargasso Sea. The three sites are teleconnected. During a prolonged positive phase of the NAO at the beginning of the 1990s, intense convective activity was observed in the Labrador Sea. Convection penetrated down to 2,300 m. In contrast, intermediate water formation slowed down in the Greenland Sea and the Sargasso Sea where waters were warmer and saltier than usually. In contrast, during the 1960s during a prolonged negative phase of the NAO, the convection in the Labrador Sea was shallow and weak, whereas in the Sargasso Sea and the Greenland Sea the convection was high, with values of several hundred metres and more than 3,000 m, respectively. Hurrell and Dickson (96) explained why such a tele-connection might occur. During a negative phase of the NAO, more storms were observed in the Sargasso Sea and the Greenland Sea, which increased the strength of the convection. Calmer conditions occurred northwards in the Labrador Sea, which enabled freshwater to accumulate. Convection was at a very low level.

The Meridional Overturning Circulation (MOC) has also been connected to the state of the North Atlantic Oscillation. This is not surprising, as this oscillation modulates the atmospheric flows that are important in controlling MOC. The Labrador Sea Water is a major component of the North Atlantic Deep Water, which I recall is part of the MOC. The NAO has also been associated with the decrease in subsurface salinity observed in the Nordic seas since the 1970s. This was associated with the NAO, which modulates the strength and the expansion of the Subpolar Gyre. Since the end of the 1990s, the Subpolar Gyre has retracted because of the influence of lower NAO indices, which has led to the pene-tration of more saline water in the Nordic seas (98). Some of this variability has also been associated with another index of teleconnections called the East Atlantic Pattern (99).

Although the NAO is the only teleconnection pattern identified for all months of the year, some authors distinguish the winter NAO from the summer NAO (100). The summer NAO is located more to the north than the winter NAO and has a smaller spatial extent.

It also influences temperature, rainfall and cloudiness. This index has been much less used in marine bioclimatology.

2.6.1.4 The El Niño Southern Oscillation

An El Niño event corresponds to a warming of the eastern Pacific along the northern part of the Peruvian coast and the southern part of Ecuador. This phenomenon has been known for many centuries. Fishermen noticed the arrival of a southward warm current along the Peruvian coast, associated with a reduction of trade winds. As the event appeared every year around the Christmas season, this warm current was named El Niño, as a reference to the Christ child. This warm current announced the end of the fishing season from January to March. Because the intensity of the warm current varied from year to year, the term El Niño was later used to qualify the year when this phenomenon was particularly pronounced. During the 1960s, scientists showed that this event was part of a large-scale phenomenon, a basin-scale coupling between the ocean and the atmosphere (101).

The Southern Oscillation (SO), representing the atmospheric part of the phenomenon, was first identified by Sir Gilbert Walker (102, 103). The British meteorologist observed that when the atmospheric pressures increased in the western part of the Pacific (Darwin, the Indonesian low-pressure centre), it decreased in the eastern part (Tahiti, the south-east Pacific high-pressure centre). He called this see-saw effect the Southern Oscillation (Plate 2.13).

The SO is therefore a year-to-year fluctuation in mean sea level pressures between the western and the eastern tropical Pacific. This see-saw effect reflects the two modes of Walker's circulation. An index of the Southern Oscillation has been created to monitor this phenomenon. It is calculated by subtracting the Mean Sea Level Pressure (MSLP) at Tahiti by MSLP at Darwin and by normalising the difference by both the long-term mean and the standard deviation of the MSLP difference. This index is available from 1860, but data in Darwin are more consistent prior to 1935 than data in Tahiti (55). Therefore, the MSLP at Darwin is often used alone to characterise the state of this oscillation. The SO controls the type of atmospheric circulation in the tropical Pacific, influencing the strength of the trade winds and precipitation.

As oceanographers and meteorologists showed that El Niño was coupled to the SO, the phenomenon was subsequently referred to as the El Niño Southern Oscillation (ENSO) (104). The ENSO largely controls the climatic variability of the tropical Pacific and also has large meteorological and climatological implications in many parts of the globe with consequences for ecosystems and their biodiversity, human health and socio-economic systems (105, 106). The super El Niño event of 1997–1998 increased the global temperature by ~0.2°C, about one-third of the magnitude of the global warming between 1905 and 2005 (107, 108).

Much progress was made on ENSO during the international programme TOGA (Tropical Ocean – General Atmosphere; 1985–1994). Among others, George Philander from Princeton University and Jacques Merle from IRD (Institut de Recherche pour le Développement, France) summarised well the level of understanding gained by the scientific community on this meteo-oceanic phenomenon (68, 104, 109). During normal conditions, trade winds flow towards the Australian–Indonesian centre of low pressure characterising the warm pool in the equatorial western Pacific (Plate 2.6B). The south-easterlies push the warm water in the warm pool, elevating the sea level by 50 cm in comparison to the eastern side of the

Pacific (Plate 2.5). Convergence occurs and triggers ascending movements at the origin of high concentration in cloudiness and precipitation (Plate 2.14).

This region is the main area of convergence and convection of the planet. The air then flows at altitude towards the centre of high pressure on the eastern side of the Pacific Ocean, completing the cell. This zonal motion of air masses is Walker's circulation, or Walker's cell (Plate 2.14).

When Walker's circulation is high, the Hadley circulation (role of meridional transport in energy) slows down and inversely. The thermocline, thin layer with great vertical thermal gradient, is much deeper (about 200 m) in the western part than in the eastern part of the Pacific (~50 m): the higher the temperature of the sea, the deeper the thermocline. The sharp thermocline is therefore inclined along the equatorial Pacific, being a direct consequence of the trade winds that flow westwards (Plate 2.6B).

The definition of El Niño was revised in 2003 by the National Oceanic and Atmospheric Administration (NOAA) as 'a phenomenon in the equatorial Pacific Ocean characterised by a positive sea surface temperature departure from normal (for the 1971–2000 base period) in the Niño 3.4 region (5°S to 5°N and 170°W to 120°W) greater than or equal in magnitude to 0.5°C, averaged over three consecutive months' (Figure 2.18).

The Niño 3.4 region encompasses the equatorial cold tongue. This region, which extends from the eastern to the central part of the equatorial Pacific between 5°S to 5°N and from 170°W to 120°W, is important because it determines patterns of tropical rainfall, and influences jet streams and patterns of temperature in many parts of the world.

As the area covered by the phenomenon is large (10% of the earth's surface), the consequences of an El Niño on the global climate are important. Conversely, a La Niña event is defined by a sea surface temperature anomaly lower than or equal to –0.5°C. An El Niño event typically takes place every three to six years (*110*), although this range of occurrence varies among authors. It has been proposed that the timescale of an El Niño event is linked to the time needed for the warm water to accumulate in the western

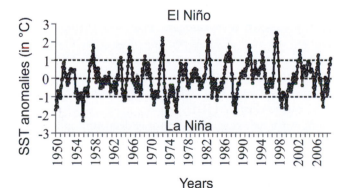

Figure 2.18 Year-to-year changes in the seasonal ERSST.V3B SST Niño 3.4, also called the oceanic Niño index. This index is the three-month average of sea surface temperature in the geographical box from 5°S to 5°N and from 170°W to 120°W. A positive anomaly greater than or equal to 0.5°C defines an El Niño event. A negative anomaly lower than or equal to 0.5°C defines a La Niña event. The index was normalized using the period 1971–2000.

Source: Data from the Climate Prediction Center of the National Oceanic and Atmospheric Administration.

part of the Pacific Ocean and the time that is necessary for the phenomenon to develop, a hypothesis termed the recharge oscillator hypothesis (*111*).

Philander (*104*) drew attention to the possible existence of precursors. A first precursory phase, a required but not sufficient condition, is the eastward movement of the ascending branch of the Walker cell to a zone located between New Guinea and the International Date Line. In the zone, the surface water warms and the pressure difference between Tahiti and Darwin decreases. This is observed between October and November, prior to the onset of an El Niño. However, Philander stressed that this precursory phase does not always lead to an El Niño, as the phenomenon appeared in 1979 and 1980 without triggering an El Niño. The second precursory phase occurs in the eastern part of the equatorial Pacific. Normally, the ITCZ oscillates between 10°N in August and September and 3°N between February and March. Before the end of the 1970s, the El Niño event was often associated with the southward migration of the Intertropical Convergence Zone (ITCZ) at the equator or even further south. This amplified the effects of the seasonal cycle, decreasing further the intensity of the trade winds and deepening the thermocline off the coasts of Peru and the Ecuador. However, this precursory phase was not observed during the super El Niño event of 1982–1983.

When the pressure difference between the western and eastern part of the Pacific decreases due to the two previous mechanisms, the intensity of the trade winds can reduce to such an extent that the reserve of warm water of the western Pacific starts to flow eastwards (Plate 2.15A–B). This further moves the low-pressure centre towards the central Pacific, which emphasises the weakening in the pressure difference between the east and the west, and thereby the Southern Oscillation. This involves a positive retroaction leading to the amplification of the phenomenon. The propagation of the warm waters thereby accelerates and the warm waters reach the eastern part of the equatorial Pacific a few months after the beginning of the phenomenon (Plate 2.15C). This deepens the depth of the thermocline and the trade winds that continue to flow eastwards along the Peruvian coast. Winds are not strong enough to reach the thermocline and trigger upwelling. The movement of the low-pressure centre towards the central Pacific reduces precipitation over the northern part of Australia and Indonesia where severe droughts take place (Plate 2.15D). The interaction between the ocean and the atmosphere is permanent, and the movement of warm waters triggers mechanisms of atmospheric convection along its way. The Hadley cell intensifies transporting more energy towards the poles. The progression of the water eastwards implicates an increase in sea level that is detected by the networks of tide gauges set up along the equatorial Pacific (Plate 2.15D).

The outgoing long-wave radiation, an indicator of deep atmospheric convection, can be used to track the consequences of an El Niño. The value of this climatological parameter decreases when deep convection occurs because clouds such as cumulonimbus are cold at their tops. When the convection is reduced, the value of this parameter increases. The value is maximal in the absence of cloud. Convection transfers energy from the ocean to the atmosphere because energy is needed to evaporate water. Such energy comes from the ocean. When condensation occurs, this energy is released to the atmosphere. We therefore understand how changes in sea surface temperatures can affect atmospheric circulation and patterns of precipitation, redistributing the energy over the tropical Pacific (Plate 2.15). The increase in the intensity of the Hadley cells during an El Niño is an important mechanism by which the effects of an El Niño propagate polewards.

Prior to the climate shift of 1976–1977, the event occurred at intervals of about five years (*104*). The weakening of the easterly winds triggered the eastward advection of the warm waters of the Western Pacific. However, the warming peak off the South American coast took place in the spring of the El Niño year. After the climate shift, the El Niño event initiated from the west by a movement of the centre of low pressure towards the International Date Line. The warming propagated eastwards and the maximum temperature was observed in the boreal spring one calendar year after the beginning of El Niño (*106*).

El Niños characterised by a warming of the eastern equatorial Pacific have been termed canonical El Niños. These events correspond to the recent definition of an El Niño proposed by the WMO. However, a new type of El Niño has become more frequently recorded since the 1980s, and has been termed El Niño Modoki by Toshio Yamagata from the University of Tokyo (*112*). The Japanese word Modoki means 'similar but different'. The event has also been called pseudo-El Niño or Central Pacific El Niño. While the super El Niño episode of 1997–1998 was a canonical El Niño (or East Pacific El Niño), the El Niño events of 1980, 1994 and 2002 were classified as El Niños Modoki (*112*). This type of El Niño is characterised by a horseshoe pattern. Such events are identified by a warming of the central part of the equatorial Pacific at about the International Date Line surrounded by negative anomalies of temperatures at the west and the east. A recent study showed that since the 1850s, the conventional El Niño occurred 32 times and the El Niño Modoki seven times (*113*). More importantly, the study provided compelling evidence that the frequency of the events increased after 1990 from 0.01 per year to 0.29 per year. Furthermore, the same study suggested that this type of El Niño should be five times more frequent under global warming (*113*). The authors showed how this type of El Niño might affect patterns of precipitation, the energetic distribution along the equatorial Pacific and teleconnections with extratropical regions of the North Pacific. For example, there are more precipitations in the eastern part of the Pacific during the canonical El Niño than during an El Niño Modoki. The location of the low-pressure centre in the North Pacific varies, changing the way the equatorial Pacific warming propagates polewards. For example, the warming due to southerly wind along North America (including Alaska) observed during a conventional El Niño is not detected during an El Niño Modoki due to the migration of the Aleutian low-pressure centre to the south-east.

The cold phase of ENSO is called La Niña. Such an event can arise before (La Niña of 1995–1996) or after an El Niño episode (La Niña of 1998–2000). Its consequences can be important for biodiversity, but are much less documented than the consequences of an El Niño. Jacques Merle mentioned a water with a temperature below 17°C at the equator near the Galapagos during the La Niña event of 1998–2000 (68). During a La Niña event, the easterly winds reinforce, increasing the intensity of upwelling along the South American coast. They also push warm waters towards the Western Pacific. During such years, the droughts are more severe in Peru and Chile. Exceptional precipitations occur around Indonesia, the northern regions of Australia and New Guinea. By increasing the depth of the thermocline and surface height of the warm pool, a La Niña event can set an El Niño when the easterly wind relaxes (e.g. La Niña of 1995–1996).

A La-Niña-like decadal cooling has taken place since 1999 in the equatorial Pacific (Plate 2.16). This situation is clearly detected when a principal component analysis is performed on sea surface temperature anomalies (first principal component; Plate 2.16A). However, this is not exactly like a La Niña condition because episodes of El Niño were also observed.

The second principal component reveals the two super El Niño events of 1982–1983 and 1997–1998 (Plate 2.16B) and the third principal component shows a warming of the south tropical Pacific and a cooling of the East Pacific (Plate 2.16C). The cooling of the Pacific is currently limiting the magnitude of global warming (114). Without the cooling of the East Pacific, the increase in global temperature would be close to ~1°C instead of ~0.7°C.

The cooling of the region may be related to an increase in Walker's circulation and the associated intensification in the easterlies (Plate 2.17). This intensification in the easterlies may have been caused by the strengthening of the contrast between high and low pressure cells in the high latitudes of the East Pacific in both hemispheres. This may have contributed to cool the water masses of the eastern and central parts of the Pacific along the hemisphere. Cooling may occur by reinforcing upwellings along the equator and American coasts and by direct air transfer from the poles to the equator. It is therefore likely to be related to natural internal variability.

Many aspects of ENSO are not yet fully understood. For example, temporal changes in the amplitude of El Niños, the negative feedback involved in the termination of the phenomenon and the aperiodic nature of the ENSO remain elusive (115, 116). Furthermore, the identification of a new type of El Niño suggests that the warm phase of the ENSO may not have the same impact on regional biodiversity.

Many hypotheses have been invoked to explain the origin, the development and the termination of the warm phase of ENSO. This latter phase requires a negative feedback. For example, the recharge oscillator hypothesis states that the system oscillates between a charge phase (La Niña) when the warm waters accumulate gradually in the equatorial western Pacific and a discharge phase (El Niño) when the warm pool flushes eastwards (111). In this model, both depth of the thermocline and wind stress are important parameters. The delayed oscillator hypothesis (117) is based on the different properties of the ocean and the atmosphere. The ocean takes more time to adjust to atmospheric changes (several months), while the response of the atmosphere to oceanic forcing is quicker (a few weeks). The delay between phases is explained by sea temperature dynamics and the time needed for upwelling **Rossby waves** to propagate westward, to be reflected at the western boundary as (upwelling) **Kelvin waves** that in turn propagate eastward (negative feedback).

Another interesting hypothesis, different from the two previous ones, has been tested. In this hypothesis, there exists a stable state that can be disrupted by stochastic atmospheric forcing. George Philander assimilates the succession of El Niño and La Niña to 'a swinging pendulum that is subject to modest blows at random times' (115). El Niño would be the result of a stable mode, also called a damped mode, forced by stochastic atmospheric variability. Contrary to the oscillator hypotheses, this hypothesis explains the observed variability in the intensity of ENSO events and the irregularities between episodes. In this model, a random perturbation such as a slight relaxation of easterly winds can trigger an El Niño. Indeed, a small attenuation of these winds can involve a displacement of warm water from the equatorial western Pacific, which in turn weakens further the easterlies. The phenomenon amplifies and initiates an El Niño episode.

2.6.1.5 North Pacific decadal variability

The North Pacific decadal variability is strongly influenced by the mean location and intensity of the winter Aleutian Low Pressure Centre, which in turn influences large-scale patterns in sea surface temperatures. Several large-scale climate indices have been proposed to monitor

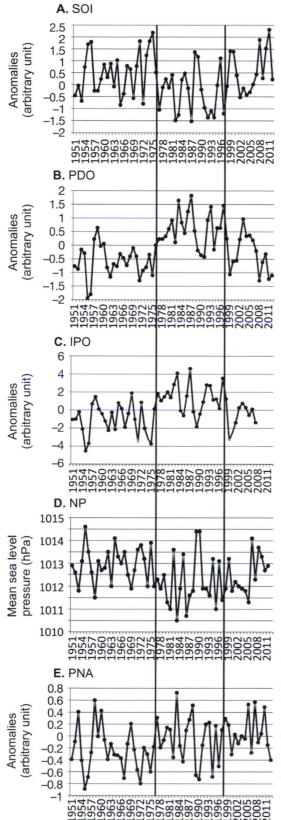

Figure 2.19
Long-term changes in the Southern Oscillation Index (SOI; A), the Pacific Decadal Oscillation (PDO; B), the Inter-Decadal Pacific Oscillation (IPO; C) the North Pacific index (NP; D) and the Pacific/North American Index (PNA; E). The two vertical lines denote the shift observed for some indices in 1976 and 1996–1998.

the North Pacific climatic variability. The Pacific Decadal Oscillation (PDO) is the first principal component obtained from a principal component analysis performed on residuals of sea surface temperature above 20°N in the Pacific Ocean (118) (Figure 2.19B).

The North Pacific climate has oscillated between cooler and warmer regimes. Climate reconstructions from tree-rings and Pacific coral suggest that this has occurred from at least 1600 (119). The PDO exhibited pronounced cool regimes from 1890 to 1924 and from 1947 to 1976. A warm regime prevailed from 1977 to at least the mid-1990s (119).

During a warm phase of the PDO, there is a reinforcement of the Aleutian Low Pressure Centre, which increases the cyclonic circulation, decreasing the temperature at the centre of the North Pacific and increasing the temperatures along the North American coast because of the advection of warmer air. This affects precipitation patterns in many parts of the Pacific. Dry conditions occur in the eastern part of Australia, Japan and the central part of Alaska, while wetter conditions occur in the coastal regions of the Gulf of Alaska. A negative phase of the PDO is characterised by the opposite climatic conditions. The PDO has strong consequences for marine biodiversity that I will detail latter in the book. The PDO is mainly distinguished from ENSO because of three characteristics. First, a phase of the PDO remains for about 20–30 years, while a phase of the ENSO lasts between 6 and 18 months. Second, exact mechanisms that control the PDO are not yet known, although recent studies suggest a strong control by the atmosphere. Third, the ENSO has its main signature in the tropics, while the PDO has its signature mainly in the extratropical part of the North Pacific.

The Inter-Decadal Pacific Oscillation (IPO) is calculated by performing a principal component analysis on global SST (Figure 2.19C). The first principal component is said to reflect mainly the changes taken place in the tropical Pacific. However, it has been stressed that the PDO and the IPO exhibit nearly identical temporal variability (120). The decadal variability of the PDO/IPO is considered to be largely influenced by ENSO (Figure 2.19A for the Southern Oscillation Index). However, the IPO can independently affect the location of the ITCZ. Both ENSO and PDO changed significantly after 1976 (Figure 2.19). The shift observed after 1976 in the PDO may be of tropical origin (121). Some results suggest that the PDO could be controlled by processes operating in the tropics, but much remains to be known about this pattern of oceanographic variability. The PDO may be the resulting emerging pattern forced by the Aleutian Low, the ENSO and the Kuroshio-Oyashio extension. Therefore, this is the pattern in atmospheric pressure that should be monitored, and not the pattern in sea surface temperature. Two main indices have been used. The first is the North Pacific Index (NP; Figure 2.19D). NP is the mean sea level pressure anomalies in the Aleutian Low over the Gulf of Alaska (30°N–65°N, 160°E–140°W) (122). The second is the Pacific/North American Index (PNA). The PNA index is the difference between the average height anomalies at 500 hPa in the two regions centred on 20°N, 160°W and 55°N, 115°W and the average height anomalies at 500 hPa in the two regions centred on 45°N, 165°W and 30°N, 85°W (123) (Figure 2.19E). It should be noted that the 1976 and the 1996–1998 shift in both the PDO and the IPO is also evident in the SOI. The second shift is also evident in tropical SST anomalies (Plate 2.16).

2.6.1.6 The Madden–Julian Oscillation

The Madden–Julian Oscillation (MJO) is the dominant mode of intraseasonal variability in the tropical troposphere (124). This oscillation was discovered by Roland Madden and

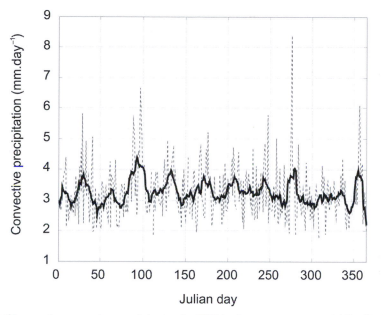

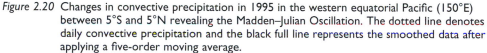

Figure 2.20 Changes in convective precipitation in 1995 in the western equatorial Pacific (150°E) between 5°S and 5°N revealing the Madden–Julian Oscillation. The dotted line denotes daily convective precipitation and the black full line represents the smoothed data after applying a five-order moving average.

Source: Convective precipitation data originated from the ERA-INTERIM data set.

Paul Julian in 1971, and it is now seen as a major source of weather variability at the equator. The dynamical and convective perturbation associated with the oscillation propagates eastwards along the equator at the speed of 5 m·s^{-1}. A complete cycle lasts between 30 and 60 days (Figure 2.20).

Despite its planetary scale, the oscillation has been clearly identified by its pattern in rainfall in the warm region of the Indian Ocean and the western and central part of the tropical Pacific. When the oscillation reaches the eastern part of the Pacific, it decreases in intensity because of the cooler temperatures. However, when the MJO reaches the equatorial part of the Atlantic, it reinforces again. The climatic parameters that are modified by the oscillation are clouds, pressure, moisture, deep convection and winds.

A research team led by Jérôme Vialard from Paris (Institute for Research and Development) showed that two MJO events that happened between November 2007 and February 2008 led to an intraseasonal variability in the range of 1.5–2°C at a mooring site located off the Chagos Islands in the Indian Ocean (*125*). They calculated that the net surface heat flux in the ocean increased from 15 W·m^{-2} during active phase of the MJO (i.e. strong convective activity) to 105 W·m^{-2} during suppressed phase of the MJO (i.e. no convective activity). They therefore showed the strong interaction between the atmosphere and the ocean during an MJO event.

A lot of uncertainties remain about the initiation, development and impact of the MJO (*126*). However, this oscillation may influence tropical cyclone activity, the onset and break activity of the Asian–Australian monsoon system and extreme flooding events in the western coast of America (*127, 128*). This oscillation also modifies the structure of the thermocline in the eastern part of the equatorial Pacific by triggering westerly wind bursts, which may in turn play a critical role in triggering ENSO events. How this may affect the biodiversity remains poorly known, and only a few publications exist on the subject. In particular, there are suggested links between the MJO and the formation of tropical cyclones, which can significantly affect coral reef ecosystems.

2.6.1.7 The Southern Annular Mode

The Southern Annular Mode (SAM), once called the Antarctic Oscillation, represents the main source of atmospheric variability in the southern hemisphere between the atmospheric masses of the middle and the high latitudes. It therefore determines the strength of the Subpolar westerly winds and thereby the strength of the Antarctic Circumpolar flow through the Drake Passage, which in turn may alter the Meridional Overturning Circulation. When a PCA is performed on global atmospheric mass, the phenomenon accounts for 10% of the total variance (129). The oscillation reflects jet stream variability and storm tracks, and is therefore primarily controlled by the internal dynamics of the atmosphere. The SAM influences the preferred path of synoptic-scale weather systems (130). During a positive phase of this oscillation, westerly winds are reinforced over the subarctic regions, while the opposite pattern is observed during a negative phase. The SAM has increased since 1958. This trend may explain, in part, the warming of temperatures observed in the Antarctic Peninsula. Indeed, a positive phase of the SAM is characterised by the location of a low-pressure centre west of the Peninsula, which tends to bring warmer air from the south. This warmer air reduces ice coverage mainly between December and May. The positive phase of the SAM also brings more precipitation around Antarctic coasts. The ENSO may also affect the SAM during the austral summer.

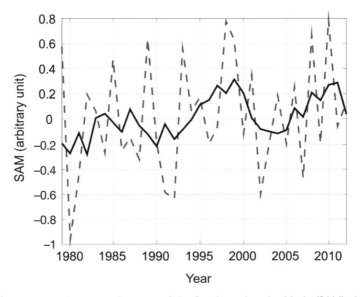

Figure 2.21 Year-to-year changes in the state of the Southern Annular Mode (SAM). A two-order moving average is superimposed (in black).

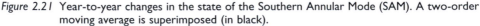

Source: SAM data are from the Climate Prediction Center.

An index of the SAM is calculated by performing a principal component analysis (PCA) of monthly mean of 850 hPa height poleward of 20°S (Figure 2.21).

Although strong year-to-year variability exists, the SAM shifted from a predominantly negative to predominantly positive phase between the middle and the end of the 1990s. The extent to which this may be connected to the cooling of the East Pacific is an important issue. The reinforcement of the low-pressure cell to the west of the Antarctic Peninsula and the South-East Pacific High may bring more cool air towards the east tropical part of the Pacific Ocean (Plate 2.17). The same phenomenon may have happened in the North Pacific, as clearly indicated by the PDO and the IPO.

2.6.1.8 The Indian Ocean Dipole

The Indian Ocean Dipole (IOD), also called the Indian Ocean Zonal Mode (IOZM), is characterised by the contrast in sea surface temperature between the eastern (Sumatra) and the western (Somalia) part of the Indian basin. This phenomenon is particularly pronounced during the boreal autumn when cooling occurs off Sumatra and warming off Somalia. An index of the IOD is calculated from a measure of the Dipole Mode Index (DMI) in the south-eastern equatorial Indian Ocean (90°E–110°E and 10°S–0°N) (Figure 2.22). When the DMI is positive, it indicates a positive phase of the IOD, and vice versa.

The IOD is not thought to be independent of ENSO. For example, the strongest IOD ever monitored took place during the super El Niño event of 1997–1998 (Figure 2.22).

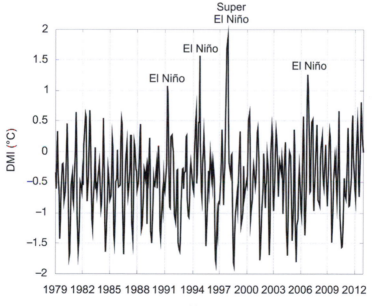

Figure 2.22 Year-to-year changes in the state of the Indian Ocean Dipole as revealed by the Dipole Mode Index (DMI) calculated here from SST ERA INTERIM data. El Niño events that coincide with strong positive values of DMI are indicated. La Niña events are not indicated, although some coincide with sustained negative values of DMI (e.g. 1996 and 1999).

Catastrophic flooding was observed in the eastern part of Africa. This meteo-oceanic pattern influenced the spatial distribution and abundance of some large pelagic fish (yellowfin and bigeye tuna).

2.6.1.9 The Atlantic Multidecadal Oscillation

The Atlantic Multidecadal Oscillation (AMO) is a large-scale oceanic phenomenon, a source of a natural variability in the range of 0.4°C in the oceanic regions of the North Atlantic (*131*). The AMO is also detected in palaeoclimatic reconstructions with a cycle ranging from 60 to 100 years (*132*). Knight and colleagues (*133*) also showed that a 1,400-year control simulation of the atmosphere-ocean general circulation model HadCM3 captured a similar pattern with both a period and an amplitude similar to the AMO. Although at present the nature of the oscillation remains unknown, it has been hypothesised that it primarily arises from changes in MOC.

A simple index can be constructed by calculating a 10-year running mean of Atlantic (detrended) sea surface temperature anomalies north of the equator (*131*) (Figure 2.23).

The AMO exhibits a cycle between 70 and 80 years when cool phases of about 30 years (e.g. 1905–1925; 1970–1990) alternate with warm phases (e.g. 1930–1960; mid-1990s onwards) (*131*, *134*) (Figure 2.23).

Other indices are based on Extended Reconstruction SST (ERSST) data, averaged in the area of 25–60°N and 7–75°W, minus the linear regression on global mean temperature. An interesting variation of the AMO index is accomplished by averaging annual sea surface temperature anomalies in the area ranging from 0° to 60°N and from 0° to 80°W and by subtracting the values by the average of annual SST anomalies for the whole global

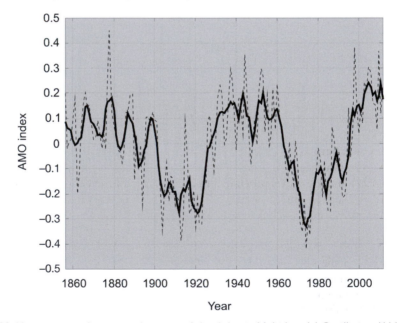

Figure 2.23 Year-to-year changes in the state of the Atlantic Multidecadal Oscillation (AMO). A two-order moving average was applied on the AMO index (heavy line).

Source: Unsmoothed AMO data were downloaded from the Climate Prediction Center.

ocean between 60°S and 60°N. This may remove the influence of global anthropogenic changes (*135*).

This oscillation influences hurricane activity and the frequency of major hurricanes in the North Atlantic, rainfall variability over North America and Europe, and may also be involved in Sahel droughts. For example, during the warming phase of the AMO, rainfall decreases over North America. Between the warm and cool phases of the AMO, a modification of 10% of the outflow of the Mississippi River outflow is observed (*131*). The recent warming phase of the AMO was characterised by an increase in sea temperature of 0.3°C compared to the cool phase of the period 1970–1990. During a negative phase of the AMO, the hurricane activity diminishes in the North Atlantic (*135*). Although uncertainties remain, experiments using general circulation models have shown that the Meridional Overturning Circulation (MOC) plays an important role in controlling the AMO. The AMO has a significant influence on marine biodiversity of the North Atlantic, from plankton to fish (*136*).

2.6.1.10 Strength and weakness of large-scale hydro-climatic indices

We reviewed here only a fraction of the existing meteo-oceanic indices. These indices are very popular to identify potential links between climate and marine ecological changes (*137–139*), and many empirical relationships have been discovered: for example, relationships between the NAO and the Atlantic cod (*137*), the Pacific Decadal Oscillation (PDO) and the ecosystem state of the North Pacific (*59*), the Atlantic Subarctic index and the ecosystem state in the eastern North Atlantic (*140*), and the North Pacific Gyre Oscillation (NPGO) and the ecosystem state in San Francisco Bay (*141*). However, these relationships are unlikely to remain stable. One reason is that similar values of an index exist for different climatic situations. For example, a positive SOI may occur for a canonical or a Modoki El Niño event, which have distinct consequences on the regional climate (e.g. sea surface temperature, depth of the thermocline). In the same way, two similar values of the NAO can mask very distinct climatic situations, depending on the location of the Icelandic Low or the Azores-Bermuda High. For example, on the eastern side of the Icelandic Low, southerly winds prevail, which tends to bring warmer air and moisture. The opposite situation is true on the western margin. The spatial pattern of the NAO can undertake subtle changes in location that can have sustained consequences on physical and biological systems. Hurrell and Dickson (*96*) illustrated this point well. They explained the changes in convection that happened in the central part of the Labrador Sea between the period 1993–1995 and 1999 by a difference in the location of the centres of action. In the former period, the spatial pattern of the NAO was conformable, characterised by a centre of low pressure over Iceland and a centre of high pressure located over the Azores. The atmospheric circulation was such that dry and chill north-westerly winds prevailed in the Labrador Sea, triggering deep convection. In 1999, the high pressures extended to Newfoundland and the southern part of the Labrador Sea. Calmer conditions prevailed, explaining the absence of deep convection.

The chaotic nature of the atmospheric circulation makes it that, on a daily basis and at a synoptic scale (i.e. the scale of the weather), the NAO is rarely observed. This is, however, the scale at which atmospheric variability affects marine biodiversity. The causality is therefore difficult to establish. Furthermore, hydro-climatic indices are often a summary of the state of a hydro-climatic subsystem, and we know that climate operates globally to redistribute energy. Therefore, it is likely that some indices are interconnected. We have seen coincidental shifts between the middle and the end of the 1990s for the PDO, IPO, SOI and SAM (AAO) (Figure 2.19).

Large-scale indices are useful to identify teleconnections or global synchrony among remote marine populations or communities, but to be able to understand processes at work and thereby develop predictive capabilities more conventional meteo-oceanic or hydro-climatic parameters are needed. Indeed, the connection between meteo-oceanic or hydro-climatic indices and biodiversity takes place through the effects of atmospheric circulation on oceanic currents, the structure of the water column and sea surface temperatures. It is therefore recommended to fully examine the mean state of the atmosphere using variables such as a mean sea level pressure that influences both atmospheric and oceanic circulations (i.e. winds, oceanic currents, upwelling and thermocline) and sea temperature that directly affects species physiology and ecology to clearly understand the climatic influence on biodiversity.

Changes in the atmospheric circulation influence storm tracks. Storm tracks have been rarely used to estimate the potential influence of weather on biodiversity. The Climate Prediction Center of the National Weather Service of the United States of America measured storminess by the minimum in Sea Level Pressure. Atmospheric circulation influences the weather regime. A weather regime is a persistent and recurrent pattern of low-frequency (suprasynoptic) intraseasonal variability, longer than the baroclinic-eddy life cycles and lasting between 10 and 100 days. Statistical methods (e.g. multivariate analyses on geopotential height) can be applied to detect these weather regimes. Each regime is characterised by a type of circulation and temperature at a large scale. Some authors have thereby classified the winter weather of the North Atlantic into four regimes (53):

1   A zonal regime identified by strong zonal wind (westerly winds) in oceanic areas south of Greenland bringing warm air over the northern countries of Europe (corresponding to the positive phase of the NAO). The jet stream crosses the whole North Atlantic.
2   A blocked regime identified by the presence of a high-pressure centre over the western part of Europe. The wind is stronger over Iceland and temperatures are colder over Europe. In the southern countries of Europe and the northern part of Africa, cyclonic activities are associated with a jet stream.
3   A regime influenced by the high-pressure centre over Greenland. This regime is zonal, but the jet stream is more located to the south.
4   A weak regime characterised by a very weak zonal circulation.

The transition from one regime to another is difficult to forecast, and the duration of a regime is variable. This way of characterising the climate seems more powerful than the use of indices such as the NAO. It may provide more elements of interpretation to understand the true nature of a link between climatic variability and the state of marine biodiversity.

Chapter 3

Large-scale biogeographic patterns

In this chapter, we examine how biodiversity is partitioned at a global scale, which is the scope of biogeography. Biogeography is defined by Lomolino and colleagues as 'the science that attempts to document and understand spatial patterns in biological diversity' (10). The authors pursued this by stating that one of the most fundamental questions in biogeography is to understand how biodiversity varies on the earth's surface. Indeed, species are not uniformly distributed, nor organised. Of the 1.2 million species living today, many are restricted to small regions of the ecosphere because they can only tolerate a narrow range of environmental conditions. Life has evolved to invade the large diversity of habitats or environments available on our planet.

3.1 Biogeography: from the terrestrial to the marine realm

Understanding how life is organised occupied early scientists such as Carolus Linnaeus (1707–1778) and Georges-Louis Leclerc, Comte de Buffon (1707–1788). Johann Reinhold Foster (1729–1798), Karl Ludwig Willedenow (1765–1812) and Alexander von Humboldt (1769–1859) showed that different plant associations were linked to the local climate (10). Augustin-Pyranus de Candolle (1778–1841) provided evidence that organisms were not only related to the climatic regime, but also competed for the resources. Alexander von Humboldt noticed that mountainous plants exhibited a zonation as a function of the altitude. At that time, Johann Reinhold Forster had already noticed latitudinal biodiversity patterns.

3.1.1 First biogeographic partitions and proposed terminology

3.1.1.1 Early classification schemes

Many classification schemes have been proposed. Augustin-Pyranus de Candolle proposed to divide the ecosphere into 20 vegetational regions in 1820 and 40 in 1838. Using mammals, the zoologist James Cowles Pritchard distinguished seven regions in 1826. Based on the study of angiosperms, the British researcher Ronald Good proposed six kingdoms while Charles Smith, using a multivariate statistical analysis on mammals, found four.

Biogeographic realms are defined as large-scale regions within which ecosystems have a high level of homogeneity due to similar evolutionary history. Current biogeographic realms are often based on the work of the eminent British ornithologist Philip Lutley Sclater. Sclater (1858) and Alfred Russel Wallace (1876) based their classification on birds, and divided the ecosphere into eight biogeographic realms (Figure 3.1).

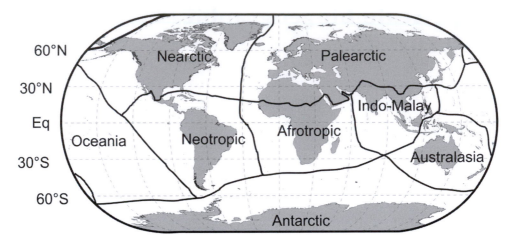

Figure 3.1 Division of the biosphere into biogeographic realms.

Source: From the Sclater and Wallace classification of faunal regions.

Early biogeographic partitions remain highly heterogeneous from a biological point of view, and the division of the biosphere has also been undertaken by biome. The word biome was coined by the two American biogeographers Frederic Clements and Victor Shelford to designate the biotic formation, a basic community unit, primarily controlled by climate (*142*). Nowadays, the term biome is also used to qualify large regions of the **ecosphere** that have the same type of climate, soils or vegetation (*10*). Heinrich Walter defined the term biome as 'habitats which correspond to a concrete uniform landscape'. He then added that the term was employed 'for the fundamental unit of which larger ecological systems were made up'. The oceanographer Alan Longhurst, in his outstanding book *Ecological Geography of the Sea*, defined the biomes as the primary partition of the biosphere (*143*). However, he used the word to refer to large-scale ecosystems (compartments). Therefore, the term has not exactly the same meaning according to all authors. For some, it refers to a primary **biocoenosis** in equilibrium with climate, and for others to a primary ecosystem in equilibrium with climate. Other terms have also been employed: for example, ecozone (*144*), ecoclimatic regions or ecoregions (*10, 145*).

3.1.1.2 A proposed terminology

The first level of partition we might distinguish is the ecogeographic/biogeographic realm. Then, I propose to use the term ecome when the ecosphere (i.e. the global ecosystem) is partitioned and biome when the biosphere (i.e. the global biocoenosis) is divided (Figure 3.2).

Ecomes are large-scale primary compartments composed of distinct vegetation or animal type called biomes that interact with their environment. Ecomes can be divided into **provinces** (*sensu* Longhurst) or **biohydrocenes** (*sensu* Walter) to give a second partition of the ecosphere. I will use the term province, which was originally proposed by phytogeographers. A province has been defined as an area characterised by some level of endemism, with species sharing a common history (*146*). In addition, a province has been defined as

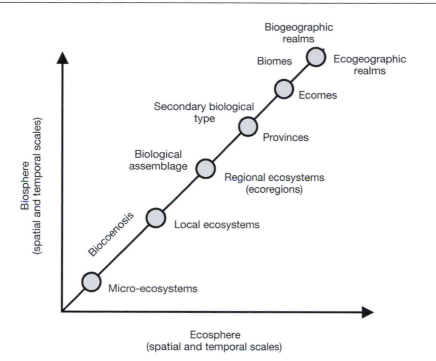

Figure 3.2 Relationships between ecological and biological systems at some spatial and temporal scales.

an association of ecosystems that may change over time in the same way. Provinces are also sometimes divided into ecoregions (*147*). Ecoregions were defined by Spalding and colleagues (*147*) as:

> areas of relatively homogeneous species composition, clearly distinct from adjacent systems. The species composition is likely to be determined by the predominance of a small number of ecosystems and/or a distinct suite of oceanographic or topographic features. The dominant biogeographic forcing agents defining the ecoregions vary from location to location but may include isolation, upwelling, nutrient inputs, freshwater influx, temperature regimes, ice regimes, exposure, sediments, currents, and bathymetric or coastal complexity.

For the authors, endemism was not a key determinant in the establishment of the Marine Ecoregions of the World (MEOW).

The marine and terrestrial ecospheres (or biospheres) are so fundamentally different that they are often separated. Heinrich Walter propose to call the geobiosphere what I call the terrestrial ecosphere and the hydrobiosphere what I term the marine ecosphere. Note that Jürgen Schultz separates the two realms into terrestrial and marine ecozones (*144, 148*).

3.1.2 Partition of the terrestrial ecosphere

We now briefly see how the present-day terrestrial ecosphere is partitioned because many more works have been carried out in this realm. The terrestrial biosphere is mainly influenced

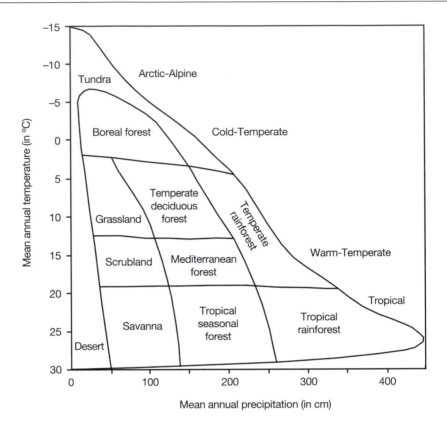

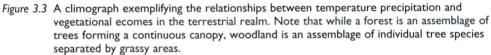

Figure 3.3 A climograph exemplifying the relationships between temperature precipitation and vegetational ecomes in the terrestrial realm. Note that while a forest is an assemblage of trees forming a continuous canopy, woodland is an assemblage of individual tree species separated by grassy areas.

Source: Modified after Whittaker (*149*).

by two climatic factors: temperature and precipitation. This is well exemplified by the classification of the ecomes by the American plant ecologist Robert Whittaker (Figure 3.3).

When average annual temperatures are lower than –5°C, tundra is the only ecome. When average annual temperatures are greater than 20°C, ecomes ranging from desert to the tropical rainforest arise. Temperature seems to discriminate ecomes slightly better than precipitation. However, inside a high thermal regime, precipitation discriminates well the ecomes. Both high temperatures and high precipitations enable the occurrence of the most diverse ecome of the planet: the tropical rainforest. This ecome, which covers only between 2 and 6% of the total land masses, encompasses an estimated 50% of the known planetary biodiversity. The examination of the climograph shows how important the two parameters temperature and precipitation are for explaining patterns in global terrestrial biodiversity. The parameters delineate the **bioclimatic envelope** of each terrestrial ecome, explaining in turn their geographical range. When the different vegetational ecomes of Walter (*148*) are superimposed on the average planetary atmospheric circulation, the relation between biomes and climate becomes patent (Figure 3.4).

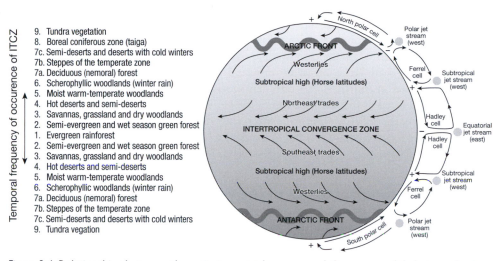

Temporal frequency of occurence of ITCZ

9. Tundra vegetation
8. Boreal coniferous zone (taiga)
7c. Semi-deserts and deserts with cold winters
7b. Steppes of the temperate zone
7a. Deciduous (nemoral) forest
6. Scherophyllic woodlands (winter rain)
5. Moist warm-temperate woodlands
4. Hot deserts and semi-deserts
3. Savannas, grassland and dry woodlands
2. Semi-evergreen and wet season green forest
1. Evergreen rainforest
2. Semi-evergreen and wet season green forest
3. Savannas, grassland and dry woodlands
4. Hot deserts and semi-deserts
5. Moist warm-temperate woodlands
6. Scherophyllic woodlands (winter rain)
7a. Deciduous (nemoral) forest
7b. Steppes of the temperate zone
7c. Semi-deserts and deserts with cold winters
9. Tundra vegation

Figure 3.4 Relationships between the main terrestrial ecomes and the average global atmospheric circulation of the earth. The different terrestrial ecomes are from Walter. The ecomes 7a–c are also influenced by the continentality or the proximity of mountains. Deciduous forests (7a) are mainly found at the margin of continents. The steppes of the temperate zone (7b) are mainly found at the centre of continents or close to mountains. The semi-deserts and deserts with cold winters (7c) are mainly found to the eastern or northern sides of mountains at the centre of continents. The symbols + and – denote the processes of convergence and divergence, respectively.

Some ecomes such as the steppes of the temperate zone and the semi-deserts and deserts with cold winters are, however, better explained by secondary factors (e.g. continentality and proximity of mountains) that modify the influence of climate zonally. Inside provinces, other secondary factors may control the type of vegetation and therefore the distribution of animals. For example, the pH, the altitude, the sun exposition, the type of soil (also influenced by climate) and the proximity of a river or a lake may have a strong influence on the biocoenosis.

3.1.3 Terrestrial versus marine biogeography

To my best knowledge, a similar climograph (Figure 3.3) has not been outlined yet for the marine ecosphere. Progress in marine biogeography has been more difficult to achieve because of the inherent nature of the marine environment, and this despite the fact that the oceans cover about 71% of the earth's surface. The marine realm is fundamentally structured in three spatial dimensions (depth, latitude, longitude). The oceanic hydrosphere has an average depth of 3,800 m and about 88% of the oceans are deeper than 1,000 m (*150*). Some areas, called trenches, can be as deep as 11 km (e.g. the Mariana Trench, with a depth of 11,033 m). From an ecological perspective, the difference between the marine and the terrestrial ecospheres lies in the presence of water, which influences light penetration, the distribution of all forms of energy (**exosomatic** and **endosomatic** energy) and oxygen. Most oceanic regions are aphotic, which means that only a weak amount of energy exists (carbon exportation takes place from the euphotic zone) to sustain biodiversity. Hydrothermal vents and cold seeps (also called cold vents), based on chemosynthesis, are perhaps the only exception.

From a biogeographical perspective, environmental conditions are more consistent in the marine realm with a priori much less microclimates, especially in the pelagic realm. Therefore, the distributional range of organisms is probably more predictable than in the terrestrial realm. This is less the case in some ecosystems such as rocky shores where microclimatic refugees occur. In contrast to terrestrial biogeography, marine biogeography has been mainly based on animals (150), although recently rigorous biogeochemical procedures have been developed (143). Delineation of biomes is more difficult than in the terrestrial realm because of the much shorter generation time of organisms that populate this sphere (e.g. a few days for phytoplankton in comparison to decades for trees) and the absence of a strict physical barrier.

3.2 Primary compartments of the marine ecosphere

3.2.1 Potential factors affecting species spatial distribution

3.2.1.1 Climatic factors

As we have seen in Chapter 2, multiple climatic parameters may influence biodiversity. Atmospheric circulation has a strong influence on surface currents, upwelling and the structure of the water column (thermocline, halocline and the resulting pycnocline). Wind intensity also affects prey-predator encounter rates (151) by its effects on oceanic turbulence, and nutrient supply rates by its effects on vertical mixing (152). Wind direction, by its control on the dispersal of meroplanktonic species, might strongly affect recruitment of some benthic organisms (153). Incident solar radiations are also important because they provide the energy that influences sea surface temperature and the structure of the water column. Photosynthetically active radiation (PAR), solar radiation spectrum in the wavelength range of 400–700 nm, is also relevant because it represents the level of energy that can be assimilated by photosynthetic organisms (154). PAR regulates both the composition and evolution of marine ecosystems, influencing the growth of phytoplankton, and in turn the development of zooplankton and fish. This factor is also strongly influenced by cloudiness. As we will see (see also Chapters 4–7 and 11), temperature is a key variable that influences the spatial distribution of species. The parameter is largely influenced by atmospheric circulation, incident solar radiation, oceanic circulation and cloudiness. In contrast to the terrestrial realm, precipitation or any measure of water availability has, in general, a small influence on the marine environment. Precipitation may become important in some coastal regions (155).

3.2.1.2 Marine environmental factors

Many environmental factors influence the arrangement of life in the oceans (Plate 3.1). Dissolved oxygen must remain high enough to support respiration (157, 158) (Plate 3.1A). Mixed layer depth (MLD) is an important parameter for phytoplankton production and controls the spatial distribution of many plankton species (152, 159) (Plate 3.1B). Both bathymetry (Plate 3.1C) and local spatial variability in bathymetry are key determinants of the marine pelagic biodiversity (160). At the bottom, the variation in the bathymetry is also responsible for the creation of very distinct habitats (160). Bathymetry influences light penetration, all parameters enunciated above, including pressure that directly influences plant and animal anatomy and their life history traits. Salinity is an important factor. Capacity of dispersal can be influenced by salinity through the ability of marine species to osmoregulate (*161*). Salinity is high over subtropical gyres and in the Mediterranean Sea (Plate 3.1D).

Sediment type is also an important parameter for benthic biodiversity. Oceanic pH influences calcifying organisms such as coccolithophores, foraminifers, corals and pteropods (*162, 163*). We will hierarchise the influence of these parameters on the biodiversity in the next sections and chapters. Although apparently uniform, the oceans are therefore highly heterogeneous at all spatial scales, creating a variety of habitats, which also explains why marine biodiversity is so diverse.

3.2.1.3 Macronutrients

Nutrients also limit primary production (*164*). Plate 3.2 shows the spatial distribution in some macronutrients (nitrates, phosphates and silicates). Large concentrations in macronutrients are found in extratropical regions (especially in polar regions) and to a lesser extent along the equator. Subtropical gyres are poor in macronutrients.

3.2.1.4 High nutrient low chlorophyll areas

High nutrient low chlorophyll (HNLC) areas are oceanic regions where phytoplankton standing stock is low despite high concentrations of macronutrients (e.g. nitrate, phosphate, silicic acid). Small phytoplankton species are more common than diatoms and spring **blooms** are never observed. These regions, which cover about 20% of the world's oceans, are mainly the equatorial Pacific Ocean, the subarctic Pacific Ocean and the Southern Ocean (Plate 3.3). Two hypotheses have been proposed to explain the existence of these areas. The first, the iron hypothesis, proposes that macronutrients are never depleted because other micronutrients (e.g. iron) control phytoplankton proliferation (*165*). Concentrations in bioavailable iron in these regions are significantly smaller than in other oceanic provinces. The second hypothesis, the grazing control hypothesis, stipulates that phytoplankton cannot reach high standing stocks because they are limited by grazing (*166*). Although grazing may be important in some regions, iron plays a very important role. Iron is mainly abundant in coastal regions. In contrast, open-ocean waters have much less iron and are said to be infertile in this respect (*165*). Iron is delivered to the pelagic oceans by dust storms from arid lands, which contain between 3 and 5% iron.

Although ocean iron experiments have given contrasting results, some have shown that the addition of iron in some HLNC regions stimulated phytoplankton and rapidly diminished nitrate concentrations. During IRONEX I, an area of 64 km^2 in the eastern equatorial Pacific was fertilized with an iron sulphate solution (*167*). The amount of iron was multiplied by 20 and was at the origin of a doubling in chlorophyll concentration at the surface. However, the biological response was short lived, and after five days phytoplankton growth returned to values observed before the beginning of the experiment because iron subduction took place at a depth where light limits phytoplankton proliferation. This experiment showed that iron is therefore an important micronutrient for phytoplankton growth and photosynthesis.

3.2.2 Partitions of the marine ecosphere

3.2.2.1 The diversity of biogeographic approaches

Dinter (*150*) reviewed a number of elements that have been used by biogeographers to partition the oceans and seas. First, each biogeographer tends to focus on a specific domain or kingdom, and many based their division on a single phylum, class, order, family or genus. Second, the spatial scale considered in the classification is neither always global, nor does it incorporate all depths of the pelagic realm or substrates or types of ecosystems for the benthic realm. Third, the methods used to divide the biosphere are numerous. Some biogeographers even stated that 'the biogeographic method does not exist, or there are as many methods as biogeographers' (*143, 168*). Past divisions have been mainly based on expert knowledge. Although current divisions tend to be based on statistical techniques (*169*), the expert knowledge continues to be used (*143*). Fourth, partitions are also influenced by the use biogeographers want to give them. In the past, divisions were exclusively dictated by biological or ecological issues (identification of biocoenoses, provinces and biomes) but recent partitions have been established for management purposes.

3.2.2.2 First partitions of the marine ecosphere

Marine organisms have virtually colonised all **biotopes** from the intertidal to the subtidal benthic realm, which range from the continental shelf to the hadobenthic zone (Figure 3.5). In the pelagic realm, living creatures have established in both neritic and oceanic regions from the epipelagic (0–200 m) to the hadopelagic zone (i.e. bathymetry > 6,000 m). Even if similar environmental conditions occur in different oceanic and neritic regions, different species may be present, according to **Buffon's law**, also known as the first principle of biogeography (*10*). This law, established in the terrestrial domain, is based on the observation that environmentally similar regions have distinct assemblages of mammals and birds. This law was subsequently extended to other taxonomic groups.

Many partitions of the marine ecosphere have been proposed, and it is not possible to enumerate all of them in this book. Mark Spalding and colleagues listed the work of Forbes (1856), Ekman (1953), Hedgpeth (1957), Briggs (1974) and Bailey (1998) (*147*). More recently, partitions based on 'thermogeography' have been proposed for benthic marine algae (*170*). Temperature variability over large timescales explained well the partition of Briggs (*171*). Briggs's partition also considered endemism, each province being based on 10% endemism. An improvement of this partition has been recently proposed (Plate 3.4). The marine ecosphere was divided into three main ecomes: (1) cold regions (Arctic and Antarctic); (2) cold-temperate; and (3) warm-temperate regions. Provinces were also identified.

3.2.2.3 Classifications for ecosystem management

Classifications have been proposed to improve ecosystem management (e.g. Large Marine Ecosystems, ecoregions). Large Marine Ecosystems (LMEs), implemented by Kenneth Sherman and colleagues, are large regions (i.e. ≥200,000 km²) based on their: (1) bathymetry; (2) hydrography; (3) productivity; and (4) trophically dependent populations (*173*). A total of 66 LMEs have been proposed so far (Figure 3.6). They were originally designed to tackle environmental issues such as fisheries management, and only concern currently large continental shelves.

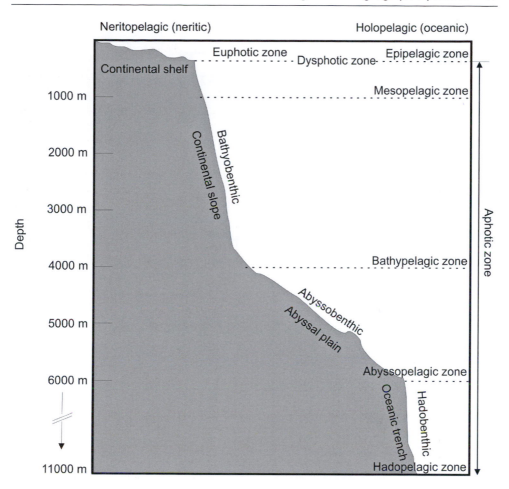

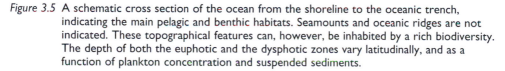

Figure 3.5 A schematic cross section of the ocean from the shoreline to the oceanic trench, indicating the main pelagic and benthic habitats. Seamounts and oceanic ridges are not indicated. These topographical features can, however, be inhabited by a rich biodiversity. The depth of both the euphotic and the dysphotic zones vary latitudinally, and as a function of plankton concentration and suspended sediments.

Another partition has been more recently proposed by Mark Spalding and colleagues (*147*). The partition was mainly based on a review of the literature (230 works published in peer-reviewed journals or in reports), expert knowledge and utilized criteria such as evolutionary history, patterns of dispersal and isolation. They divided the continental shelves into 12 realms, 62 provinces and 232 ecoregions (Figure 3.7).

3.2.3 The pelagic realm

The principal difficulty in partitioning the marine pelagic ecosphere is related to the dynamic movement of water masses and the locations of surface features, which are influenced by atmospheric conditions. This difficulty led the biogeographer van der Spoel

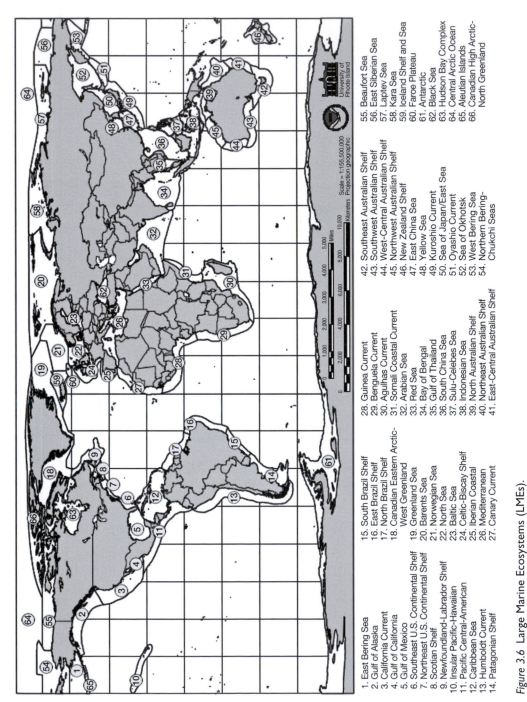

Figure 3.6 Large Marine Ecosystems (LMEs).

Source: Courtesy of Kenneth Sherman.

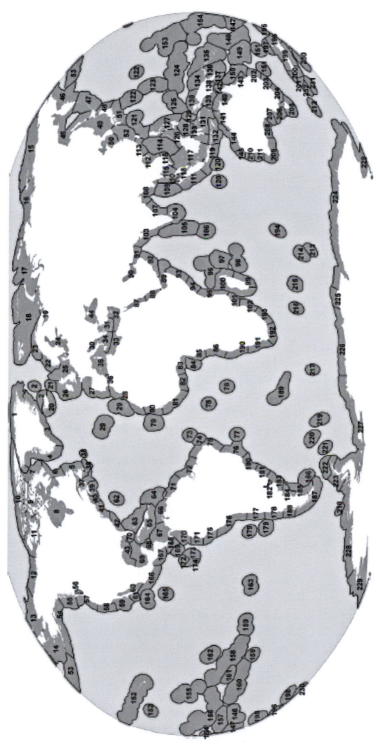

Figure 3.7 Marine Ecoregions of the World (MEOW).

Source: Courtesy of Mark Spalding.

(*174*) to separate the biotope of pelagic ecosystems into two components: (1) a stable-biotope component (geographically stable) in which a primary related community lives; and (2) a substrate-biotope component (depending on water mass) characterised by a secondary related community (mixed primary community) (*175*). An ecosystem is mainly characterised by a primary related community linked to a stable-biotope component, whereas an ecotone is more distinguished by a secondary related-community depending on water masses. It is also known that an ecotone can also be characterised by its own biological composition (*176–178*). The distinction van der Spoel made is fundamental to correctly understand how the pelagic ecosphere can be divided.

3.2.3.1 Biogeochemical partitions

Biological partition is rarely achievable with great precision at a global scale because the spatial distribution of species is poorly known. This is perhaps why some authors have proposed new partitions based on biogeochemical parameters. The development of satellite technology and the globalisation of environmental data sets have enabled the establishment of global biogeography. Plate 3.5A shows the division of the marine ecosphere into biomes and provinces by Alan Longhurst (*152*). Four primary biomes (Polar, Westerlies, Trades and Coastal) and 56 secondary provinces were identified (Plate 3.5A). This partition of the marine ecosphere by Longhurst was mainly based on the characterisation of the seasonal cycle of primary production (*152*). Variables used to establish the partition were chlorophyll concentration, mixed layer depth, nutrients, the Brunt-Vaisala frequency, the Rossby radius of internal deformation, photic depth, algal biomass and primary production. These variables allowed the identification of a number of ecological situations: (1) polar irradiance-limited production peak; (2) nutrient-limited spring bloom; (3) winter-spring production with nutrient limitation; (4) small amplitude response to trade wind seasonality; (5) large amplitude response to monsoon reversal; and (6) various responses to topography and wind-stress on continental shelves, including coastal upwelling (*179*). Using four parameters (bathymetry, chlorophyll a concentration, surface temperature and salinity), Reygondeau and co-workers applied a procedure based on the Non-Parametric Probabilistic Ecological Niche model (*180*) to propose a more dynamical partition of biogeochemical provinces of Longhurst. The average contour of the provinces was, in general, in good agreement with those originally proposed by Longhurst (Plate 3.5B).

3.2.3.2 Biological partitions

Biogeographical partitions based on species distribution have been proposed by many authors. Mary Somerville (1780–1872), in her book about physical geography, divided the marine ecosphere into homozoic zones. Based on Mollusca, Edward Forbes (1815–1854) established nine homozoic zones and related them mainly to marine isotherms. It is unfortunate, however, that the spatial distribution of so many species remains poorly identified because this would allow a better partition of the biosphere.

Taxonomic biogeography has been said 'to belong to the family of intractable scientific problems' (*143*). However, I think that basing the biogeography on only a few biogeochemical parameters may lead to a too simplistic scheme because most species are very sensitive to the environment. I illustrate my point here focusing on pelagic species, but I expect the discussion to remain valid for benthic species. Pelagic species can be stenograph (i.e. local spatial distribution, ecotone species) or eurygraph (i.e. large spatial distribution) due to the large clines in environmental conditions and the absence of geographical barriers. Because their spatial distribution integrates many environmental parameters, they may be more powerful to partition the ecosphere (Plate 3.6).

Plate 3.6 shows different types of spatial distribution of copepods in the North Atlantic Ocean. Unfortunately, the area under investigation cut the spatial distribution of some species. While we can only see the southern edge of the spatial distribution of arctic and subarctic species, we can only visualise the northern edge of warm-temperate and subtropical species on the plate. Despite this limitation, the plate shows that some species have large (the group *Para-Pseudocalanus* spp.) or more restricted (*Candacia armata*) distributional range. Some are cold-water indicators (e.g. the subarctic species *Paraeuchaeta norvegica*) and some are warm-water indicators (e.g. the genus *Clausocalanus* and *Euchaeta marina*), while others are oceanic or pseudo-oceanic (i.e. occur in both neritic and oceanic regions but preferentially along shelf-edges). *Metridia lucens* has a higher abundance in the ecotone located in the extension of the Gulf Stream and the North Atlantic Current to the west of the British Isles. This copepod is qualified of mixed-water species because it is observed at the boundary between temperate and subarctic waters. Considering the information of all species taken together, this shows that the North Atlantic Drift Province (NADR; *sensu* Longhurst) (Plate 3.5A) remains highly heterogeneous.

From Plate 3.6, we can also see that pelagic biodiversity is constrained by hydro-climatic parameters such as temperature, bathymetry and oceanic surface currents or large-scale hydrodynamic features such as the subarctic gyre (69, 160, 182). Wolfgang Dinter (150) made a list of the environmental factors that control marine species geographical range. The environmental factors, enumerated above, were thought to be the most important factors to explain species geographical distribution. Dinter also added water quality (e.g. nutrients, salinity and turbidity) that can influence pelagic ecosystems over large areas (e.g. HNLC areas). At more local scales, he listed factors such as tidal currents, types of substratum, harbour effect and freshwater inputs that can be relatively important.

Many marine partitions have been proposed. Developments of remote sensing and large-scale ship-based surveys have allowed a better demarcation of the biomes occupied by various taxonomic groups such as coccolithophores (183), N_2 fixers (184) and picocyanobacteria (185). Tuna and billfish species are oceanic top predators that are important for both ecological and economic reasons. Reygondeau and colleagues (169) showed that both tuna and billfish species formed nine well-demarcated assemblages at the global ocean. Each assemblage was characterised by specific environmental conditions and showed a specific species composition. We found high similarity (68.8% homogeneity) between the spatial distribution of the communities of tuna and billfish and Longhurst's biogeochemical provinces, which provide evidence of a solid relationship between these top-predators and the physical and chemical characteristics of the global ocean, in spite of their high tolerance for a wide range of environmental conditions.

Relatively little is known about deeper pelagic realms such as the mesopelagic and the bathypelagic zones. Scientists have only sparse information on these domains. In these deeper waters, the energy on which the food web is based generally comes from organic materials (particulate organic carbon) exported from the epipelagic zone.

3.2.4 The benthic realm

3.2.4.1 Vertical structure of the benthic realm

Although poorly known, some scientists speculated that the seabed might contain about 98% of the total number of marine species (*186*). This realm is structured vertically (Figure 3.5) and horizontally. Vertically, four regions can be recognised: (1) the seabed of the continental shelf; (2) the seabed of the continental slope or the bathyobenthic zone; (3) the abyssobenthic zone (below 4,000 m, over the abyssal plain); and (4) the hadobenthic zone (below 6,000 m). Some authors locate the abyssobenthic zone from 3,000m and subdivide it into an upper-abyssal (3,000 and 4,500 m) and a lower-abyssal layer (4,500 and 6,000 m). Abrupt shifts in species composition have been noticed between 400 and 600 m, at about 1,000m and between 1,400 and 1,600m (*187*).

3.2.4.2 Horizontal structure of the benthic realm

The large-scale horizontal structure of the benthic realm is less understood than the pelagic realm because biological data are scarcer due to limiting equipment and funds for ship time (*146*). The first author who provided a comprehensive division of the abyssobenthic and hadobenthic zones was Vinogradova (*188*). A partition for areas of the deep ocean floor greater than 800 m was proposed by Watling and colleagues (*146*). The latter authors improved the Global Open Ocean and Deep Sea (GOODS) classification outlined during an expert consultation workshop in 2009. Using an approach similar to Longhurst, they used the three parameters – temperature, salinity and particulate organic carbon (POC) – to delineate 14 lower bathyal (801–3,500 m), 14 abyssal (3,500–6,500 m) and 10 ultra-abyssal (or hadal; >6,500 m) provinces (Plate 3.7).

The authors cautioned, however, that the proposed provinces should be considered as hypothetical because they should be tested against observed biological data. These provinces may also be subsequently subdivided. However, they found good agreement between some of their provinces and biological data (e.g. demersal fish for the lower-abyssal provinces; protobranch bivalves and tunicates for abyssal provinces).

3.2.4.3 Neritic regions

Over neritic regions, some typical ecosystems can be recognised: mangroves, kelp forests, seagrasses, coral reefs, intertidal zones and estuaries. Mangrove forests are composed on many halophytes that grow on saline coastal sediments (swampy clay and muddy substratum). About 110 species are representative of these ecosystems (e.g. the genus *Rhizophora* and *Avicennia*). They are places where birds nest and where many crustaceans (and molluscs) and fishes reproduce. They cover 18 million hectares (*189*). Coral reefs occupy about 600,000 km^2, occurring in the waters of more than 100 countries. Their biological diversity is just behind that of tropical forests. We will examine neritic biodiversity in the next chapters of this book.

3.2.4.4 Deep-sea regions

Over the deep sea, many ecosystems can be distinguished: seamount sand knolls, deep-sea sediment habitats, hydrothermal vents and cold seeps (ecosystems based on chemosynthesis),

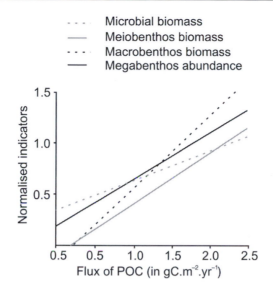

Figure 3.8 Relationships between the structure of the abyssal ecosystems and the flux of POC.
The slopes were derived from regression analysis based on seven pairs of points for
microbial biomass (r^2 = 0.58), five for meiobenthos (nematode) biomass (r^2 = 0.92),
seven pairs of points for macrobenthos biomass (r^2 = 0.96) and five pairs of points for
megabenthos abundance (r^2 = 0.94).

Source: Simplified from Smith and colleagues (*191*).

the latter being located close to volcanically active places along mid-ocean ridges. Deep-
sea regions (i.e. depth greater than 2,000 m) cover about 60% of the earth's surface (*190*).
Less than 1% of deep-sea regions are represented by mid-ocean ridges where ecosystems
are based on chemosynthesis (hydrothermal vents). The rest is covered by soft-sediment
ecosystems, which are fuelled by photosynthetic carbon produced in the **euphotic zone** (Figure
3.5). These systems are therefore limited by the level of energy, and especially the level of
endosomatic energy (i.e. biochemical energy contained in the biomass) that arrives as
phytodetritus, or more generally as POC (Figure 3.8).

The total species abundance on abyssal plains might only represent 1% of the abundance
found on the continental slopes (*191*). Diversity is also thought to be positively related to
POC concentration, peaking over continental slopes, being higher below upwelling regions
(e.g. equatorial upwelling) and lower below large oceanic gyres.

3.2.4.4.1 DEEP-SEA SOFT-SEDIMENT ECOSYSTEMS

Soft-sediment biodiversity is much higher than previously thought. Grassle and Maciolek
(*192*) collected 233 box-core samples off the coast of New Jersey and found 798 species in
a prospected area of 21 m^2 (Table 3.1). Subsequently, by extrapolation, the authors estimated
that 300 million of benthic macrofaunal species might occur in the deep-sea, but they
suggested a conservative estimate of 10 million (*12*). By adding the meiofauna (nematodes),
Lambshead (*194*) speculated that this could raise the number to 100 million. These estimates
should be considered with care because deep-sea biodiversity remains poorly sampled.

Table 3.1 Biodiversity of macrofaunal animals
off New Jersey at depths between
1,500 and 2,500 m.

Phylum	Number of species
Annelida	385
Arthropoda	185
Mollusca	106
Echinodermata	39
Nemertea	22
Cnidaria	19
Sipunculida	15
Pogonophora	13
Echiura	4
Hemichordata	4
Brachiopoda	2
Priapulida	2
Bryozoa	1
Chordata	1
Total	798

Note: A total of 233 samples were collected,
enabling a total of 90,677 animals to be identified.

Source: Data from Grassle and colleagues (193).

Another open question is whether the deep-sea soft-sediment ecosystems are composed of many cosmopolitan species and whether biodiversity hot spots exist. For isopods, foraminifers and polychaetes, some studies indicate the existence of cosmopolitan species. For example, the foraminiferan species *Epistominella exigua* occurs from the Arctic Ocean to the Weddell Sea. Genetic studies provided evidence for gene fluxes between these two regions. However, for isopods, many species are found exclusively over a unique abyssal region. Furthermore, for the same taxonomic group, cold spots are observed over the North Atlantic, whereas hot spots are detected in the Atlantic section of the Southern Ocean (191). This pattern has been attributed to higher fluxes of Particulate Organic Carbon.

3.2.4.4.2 HYDROTHERMAL VENT ECOSYSTEMS

Discovered in 1977 over the Galapagos Spreading Center (2,500 m depth), hydrothermal vent ecosystems cover large regions of the deep sea. Using Multivariate Regression Tree (MRT) analysis, Bachraty and colleagues (195) clustered together 63 hydrothermal fields into six provinces on the basis of their species and genera composition. The six provinces were: (1) Northwest Pacific (Izu-Ogasawara arc, Okinawa Trough, Okinawa arc, Japan ridge); (2) Southwest Pacific (Mariana, Manus, North-Fiji and Lau back-arc basins, Kermadec arc, Tabar-Feni Volcanic Fore arc, Intraplate seamount, Central Indian ridge); (3) Northeast Pacific (Juan de Fuca, Explorer, Garda ridges); (4) Northern East Pacific rise; (5) Southern East Pacific rise; and (6) Northern Mid-Atlantic ridge.

A. Seamounts

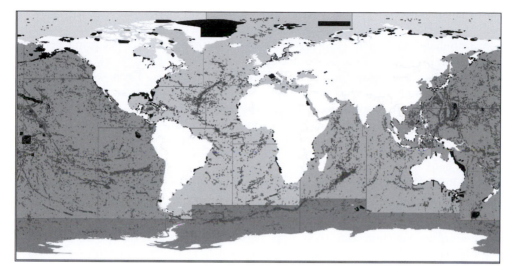

B. Knolls

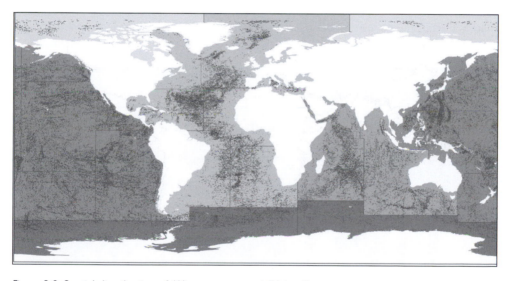

Figure 3.9 Spatial distribution of (A) seamounts and (B) knolls.

Source: Modified from Yesson and colleagues (*196*). Courtesy of Dr Chris Yesson, Zoological Society of London.

3.2.4.4.3 SEAMOUNTS

The deep-sea floor is also interrupted by topographic structures called seamounts (topographic rise of the seabed >1,000 m), submarine knolls (between 500 and 1,000 m) and hills (<500 m). Seamounts and knolls are widely distributed in the marine ecosphere (Figure 3.9).

Chris Yesson and colleagues (*196*) estimated that the numbers of seamounts and knolls were 19,620 and 10,972, respectively. These topographic elevations of the seabed have a strong influence on the benthic biodiversity. They provide a large variety of habitats and host a large number of endemic species. Shallow seamounts may penetrate to the euphotic zone. On such seamounts, the abundance of phytoplankton is high, and macroalgae or zooxanthellate corals may be observed. Middle-depth seamounts, occurring between 1,500 m and the basis of the euphotic zone (i.e. the dysphotic zone), may interact with zooplankton and increase the productivity of demersal fishes. Deep seamounts have biodiversity mainly influenced by the type of substratum, currents exposure and slope. Suspension feeding epibenthic species such as hydroids and cold corals may be observed on hard substrata, whereas species such as sea pens or xenophyophores may occur on soft substrata.

3.2.4.4.4 BIOLUMINESCENCE

Deep-sea biodiversity is peculiar. The most noticeable feature of pelagic deep-sea life is the frequent presence of bioluminescent species bearing complex photophores or light organs. For example, many crustaceans (both hoplophorid and sergestid prawns and euphausiids), starfish, sea cucumbers, octopods and fishes produce light (12). These organisms emit visible light by oxidising a molecule called luciferin by means of an enzyme named luciferase. The biochemical molecule emits light when it is exited, and in some cases transfers its energy to another fluorescent molecule characterised by a specific wavelength. At depth greater than 500 m in the eastern part of the North Atlantic, Peter Herring estimated that 70% of fish and 90% of individuals were bioluminescent. An estimated 80% of decapods below

Figure 3.10 Photo of the giant isopod *Bathynomus giganteus*.

Source: Courtesy of the National Oceanic and Atmospheric Administration (NOAA Photo Library).

500 m are luminous. Most (99% of 87 species) euphausiids, 20–30% of all copepods and most ostracods are luminous in the upper 1,000 m. Some bacteria (genus *Photobacterium*, *Vibrio*, *Shewanella*) are bioluminescent. Some deep-sea species such as the anglerfish *Chaenophryne ramifera* cultures bioluminescent bacteria (luminous symbionts) in lures to attract prey. The function of the bioluminescence in the deep sea is probably related to species interactions (e.g. predator-prey interaction, schooling and sexual display). However, this subject remains extremely difficult to tackle with our current level of technology and the paucity of funding.

3.2.4.4.5 GIGANTISM

Another interesting feature in the deep sea is gigantism. Giant isopods and amphipods are common. For example, the amphipod *Eurythenes gryllus* can grow up to 14 cm, and the swimming isopod *Bathynomus giganteus* can reach 50 cm (Figure 3.10).

Gigantism also occurs in the protozoan group called xenophyophores (*197*). This single cell can be bigger than some metazoans. For example, the genus *Stannophyllum* can reach a diameter of 25 cm and can be as thick as 1 mm (*197*). It remains unclear why these large organisms occur in such an oligotrophic environment. This could be a manifestation of the temperature–size rule, which we will examine in Section 3.3.1.2.

3.3 Ecogeographic patterns

Here, I describe briefly some ecogeographical patterns that may provide some clues on how marine biodiversity is organised. It is important to recall that the generality of these ecogeographic patterns have been debated – that all have exceptions, and that the causal mechanisms that may generate them remain elusive. Although they have been qualified as rule, I prefer to call them as ecogeographical patterns. Latitudinal biodiversity gradients are examined in great detail in Chapter 4.

3.3.1 Bergmann's pattern

Carl Bergmann found in 1847 that the biggest bird and mammal species were often located at higher latitudes. Bergmann worked at the inter-specific level. To explain this eco-geographical pattern, he proposed that this enabled the largest endotherms to reduce heat loss in cold regions. Many studies subsequently documented this pattern for endotherms, and the ecogeographic pattern was termed Bergmann's rule. This pattern has been shown for many **homeotherms** (*198*), although exceptions have been reported (*199*). Kyle Ashton (*200*) stressed that on 110 examined mammal species, 78 exhibited Bergmann's pattern. For birds, on 100 examined species, 72 were conformed to predictions. The pattern has also been detected for some ectotherms and marine taxonomic groups. However, many exceptions have been reported (*199, 201*). For example, the body size of insects generally decreases with temperature (*202*). Timothy Mousseau found that crickets (*Melanoplus sanguinipes*) reared in cold environments were smaller than their counterparts reared in warmer environments. Kyle Ashton (*200*) mentioned that on 23 turtle species, 19 species followed the ecogeographic pattern; on 82 squamate reptiles (lizards and snakes), 60 did not follow the ecogeographic pattern. Using an extensive data set of marine bivalvia composed of 4,845 species in 59 families, Sarah Berke and colleagues (*203*) found many trends that did not conform to Bergmann's pattern. It therefore appears that the pattern is more valid for endotherms than ectotherms, although exceptions are likely to be frequent for ectotherms as they compose 99.9% of animals on earth.

An intense debate has taken place recently on Bergmann's pattern (199, 200, 204, 205). The debate has been particularly intense on three points (204). First, Bergmann's pattern is often considered as an ecogeographic pattern (i.e. the tendency of increasing body size towards the poles) and a process (i.e. heat retention mechanism for homeotherms). Second, although originally formulated for homeotherms, Bergmann's pattern has also been extended for ectotherms. The mechanisms are likely to be different, and some researchers have questioned whether Bergmann's pattern should be restricted to birds and mammals. Third, Bergamnn's pattern has been tested at different biological levels: intraspecific, interspecific and at the assemblage levels. Note that Bergmann's pattern at the intraspecific level is sometimes called James's rule, in recognition of the work of the professor who translated the original paper of Bergmann (1847) and noted that the intraspecific level represented a substantial change of the pattern formulated by Bergmann (interspecific level) (206). I think that we should remain as general as possible, however, taking into account the years of research devoted to Bergmann's pattern, and propose to define the pattern as the tendency for the body size or body mass of individuals, species and assemblages to increase with latitude. Note that this definition is close to the definition proposed by Tim Blackburn and colleagues (206): 'the tendency for a positive association between the body mass of species in a monophyletic higher taxon and the latitude inhabited by those species'. I, however, follow Shai Meiri by removing the constraint on the biological level (204). The definition I propose allows a clear separation of the pattern and the processes, and allows testing to be carried out at all biological levels.

Isabelle Rombouts and I tested Bergmann's pattern on copepods at the intraspecific level and found that minimum size of individual copepods increased polewards. At the assemblage level, we also clearly detected Bergmann's pattern on North Atlantic copepods, the mean size of copepod assemblages being smaller south of the Oceanic Polar Front than in the Atlantic Polar Biome (Plate 3.8).

Different hypotheses have been proposed to explain Bergmann's pattern (206).

3.3.1.1 The heat conservation hypothesis

The heat conservation hypothesis (or thermal independence hypothesis) stipulates that bigger species may inhabit colder habitats because their size increases surface-to-volume ratios and limits heat loss (206). However, for ectotherms, this hypothesis may not apply because bigger size may also slow down heat gain.

3.3.1.2 The temperature–size rule

The temperature–size rule (TSR) may explain Bergmann's pattern observed for ectotherms at the intraspecific level. The link between climate and body size has often been found for both endotherms and ectotherms. For ectotherms, the TSR may explain Bergmann's pattern, at least at the intraspecific level. TSR is defined as the phenotypic plastic response of species' size to temperature (208); individuals reared at colder temperatures mature as larger adults than at warmer temperatures (209). TSR has been observed for many aquatic ectotherms. On 61 studies, 90% (55 studies) reported a reduction in the mean size of species when temperature increased, and 10% found the opposite (210).

3.3.1.3 Migration ability hypothesis

Small species have lower dispersal capability, and this may prevent them from occurring in high-latitude regions (206). I think that it is extremely unlikely that this hypothesis holds for many taxonomic groups. In the marine realm, plankton are, in general, represented by small organisms (some exceptions are jellyfish) that can disperse over large areas.

3.3.1.4 Starvation resistance hypothesis

Bigger species have more reserves, which allow them to withstand periods of starvation. This may allow them to survive in high latitudes, which fluctuate strongly at a seasonal scale. Starvation resistance may also explain some exceptions to the pattern. The fact that many insects overwinter in dormant stages may explain why they do not need to increase their body size. However, in the marine environment, copepods that also overwinter display Bergmann's pattern.

3.3.1.5 Resource availability hypothesis

Higher mortality polewards, related to severe climatic conditions, allows more resources to become available for other surviving individuals. Competition is alleviated and individuals can reach larger body size. This hypothesis is, however, unlikely. Some experiments carried out in a laboratory with the same level of food have shown that individuals that grow at high temperature are smaller than individuals that were reared at colder temperatures.

3.3.1.6 Competition, predation and metabolic interaction hypothesis

This interesting hypothesis is not unrelated to the previous one. Interspecific competition favours small species, whereas intraspecific competition favours bigger ones. Small species are expected to divide the resource more finely than bigger ones (211). Diverse communities should therefore be composed of smaller species. These predictions were confirmed for copepods in the North Atlantic, not only at the spatial but also at the seasonal scales (60, 207).

However, this explanation does not consider the implications of temperature, a parameter often invoked in Bergmann's pattern. Eric Edeline and co-workers (212) added a metabolic mechanism to the competition hypothesis. An increase in temperature elevates the metabolism and the energetic demand. This can be compensated by an increase in food uptake, which increases competition. Net energy gain, which is energetic gain through feeding minus maintenance rate, increases faster for smaller organisms and gives them a competitive advantage under warm conditions.

Predatory interactions may also be influenced by the effects of temperature on metabolism. Because warming increases metabolism of predators, they should attack their prey more often. Predation pressure may select for earlier maturation and may therefore diminish the mean size of preys.

3.3.2 Allen's pattern

Allen's pattern states that the length of the protruding parts (limbs, wings, ears) is shorter polewards (213). The main explanation for this pattern is that longer appendages allow the species to more efficiently dissipate excess of heat in the body (role of thermoregulation). This pattern has been shown for terrestrial species, especially endotherms (foxes and rabbits are well-known examples). As with Bergmann's pattern, Allen's pattern does not seem to apply to insects (*214*). Allen's pattern has been much less tested than Bergmann's pattern, and only a few studies have investigated this pattern in the marine domain (e.g. the intertidal zone).

3.3.3 Rapoport's pattern

Rapoport's pattern states that the mean latitudinal range of a species increases with latitude (*215*), which led Stevens to the conjecture that global-scale biodiversity gradients may be the result of this pattern. The rationale of this pattern is based on the relationship between the ecological niche of a species and its distributional range. Because the environment is more variable towards the poles, the species need to have larger environmental tolerances, which implicate larger spatial range polewards. As with Bergmann's pattern, Rapoport's pattern has many exceptions. For example, the pattern has not been confirmed for marine fish (*216*). We will examine deeper this ecogeographic pattern at a global scale in Chapter 11. We demonstrated that although there is a Rapoport effect, this pattern cannot be observed for all taxonomic groups because Rapoport's pattern depends on the characteristics of the species ecological niche within a taxonomic group (*217*). These results explain the lack of universality of the pattern.

3.3.4 Thorson's pattern

Thorson's pattern is a tendency for marine benthic species to have a more direct development polewards (*218*). Gunnar Thorson observed that planktotrophic larvae (i.e. meroplanktonic larvae with poor reserve) accounted for 5% in high arctic seas, 55–65% in boreal seas and 80–85% over tropical shelf seas. **Lecithotrophic** larvae, originating from large yolky eggs and occurring in small numbers, were absent from Arctic seas.

Thorson provided a number of key observations that may explain this ecogeographic pattern. First, benthic species often require a higher thermal regime during development. Second, some species require more food (5 to 10 times more during the developmental than the adult stage). Third, benthic species tend to put their eggs in highly saline waters. Some food accumulation is sometimes needed to trigger reproduction. Thorson proposed that these ecological requirements might explain the need for some species to migrate during the developmental stages in some ecosystems. Larval development in higher thermal regimes, with more food concentration, allows the species to mature quicker and to escape predators faster. In the Arctic, larval development is difficult because of colder temperatures, lower salinities and the frequent presence of sea-ice.

3.3.5 Other biogeographic patterns

3.3.5.1 Jordan's pattern

This ecogeographic pattern posits that the number of vertebrae in fish increases from the equator to the poles. This morphological pattern is probably related to Bergmann's pattern.

3.3.5.2 Foster's pattern

Foster's pattern (*219*), or island pattern, is the tendency for small mammals (e.g. rodents) to be bigger and the tendency for large mammals (e.g. deer) to be smaller on small islands than on the continent (*220*).

3.3.5.3 Guthrie's or Geist's pattern

This pattern states that the seasonal concentration of food influences the body size of large mammals. An explanation is that when the concentration of food is high, an organism can complete quicker its life cycle and develop a bigger body.

Chapter 4

Large-scale biodiversity patterns

The latitudinal gradient in species diversity is the tendency for biodiversity to be higher equatorwards. Turner mentioned that this gradient is among the oldest observed ecogeographic patterns, already pointed out by Forster in 1778 and von Humbold in 1808 (221). In the marine realm, this ecogeographic pattern has been documented for ostracods, euphausiids, foraminifers, decapods and fish (37, 39, 222, 223). This pattern does not appear to be a perfect latitudinal cline, however. For some taxonomic groups, rapid and pronounced spatial changes occur at the middle latitudes (69, 224). For example, at a global scale, Rutherford and colleagues (225) revealed that foraminifera biodiversity showed a maximum in middle latitudes, a minimum in high latitudes and was intermediate at the equator. This pattern is clearly observed when foraminifera biodiversity is zonally averaged (Figure 4.1). Such a hump-shaped relationship has also been observed in the terrestrial realm for birds. A pronounced discontinuity in the diversity of birds was found between the Hadley and the Ferrel cell (221).

4.1 The search for a primary cause

Causes for the generation of the latitudinal biodiversity gradient have been vigorously debated (227–229). Factors that contribute to biodiversity in the marine realm are numerous and belong to a large range of temporal and spatial scales. Geological events that have involved modification in the distribution of continents, open or closure of seaways and changes in the general thermohaline circulation have led to speciation. Climatic oscillations have involved modifications in the geographic distribution of species and in sea levels, which has been shown to contribute to evolution in certain circumstances. At a smaller scale, ecological factors act on the ranges of species tolerance, contributing to the spatial-temporal regulation of diversity. All these factors synergistically have contributed to set up or maintain species and to shape the present-day biodiversity. These large-scale events should still be considered in the explanation of contemporary patterns of biogeography.

Many authors have made significant attempts to identify the primary factor(s) involved in global biodiversity patterns, and a large number of explanations have been propounded (227). In this chapter, I review factors, theories or hypotheses that have been proposed to explain global biodiversity patterns, and outline a unifying scheme, showing the hierarchy between factors and mechanisms and their relationships. I believe that climate is the primary factor regulating diversity, acting mainly through temperature (marine and terrestrial realms) and precipitation (terrestrial realm). However, the rhythm of energy injection and

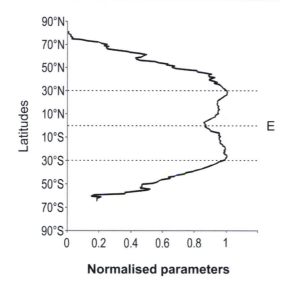

Figure 4.1 Latitudinal pattern in foraminifera biodiversity.

Source: Data from Prell and co-workers (226).

its temporal variation is also probably a key parameter. Secondary parameters, which can influence biodiversity, are also numerous. I will frequently refer to the terrestrial realm because I believe that to understand global scale biodiversity patterns and to go towards a unifying theory, we should consider the whole **ecosphere**. In Chapter 11, I propose the Macro-Ecological Theory on the Arrangement of Life, called the METAL theory, which will be based on some mechanisms developed in this chapter and will explain why some taxonomic groups (e.g. pinnipeds) show an inverse global biodiversity pattern.

4.2 Neutral and null models or theories

Gotelli and McGill defined a neutral model (or theory) as a model considering that the variability in speciation and extinction rates and dispersal are random (230). Neutral models can explain observed patterns in species abundance (e.g. the neutral theory of diversity and biogeography). In contrast, they defined a null model as 'a pattern-generating model that is based on randomization of ecological data or random sampling from a known or specific distribution' (e.g. the Mid-Domain Effect) (230). Although I do not think that such models or theories can explain the cause of global biodiversity patterns (see the next sections and Chapter 11), I think they are an important starting point.

4.2.1 The neutral theory of diversity and biogeography

This theory, developed by Hubbell (231), is based on the theory of island biogeography and can be compared to the **genetic drift** (232). The theory attempts to explain diversity patterns, both abundance and distribution of organisms, assuming that demographic changes in individuals are neutral. The theory focuses on species that have similar trophic characteristics, occurring in sympatry and competing for similar resources. In contrast to species assembly models, the theory asserts that the diversity of such trophic guilds is simply a function of stochastic processes. A species is created by neutral processes within a metacommunity (i.e. a group of local communities). The diversity of the local community is a function of migration, births and deaths, all governed by stochastic variability. All species have the same competitive ability. If the theory is right, the latitudinal biodiversity pattern would only occur by chance, which we will see later is unlikely. A species lives and dies by a process called ecological drift.

A surrogate measure of the species richness, called the fundamental biodiversity number θ, is given by:

$$\theta = 2J\nu \tag{4.1}$$

Here J is the total number of individuals and ν the speciation rate. Because J is a function of the density of individuals σ and the area A, Equation 4.1 can be transformed as follows (*231*):

$$\theta = 2\sigma A\nu \tag{4.2}$$

From Equation 4.2, it is clear that diversity should be a function of the production of individuals (see the production theory), the area and the speciation rate. The theory has been criticised. Brown (2001) stated 'how can we have a neutral theory of biodiversity that not only ignores, but seems to contradict, the single most pervasive feature of life: the incredible variety of size, form and function' (*232*). I tend to concur, and add that the theory is invalidated by compelling evidence suggesting that climate is a major determinant of species richness and that temperature is an important controlling factor of the structure of communities (see next sections and Chapters 2, 5–7 and 11). The climate influence on species through their ecological niche alters their physiology and their competitive ability. Therefore, the assumption that all species have the same competitive ability is unrealistic. The relative success of this theory in explaining observed biodiversity patterns may be explained by the density of individual σ, which is probably strongly related to climate.

4.2.2 The Mid-Domain Effect

The Mid-Domain Effect (MDE) was first detected by Colwell and Hurtt (*233*), but was fully described subsequently by Colwell and Lees (*234*). Along a geographical (e.g. latitudinal) domain, when species are randomly generated, modelled species richness becomes inevitably higher at the centre of the spatial domain (Figure 4.2).

Consequently, biodiversity peaks at the centre of an area. If the meridional domain is fixed between the two poles, diversity peaks at the equator. This led Colwell and Lees to propose that high diversities observed at the equator may be largely influenced by this geometric effect. Along an elevational gradient, a peak in species richness is also observed at intermediate elevations. The paper of Robert Colwell and George Hurtt was inspired by the research of Stevens, who proposed that Rapoport's pattern explained the latitudinal pattern in biodiversity (*215*). As we saw in Chapter 3 the Rapoport's rule states that the mean latitudinal range of a species increases with latitude. Stevens hypothesised that the increasing species richness equatorwards may be the result of Rapoport's rule. However, Robert Colwell and his colleague showed, by applying their null model, that the ecogeographic biodiversity gradient was unlikely to result from Rapoport's rule. We also found the same result with the METAL theory (Chapter 11).

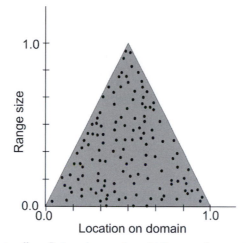

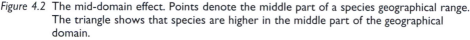

Figure 4.2 The mid-domain effect. Points denote the middle part of a species geographical range. The triangle shows that species are higher in the middle part of the geographical domain.

Source: Simplified from Colwell and Lees (*234*).

However, the MDE has an important drawback. Indeed, the model neglects the longitudinal dimension (*235*). If this second dimension is considered, a peak of richness should be detected at the middle of a continent or an ocean. This is, however, not generally observed. As an example, in the Pacific Ocean, there is no sign of such a two-dimensional biodiversity peak. The second objection against the MDE, as envisioned originally by Colwell and Lees, is the impossibility to connect this model with the biogeographical observations linking the locations of the main ecomes and biomes with the climatic regime. The MDE is very interesting, but I think that this model should have been done not in the geographical space, but in the Euclidean space of the ecological niche (*236*). We will examine this point later in Chapter 11.

4.3 The area hypothesis

The relationship between the area and biodiversity was well revealed by Darlington (*237*), who studied the herpetofauna (amphibians and reptiles) of the West Indies. The author found that species richness was divided by 2 when the area of an island was decreased by a factor of 10. The species–area curve is one of the main stones of the theory of Island Biogeography (*238*). The species–area relationship was expressed by MacArthur and Wilson as follows:

$$S = CA^z \tag{4.3}$$

S is species richness of the island, A the area of the island expressed in square miles. C is a parameter depending on the taxonomic group and the region, and z ranges usually between 0.2 and 0.35 for a small island and between 0.12 and 0.17 for a continent or a large island. Many regression analyses have revealed relationships between the area and

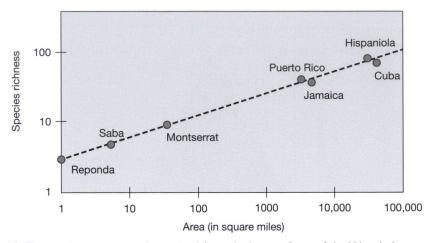

Figure 4.3 The species–area curve determined from the herpetofauna of the West Indies.

Source: Redrawn from MacArthur and Wilson (*238*).

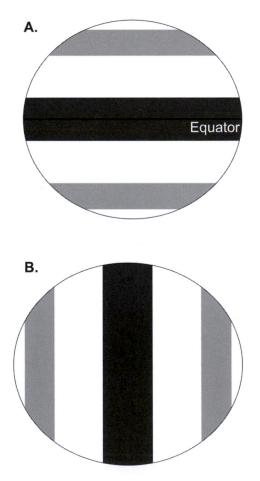

Figure 4.4
Planetary area and biodiversity. (A) The area hypothesis stipulates that the tropics have high diversity because they are large areas. The poles have low diversity because they are smaller. (B) If the area was the primary factor of the gradient, we should also expect large diversity along meridians and low diversity at their 'poles'.

Source: Redrawn from Turner and Hawkins (*221*).

species richness. Terborgh (*239*) was among the first scientists to suggest that the latitudinal gradient could be explained by the area of a latitudinal zone. MacArthur and Wilson stated that area accounted for a large part of the variability in species richness among islands (Figure 4.3).

Rosenzweig (*240*) attempted to demonstrate, in his book entitled *Species Diversity in Space and Time,* that the area was the primary factor that explains the polar-tropical difference in diversity. He suggested that the large area of the tropics explains their high diversity. This happens because the two tropical belts join at the equator and because each tropical belt is wider than other zonal regions of the planet. However, Turner and Hawkins (*221*) stressed that if area was the primary factor, not only would the tropics have high diversity, but meridians would also be characterised by greater biodiversity (Figure 4.4).

I recall that the area effect is included in the Neutral Theory of Biodiversity and Biogeography (Equation 4.2). Although I think that there is a clear relationship between the size of an island and its biodiversity, I do not think that this relationship should be extrapolated to explain latitudinal biodiversity gradients.

4.4 History

Historical time is an important factor. This parameter enables a full colonisation of potential niches by species (species niche saturation) and allows more speciation to occur (*241*). Tectonic events have had strong effects on marine biodiversity. For example, the physical isolation of Antarctica, the closure of the Tethyan Ocean in the Middle East, the collision of Australia with Southeast Asia and the uplift of the Central American Isthmus all altered the oceanic circulation from primarily equatorial to meridional or gyral (Chapter 5). The net effect was a global cooling. The opening of the Drake Passage (36–23 million years BP) led to the establishment of the Antarctic Circumpolar Current and ultimately the formation of the Polar Frontal Zone.

There is compelling evidence that the diversity gradient has reinforced through time during the last 150 million years (Figure 4.5).

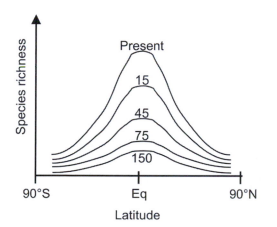

Figure 4.5 Theoretical reinforcement of the latitudinal gradient in marine biodiversity through time. Time is indicated in million years.

Source: Redrawn from Alistair Crame (*242*).

Taxonomic groups such as neogastropods, heteroconch bivalves, decapods and teleost fish radiated. Alistair Crame (242) proposed that the radiation of these groups took place mainly over the tropics during a period of global cooling.

The opening or closure of seaways also causes **vicariance** events. Planetary cycles such as the Milankowitch-Croll (Chapter 5) may also have an effect on regional biodiversity. These planetary cycles, which triggered interglacial and glacial periods, have probably undersaturated ecological niches polewards. Although history is a key parameter when interpreting present-day biodiversity patterns, I do not think this is the primary factor. Observations of invasive species in Europe and North America showed that latitudinal biodiversity gradients of exotic species developed in just a few decades (243).

4.5 Hypotheses based on fundamental processes

4.5.1 The equilibrium/nonequilibrium hypothesis

It has been proposed that the tropics have more species because they have reached an equilibrium state, whereas polar ecosystems have been disturbed several times by the repeated glaciations of the Pleistocene and remain ecologically young because of niche desaturation (241, 244). Their present flora and fauna might not yet be in equilibrium in contrast to the biodiversity from tropical regions. Non-saturation is common, and it has been argued that between 12 and 54% of the available niches may be vacant in the marine realm (245). Again, gradients in exotic species that develop within decades suggest that nonequilibrium is unlikely to be the main mechanism involved in global biodiversity patterns (243), although it plays a role (217) (Chapter 11).

4.5.2 Metabolic theory of ecology

Biodiversity encompasses species that cover over 21 orders of magnitude in body mass from the smallest microbes (10^{-13} g) to the largest mammals (10^8 g) (246). The metabolic theory of ecology (MTE) uses some key principles of allometry and kinetics to establish predictions of metabolism, biological rates and ecological processes (247). The metabolic rate is related to: (1) body size by using **Kleiber's Law**; and (2) temperature by using the Boltzmann factor or the Van't Hoff-Arrhenius relation.

Organism metabolic rate scales with the ¾ power of body mass:

$$I = I_c M^{3/4} \tag{4.4}$$

This relationship, discovered by Max Kleiber in 1932 (248) and subsequently documented by Robert Peters (249), has been termed Kleiber's rule or Kleiber's Law. An explanation for this rule was given by West and colleagues, who demonstrated that the ¾ exponent was related to the fractal dimension of networks of branching tubes (e.g. plant vascular systems, insect tracheal tubes, mammalian circulatory systems, bronchial trees) that allow molecules to be transported throughout the entire organism (246, 250). The fractal network must irrigate the whole n-dimensional volume, which gives an exponent equal to $n / (n+1)$ = ¾; n = 3 because organisms are three-dimensional. This increases the surface-to-volume ratio.

Boltzmann's factor is part of the equation that relates the individual metabolic rate of species. Metabolism is a process allowing energy and materials to be transformed within an organism and to be exchanged between the organism and the environment. Temperature influences metabolism through its effects on biological rates and biochemical reactions. Reaction kinetics varies with temperature according to Boltzmann's factor:

$$\exp\left(-E / kT\right) \tag{4.5}$$

Here E is the activation energy, k is Boltzmann's constant (8.62×10^{-5} eV K^{-1}) and T the absolute temperature (K). Metabolic rate is the product of many different reactions, which are controlled by three major variables: (1) concentration of reactants; (2) fluxes of reactants; and (3) kinetic energy of the system.

Gillooly and colleagues (251) related the individual Metabolic Rate B to temperature (T, Kelvin) and body mass (M) as follows:

$$B = b_0 M^{3/4} \exp\left(-E / kT\right) \tag{4.6}$$

Here b_0 is the normalisation constant independent of body size and temperature, E the activation energy (in eV) and k Boltzmann's constant. This is valid within the limited range of biologically relevant temperatures, between 0 and 40°C. The metabolism is therefore a function of the mass of a given species and temperature. Small animals cool and warm quicker than larger animals, and therefore their metabolism is higher than bigger animals (Chapter 6). The same result was found earlier by only considering the surface of an individual species (e.g. the Surface Law).

Andrew Allen and James Gillooly (252) stated that 'the MTE has been formulated based on the premise that the structure and dynamics of ecological communities are inextricably linked to individual metabolism'. Equation 4.6 was subsequently extended to many biological rates (Table 4.1):

$$Y = Y_0 M^b \tag{4.7}$$

This theory has provided very good insights on mechanisms that influence biological rates and processes and has become a major theory, even if some assumptions remain debated.

Table 4.1 Values of the exponent b in Equation 4.7 for some anatomical structures or processes.

Processes	Parameter b
Metabolic rate	3/4
Lifespan	1/4
Growth rate	−1/4
Heart rate	−1/4
DNA nucleotide substitution rate	−1/4
Length of aortas and heights of trees	1/4
Radii of aortas and tree trunks	3/8
Cerebral grey matter	5/4
Densities of mitochondria, chloroplasts and ribosomes	−1/4
Concentrations of ribosomal RNA and metabolic enzymes	−1/4

Source: Brown and colleagues (247).

Andrew Allen and colleagues (253) proposed a mechanistic framework based on the MTE to explain large-scale biodiversity patterns. Compared to most hypotheses proposed to explain diversity gradients, the MTE is a process-based model that is testable against field data. The model developed by Allen and colleagues is based on the **energetic equivalence rule**. This rule states that the total energy flux of a population per unit area B_T is invariant of body size. Species of different size have similar values of B_T because species metabolic rates B_i increase with body size M_i:

$$B \propto M^{3/4} \tag{4.8}$$

Because population density per unit area N_i decreases with body size as:

$$N \propto M^{-3/4} \tag{4.9}$$

Energy flux of a population per unit area is therefore:

$$B_T \propto B_i N_i \propto M_i^{3/4} M_i^{-3/4} = M^0 \tag{4.10}$$

They extended the energetic equivalence rule to include temperature by incorporating the biochemical kinetics of metabolism:

$$B_T = N_i B_i = N_i b_0 M_i^{3/4} \exp(-E/kT) \tag{4.11}$$

b_0 is the normalisation constant independent of size and temperature ($b_0 \approx 2.65 \times 10^{10}$ W g$^{-3/4}$); E is the activation energy ($E \approx 0.78$ eV or 1.25×10^{-19} J); k is Boltzmann's constant (8.62×10^{-5} eV K^{-1}); and T is the absolute temperature (K). For ectotherms, the temperature of a species can be approximated by the environmental temperature, and for endotherms T is close to 40°C. For testing, Allen and colleagues used a mass-corrected version of Equation 4.11:

$$N_i M_i^{3/4} = \frac{B_T}{b_0} \exp\left(\frac{E}{kT}\right) \tag{4.12}$$

$$\ln\left(N_i M_i^{3/4}\right) = \frac{E}{kT} + \ln\left(\frac{B_T}{b_0}\right) \tag{4.13}$$

The authors multiplied by 1,000 and with $C_0 = \ln(B_T / b_0)$, Equation 4.13 becomes:

$$\ln\left(N_i M_i^{3/4}\right) = \frac{E}{1000k} \frac{1000}{T} + C_0 \tag{4.14}$$

Therefore, the mass-corrected population density was tested as a function of temperature. Using the same reasoning, the temperature-corrected population density was also tested. They therefore demonstrated the temperature invariance of the total energy flux of a population per unit area B_T. Because $C_0 = -19.63$ (intercept) and $\ln(C_0) = B_T / b_0 = \exp(-19.63)$ and $b_0 = 2.65 \times 10^{10}$ Wg$^{-3/4}$, B_T was estimated to be close to 80 W·km^{-2}.

The average population density $\bar{N}$ in a community composed of J individuals and S species is:

$$\bar{N} = \frac{J}{AS} \tag{4.15}$$

Here A is the area delimiting the community. The individuals are calculated as follows:

$$J = \sum_{i=1}^{S} N_i A \tag{4.16}$$

Therefore, the average population density $\bar{N}$ in a community becomes:

$$\bar{N} = \frac{\sum_{i=1}^{S} N_i A}{AS} = \frac{\sum_{i=1}^{S} N_i}{S} \tag{4.17}$$

The average metabolic rate of an ectotherm is expressed as follows:

$$\bar{B} = \bar{B}_0 \, \exp\left(- E / kT_{env}\right) \tag{4.18}$$

Where

$$\bar{B}_0 = b_0 \bar{M}^{3/4}$$

and T_{env} is the environmental temperature.

Holding A constant across community community samples,

$$\bar{B}_T = \bar{N} \, \bar{B} = \left(\frac{J}{AS}\right) \bar{B}_0 \, \exp\left(\frac{-E}{kT_{env}}\right) \tag{4.19}$$

Prediction of changes in the diversity of ectotherms along a temperature gradient is calculated as follows:

$$S = \frac{J}{A} \frac{\bar{B}_0}{\bar{B}_T} \exp\left(\frac{-E}{kT_{env}}\right) \tag{4.20}$$

For testing, Equation 4.20 becomes:

$$\ln(S) = \frac{-E}{kT_{env}} + \ln\left(\frac{\bar{B}_0}{\bar{B}_T} \frac{J}{A}\right) \tag{4.21}$$

By multiplying by 1,000 the first term and C_1 = second term, Equation 4.21 becomes:

$$\ln(S) = \frac{-E}{1000k} \frac{1000}{T_{env}} + C_1 \tag{4.22}$$

With

$$C_1 = \ln\left(\frac{\bar{B}_0}{\bar{B}_T} \frac{J}{A}\right)$$

$\ln(S)$ is a linear function of $1000/T_{env}$ for ectotherms and the theoretical slope is: $-E / 1000k$ = -9 K (including aquatic taxa). C_1 is assumed to be independent of temperature, which means that both J / A and $\overline{M}^{3/4}$ are not affected by temperature. This assumption was said to be supported for trees. Although Allen and co-workers confirmed the expected theoretical slope (-9 K) for both terrestrial (Costan Rican trees, North American amphibians, Ecuadorian amphibians) and marine taxonomic groups (prosobranch gastropods, ectoparasite species of marine teleost fish), other studies did not find values predicted by the MTE (254). In this paper, we tested the MTE on copepod biodiversity using three data sets: a global-scale data set (Atlantic Ocean and two regions of the Pacific Ocean), data from the **Continuous Plankton Recorder (CPR)** survey (North Atlantic) and the ODATE collection data set (East Japan Sea). The test was performed using the following equation:

$$\ln(S) = -E\left(\frac{1}{kT}\right) + C_1 \tag{4.23}$$

With C_1 corresponding to Equation 4.22.

Therefore, we expected a slope of $-E$ (i.e. between -0.6 and -0.7 eV) (252). Even if there were strong linear relationships between copepod diversity and inverse temperature, we found lower values than expected from the theory for all data sets: -0.38 for the global-scale dataset, -0.22 for the ODATE collection and -0.32 for the CPR dataset (Figure 4.6).

In a more recent paper, Allen and colleagues (255) extended their theory by proposing that the origin and maintenance of biodiversity gradients depend upon energetic constraints of speciation-extinction dynamics. The research team found that the absolute rates of DNA evolution rose exponentially with environmental temperature in the same way as individual metabolic rates (256). They hypothesised that speciation was influenced by both the effects of individual-level variables (e.g. body size and temperature) on the rates of genetic divergence among populations and the effects of ecosystem-level variables (e.g. net primary production) on the numbers of genetically diverging populations maintained in communities (252). Extinction was assumed to be a function of both speciation rate and population abundance, in agreement with Hubbell's neutral theory.

Isabelle Rombouts and colleagues stressed that the validity of Equation 4.20 depends on two important assumptions: independence of both mean body size distribution and total number of individuals in the community from temperature (254). We showed that the mean size of copepod community varies as a function of temperature (207), and Bergmann's pattern (Chapter 3) suggests that body size independence to temperature is unlikely. Unfortunately, detailed information on the mean size of species is difficult to obtain for the marine realm. Allen and colleagues (257) argued that their model was robust to departure from these two assumptions because richness is predicted to vary exponentially with temperature, whereas richness varies less than linearly with average body size due to the ¾ exponent and total abundance due to sampling properties of species abundance distribution. They provided an example with the increase in tree diversity moving from boreal ($-5°C$) to tropical ($30°C$):

$$\frac{\exp\left(-E/k(273+30)\right)}{\exp\left(-E/k(273-5)\right)} = 50$$

Such a change should be far more important than other alterations in Equation 4.20.

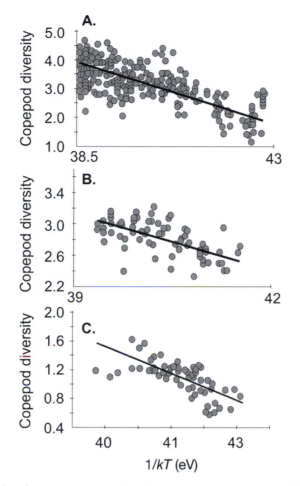

Figure 4.6 Relationships between copepod diversity and inverse temperature (1/*k*T). (A): Global-scale dataset; (B): ODATE collection; (C): CPR dataset.

Source: Modified from Rombouts and colleagues (*254*).

The MTE has been vigorously criticized (*258, 259*). In particular, Michael O'Connor and co-workers argued that it is too simplistic to transpose the individual metabolic rate equation to higher levels such as population, community and ecosystem because the processes to consider are too complex and numerous. In agreement with Bradford Hawkins and colleagues, they also stressed that Equation 4.20 is likely to be invalid. In the case of copepods, additional variation in metabolic rates along the temperature gradient could be caused by region-specific life history strategies. For example, some species exhibit diapausing stages at high latitudes that are characterised by temporarily reduced metabolic rates (*254*). By relaxing the assumption that body mass is independent on temperature, Nicholas Record and co-workers (*260*) obtained better fit of the theory to observations.

The MTE is an important theory that explains many biological processes when a large range of species body sizes are considered. James Brown and colleagues (*261*) stressed that the effects of temperature on metabolic rates should be incorporated in any theory that attempts to explain the latitudinal biodiversity gradient. I will show in Chapter 11 that although temperature is indeed a key parameter, its effects on metabolism can be ignored by working on emerging ecological properties.

4.5.3 Energy partitioning in living systems

How energy partitioning affects biodiversity remains elusive. To understand the relationship between biodiversity and energy partitioning, we need to investigate the relationships between species abundance N, species body mass M and species richness. We should also comprehend the links between those parameters and the shape of the species niche (optimum and width of the niche).

The relationship between N and M has been well investigated (262):

$$N \propto M^\beta \tag{4.24}$$

The coefficient β has been found to range between –1.16 and –0.76 for phytoplankton of 67 sites in lakes (e.g. Lake Superior) and in the ocean (e.g. North Pacific Central Gyre) with different nutrient levels (263). These results revealed that smaller species consume a larger proportion of the energy flux in stable ecosystems. Furthermore, between 91 and 98% of the ecosystem's respiration was attributed to bacteria in stable ecosystems, whereas only 9% was credited to bacteria in unstable ecosystems (264). Investigating 6,339 sea samples, Li and colleagues found that smaller phytoplankton cells were 10,000 times more abundant than larger cells in stable systems (265). The coefficient β ranged between –4/3 for stable and –1/3 for unstable systems. Ecosystems where energy is partitioned by smaller organisms use more equitably resources than ecosystems where energy is fuelled by larger organisms. We found a negative relationship between biodiversity and the mean size of North Atlantic copepods (Figure 4.7). This result suggests that both body mass and number of individuals are related to biodiversity. More remains to be done in this interesting research area. This result also reveals that it is unlikely that mean body size distribution is independent upon temperature, as assumed in the MTE (Equation 4.20).

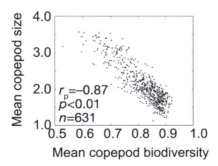

Figure 4.7 Relationships between biodiversity and the mean size calanoid copepods in the North Atlantic Ocean. A strong negative relationship was found.

Source: Modified from Beaugrand and colleagues (207).

4.5.4 The niche-assembly or the structural theory

An intuitive hypothesis stipulates that there are more species in a region because they are more niches (266). The land has more species than the global ocean because there are more potential niches on the land, and similarly for the tropics. The niche-assembly theory states that there are more species in the tropics because species are more specialised, which results in narrower ecological niches. One central aspect of this hypothesis is adaptation to climate,

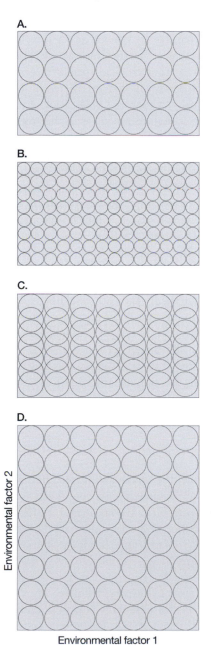

Figure 4.8
Main versions of the niche-assembly theory.
(A) Hypothetical niches in a temperate system according to two environmental dimensions.
(B) Hypothetical niches in a tropical system. The niches are narrower than in A. (C) Hypothetical niches in a tropical system. The niches have the same breadth as in A, but they can overlap.
(D) Hypothetical niches in a tropical system. The niches have the same size as in A, but the gradients along the two environmental factors are larger.

Source: Redrawn from Turner and Hawkins (221).

but, perhaps more importantly, to other species (*221*). Different forms of the niche-assembly theory exist (Figure 4.8).

The first simple alternative is that tropical species have narrow ecological requirements (Figure 4.8A–B). If the resource remains constant, this can be accomplished by speciation, which results in a better sharing of resources. This is sometimes called species packing. As more species share the same amount of resource, the mean production of a species diminishes. The second alternative (Figure 4.8C) is that tropical species have the same niche breadth as temperate species, but that overlapping of the niches is possible. Here also, species pack along the resource gradient. This can be made feasible by species interaction (e.g. mutualism), which tends to be more pronounced in the tropics. The third alternative (Figure 4.8D) is that the gradient in the resources is larger so that tropical species are more numerous. Many researchers think that, at least in the terrestrial domain, there are some indications that the resource gradient is larger in the tropics than the poles. Ricklefs and Miller (*267*) recalled that there is a larger diversity of ecological functions in the tropics, with an increase in the number of nectarivorous or insectivorous species. For example, the higher diversity of terrestrial mammals in the tropics is mostly due to nectarivorous bats. However, this alternative version of the species-assembly theory could be a direct consequence of the increase in diversity, and therefore may not represent the primary cause.

The species-assembly theory has never been demonstrated. In particular, why the tropics should have narrower niches than the higher latitudes is not known. Climate is clearly an important factor, but how could climate potentially interact with species ecological requirement? Climate stability in the tropics has often been invoked to explain the higher degree of specialisation of tropical species. Climate seasonal stability in the tropics implicates a stability of resource availability throughout the year, and the year-to-year variability ensures a long-term stability in the extinction rate. However, the theory currently does not provide any explanation on why niches would overlap or why niches would be narrower. Furthermore, testing this hypothesis is not straightforward because it is not easy to delineate the niche and select the appropriate number of ecological dimensions (*160*). It is clear that some dimensions are more important than others. A large number of dimensions may arise from biological interactions, which cannot be realistically accounted for in models. A further complexity emerges from the likely interactions between dimensions. We will come back to these points later (Chapter 11).

4.6 The climatic influence

Despite the numerous influences organisms have to face, the fundamental role of climate in shaping biodiversity has perhaps been underestimated. Climate is often considered as the primary factor controlling the presence of primary ecological (ecomes) and biological (biomes) compartments (Chapter 3). The spatial distribution of species is more meridionally than zonally constrained (*221*), suggesting a strong climatic implication, a statement reinforced by the success of bioclimatic envelopes in reproducing spatial patterns in species distribution (*180*).

What are the mechanisms by which climate influences biodiversity? Climate controls a large part of the energy used by biodiversity, either directly or indirectly. However, climate does not control all the energy available on the planet. For example, tidal energy (e.g. currents and tidal fronts) and hydrothermal sources are not dependent on climate. In Chapter 2, we saw that climate balances the unequal distribution of incoming solar energy. By doing so, climate modulates temperature and precipitation, which have a strong influence on organism physiology. It also enables both atmospheric and oceanic fluids to move and interact (Chapter 2).

4.6.1 The species-energy theory

The species-energy theory is a climatically based theory (268). Many studies have suggested that energy is an important factor for explaining global-scale biodiversity patterns (269, 270). I examine in this section what is meant by energy. We will also see that the species-energy theory has been separated into two theories: (1) the ambient energy theory; and (2) the production theory.

4.6.1.1 What is meant by energy?

Energy is essential to all physical, chemical and biological processes. Things that move, change, grow, replicate or reproduce need energy (271). Defining energy is therefore an essential prerequisite.

First, in mechanics, energy is defined as the capacity to do work. Energy comes from the Latin *energia*, which means force in action, and from the Greek *ergon*, which means work. This term was introduced for the first time by Thomas Young in 1802 (272). Although physicists distinguish many forms of energy (e.g. electrical, chemical, thermal, solar, magnetic), there are two fundamental types of energy: kinetic energy and potential energy, mechanical energy being the sum of both kinetic and potential energies. The kinetic energy ε_k (J) of an object is linked to its mass m (kg) and its speed v (m·s^{-1}):

$$\varepsilon_k = \frac{mv^2}{2} \tag{4.25}$$

There are different forms of potential energy ε_p (J). An object possesses at rest a potential energy that is a function of its mass m (kg) and its height h (m):

$$\varepsilon_p = mgh \tag{4.26}$$

With g the constant of gravitational acceleration ($g = 9.80$ m·s^{-2}).

Second, in thermodynamics, the energy can also be defined as the capacity of a system to provide heat. Thanks to the statistical theories elaborated by James Clerk Maxwell (1831–1879) and Ludwig Eduard Boltzmann (1844–1906), the temperature can be considered as a measure of the average kinetic energy. The average kinetic energy $\bar{\varepsilon}$ of the centre of mass of an ideal gas can be deduced from the knowledge of its temperature T (K):

$$\bar{\varepsilon}_k = \frac{3}{2}kT \tag{4.27}$$

With $k = 1.38 \times 10^{-23}$ J·K^{-1}, the value of Boltzmann's constant. I recall that this constant can also be expressed as $k = 8.617 \times 10^{-5}$ eV·K^{-1}. The average kinetic energy of an air mass of 15°C is therefore:

$$\bar{\varepsilon}_k = \frac{3}{2}8.617 \times 10^{-5} (273.15 + 15) = 0.0372 \text{ eV}$$

Or:

$$\bar{\varepsilon}_k = \frac{3}{2} 1.38 \times 10^{-23} (273.15 + 15) = 5.96 \cdot 10^{-21} \text{ J}$$

Biological systems follow the laws of thermodynamics. The zeroth law of thermodynamics states two systems in equilibrium with a third one are in equilibrium with each other. The first law of thermodynamics stipulates that the total amount of energy in the universe remains the same; no energy can be created or destroyed. The second law affirms that an isolated system evolves towards a final state of thermodynamic equilibrium that maximises entropy (i.e. a measure of the amount of energy in a system unavailable for doing work or a measure of disorder) and that heat is the most degraded form of energy. The third law of thermodynamics states that the production of entropy is a function of temperature. The entropy is null at absolute zero (−273.15°C). A tentative fourth law of thermodynamics has been proposed by Sven Erik Jorgensen (273). This law states that all systems that receive high-quality energy (or exergy) exploit this energy to remain far from the thermodynamic equilibrium. From these laws, we understand better that heat is a form of energy (the most degraded form of energy). Larger objects (or species) contain more heat than smaller object (or species) with the same temperature.

Third, in meteorology, the link between energy and temperature can also be better understood by considering the Ideal Gas Law (53). From Equation 2.4 (Chapter 2), the Ideal Gas Law can be expressed as follows:

$$pV = NkT \tag{4.28}$$

With p the absolute pressure (Pascal), V the volume of the gas (litre), N the number of molecules, k the Boltzmann constant and T the temperature (Kelvin). This equation clearly shows the relationship between the pressure, the temperature and the density of molecules. By combining Equation 2.4 and Equation 4.28, the average kinetic energy of molecules can be assessed from the knowledge of the atmospheric pressure:

$$\bar{\varepsilon}_k = \frac{3pV}{2N} \tag{4.29}$$

Or:

$$\bar{\varepsilon}_k = \frac{3kT}{2} \tag{4.27}$$

Therefore, the assessment of temperature or atmospheric pressure reflects the average kinetic energy of molecules. It is by no means the total energy of the system. When the atmosphere is barotrope (i.e. the isopycns follow the isobars, or in practice the isotherms follow the isobars), pressure and temperature should reflect well the energetic state of the system and would give the same correlation with diversity. However, in regions where the atmosphere is barocline (i.e. a strong temperature gradient for a same isobar), this would be different, and temperature should be preferred. However, pressure is an important parameter for other variables of importance for biodiversity: atmospheric circulation and precipitation (Chapter 2). Air temperature can therefore be considered as an indicator of the energetic level of the atmosphere, resulting from heat exchanges between all units of the climate system (52).

Hawkins and colleagues separate the species-energy theory into two theories: (1) the ambient energy theory; and (2) the production theory (268).

4.6.1.2 The ambient energy theory

This theory stipulates that biodiversity is controlled by the level of **exosomatic energy**. Gradients in woody plant biodiversity are closely related to water–energy dynamics (274). The water–energy regime maximises directly usable chemical energy, which in turn becomes available for higher trophic levels. Across South Africa, both the energetic state inferred from heat and light and water were highly correlated to woody plant biodiversity at the species, genus and family levels. In the marine realm, as water is no longer a limiting factor, temperature is often a major determinant of marine biodiversity (see Section 4.7).

At a global scale, the positive covariation between solar radiation and temperature makes their influence on biodiversity difficult to separate. However, the fact that solar radiation increases with altitude while both temperature and diversity decrease suggests that temperature is the primary factor (275).

Rombouts and colleagues (69) found a positive relationship between salinity and copepod diversity. The lower copepod biodiversity at the equator in comparison to the tropics was associated with lower salinity related to a larger amount of precipitation due to the presence of the Intertropical Convergence Zone (ITCZ). Sea temperatures decrease at the equator not only because of equatorial upwellings, but also because the amount of incoming solar energy is reduced due to an increased cloudiness associated with the presence of ITCZ (Chapter 2). On average, the equator receives less solar energy (700 $Kj \cdot ha^{-1}$) than the tropics (800 $kJ \cdot ha^{-1}$) (276). Therefore, the positive relationship between salinity and biodiversity was indirect and revealed an effect of the temperature (or the energy) on biodiversity.

4.6.1.3 The production theory

This production theory, also improperly termed the productivity theory, claims that biodiversity is modulated by the rate of carbon fixation through photosynthesis (266). The emphasis is made here on the production of **endosomatic energy**. The carrying capacity not only depends on the biomass (i.e. the biochemical energy available), but also on the **production**, the quantity of biochemical energy (also called the Gibbs free energy) produced in space and time. This endosomatic energy represents the basis of the functioning of both marine and terrestrial ecosystems. Production (i.e. biomass produced per unit of time) or **productivity** (i.e. the production normalised by the biomass) varies among ecosystems, depending on climate, nutrient availability and anthropogenic influences. The **exosomatic energy** is stored as endosomatic energy by primary producers and then propagates to higher trophic levels, determining the diversity of all trophic groups. About 5% of the solar energy that arrives at the earth's surface is caught for photosynthesis (277). As a consequence of the second law of thermodynamics, the **carrying capacity**, which can also be defined as a measure of energy usable for an organism or a trophic group, decreases from primary producers to higher trophic levels. Between 0.1 and 10% of the energy can be transferred from one trophic group to another (*10*). Secondary production should therefore depend on primary production.

The development of the production theory originated from the terrestrial realm. On land, diversity tends to be higher in places where production is elevated. The production is assumed to increase the total number of organisms in an area (the density), providing that the size of organisms is spatially constant. That is why Turner and Hawkins (*221*) proposed that biodiversity predicted from the neutral theory of biodiversity and biogeography of Hubbell (*231*) should indirectly depend on primary production (Equation 4.2): the higher the production, the higher the density of total individuals, and therefore the total number of species. On land, production is mainly influenced by water availability, Photosynthetically active radiation (PAR) and ambient temperature. These climatic parameters have a clear latitudinal distribution that may explain well the latitudinal gradient. Because of the positive relationships between temperature, water availability and productivity, some researchers have attempted to find support for the theory by examining the broad-scale relationships between fluctuations in diversity and measures of production, energy or water availability.

Primary production in the oceanic realm strongly depends on both temperature and the spatial distribution of both micronutrients (e.g. iron) and macronutrients (e.g. nitrates and phosphates). The marine ecosphere being more three-dimensional than the terrestrial realm, the quality and quantity of light penetrating at depth is also an important parameter. Both quality and quantity of light are influenced by turbidity and plankton concentration (shelf-shading). The production theory does not hold in the marine realm because the relationship between production and diversity is negative, contrary to what is generally observed in the terrestrial domain (*58*). Plate 4.1 shows that the subtropical regions, characterised by high diversities, have low net primary production.

Furthermore, Hawkins and Porter (*279*) tested the hypothesis that plant or animal biodiversity was positively related to production. The researchers found an absence of congruence between plant and butterfly biodiversity and suggested that both were influenced by a common environmental factor.

The paradox of nutrient enrichment (*280*) stipulates that in a system composed of a predator and a prey, the increase in the carrying capacity of either the prey or the predator leads to the disappearance of one of the species or both. Although rarely observed in real systems (*281*), when there is an enrichment the resulting increase in biomass often leads to a reduction in diversity. For example, eutrophication is often accompanied by a reduction in diversity. Another important point is that the production theory cannot be considered as the primary factor that controls global-scale biodiversity patterns because primary production is also influenced by environmental fluctuations.

A variant of the production theory, called the Slope–Abyss Source–Sink (SASS) hypothesis, has been proposed to explain the low species richness of bivalves and gastropods in some deep-sea soft-sediment ecosystems (*282*). The hypothesis is based on the fact that the abundance of the macrobenthos diminishes substantially from the bathyal region to the abyss (Figure 4.9).

The decline in abundance towards the abyssal plains might be related to severe food limitation, which is thought to approach adaptive limit (*191*). Because most species are **plantotrophic**, they may disperse from bathyal regions (source area) to the abyss (sink area) over long distances. The low abundance observed in the abyss is thought to make

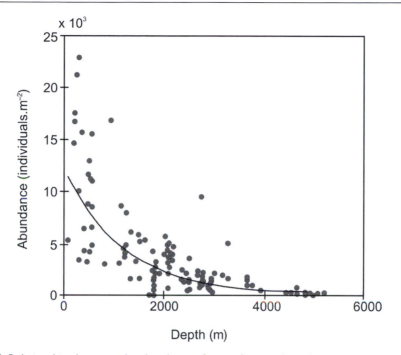

Figure 4.9 Relationships between the abundance of macrofauna and depth.

Source: Redrawn from Rex and co-workers (*282*).

reproduction more difficult because of density-dependence reproduction failure (**Allee effect**). However, the hypothesis remains to be evaluated for organisms other than bivalves and gastropods.

4.6.2 Selection of climatic parameters

In this section, I review the main climatic parameters that could be used to test the relationship between biodiversity and **exosomatic energy**.

4.6.2.1 Downward radiation

First, incoming solar radiation and its average and temporal variability are probably important factors. Incident solar radiation is available for the top of the atmosphere or at surface (clear sky) in many data sets (e.g. ERA data sets from the European Centre for Medium Range Weather Forecasts). The level of solar irradiance influences strongly temperature. Downward solar radiation flux at surface ($W \cdot m^{-2}$) is the solar radiation that is not intercepted by the atmosphere and touches the ground (Plate 4.2). Downward solar radiation flux at surface varies strongly in extratropical regions, and to a lesser extent around the equator. At a regional scale, the amount of radiation is highly dependent on the density and type of clouds present in the troposphere.

Incident solar radiation at the surface is not the only source of energy that warms the oceans, however. Downward long-wave radiation also has an effect and contributes to drive weather and climate. This energy comes from incident solar energy that is absorbed by the atmosphere and subsequently emitted as long-wave radiation (Plate 4.3). The plate shows the strong seasonal variability in downward longwave radiation in the extratropical regions. Expectedly, the amount of radiation is highly correlated to downward solar radiation flux at surface. Downward long-wave radiation varies with cloudiness, temperature and humidity.

4.6.2.2 Upward radiation

This incoming (short-wave and long-wave) energy can be stored in the oceans, but some return to the atmosphere as upward long-wave radiation as a direct consequence of Planck's law or Stefan-Boltzmann's law (Plate 4.4; see also Chapter 2). The seasonal variability in upward long-wave radiation is much smaller over the oceans than the continents. Upward long-wave radiation is highly correlated to skin (or surface) temperature.

4.6.2.3 Sensible and latent heats

Heat may also be transferred by conduction, which takes place by direct contact (e.g. the ocean and the underlying air). The air can subsequently carry the heat away from the ocean by convection or sensible heat. Sensible heat is the heat needed to increase the temperature of a gas or a liquid with no phase change. The examination of the spatial distribution of sensible heat flux shows that extratropical regions are warmed by the oceans in winter and cooled in summer (Plate 4.5). In particular, the North Atlantic Current, the Gulf Stream and the Kuroshio currents warm the atmosphere in winter.

Latent heat is the energy of a given substance absorbed or released when it changes phase. The energy does not affect the temperature of the substance, only its state. Latent heat is absorbed when ice melts or water evaporates. The energy is then stored in the water molecules and released when there is condensation or freezing. At 20°C, 2,454 J are needed to shift the state of 1 g of water molecules from liquid to vapour (283). At 10°C, 2,477 J are required. At 0°C, 334 J are necessary to melt 1 g of water. The spatial distribution of latent heat in January shows high positive values over subtropical regions and the Gulf Stream (Plate 4.6). This indicates that the net process affecting these oceanic regions are condensation. In summer, the latent heat net flux decreases in these regions, showing that the part of evaporation increases. Subtropical regions in summer export latent heat and are the main source of atmospheric water of the planet. Both sensible and latent heats act on the atmosphere and contribute to control the vertical motion of air and wind.

4.6.2.4 Cloudiness

Water vapour is a key component for cloud formation. In dry air, clouds are unlikely to form. In contrast, moist air may condense to form clouds. During cloud formation, latent heat is released into the atmosphere, which warms locally the atmosphere and increase the instability (see Chapter 2; buoyancy). Depression formation is related to both sensible and latent heat fluxes that enrich energy in the lower troposphere (284). When vapour molecules become separated from the surface of the ocean, they are advected by wind and convection takes place. At high altitudes, the air is colder and condensation or freezing arises, which releases a lot of energy in the surrounding air. Heat release allows atmospheric low pressures to deepen. This mechanism leads to cloud formation, and ultimately precipitation. Cloudiness is a parameter important for marine biodiversity through its effect on parameters such as temperature (marine biodiversity) and photosynthetically active radiation (photo-autotrophs; Plate 4.7).

Cloudiness affects regional energetic balance in opposite ways. For example, cloudiness reflects directly electromagnetic radiation from the sun during daytime and therefore cools regional surface. During night-time, it retains infrared radiation and enables temperatures to remain higher than in the absence of cloud. The type and the height of cloud is also an important parameter. Thick and low clouds such as stratus tend to cool the surface because they mostly reflect electromagnetic radiation back to space. Therefore, they increase regional albedo. Cirrus clouds composed of ice crystals do not reflect solar radiation, but increase the regional greenhouse effect. This parameter should therefore be used with caution.

4.6.2.5 Mean precipitation

When cloudiness and mean precipitation are compared (Plate 4.7 versus Plate 4.8), it becomes apparent that regions with a high concentration of cloud coincide with regions with medium to large precipitation (Plate 4.8), as the total amount of cloudiness controls the water cycle. However, the comparison also reveals that both climatic parameters are not perfectly related because cloudiness merges convective (e.g. cumulonimbus) and stratiform (e.g. altostratus) clouds, low- (e.g. stratocumulus and nimbostratus), middle- (e.g. altocumulus and altostratus) and high-altitude (e.g. cirrocumulus, cirrostratus and cirrus) clouds, and not all clouds generate precipitation. For example, nimbostratus generates precipitation in contrast to cirrostratus.

4.6.2.6 Atmospheric circulation

The spatial distribution of the energy is also controlled by the global atmospheric circulation, which is an excellent indicator of its energetic state. The spatial distribution of atmospheric surface pressure controls and is the result of the atmospheric circulation (Plate 4.9). The locations of the main centres of action vary mainly with season and from a year-to-year to multidecadal scales. Plate 4.9 shows the reinforcement of the contrast between high and low surface atmospheric pressure cells (e.g. Aleutian Low versus North-Pacific High and Icelandic Low versus Bermuda-Azores High) in the extratropical regions during winter and substantial changes over Eurasia.

Large-scale atmospheric circulation is highly associated with the spatial distribution of atmospheric pressure (Plate 4.10). Atmospheric circulation reinforces in extratropical regions in winter as a result of the increasing contrast between high-latitude low and low-latitude high pressures (Chapter 2). Large-scale atmospheric circulation redistributes heat over large regions. Atmospheric circulation can export energy as latent heat (evaporation in the tropics and then condensation in the extratropical regions) and contribute to drive oceanic currents that also export energy as sensible heat.

4.6.3 The theory of ergoclines

Biodiversity is also influenced by the rhythm of injection of the energy. The theory of ergoclines, proposed by Legendre and Demers in 1985, states that both high biomass and production tend to be located at the boundary between regions characterised by different levels of energy or at periods during which there is a transition between energetic levels (285). This theory is confirmed by the occurrence of regions of high biomass that are often detected near frontal structures (e.g. tidal or thermal frontal structures). The Ushant tidal front in the Celtic Sea (North Atlantic Ocean) begins to form in April and is well established from June to September. Phytoplankton biomass is high in the front and phytoplankton production propagates to the zooplankton level (36, 286). Oceanic frontal structures are known to be highly productive places where top predators such as tunas concentrate. In the Mediterranean Sea, the Atlantic bluefin tuna (*Thunnus thynnus*), which mainly feeds on krill, has a feeding habitat well defined by frontal structures (287). The spring bloom of high-latitude regions occurs at the transition between mixed and stratified waters.

4.6.4 Synthesis

To summarise, as we have seen in Chapter 2, climate redistributes part of the latitudinal unbalance in energy created by the earth's inclination. This unbalance triggers the movement of both atmospheric and oceanic fluids in a way also driven by the rotation of the earth (Coriolis force). Solar energy is therefore converted into mechanical energy (kinetic and potential energy) and heat. Oceanic convection, which results from the combined action of temperature (low temperature) and wind (high wind intensity and wind-caused turbulence) places nutrients in the euphotic zone when most of the primary production takes place. This form of energy has sometimes been termed ancillary energy (177). Clouds modify the planetary albedo (α_a), which influence the level of incident solar radiation that reaches the earth's surface and the quality and quantity of solar radiation needed for photosynthesis.

Regional climate determines the level of **exosomatic energy** available for biodiversity. The oceans and seas absorb incoming long-wave and short-wave radiations. Most solar radiation is absorbed by the oceans, but a little fraction is reflected (the surface albedo α_s; Chapter 2). Long-wave radiations are then emitted in the way predicted by Planck's law and Stephan-Boltzmann's law (Chapter 7). Heat can be transferred by conduction to the surrounding atmosphere, and atmospheric circulation may remove the heat by convection (vertical atmospheric motion) or advection (horizontal atmospheric motion) with associated transport of sensible and latent heat. This can be summarised by the surface energy balance equation:

$$NR + H + L_v E + G = 0 \tag{4.30}$$

Where NR is the net radiation, H the sensible heat flux, $L_v E$ the latent heat flux and G the ground (here, the ocean) heat flux. The net radiation is calculated as follows:

$$NR = S\downarrow (1 - \alpha) + L\downarrow - L\uparrow \tag{4.31}$$

Where $S\downarrow$ is the net radiation (downward solar radiation) and α the oceanic surface albedo. $L\downarrow$ is the atmospheric counter-radiation (downward longwave radiation) and $L\uparrow$ the oceanic (or ground) emission (upward longwave radiation).

Thereby, there are many different types and sources of energy. Temperature is a good proxy of the energetic level of the atmosphere or the ocean because it integrates the effects of solar, downward and upward long-wave radiations, the atmospheric albedo effect exerted by clouds, the effects of oceanic currents and atmospheric circulation that redistributes heat (sensible and latent heat). However, we saw that there are some noticeable differences between temperature and energy. Photosynthetically active radiation (PAR), the amount of radiation (between 400 and 700 µm) available for the photosynthesis, is a good indicator of the energy utilised by plants and algaes to photosynthese and therefore store the **exosomatic energy** as biochemical **endosomatic energy**. The endosomatic energy then propagates from lower to higher trophic levels. At each trophic level, a large part of the energy is lost by respiration to ensure the metabolism (typically between 0.1 and 10%). The type of radiation absorbed by plants and algaes also varies as a function of their depth of occurrence and photosystem types.

4.7 Temperature

We have seen above that among climatic parameters, temperature is probably the most important parameter for marine biodiversity. Although many factors can be invoked to explain spatial patterns in biodiversity, many recent studies suggest that temperature is strongly implicated in the spatial pattern of biodiversity observed from regional to global scales. Effects of temperature at smaller organisational levels are reviewed in Chapter 6.

4.7.1 Biodiversity and temperature at a basin scale

In the region covered by the **Continuous Plankton Recorder (CPR) survey**, we showed that spatial patterns in plankton biodiversity were highly related to both mean and variation in temperature (*182, 288*). The positive correlation between copepod biodiversity (marine crustaceans; 109 species or genera) and mean sea surface temperature was elevated (Plate 4.11). For example, the low diversity of the subarctic gyre coincided with low mean sea surface temperatures observed north of the Oceanic Polar Front (Plate 4.11).

Biodiversity patterns are not similar among plankton groups. We investigated the spatial distribution in the biodiversity of two phytoplanktonic groups: diatoms (35 genera) and dinoflagellates (the genus *Ceratium*; 47 species). We found a lack of congruence among these two groups (Plate 4.12). Although *Ceratium* biodiversity was similar to the spatial pattern exhibited by calanoid biodiversity (Plate 4.15), diatom biodiversity showed a distinct geographic pattern.

We tested whether calanoid biodiversity (109 species, mostly), diatom (35 genera) and *Ceratium* (47 species) biodiversity was related to maximum sea surface temperature (SST) and an index of annual variability in SST calculated as the coefficient of variation of monthly SST. Among the indicators of temperature, maximum SST is one of the best predictors of the spatial distribution of some marine organisms in the North Atlantic (*60, 289*). The annual variability in SST was also an important factor for the biodiversity of the three plankton groups (*290, 291*). Both maximum SST and the index of annual variability in SST were significantly related to the biodiversity of the three plankton groups (Plate 4.13).

For copepods and the genus *Ceratium*, biodiversity was higher when temperature increased and when annual variability in SST diminished. This was not the case for diatoms. Instead, this group exhibited higher biodiversity for a certain range of temperatures and an intermediate level of annual SST variability (or thermal annual stability). Although the relationships between temperature, annual variability in SST and biodiversity differed among taxonomic groups, these results showed that temperature has a cardinal influence on plankton biodiversity at the scale of an oceanic basin. The difference in the influence of temperature and the lack of congruence among taxonomic groups reflect the degree of specialisation of the groups for a specific habitat. The METAL theory, exposed in Chapter 11, allows the lack of congruence between taxonomic groups to be understood (*217*).

4.7.2 Copepod biodiversity and temperature on a global scale

The strong effect of temperature on copepod biodiversity has also been found at a global scale (*58, 69*). In 2009, Rombouts and colleagues found that oceanic temperatures accounted for 58% of the spatial variability in the biodiversity of copepods (Figure 4.10).

The researchers subsequently tested the influence of a total of 11 ecogeographic variables: bathymetry, oxygen concentration, temperature, salinity, nitrate, phosphate and silicate concentration, net primary production, mixed layer depth, surface oceanic currents and chlorophyll a concentration (*58*). Of the 11 factors, temperature and its environmental correlates (e.g. oxygen and nutrient concentration) were the variables that best explained spatial patterns in copepod biodiversity.

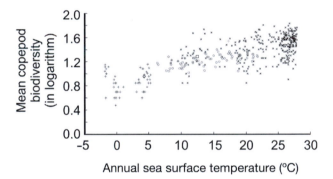

Figure 4.10 Positive relationship between copepod biodiversity (log of taxonomic richness) and temperature. The zooplankton data sets used for the analysis are identified by different marker codes. Thin plus: Barents and Kara Sea; square: Eastern English Channel; circle: east of Japan; cross: tropical and South Atlantic; diamond: White Sea; thick plus: Point B (the Mediterranean Sea); star, Station P and adjacent sites.

Source: Modified from Rombouts and co-workers (*69*).

4.7.3 Relationships between marine biodiversity and temperature from plankton to marine mammals

Tittensor and colleagues (*292*) compiled data on the global distributional range of 11,567 species belonging to 13 taxonomic groups (coastal and oceanic) and spanning 10 orders of magnitude in body mass (Plate 4.14). They found three spatial patterns. First, primary coastal taxonomic groups, other than pinnipeds, had higher diversity over the Indo-Pacific and exhibited a clear latitudinal gradient of increasing diversity towards the equator along continents. Second, primary oceanic species showed higher diversity at mid-latitudes. Third, pinnipeds showed an inverse latitudinal gradient, with higher diversity towards higher latitudes.

They used six ecogeographic variables – annual sea surface temperature (SST), SST slope, SST range, coastline length, primary productivity and oxygen stress – to show that the only factor that explained the spatial patterns of all taxonomic groups was annual SST. The relationships were, in general, positive, with the exception of pinnipeds. However, as Figure 4.10 shows, the relationships between temperature and diversity were not straightforward. If temperature is the common factor at the origin of the diversity pattern, how can we explain that some taxonomic groups (e.g. pinnipeds) are negatively related to temperature? Again, this will be explained in Chapter 11.

4.8 Environmental hypotheses

4.8.1 Environmental stability

The environmental stability hypothesis was developed by Joseph Connell and Eduardo Orias (266). The authors quoted this sentence from Margalef (293): 'the energy required to maintain an ecosystem is inversely proportional to its stability'. In a stable environment, organisms can allocate more energy to growth and reproduction. This energy becomes stored as organic matter and thereby accessible for the whole food web. This hypothesis is therefore related to the production theory. When random or cyclical variations are common, more energy is needed to compensate for perturbations and ensure **homeostasis**. Connell and Orias proposed a theoretical model to understand how environmental stability may operate on biodiversity (Figure 4.11).

When climate, or more generally hydro-climatic variability, is constant, the environment is more stable and the energy captured by photosynthesis can be invested into organic matter, increasing population size. The number of individuals into a population increases the intraspecific variability and, through evolutionary processes, species richness. Diversification contributes to a better cycling of nutrients, which in turn reinforces the production of organic matter (Figure 4.11). Diversity increases the number of links between species and alleviates the influence of environmental fluctuations (Chapter 11).

However, in a second time, overspecialisation, by reducing the number of individuals per population, may reduce the within-population genetic variability, and therefore both the **resistance** and the **resilience** of the species. Overspecialisation can also fragment populations and reduce mobility of organisms. Reproduction may become more difficult, especially if climate becomes more variable. In this theoretical framework, the production of a population can also be controlled by a limiting factor (Liebig's law), directly influencing **production**. This theory is highly dependent on climatic fluctuations. It explains how climate may influence initially the development of regions of high diversity and may make them later more fragile to perturbations. However, the link between primary production, population size and species richness has proved to be inconsistent from one system to another (see the discussion about the production theory above). This hypothesis is, however, quite interesting.

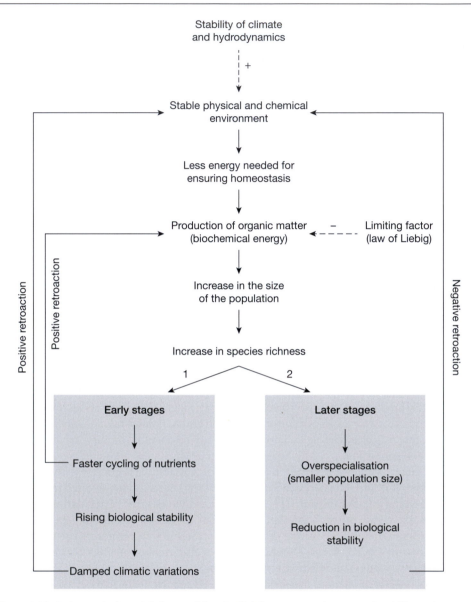

Figure 4.11 A theoretical model that explains the link between environmental stability and diversity. A '+' and a '–' indicate a positive and a negative influence, respectively.

Source: Redrawn from Connell and Orias (266).

4.8.2 Environmental harshness

It has been proposed that environmental harshness may be at the origin of global-scale biodiversity patterns because relatively few species can adapt to rigorous conditions (239). Although a clear mechanistic explanation is lacking, environmental harshness should influence the ecogeographical pattern in biodiversity (Chapter 11).

4.8.3 Disturbance

Many types of disturbance exist in nature. In the terrestrial domain, these disturbances include volcanic eruptions, fires, overgrazing, drought and tropical cyclones. Fires are, for example, important to kill invading shrubs or trees in grassland. In the marine domain, the physical agents of disturbance range from oceanic currents, variation in sea temperature, tropical cyclones, extratropical storms and unstable seabed. The intensity of a disturbance is important, and ranges from small to high. Strong disturbances can lead to catastrophic changes, but most frequent disturbances are moderate or take place periodically. Moderate disturbances delay competitive exclusion and increase diversity. Periodic disturbances are an important source of biodiversity regulation. Many studies have shown that some species depend on these disturbances to occur in a region.

4.8.3.1 Rocky shores

Natural physical disturbances may affect intertidal biocoenoses, coral reef and animal soft-sediment communities. The spatial distribution of organisms on rocky shores is limited by physical factors such as length of emersion, which influences desiccation and temperature changes (294). Some species are limited on the upper part of the shore by some events that take place a few times a year. For example, the algae *Fucus spiralis*, which is found on the upper part of a rocky shore, can be damaged when sunny weather coincides with neap tide. In such a case, the sunny weather exacerbates the impact of desiccation. Schonbeck and Norton (295) showed that five events that took place between May and September 1975, damaged *Fucus spiralis* and two other algaes *Ascophyllum nodosum* and *Pelvetia canaliculata*. Their study showed how a few critical days may limit species spatial distribution.

4.8.3.2 Ice disturbances

Ice disturbance occurs every year on several high-latitude continental seabeds (296). On these polar coasts, ice scouring (icebergs and broken shelves) represents an important source of predictable disturbance eroding both soft and hard substrates. Ice scouring affects 149,000 km of coasts worldwide – the coasts of the Beaufort and Barents Seas, the coasts of Greenland and Antarctica. The phenomenon has deep consequences, and some authors have estimated that seabeds with a depth between 6 and 14 m might be entirely impacted within 50 years in the Beaufort Sea (297). Ice scouring can alter benthic habitats as deep as 500 m in Antarctica.

4.8.3.3 Level of a disturbance

Disturbances may have an opposite influence according to the degree to which they affect biodiversity. A small disturbance is thought to increase diversity locally by removing dominant competitors and by creating clear patches that can be rapidly colonised by animals such as worms, urchins, sea spiders and fish. However, an increase in the intensity or the frequency of this phenomenon may affect seabed animal communities. This type of disturbance exerts a differential effect on trophic guilds. It eliminates large animals such as kelp, sea urchins and bivalves, and decreases the number of large primary producers such as kelp and carnivores, while increasing the number of deposit feeders and scavengers (296).

These authors showed an increase in spionid and capitellid polychaetes, which tolerate a larger degree of disturbance.

4.8.3.4 Biogenic disturbance

Disturbance can also be biogenic, originating from biotic foraging activity, bioturbation or a species bloom. Investigating a rocky shore, Paine (298) showed that when the starfish *Pisaster ocraceous* bloomed, it decreased the number of mussels (*Mytilus californianus*), creating gaps that enabled other species to settle (algae and invertebrates), increasing local biodiversity.

4.8.3.5 Summary

Although the disturbance can explain differences in regional biodiversity, it is unlikely to represent the primary cause of global-scale biodiversity patterns. However, locally and even regionally, it may strongly affect biodiversity. Furthermore, the sign of the relationships between disturbance and biodiversity may depend upon the magnitude and frequency of the peturbations.

4.8.4 Habitat heterogeneity

Many studies have shown that biodiversity increases with habitat heterogeneity. I recall here that a habitat is where a species lives. A habitat is often occupied by many species. The habitat can be inhabited by a species only if environmental conditions are compatible with its environmental tolerance and requirements (in other words, with its ecological niche) (*sensu* Hutchinson; Chapter 11); the more heterogeneous the habitat is, the more niches it contains, and therefore the more species it holds. The effect of habitat heterogeneity can be observed at large spatial scales. For example, species richness is higher over the continental slope, a pattern that may be attributed to habitat heterogeneity (288). I think that habitat heterogeneity is a key factor at a regional scale and that it is axiomatic that habitat heterogeneity increases biodiversity for the reasons invoked above.

4.8.5 Hydrodynamics

Ruddiman (299), investigating North Atlantic foraminifera biodiversity, stressed that the diversity gradient was virtually erased at a regional scale by the strength of the diverse Subtropical North Atlantic Gyre. Beaugrand and co-workers (288) also showed the strong influence of regional hydrodynamic features on North Atlantic copepod biodiversity (Plate 4.15). Results revealed an east-to-west asymmetry in the spatial patterns of copepod bio-diversity that was also apparent in mean sea surface temperatures. Environmental fluctuations can be induced by the **stable-biotope** (e.g. solar radiation, bathymetry) and **substrate-biotope** (e.g. warm and cold surface current, eddies) **components** (Chapter 3). The North Atlantic Current, the Intermediate Shelf Edge and Mediterranean Intermediate Currents, transporting heat towards Europe (300, 301), are important in maintaining pelagic diversity. The influence of hydrodynamics was observed on the spatial variance in calanoid biodiversity (Plate 4.16).

Biodiversity varied as small spatial scales in the path of the North Atlantic Currents and varied at larger spatial scales over the subarctic gyre and the northern part of the subtropical gyre. Hydrodynamic features can therefore reshape strongly global biodiversity patterns at a regional scale.

4.9 Evolutionary rate

The theory posits that evolution takes place quicker in more productive ecosystems, leading to greater speciation rates and species accumulation (227). Because molecular evolution is a critical condition for allopatric speciation and for reproductive isolation in sympatric speciation, higher rates of molecular evolution increase the speciation rate (302). Productive environments may produce greater rates of molecular evolution because metabolic rates are higher and because the rate of nucleotide substitution positively covaries with the metabolic rate or because generation times are shorter (303). Some studies have related rapid rates of molecular evolution to species accumulation (304, 305). Len Gillman and Shane Wright (302) stressed that there is some evidence for a positive relationship between rates of molecular evolution, biodiversity and productivity. If the theory is true, why did Sax (243) observe that invasive species exhibit a latitudinal pattern of increasing diversity equatorwards? Although this observation does not contradict the theory, it does show that other important factors or mechanisms operate.

4.10 Biotic interactions

Biotic interactions are likely to increase the complexity of biodiversity patterns. I think that biotic interactions inflate local species richness but are not the primary cause of global-scale biodiversity patterns. However, biotic interactions are fundamental in the origination of biodiversity through speciation.

4.10.1 Competition

Competition is defined as the reduction in resource availability by an individual, which is detrimental for others (267). Competition leads to the decrease in the environmental resource, which is often limited. Competition can take place for food, nutrients (nitrogen), light (shelf-shading) and space. It can occur between individuals of the same species (intraspecific competition or density dependence). Competition also happens between species (interspecific competition). Two predators can be in competition for the same prey. Competition occurs through a direct exploitation of the resource (passive competition) or by interference (active competition). Some corals use chemical substances to push away other species, a process called **allelopathy**.

Clear evidence for competition for space has been found in the benthic realm. On a rocky shore (Isle of Cumbrae, Scotland), Norton (306) investigated the competition of *Fucus spiralis* and *Pelvetia canaliculata*. In the upper part of the shore where both algaes can settle, the researcher cleared *Fucus spiralis* regularly, which allowed the algae *P. canaliculata* to colonise the area. In the absence of clearing, *P. canaliculata* was unable to colonise the area because of the rapid growth of *Fucus spiralis*. Little and Kitching (294) reported several investigations on the Isle of Man that suggested that many fucoids would have a larger vertical distribution on the rocky shores in the absence of their competitors. They showed that when the species *Fucus vesiculosus* normally located above *Fucus serratus* and *Laminaria digitata* below *F. serratus* were removed, the species expanded in both directions.

4.10.2 Predation

Predation is the consumption of an individual (the prey) by another (the predator) (*291*). As the predator must be where the prey is, it must therefore have an ecological niche (*sensu* Hutchinson) that corresponds at least partially to the niche of its prey, although this figure depends on the degree of predator specialisation. If the predator has many preys, it is likely to have a larger niche than at least one of its preys. In the case of a specialised predator, the prey and the predator can have a similar (or nearly identical) ecological niche. As the prey can also be exposed to predation according to a specific rhythm, the predator niche can sometimes be narrower than the prey niche.

Predator–prey interaction strongly influences spatial and temporal biodiversity patterns at different scales. The prey tries to escape to predation by occupying microhabitats not generally visited by the predator. Some prey can become nocturnal to escape diurnal predators. The prey can become cryptic and look like its environment. Slow-moving animals are protected against predators by shells. This is the case for many gastropods. Marine copepods are known to undertake diel vertical migration. The organisms are more frequent at the surface during night where they graze on phytoplankton. When daylight arises, they migrate at depth in the epipelagic zone (*307*). Graeme Hays showed that diel vertical migration was more intense when the species was pigmented (*308, 309*). He proposed that when species pigmentation is high, this makes the prey more detectable by visual predators. Using data from the CPR survey, we showed how this process influences large-scale biodiversity patterns of copepods (*182*). Copepod biodiversity, measured as the mean number of species per CPR sample, varied strongly in surface water (depth of sampling 6.5 m) as a result of their diel vertical migration (Figure 4.12). Biodiversity was higher during dark periods and lower during daylight periods. This example shows how predation can lead to behavioural adaptation and how it influences spatial and temporal biodiversity patterns at a basin scale.

Predation is also a fundamental strength for speciation, and therefore for biodiversity origination. Predators develop anatomic structures to detect (sensory mechanisms) and catch their prey. The choanocytes of sponges, the nematocysts of cnidarians, the filtering system of bivalves, the tentacles of cephalopods, the multiple arms of sea stars, the prominent pharyngeal slits function of tunicates and the **gill rakers** of filter feeding fish such as herring are a few examples of anatomic structures developed by animals for predation. The blue whale (*Balaenoptera musculus*) feeds on euphausiids. During the feeding season, the animal can consume up to 3,600 kg of these crustaceans, which represents about 40 million krill a day. The animal has a series of 260–400 fringed overlapping plates made in keratin on each side of the upper jaw. The plates fray out into fine hair towards the tongue where the krill is filtered.

Prey–predator interactions can reverse during the life cycle. For example, adult sprat (*Spratus spratus*) feed on cod eggs, whereas adult cod (*Gadus morhua*) feed on adult sprat. This leads to a positive feedback, which is probably one of the causes implicated in the alternation of species dominance in the Baltic Sea (*310*). When cod becomes abundant, sprat decreases due to predation by adult cods. The environment influences cod at the egg

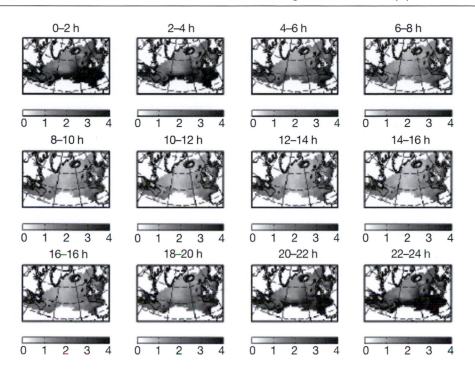

Figure 4.12 Diel changes in the diversity of calanoid copepods showing how the effect of predation on calanoids affects the spatial distribution of the diversity in surface at the scale of an oceanic basin.

Source: From Beaugrand and co-workers (*182*).

stage through a control by low salinity and oxygen concentration, and at the larval stage for sprat through a control by zooplankton.

The environment strongly influences prey-predator interaction. For example, it has been shown that wind-induced turbulence may play an important role. Turbulence has a positive effect on zooplankton, particularly on copepods. Turbulence allows an increase in the contact rate between prey and predator (*151*). However, high turbulence may become a problem and prevents the predator from catching its prey. Here also, a trade-off exists. If species interactions are modulated by hydro-climatic variability, they cannot represent the primary cause for global biodiversity patterns.

4.10.3 Herbivory and grazing

Herbivory is a form of predation. The difference is that grazers only take a part of an individual and do not generally kill their prey. This form of interspecific interaction is also an important factor to explain species spatial distribution. In a laboratory experiment mimicking a rocky shore (*311*), some inserted marine prosobranch gastropods distributed at low tide in the same way as they would be in the rocky shore. However, the winkle *Littorina obtusata* placed itself at the high tide mark of the artificial shore until it died from desiccation. When fucoid seaweed was inserted into the experiment, the herbivorous winkle placed itself in the position similar to the one it occupies on the natural shore. This example shows how interspecific interaction modulates species spatial distribution.

On the other side, the environment strongly influences the plant (or phytoplankton)/ herbivores interaction. For example, small-scale turbulence changes the behaviour of marine pelagic copepods (*312, 313*). In particular, the intensity and pattern of swimming behaviour changes, and it has been shown that the time allocated to 'slow swimming' (i.e. feeding) increases with turbulence even when food is absent (*314, 315*).

4.10.4 Symbiosis

Symbiosis is defined here as a close interaction between two species that can be neutral, harmful or beneficial (*316*). It includes parasitism, mutualism, commensalism and amensalism. This contrasts with the definition that restricts symbiosis to close mutualist relationships (*317*). Symbiosis is an important selective force involved in biodiversity evolution.

4.10.4.1 Mutualism

A third form of species interaction that can influence biodiversity is mutualism. Mutualism occurs when two species interact in such a way that the relationship is beneficial for both in term of fitness (e.g. resource, services, habitat or defence against predators or parasites). Mutualist organisms compose a large part of the biomass (*291*). There are different degrees of mutualism from facultative to obligate.

In the terrestrial realm, dispersal mutualism concerns the transport of pollen or seed. Seed dispersal mutualism does not represent a high degree of mutualism. Pollen dispersal mutualism is more specialised because the plant has to ensure that the other organism is attracted by its own species to maximise the probability of pollination. In the marine realm, this type of mutualism is perhaps less frequent, probably because oceanic current plays an important role in gametes or eggs dispersal. Indeed, many examples of the influence of oceanic currents for egg transport can be found in the literature from South Africa to the English Channel, from the pelagos to the benthos and from plankton to fish.

Defensive mutualism is well exemplified by the clown fish (*Amphiprion ocellaris*) that lives close to sea anemones in coral reef. When a predator arises, the fish withdraws close to the tentacles of the cnidarian. The anthozoan offers a protection against a potential predator and the clown fish protects the cnidarian against some fishes feeding on sea anemones. Another type of defensive mutualism is represented by cleaner-client interaction. Among all marine fishes described, 45 fish species are known to be involved in cleaner-client interaction (*318*). For example, the cleaner wrasse *Labroides dimidiatus* remove a gnathiid isopod parasite from the client fish *Hemigymnus melapterus*. An experimental study provided compelling evidence that this cleaner fish decreased by a factor of 4 the number of parasites on the client fish within 12 hours (*319*). However, the interaction between these two species depends on the proportion of mucus collected by the cleaner. If the cleaner takes too much mucus, the client may chase the cleaner (*320*).

At the community level, the cleaner–client interaction has been shown to increase fish biodiversity at the Ras Mohammed National Park in Egypt (Red Sea) (*321*). When the cleaner wrasse was absent or removed artificially from a reef site, fish biodiversity decreased by about 10–15% after a minimum of four months. In contrast, when the cleaner wrasse was introduced or arrived to a new reef site, biodiversity increased by more than 20% after a few weeks. This example clearly illustrates that mutualism can locally increase biodiversity.

Trophic mutualism occurs when two species interact for energy or nutrients. This type of mutualism involves more intimate relationships between two species. The lichen, which results from a mutualist association between a fungus and a photosynthetic alga, is among the well-known terrestrial examples. This interaction is so strong that it resulted in the establishment of a new taxon. In the marine domain, the intimate relationship of the photosynthetic algae zooxanthellae and corals is another example of strong mutualism.

4.10.4.2 Commensalism

Commensalism is a type of interspecific relationship where a species uses another. The second species is not positively or negatively affected. There are several types of commensalism. First, phoresy is a type of commensalism where a species is transported by another. A clear example is barnacle, which can attach to a scallop shell. Phoresy can be either facultative or obligate. Second, inquilinism occurs when a species uses another for housing. This form of commensalism is observed between some fish and corals in the marine realm and between epiphytic plants and trees (or the bird nest in a tree hole) in the terrestrial realm. This form of commensalism arises from the presence of an **ecosystem engineer** (e.g. trees in the terrestrial domain). In the marine realm, examples of ecosystem engineers are seaweeds, kelps and corals. Ecosystem engineers modify the habitat and create new biotopes that enable the establishment of new species. We therefore see here some forms of positive retroaction, which tend to inflate biodiversity. Third, metabiosis occurs when a species uses another when dead. The hermit crab uses the shell of a gastropod to protect itself against predation.

4.10.4.3 Amensalism

Amensalism is a type of interspecific interaction where a species has a negative effect on another. This often occurs when a species reject a chemical compound that has a stressful or harmful influence on another. A classic example of such an interaction is the secretion of penicillin by the ascomycetous fungi of the genus *Penicillium* that kills most bacteria. Amensalism is probably common in the marine environment, as attested by the many examples of allelopathy.

4.10.4.4 Parasitism

Parasitism is a type of interspecific interaction where an organism called a parasite feeds on another species called a host. The parasite can be outside (ectoparasite) or inside the host (endoparasite). Parasitism is distinguished from parasitoidism, which is a more severe form of parasitism leading to host death. The parasitoids generally have larvae that develop inside the host. This form of parasitism is very close to predation. Parasitism is a very important form of species interaction. Some authors assessed that about 30% of the total number of species may be parasite (316). The acanthocephalan worm of the genus *Polymorphus* is a parasite living at the adult stage in the intestine of the eider duck (*Somateria mollissima*) causing gut ulceration. The parasite uses as intermediate hosts decapod crustaceans and modifies its behaviour in such a way that it spends more time outside water. This modification increases the probability of the crustacean to be eaten by the seabird and enables the parasite to complete its life cycle. Parasites tend to have a narrower niche than their host and can accomplish a large part of their life cycle on a single individual.

4.10.5 Are biotic relationships important to explain global biodiversity patterns?

Although evidence suggests that climate (mainly temperature and precipitation for terrestrial biodiversity and temperature for marine biodiversity) is the primary factor that explains global biodiversity patterns (see also Chapters 5–7 and 11), biotic interactions have a significant influence, complexifying the primordial (or physically shaped) latitudinal pattern. Biotic interactions create local environmental conditions that are virtually unlimited. Engineer species create new conditions (microhabitat and microclimate) that radically change the influence of the local environment that controls species niche. There is no need to have an engineer species to modify the environment, however. Most species, by their presence, can create microhabitats that modify or temper environmental influences. In the terrestrial domain, a tropical tree, by its shadow, can create cooler condition propitious to the

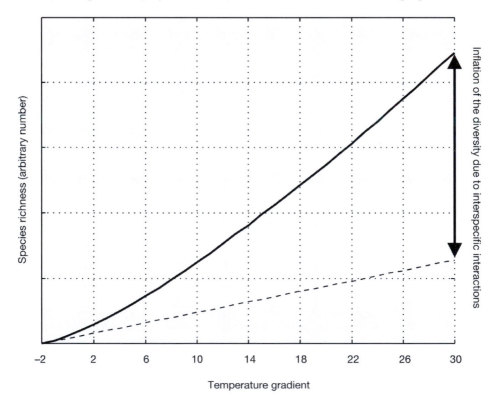

Figure 4.13 Potential influence of biotic interactions (the black curve) on the primordial relationship between biodiversity and climate (here, a linear temperature gradient; dashed line). The magnitude of the effect of biotic interactions is inflated.

establishment of a less thermophile species. It can also serve as a support for others. A plethora of examples also exists in the marine realm. Corals, macroalgae and mussels are among species that can alter locally the environment. This is also observed in pelagic ecosystems (e.g. shelf-shading), although examples are rarer.

The adaptation of species to their environment tends to trigger convergence, whereas biotic interactions involve diversification (267). Even if organism physiology is climatically constrained, species live in a network of biotic interactions. Therefore, Hutchinson's concept of the niche can only determine the potential distribution of a species because species have to compete with others to balance their energetic metabolism, face predation and be resistant to parasitism. However, all of these interspecific interactions are also mediated by climate. When climate pushes a species outside its ecological niche, it places the species in stressful conditions that not only decreases its potential growth and alters or halts its reproduction, but also makes the species less resistant to negative species interactions.

Although biotic interactions are not at the origin of global-scale biodiversity patterns, there is no doubt that they exacerbate it, adding more non-linearity to the primordial relationship between biodiversity and climate. Interspecific interactions are a function of the diversity of available potential niches (*sensu* Hutchinson). This element of circularity was mentioned by Rohde (227). More locally, biotic interactions change the habitat, create new opportunities and increase the number of niches a habitat contains. Biotic interactions alter regionally patterns of biodiversity and are involved in the creation of new species – the more species, the more possibility of creation of new microhabitats (Figure 4.13).

Chapter 5

Marine biodiversity through time

In Chapters 3 and 4, we saw that climate is important in the interpretation of large-scale biogeographic and biodiversity patterns. In this chapter, temporal fluctuations in marine biodiversity are examined at different spectral bands ranging from millions of years to single-day events (Table 5.1).

Again, we will see that climate tends to be an important parameter controlling biodiversity changes, even if it is not always the primary cause. When space/time adjustments of ecosystems and their species are not possible, local extirpations or global extinctions take place. I therefore review the five major extinctions that have occurred since life appeared 3.8 billion years ago, and examine environmental parameters and mechanisms at the origin of these extinctions.

Geological time is split into eons, eras, periods and epochs (Figure 5.1). The Precambrian is divided into three eons: Hadean, Archaean and Proterozoic eons. The Precambrian terminated 544 million years ago. The Phanerozoic is split into the Palaeozoic, Mesozoic and Cenozoic Era. The Palaeozoic stopped at the End Permian extinction, and the Mesozoic is separated from the Cenozoic by the End Cretaceous extinction. The Cenozoic Era is split into the Tertiary and the Quaternary period, the last period including the Pleistocene and Holocene epochs.

Surface temperature has greatly varied for 4.5 billion years. Global temperature decreased by about 60°C over the full earth's history (323). Over the 544 million year (Ma) timescale of the Phanerozoic, global temperatures have fluctuated by ~10°C (324). During the Pleistocene, and especially the last 800,000 years, temperatures varied between glacial maxima and interglacial by 13°C over Antarctica, and by 4–5°C at a global scale (325, 326). In the last 2,000 years, northern hemisphere temperatures varied in the range of 1°C due to a combination of solar and volcanic forcing (327, 328). At shorter timescales, temperature is also influenced by meteo-oceanic oscillations (Chapter 2) (329). Even at timescales of less than a year, large temperature changes may be observed.

5.1 Palaeoclimatic changes

Understanding palaeoclimate and its interactions with the palaeoecosphere/palaeobiosphere is an important prerequisite for better anticipating both present and future effects of climate change on biodiversity.

Table 5.1 Temporal fluctuations in marine biodiversity at timescales ranging from the tectonic to the intra-annual frequency band.

Timescale	Duration	Forcing type	Biodiversity response and example
Tectonic	>1,000,000 years	Wilson's cycle Tectonics	Creation of new ecomes and biomes Speciation Major extinction Biocoenosis disruption
Orbital	10,000–1,000,000 years	Milankovitch cycles Trace gases Tectonics	Ecome/biome shifts Biocoenosis disruption Few extinctions Fewer speciations (although controversial at present) Creation of ecotypes
Millennial	1,000–10,000 years	Trace gases Insolation Ice sheets	Species and ecome/biome shifts Biocoenosis disruption
Secular	100–900 years	Volcanic eruptions Solar activity	
Decadal	10–90 years	PDO, AMO	Changes in species dominance (e.g. herring/sardine fluctuations) Subtle ecome/biome shift
Year-to-year annual	1–9 years	ENSO events NAO	Coral bleaching Fish stock collapse Fluctuation in dominance
Intraseasonal	<3 months	Heatwave	Massive mortality of intertidal species
Synoptic	One to several days	Storm, hurricane	Coral reef and mangrove destruction

Source: Modified from Overpeck and co-workers (*322*).

5.1.1 Biodiversity and climate coevolution

The relationships between climate and biodiversity may be traced back billions of years. Solar activity has changed since the earth's formation (*330*). The sun increases its luminosity by 1% every 100 million years because of the thermonuclear transformation of hydrogen into helium at the star's core (*54*). Model computations estimate that the sun was 25 ± 10% less luminous 4 billion years ago than today (Holocene). Theoretical calculations indicate that if greenhouse gas concentrations are fixed at a pre-industrial level, earth's surface temperature should have been <0°C. Positive feedback due to the ice albedo effect would have completely glaciated the planet 1 billion years ago, according to the snowball earth hypothesis (*331, 332*). However, life has existed on the planet for 3.8 billion years, which requires liquid water. The earth was not, therefore, entirely glaciated. This apparent discrepancy, termed the 'faint young sun paradox', only holds if greenhouse gas concentrations are similar to today's concentrations. The same theoretical considerations indicate that solar luminosity might have increased global temperatures by ~20°C, although the current global temperature is ~14.5°C. Palaeoclimatologists are currently trying to resolve this conundrum.

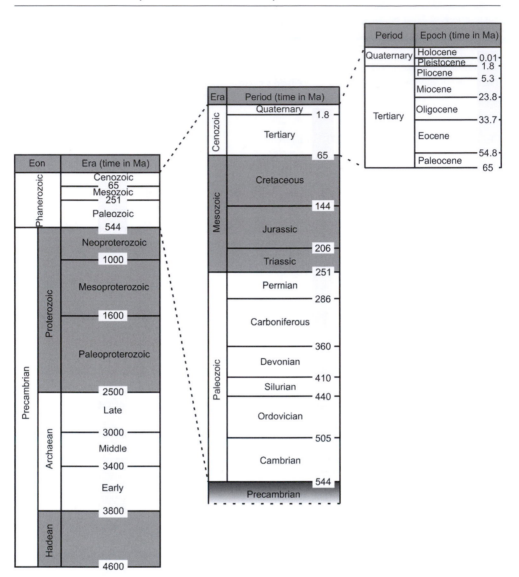

Figure 5.1 Main divisions of geological time.

Source: Redrawn from Kump and colleagues (54).

James Lovelock and Lynn Margulis postulated in the Gaia hypothesis that the earth's biodiversity contributes to maintain the planet's environment constant and habitable by self-regulating feedback mechanisms (2). Thus, living organisms might have reduced CO_2 concentration to counteract the increase in solar luminosity and enable the planet to maintain a thermal regime compatible with life requirements. This theory supposes that both climate and the ecosphere have coevolved with the existence of strong interactions and feedback mechanisms (333, 334). In 5 billion years (the expected lifetime of the sun), solar luminosity should be two to three times more elevated than it is at present. This raises the question of how much level of disturbance ecosystems can buffer.

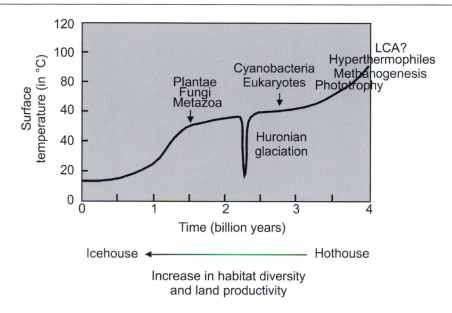

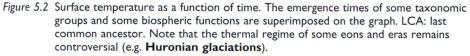

Figure 5.2 Surface temperature as a function of time. The emergence times of some taxonomic groups and some biospheric functions are superimposed on the graph. LCA: last common ancestor. Note that the thermal regime of some eons and eras remains controversial (e.g. **Huronian glaciations**).

Source: Redrawn from Schwartzman and Lineweaver (335).

David Schwartzman and Charles Lineweaver postulated that the earth's ecosphere has evolved almost deterministically as a self-regulating system (335). The idea was not new, and the authors cited, for example, Conway Morris, who wrote that 'biotic history is constrained and that not all things are possible'. Schwartzman and Lineweaver used the term 'quasi-deterministic' because randomness (e.g. multiple climate attractors, meteorite or comet collision) also plays a role. They also considered that 'determinism likely breaks down at finer levels'. They argued that temperature has constrained each major biological advance such as the appearance of cyanobacteria (oxygenic photosynthesis) because organisms operate within thermal limits fixed by their physiology (Figure 5.2). Figure 5.2 shows that surface temperatures were greater than 50°C from 4 to 1.5 Ga (giga-annum) and that temperatures started to decrease substantially towards the end of the Proterozoic. In parallel, life evolved from hyperthermophile archaea and bacteria to cyanobacteria and eukaryotes. Plantae, fungi and metazoa appeared at about 1.5 Ga (mesoproterozoic). Since the beginning of the Phanerozoic, global temperature is thought to have varied in the range of 5–10°C from warm to cold regimes (Figure 5.3).

Some studies suggest that temperatures varied in phase with CO_2 concentrations. Dana Royer and colleagues (324) found that long-lived glacial periods coincided with low (<500 ppm) CO_2 concentrations, whereas warmer periods corresponded to higher CO_2 concentrations (>1,000 ppm). Although a study showed a closer link between cosmic ray flux and temperature than between CO_2 and temperature (336), Royer found that during the Phanerozoic, all cool events coincided with CO_2 concentrations lower than 1,000 ppm by examining 490 published CO_2 proxy records from the Ordovician to the Neogene (337).

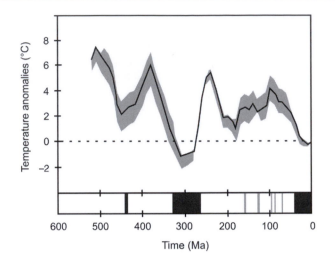

Figure 5.3 Historical changes in mean global temperature for the Phanerozoic. Changes in temperature anomalies (past minus present temperature) were assessed from $\delta^{18}O$ carbonate with a correction to account for changes in pH due to the alteration in the concentration of seawater calcium and CO_2 (model GEOCARB III). The grey envelope is the result of a sensitivity analysis by keeping constant the concentration of calcium in seawater (lower band) or by enabling temporal changes in the saturation state of calcium carbonate ($CaCO_3$; upper band). Periods of glacial (dark grey) and cool (light grey) climate are indicated at the bottom of the figure.

Source: Simplified from Royer and co-workers (*324*).

5.1.2 The tectonic frequency band (> 1,000,000 years)

At this scale, many forcing types, referred to as TECO events by palaeontologists are at work. In TECO, T stands for Tectonics, E for Eustatic (sea level variations), C for Climate change and O for Orogeny (mountain formation).

5.1.2.1 Plate tectonics

As we saw in the introduction, at the end of the 1910s, the German meteorologist Alfred Lothar Wegener (1880–1930) proposed the theory of continental drift (*338*). The tectonic frequency band incorporates plate tectonics and the **Wilson cycle** (*339*), also called the supercontinent cycle (Plate 5.1). The supercontinent Rodinia, which formed 1,100 Ma ago, broke up in the late Proterozoic (650 Ma). In the late Cambrian (514 Ma), the supercontinent Gondwana formed and was located close to the South Pole. In the early Devonian (390 Ma), the Palaeozoic oceans regressed forming the pre-Pangaea. The Pangaea (meaning the entire earth) formed in the late Palaeozoic (Permian, 255 Ma) and lasted up to the Triassic (251–206 Ma). It was composed of Gondwana in the south and Laurasia in the north. A unique circum-global ocean called Panthalassic Ocean surrounded Pangaea. At the end of the Mesozoic, India moved towards Asia and Australia remained attached to Antarctica. Other land masses were close to their modern configuration. Plate 5.1 also shows how the continents are expected to move in the future, eventually returning to a Pangaea

A. Extreme warm phase

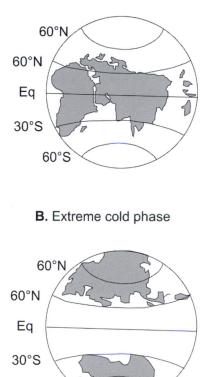

B. Extreme cold phase

Figure 5.4
Relationships between the position of continents and global climate from Lyell's theory (*340*). When most continents are close to the equator, global climate is in a warm phase and when continents are close to the poles, global climate is in a cold phase.

Source: Simplified from Lomolino and co-workers (*10*).

stage called 'Pangaea Ultima'. During periods when plates were divided by shallow seas, it is thought that the numerous shallow seas favoured speciation. Isolation promoted **allopatric speciation** while opening of passage between seas or ocean basin favoured biotic exchanges (*10*). Plate tectonics have therefore been associated with both gradual and sometimes more rapid (closure/opening of seaways) changes in marine biodiversity.

5.1.2.2 The Wilson cycle and climate

The Wilson cycle also affects climate, which in turn alters biodiversity. The drift of continents had potential large consequences for both oceanic circulation and mean atmospheric variability. According to Lyell's theory (*340*), when most land masses are close to equator, the earth is in a warm phase, whereas when continents are located towards the poles, the earth is in a colder phase as ice sheets are able to form (Figure 5.4) (*10*). The planet is currently in a cold phase.

The separation of Antarctica from South America provides another example of the impact of plate tectonics on global climate. The Drake Passage is thought to have opened between 49 and 17 Ma. Recent evidence suggests that it formed 41 Ma ago and that subsequent deepening of the passage was associated with the establishment of the Antarctic Circumpolar Current (ACC), which contributed to alter the greenhouse climate in the early Cenozoic and initiated the permanent glaciation of Antarctica (*341*). The formation of the Drake Passage enabled seawater interoceanic exchange and an isolation of Antarctica from poleward heat transport. Scher and Martin (*341*) provided evidence that an abrupt increase in oceanic productivity paralleled the deepening of the passage because of a widespread intensification of upwellings. The resulting increase in carbon export from the ocean surface to the bottom triggered a reduction in atmospheric CO_2 concentration that also contributed to accelerate the greenhouse-icehouse transition. At the late Eocene, the ACC was fully established.

5.1.2.3 Eustatic variations

Global sea level change can also trigger variations in marine biodiversity, including mass extinction. However, many uncertainties remain on potential links between radiation/extinction and eustatic fluctuations. Hallam (*342*) conjectured that ammonite biodiversity was positively correlated with the area covered by **epeiric seas** during the Mesozoic and the beginning of the Cenozoic (Figure 5.5).

Figure 5.5 shows similar changes between the areal extent of epeiric seas and ammonite biodiversity. To what extent this result was influenced by regions favourable to fossil ammonite formation is, however, poorly known.

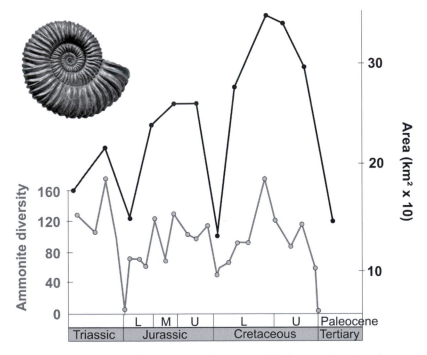

Figure 5.5 Relationships between fluctuations in the genus and subgenus diversity of ammonites (in grey) and the area of epeiric seas (in black).

Source: Simplified from Hallam (*342*) and Lomolino and co-workers (*10*).

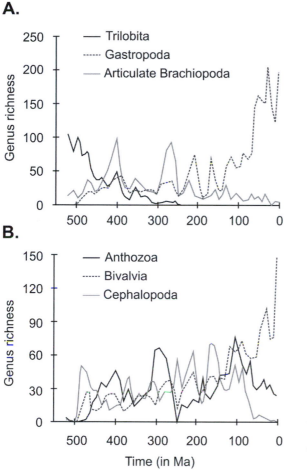

A.

Figure 5.6
Phanerozoic changes in the
biodiversity of (A) Trilobita,
articulate Brachiopoda and
Gastropoda, and (B) Bivalvia,
Anthozoa and Cephalopoda.

Source: Redrawn from Alroy (*344*).

5.1.2.4 Lack of congruence in biodiversity changes

Understanding of the dynamics of biodiversity has been impeded by uneven sampling and gaps (*343*). Using innovative procedures, Alroy (*344*) showed new estimates of the dynamics of Phanerozoic marine biodiversity. He found differences in long-term changes of biodiversity among taxonomic groups – what is commonly termed lack of congruence (Figure 5.6). For example, articulate Brachiopoda biodiversity diminished, in contrast to both Gastropoda and Bivalvia biodiversity.

Interestingly, Alroy also showed that patterns that control diversification of major marine taxonomic groups were not constant through time. Marshall (*343*), who introduced Alroy's results, took the example of the End Permian mass extinction that strongly affected articulate Brachiopoda and Trilobita biodiversity (Figure 5.6). In ecology, the literature abounds with examples of events that have permanent or long-lasting effect on biodiversity (e.g. long-lasting effect of hurricane on coral reef). Alroy's results also suggest that biodiversity is limited for many groups, a finding that contradicts others that suggest unlimited diversification of the marine biota (*345*).

5.1.2.5 Biodiversity changes from a greenhouse to an icehouse climate

Planktonic microfossils offer the most comprehensive biodiversity record of any taxonomic group. Among planktonic microfossils, **coccolithophores** were first detected in Upper Triassic sediments circa 225 Ma (346). Paul Bown (347) investigated long-term biodiversity changes in fossil calcareous nanoplankton, diatoms and dinoflagellate cysts (Figure 5.7). The different groups again exhibited diversity peaks at different times (i.e. lack of congruence).

Bown found a continuous increase in calcareous nanoplankton biodiversity from 225 Ma to the K/T boundary, a period corresponding to a greenhouse climatic regime. During the Cretaceous, a cooling event enabled biogeographic regions to increase and diversify, which also contributed to an increase in the number of potential habitats for calcareous nanoplankton. After the K/T boundary, the biodiversity of this group collapsed and ~93% of its species disappeared. Then, a rapid increase took place up to 51 Ma. The examination of many palaeontological proxies indicate that the early Eocene (up to 50 Ma) was the warmest period of the Cenozoic where poles were not only unglaciated, but covered by forests. This warm period is thought to have resulted from a combination of high CO_2 concentrations and an absence of a circumantarctic current (348). This period was subsequently followed by a constant decline up to the present that coincided with the establishment of an icehouse climatic regime. Long-term changes in the species richness of

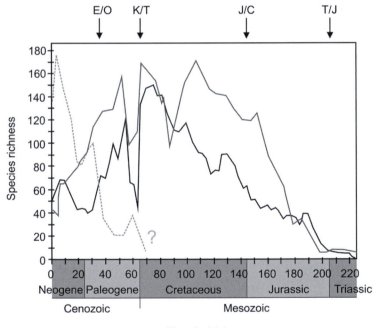

Figure 5.7 Biodiversity changes in fossil calcareous nanoplankton (black line), diatoms (pale dashed grey line) and dinoflagellates (cysts; grey line) from 225 Ma (mega-annum). T/J: Triassic/Jurassic boundary; J/C: Jurassic/Cretaceous boundary; K/T: Cretaceous/Tertiary boundary; E/O: Eocene/Oligocene boundary.

Source: Redrawn from Bown and co-workers (346).

dinoflagellate cysts (Figure 5.7) and in **foraminifera** (*347, 349*) (not shown) were roughly similar to those of calcareous nanoplankton. The diversity reduction of dinoflagellates and calcareous nanoplankton coincided with a rapid increase in diatom biodiversity from 70 Ma to present that paralleled the shift from a greenhouse (Cretaceous and early Palaeogene) to an icehouse (Oligocene) climate. Strong cooling occurred in the late Eocene, which resulted in a colder oceanic temperature regime in high-latitude regions. This created new habitats in which diatoms diversified. Biosiliceous sedimentation (siliceous **ooze**) therefore progressively replaced both nanoplankton and foraminiferan calcareous ooze. Changes in biodiversity observed for calcareous nanoplankton, dinoflagellates and diatoms can be explained by some ecological and physiological traits. Diatoms are photoautotroph and generally live in turbulent waters and unstable environments rich in nutrients (*207, 290*). These r-strategist species generally contain large vacuoles that absorb large quantities of nutrients. In contrast, the mixotrophic dinoflagellates and the photoautotrophic calcareous nanoplankton (both K-strategist species) dominate in stratified regions, which have much lower concentrations of recycled nutrients.

5.1.2.6 Global temperature, greenhouse gases and marine biodiversity

Peter Mayhew and collaborators (*350*) examined the long-term association between global temperature and both land and marine biodiversity for the last 520 million years (Figure 5.8). The authors worked on detrended time series of temperature and biodiversity (animals and protists). They found a negative correlation between global temperature and marine family richness. Both detrended origination and extinction rates were positively correlated to detrended global temperature. The authors conjectured that it might be extinction rates that drive biodiversity fluctuations (extinction rates should be greater than origination rates when detrended temperatures are positive).

The authors also discovered some relationships with CO_2 concentration, suggesting that greenhouse gases might play an important role. Furthermore, they found that 4 out of 5 mass extinctions (End Ordovician, Late Devonian, End Permian, Early Triassic and End Cretaceous) happened at a time when detrended global temperatures were high. Wignall hypothesised a potential link with rapid periods of global warming due to large igneous province eruptions (*351*). At a time of greenhouse climate, such events reinforced thermal stress on biodiversity and increased the probability of anoxic oceanic events.

5.1.3 The orbital frequency band (1,000,000–10,000 years)

The Quaternary has been a time of pronounced climatic oscillations, characterised by a sequence of cold and dry glacial climates (about 100,000 years) being suspended by shorter interglacials (between 10,000 and 30,000 years) when climate is warmer and moister.

5.1.3.1 The Milankovitch theory

The Milankovitch theory relates these climatic oscillations to the characteristics of the earth's eccentricity, obliquity and the precession of equinoxes.

The eccentricity of the earth's orbit around the sun fluctuates periodically every 100,000 and 400,000 years from a quasi-circular (eccentricity $e = 0$) to a slightly elliptic ($e = 0.06$) orbit. The eccentricity, currently $e = 0.017$, causes variations in annual incoming solar

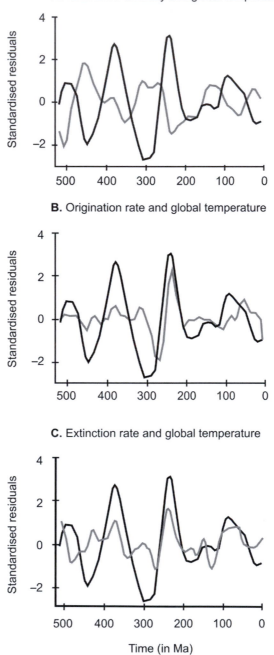

A. Taxonomic diversity and global temperature

B. Origination rate and global temperature

C. Extinction rate and global temperature

Time (in Ma)

Figure 5.8 Association between global (detrended) temperature (in black) for the Phanerozoic and detrended (A) global biodiversity (in grey), (B) origination (in grey) and (C) extinction rates (in grey).

Source: Simplified from Mayhew and co-workers (*350*).

radiation by 0.2%. The earth intercepts more solar energy when eccentricity is maximum (54). The difference is small, but amplification occurs through positive feedbacks with the other components of the climate system (e.g. cryosphere, biosphere).

The obliquity of the axial tilt changes from 22° to 24.5° (current obliquity = 23.4°) with a period of 41,000 years (283). Contrary to eccentricity, obliquity does not change the annual amount of incoming solar radiation. A higher obliquity increases seasonality in both middle and high latitudes.

The precession of equinoxes influences the location of the equinoxes and solstices. It also moves the location of the North Pole among stars. The periodicities are 19,000 and 23,000 years (54). The precession cycle does not modify the annual amount of solar energy, but the seasonality. Currently in the northern hemisphere, the earth is closer to the sun in winter (warmer winter) and farther in summer (cooler boreal summer). The seasonality is therefore attenuated in this hemisphere. The opposite occurs in the southern hemisphere, where seasonality is currently stronger.

These orbital changes affect the amount of incoming solar radiation that reaches the different regions of the planet (precession) or the whole planet (eccentricity) or the seasonal pattern in incoming solar radiation (obliquity). Milutin Milankovitch conjectured that the essential factor for the initiation of the glaciation in high-latitude regions of the northern hemisphere was related to summer insolation. When summer insolation is higher, it enables the snow that accumulates on continents in winter to melt. Interglacial periods such as the Eemian 125,000 years ago or the Holocene correspond to large positive temperature anomalies in summer. The main periodicity therefore corresponds to the eccentricity cycle. However, positive feedbacks amplify the orbital forcing. The first feedback (positive) is related to the growth of continental ice. As we saw in Chapter 2, the ice albedo is more elevated than the vegetation albedo. Ice contributes to decrease solar radiations caught by the planet and to amplify the cooling. Another positive feedback involves marine biodiversity. When the ocean becomes colder, its biological productivity increases. The exact mechanisms by which this occurs is still in debate. Organisms might become bigger (temperature-size rule; Chapters 3 and 6–7) and export biogenic carbon more efficiently from the surface to the bottom. Biodiversity changes might also contribute to increase carbon exportation, as we saw above (e.g. temperature-caused diatom/dinoflagellate fluctuations in dominance). The whole biocoenosis might indeed be more efficient when temperature is colder (207). This tends to reduce atmospheric concentrations of CO_2 and therefore amplify the orbital forcing. The importance of CO_2 concentration is emphasised by the fact that CO_2 fluctuations seem to precede any corresponding changes in temperature anomalies in the Antarctic (Figure 5.10).

5.1.3.2 Glacial-interglacial cycles

Glacial conditions occurred in 80% of the Quaternary, the remaining 20% being represented by interglacial periods during which conditions were comparable to, or warmer than today. Rapid transitions from glacial to interglacial periods have been termed terminations (352). Palaeoclimatic records from the European Project for Ice Coring in Antarctica (EPICA) at Dome C provide evidence for climate change at the orbital frequency band. The drilling at Dome C, which was brought to an end at a depth of 3,260 m, revealed a sequence of eight glacial-interglacial cycles for the last 800,000 years (Figure 5.9). Temperature anomalies in Antarctica oscillated between +5°C during interglacials and –10°C during glacial maxima.

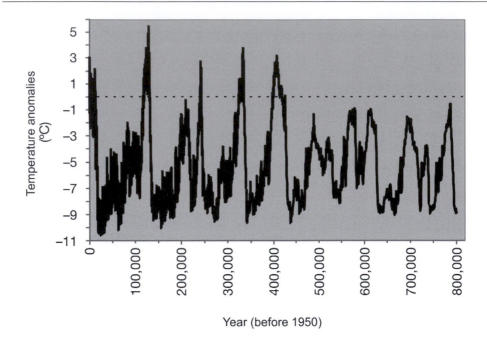

Figure 5.9 Millennial changes in temperature anomalies at Dome C (Antarctica) for the last 800,000 years. Temperature anomalies are the difference from the average of the last 1,000 years. Temperature was estimated from the ice-core Deuterium profile.

Source: Data and methods from Jousel and colleagues (*326*).

Ice-core drilling also took place at Vostok station (east Antarctica). The drilling reached a depth of 3,623 m, which enabled Petit and colleagues to focus on the last four cycles (*325*). The authors found a clear correspondence between CO_2 concentration and temperature anomalies (Figure 5.10). Glacial-interglacial cycles lasted about 100,000 years. The cycles corresponded to Milankovitch forcing, with a main periodicity matching the one of eccentricity.

The last interglacial named the Eemian (130,000–114,000 years ago) had a global mean temperature that was around 1°C warmer than the modern pre-industrial average (*353*). Greenland temperatures were 5°C greater than today's Greenland temperatures (*354*). At that time, mean sea level was 4–6 m higher than present sea level.

5.1.3.3 Higher biomass production at glacial time

During the Last Glacial Maximum 20,000 years ago, global oceanic temperature regime was lower by ~3°C in comparison to temperatures observed today (Plate 5.2). For example, average sea surface temperature (SST) was 11.28°C in 20,000 BP, whereas it reached 14.15°C during the period 2000–2008. The magnitude of temperature change was greater towards high-latitude regions, and especially over the North Atlantic Ocean where SST differences reached 10°C. However, these estimates vary between different atmosphere-ocean general circulation models.

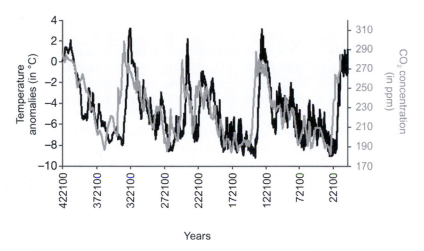

Figure 5.10 Millennial changes in temperature anomalies (black) and CO_2 concentration (grey) at Vostok (Antarctica) for the last 420,000 years. Temperature was estimated from an ice-core Deuterium profile.

Source: Data from Petit and colleagues (*325*).

During glacial periods, primary production was higher. Paytan and colleagues (*357*) found an empirical correlation between marine barite ($BaSO_4$) accumulation rate in core-top sediment samples from two equatorial Pacific transects and the estimated primary production of the overlying water column. They applied this relationship to examine glacial to interglacial fluctuations in productivity. During glacial periods over the past 450,000 years, they showed that productivity in both the central and eastern equatorial Pacific Ocean was about twice as high as during interglacial periods. This result was coherent with other evidence in the eastern and central equatorial Pacific Ocean. This area was more productive during the LGM (*358*).

Many theories have been proposed to explain changes in oceanic productivity. John Martin (*359*) conjectured that it was related to changes in oceanic iron concentration. New productivity in the Southern Ocean (7.4×10^{13} gCyr^{-1}) is limited by low iron concentration. Without this low concentration, the researcher estimated that productivity could reach between 2 and 3×10^{15} gCyr^{-1}. Iron supplies in atmospheric dust were 50 times higher during the LGM because continents were drier and wind speeds greater than today. The larger amount of iron is likely to have stimulated phytoplankton. As a consequence, the atmospheric concentration of CO_2 was much lower than at present (200 ppm).

5.1.3.4 Biodiversity and climate change

Other studies indicate that diatoms were more abundant and diverse during glacial than interglacial periods throughout the Pleistocene (*346*). Higher wind speed increases instability of the upper oceanic layer, which enables diatoms to outcompete both calcareous nanoplankton and dinoflagellates. Therefore, glacial and interglacial periods showed an alternation between time characterised by higher and lower deposition of **opal**/calcite in high-latitude regions (*360*).

Biotic responses involve ecome/biome shifts, contraction, expansion, breaking up and reorganisation. For example, Grant Bigg and colleagues showed that the potential distribution of Atlantic cod was more restricted at the LGM than today because of the reduction of mean sea level by about 120–135 m, and therefore the extent of the continental shelf. This was caused by colder sea temperatures and an increase in the size of ice sheets and glaciers (361). As a consequence, the potential habitat for marine organisms was 0.7 million km^2 at the LGM in contrast to 3.5 million km^2 today. Cod stocks from the western part of the Atlantic were isolated from the eastern part at the LGM. It is quite possible that populations remained isolated for several thousand years, and Bigg and colleagues (361) provided evidence for some degrees of genetic divergence. To what extent it might have influenced the ecological niche of the individuals is worth investigating in the future.

5.1.3.5 Speciation

The extent to which Pleistocene glaciations led to speciation remains controversial (362, 363). Some authors have posited that glacial-interglacial cycles often did not lead to the creation of new species. Barnosky and Kraatz (364), reviewing examples from the terrestrial realm, outlined two scientific strategies to examine the impact of climate change on speciation. The first strategy is to trace species population through the Pleistocene by examining fossils. For mammals, fossils that accumulated during the last 750,000–1,000,000 years can be found at Porcupine Cave in the mountains of Colorado. By taking two species characterised by different life history strategies (marmots *Marmota* sp. and the sagebrush vole *Lemmiscus curtatus*), they discovered that the effects of repeated climate changes during the Pleistocene varied among species. For marmots, which hibernate and for which microclimates are important, no different morphotypes were found. In contrast, the sagebrush vole, which does not hibernate, exhibited dental changes after the strongest climatic transitions of the period. However, the shift was only perceptible at the population level and did not engender a new species. To what extent this phenotypic change might have influenced its ecological niche remains contentious.

The second strategy assesses the divergence time (i.e. the time when two sister species share a common ancestor) of two closely related species (sister species) by examining mitochondrial DNA. Performed on North American mammals, this analysis revealed that speciation was not significantly higher during the Pleistocene (365). These results indicate that Pleistocene climate changes did not isolate populations enough to trigger speciation. Low speciation was also found by some authors for plants (366), insects (367) and fish (368). For plants, Willis and Niklas did not observe any increase in speciation rates during the Pleistocene. In mid-latitude regions, more extinction than speciation occurred in plants. Furthermore, when speciation took place, it was restricted to small regions. For insects, Coope posited that insects exhibited a high degree of stability throughout the Pleistocene.

Some findings using the mitochondrial DNA of living North American birds suggested that speciation had occurred during the Pleistocene, including the last two cycles (369). Weir and Schluter provided evidence that Pleistocene climate change enabled both the initiation and completion of speciation of New World birds, especially the boreal avifauna (363). However, Johnson and Cicero (369) obtained very large divergence time often greater than 100,000 years (their Table 1). Genetic changes probably occurred at the population level. Rates of diversification were rapid and led to new ecotypes or morphotypes at the population level.

5.1.4 The millennial frequency band (10,000–1,000 years)

This frequency band reflects mainly climatic instabilities that took place during the last glacial period from 110,000 until 10,000 years (*370*). Both marine and ice records have contributed to document these abrupt instabilities (*371*). High-latitude records suggest that these events were more pronounced and frequent than during the Holocene (Figure 5.11). Holocene climatic stability has been remarkable. We can see in Figure 5.11 that the last deglaciation was punctuated by rapid climatic fluctuations and that the time needed to leave the glacial regime and enter the Holocene was less than a century.

Two main types of events have been documented during the last glacial period and the deglaciation.

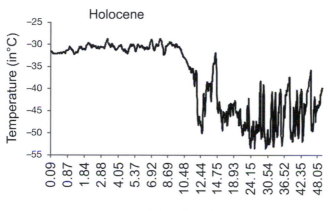

Figure 5.11 Millennial changes in temperature in the central part of Greenland (GISP2 ice core) for the last 49,000 years. Temperature interpretation is based on analyses of stable isotopes and snow accumulation.

Source: Data from Alley (*372*).

5.1.4.1 Dansgaard–Oeschger events

These events, which occurred about 23 times (23 interstadial events) during the last glacial period, were pronounced climate change characterised by an increase in temperature ranging from 8°C to 16°C over Greenland. Examination of foraminifera shells indicates that North Atlantic surface water warmed by up to 10°C in four decades, an average increase of 0.25°C per year (*371*). These events were first documented by Dansgaard and colleagues (*373*) and Dansgaard and Oeschger (*374*) by measuring oxygen-isotope ratios in Greenland ice cores. The duration of each Dansgaard–Oeschger (D/O) event was between 2,000 and 3,000 years. Each warm interstadial event was subsequently followed by a progressive cooling lasting several centuries. These events were most marked in the North Atlantic. Although they were also found in the South Atlantic, the temperature decreased in contrast to the pattern in the North Atlantic. This not-yet-fully-understood pattern has been termed the hemispheric see-saw effect.

5.1.4.2 Heinrich events

During the last glaciation, Heinrich events (HE) were episodes of major cooling that were marked by a reduction in sea surface salinity and an increase in the coarse lithic fraction (>150 μm) in sediment cores. They were first identified by the German sedimentologist Helmut Heinrich (1988) in deep-sea sediment cores in the north-eastern part of the North Atlantic in a region called Dreizback Seamount (close to the Azores). He observed some cyclical events of ice-rafted dropstones (IRD) where foraminifera were rarer. Six Heinrich events have been distinguished: H1 (16,000 years ago), H2 (22,000 years ago), H3 (27,000 years ago), H4 (37,000 years ago), H5 (46,000 years ago) and H6 (60,000 years ago). These irregular events are separated by intervals ranging from 5,000 to 14,000 years. The frequencies of these events are therefore different to the Milankovitch cycles. Helmut Heinrich conjectured that they corresponded to rapid and massive surges of icebergs. Later, it was hypothesised that these icebergs might come from the Laurentide Ice Sheet through Hudson Strait. In his book, Jean-Claude Duplessy estimated that these massive surges of icebergs represented 2% of both Canadian and European ice sheets of the time (375) or up to 10% of the Laurentide ice sheet (353). This probably increased mean sea level by 2 or 3 m. When the phenomenon started, sea level increase contributed to exacerbate the discharge of icebergs. It has been posited that the vast amount of freshwater (about 0.1 Sv; one Sverdrup = 1 million m^3 per second) perturbed the North Atlantic thermohaline circulation, and therefore the regional climate. Sediment data, for example, indicate that North Atlantic Deep Water (NADW) formation was interrupted during Heinrich events (376).

5.1.4.3 Mechanisms involved in climatic instabilities

Stefan Rahmstorf reviewed the potential role played by the ocean thermohaline circulation in these climatic instabilities (353). Three types of thermohaline circulation have prevailed. The first is a warm mode, a mode currently occurring in the North Atlantic (Figure 5.12). In this mode, NADW forms in the Nordic and Labrador Seas. An amplification of this mode is observed during a Dansgaard/Oeschger event.

The second is a cold mode where NADW forms south of Iceland (Irminger Sea). This mode was prevalent during the glacial stage. The last mode, probably observed during Heinrich events, is a mode where no NADW formation occurs (complete shutdown of the Atlantic thermohaline circulation). The event terminated when the number of icebergs diminished. The amount of heat (and saline water) stored south of 40°N started to move northwards with the return of the thermohaline circulation. The phenomenon was amplified by the strong thermohaline contrast between middle and high latitudes, which explains the rapid increase in temperature that was observed at the end of the Heinrich event. Cold periods less intense than a Heinrich event have subsequently been reported.

The Dansgaard-Oeschger events correspond to the rapid warming that is observed at the end of each Heinrich or cold event. The whole cycle has been termed a Dansgaard-Oeschger cycle. Sensitivity of the thermohaline circulation to freshwater inputs fluctuates (353). Nowadays, the thermohaline circulation is stable to freshwater inputs (Figure 5.13). Only substantial freshwater input can destabilise the system. This was not so during the LGM. The **hysteresis** loop was narrower and the amount of freshwater needed to perturb the system was much smaller (about 0.1 Sv).

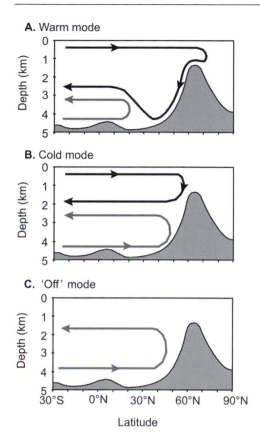

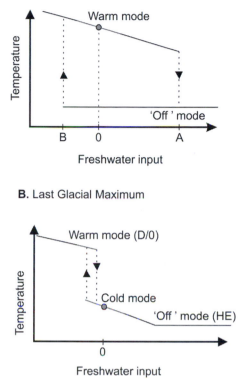

Figure 5.12
The different modes of ocean circulation in the Atlantic Ocean: (A) warm mode, (B) cold mode and (C) 'off' mode, which is characterised by a complete shutdown of the thermohaline circulation. The black line shows the North Atlantic overturning and the grey line symbolises Antarctic bottom water. Bottom topography is in grey.

Source: Redrawn from Rahmstorf (353).

Figure 5.13
Schematic of the modes of Atlantic thermohaline circulation as a function of temperature and freshwater influx.
(A) Current climate. The system has two possible equilibria. Only a large freshwater input (>A) might involve a collapse of the Atlantic thermohaline circulation. The return to initial conditions is possible if the amount of freshwater falls below B. The current system is stable because it has a large hysteresis loop.
(B) The Last Glacial Maximum was less stable because the hysteresis loop was smaller. The system was very sensitive to the quantity of freshwater amount. A small decrease in freshwater was sufficient to induce the warm mode observed during Dansgaard/Oeschger (D/O) events and a small increase in freshwater influx was enough to interrupt completely the Atlantic thermohaline circulation as observed during a Heinrich event (HE).

Source: Adapted from Paillard (377).

5.1.4.4 The Younger Dryas

Following the Bølling-Allerød warm period (14,700–12,700 years BP), the **Younger Dryas** was a cold interval that started 12,600 years BP and came to an end 11,500 years BP (Before the Present; by convention, the present is the year 1950). It is named after the alpine flower *Dryas octopetala*, which can still be found in the vicinity of Alpine glaciers and in the tundra. This cold event (also referred to as the Big Freeze) happened at a time warming was pronounced. The cooling happened in a few decades. Mean annual temperatures diminished by 12°C to 17°C in comparison to today's values. In winter, the drop was even more substantial between 22°C and 28°C, whereas it was remarkably smaller in summer, between 3°C and 6°C (352). This pattern, similar to the one observed during the first Heinrich event, reflects an increase in northern hemisphere seasonality. This event affected at least the whole northern hemisphere and possibly the whole planet. As observed for the Heinrich-1 event, some studies showed that the increase in sea-ice extent during the Younger Dryas pushed southwards the Intertropical Convergence Zone (ITCZ), reducing, for example, the Asian monsoon (378).

Duplessy stressed, however, that this stadial was distinct from a Heinrich event because foraminifera were still abundant (375). This indicates that sea surface salinity remained high, as foraminifers (as most marine plankton) would not tolerate a salinity drop of the magnitude observed during a Heinrich event. This and other evidence (e.g. absence of substantial salinity changes in the Gulf of St Lawrence) suggest that the hypothesis of an influx of freshwater from Lake Agassiz (four times the size of France) (379) was not the source behind the origin of the Younger Dryas. Jean-Claude Duplessy conjectured that Arctic icebergs might have invaded the North Atlantic and triggered a reduction or a collapse of the North Atlantic Meridional Overturning Circulation. The use of general circulation models (GCMs) has shown that an influx of only 0.1 Sv in the North Atlantic between 50°N and 70°N was sufficient to trigger a cooling in agreement with palaeoclimatic reconstructions (380).

A more recent alternative theory posits that the Younger Dryas might be related to a change in the mean position of the Polar Jet Stream as a response to the changes in topography triggered by the receding glacial ice sheets (381). The resulting increase in net precipitation over the North Atlantic was sufficient to reduce the ocean overturning circulation. Positive feedback related to increasing sea-ice amplified the phenomenon. Eisenman highlighted the fact that present net precipitation over the North Atlantic represents a flux of freshwater two to three times >0.1 Sv. Therefore, a minor change in precipitation might have a strong effect. Other theories have been proposed, but most invoked a reduction or collapse of the Meridional Overturning Circulation.

5.1.4.5 Consequences of climatic instabilities for biodiversity

Current estimates of the impact of these climatic instabilities on biodiversity remain elusive because of the lack of archives (382). Both cold and warm events had probably large consequences on middle- and high-latitude marine biodiversity. Such events would today considerably alter the structure and functioning of marine ecosystems and the services they provide. Flores and colleagues investigated the spatial distribution of large (>4 μm) coccolithophore species *Emiliania huxleyi* from the LGM to the Holocene at 11 sites distributed over the eastern North Atlantic and the western Mediterranean Sea (383). During the LGM, the distribution of cold-water species appears to have extended from the equator to the North Pole (Figure 5.14).

(A) Last Glacial
Maximum

(B) Heinrich 1

(C) Termination Ib

(D) Holocene

■ >25%	■ 1–5%
■ 6–25%	□ <1%

Figure 5.14
Spatial distribution of the large form of *Emiliania huxleyi* (>4 μm) in the eastern North Atlantic during (A) the Last Glacial Maximum, (B) Heinrich I event, (C) Termination Ib (transition between the Younger Dryas of the Pleistocene and the Preboreal period of the Holocene) and (D) the Holocene. This schematic is based on the examination of the relative abundance of the species in 11 sites in the eastern North Atlantic and the western part of the Mediterranean Sea.

Source: Redrawn from Flores and co-workers (*383*).

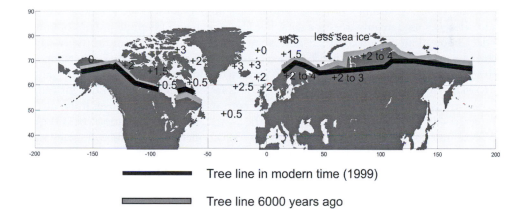

Tree line in modern time (1999)

Tree line 6000 years ago

Figure 5.15 Map showing summer temperature changes between the present day (1999) and 6,000 years ago in both the terrestrial (summer air surface temperature) and the oceanic (summer sea surface temperature) realms. The northern boundary of the boreal forest is indicated for modern time (black) and for the period corresponding to 6,000 years ago (grey).

Source: Simplified from Kerwin and co-workers (*385*).

During the cold period corresponding to the Heinrich 1 Event, *E. huxleyi* was distributed from 18°N to 75°N. The relative abundance of this large coccolithophore remained high in the western part of the Mediterranean Sea. At Termination Ib (transitional period between the Younger Dryas of the Pleistocene and the Preboreal period of the Holocene), the relative abundance of large *E. huxleyi* decreased over the whole region. In the Holocene, the species became limited to the polar biome.

The climatic optimum of the Holocene, which took place from 9,000 to 6,000 years BP (*384*), was warmer than our current period. By examining palaeoceanographic observations at 6,000 years BP in the North Atlantic, Kerwin and colleagues (*385*) provided evidence that maximum summer sea surface temperatures were up to 2–4°C warmer than those observed in the 1990s in some regions of the Norwegian Sea between Iceland, Norway and Svalbard (Figure 5.15). The boreal forest was also located more towards the pole than in the 1990s.

Koç and colleagues found that the climatic optimum was characterised by diatom assemblages that were an indicator of warmer summer sea temperatures (0.5°C and 4°C warmer 6,000 years ago than today) (*386*). Examination of other taxa such as molluscs and dinoflagellate cysts also revealed warmer conditions (*385*).

5.1.5 Sub-millennial frequency bands (<1,000 years)

5.1.5.1 Climate change

Over this band, climate change is more related to changes in volcanic and solar activity and the inherent dynamics of the climate system (Chapter 2). Two main climatic anomalies occurred: the Medieval Warm Period (MWP, between AD 1000 and 1300), also sometimes called the Medieval Climate Anomaly (MCA) and the Little Ice Age (LIA, between AD 1350 and 1830). The limits of these periods vary among authors (*83*). Anders Moberg and colleagues used mainly tree-ring data, but also pollen and borehole proxies of annual to decadal resolution, to reconstruct millennial-scale climatic variability in the northern hemisphere (*327*). From AD 1 to 1979, they found large natural variability in temperature between MWP and LIA in the range of 0.7°C with a maximum amplitude of ~1.5°C (Figure 5.16). The variability found by Moberg's group was larger than previously thought (*387*).

The transition between MWP and LIA corresponded to a change from an era characterised by persistent positive phases of the North Atlantic Oscillation (NAO) to an era of sustained average to negative NAO phases (*388*). Valérie Trouet and colleagues stated that the NAO was not the only phenomenon that changed at that time. MWP was also marked by La-Niña-like conditions and by an intensification of the Atlantic Meridional Overturning Circulation.

5.1.5.2 Consequences for biodiversity

The consequences of these changes for marine biodiversity are not well documented. A palaeoceanographic study on diatoms in some fjords of the Faeroe Islands revealed that freshwater diatoms (e.g. genera *Eunotia*, *Epithemia*, *Cymbella* and *Tabellaria*) were more abundant between 1,200 and 700 BP (*389*). This result suggests that climatic conditions were warmer and characterised by an increase in local precipitation associated with an enhanced cyclone activity over the North Atlantic. This finding makes sense as the NAO was in a prolonged positive phase at this time (Chapter 2).

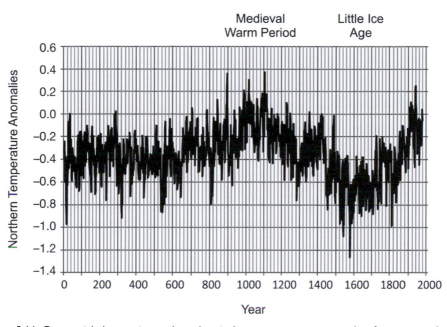

Figure 5.16 Centennial changes in northern hemisphere temperature anomalies from proxy data (tree rings, pollen and boreholes) for the period AD 1–1979. Data are anomalies based on the mean northern hemisphere temperature for the period 1961–1990.

Source: Proxy palaeoclimatic temperature data are from Moberg and colleagues (327).

In the terrestrial realm, Hadly and colleagues examined the DNA of mammalian fossils in a cave deposit (Lamar Cave in Yellowstone National Park) over the last 3,000 years (390, 391). During the MWP, some studies on terrestrial mammals have clearly shown that speciation was not possible in such a short time. However, the genetic diversity of populations of some terrestrial mammals was modified. The sign and magnitude of the change depended on the life history traits of the species. For example, populations of northern pocket gopher (*Thomomys talpoides*), an animal that spends its life close to its burrow, decreased in size. As a result of the absence of dispersal, genetic diversity diminished. The population of the montane vole (*Microtus montanus*), an animal that disperses more widely, also showed a reduction in population size. However, the genetic diversity increases because of continued immigration of individuals from other populations.

The LIA was especially cold during the Spörer Minimum (1460–1550), the Maunder Minimum (1645–1715) and Dalton's Minimum (1790–1820) (Chapter 2). As we saw earlier, palaeoclimatic studies suggest a northern hemisphere temperature reduction ranging from 0.5°C to 1°C due to summer isolation reduction (328). However, poor knowledge exists on the response of marine biodiversity during this period. In the Sicily Channel (Mediterranean Sea), increased marine productivity was inferred for the LIA (392). For a record reaching back to AD 1650, Incarbonara and co-workers (392) found a reduction in the abundance of the coccolithophore *Florisphaera profunda* associated with an increase in placoliths (shield shaped coccoliths) prior to AD 1850. This result suggests an increase in marine productivity during most of the LIA. The authors explained this result by a strengthening in northerly wind, which triggered an increase in vertical mixing and stimulated primary production. The negative consequence of an increase in vertical mixing on temperature was also reinforced by a drop in solar irradiance. Off the northern Sicilian coast, an isotopic analysis performed on a vermetid (*Dendropoma petraeum*) reef indicated that the thermal drop was close to 2°C (393).

5.2 Natural causes of extinction

Extinction is a natural phenomenon and it is part of the earth's history. David Raup even stated that:

> the extinction of species is not normally considered an important element of neodarwinian theory . . . This is surprising in view of the special importance Darwin attached to extinction, and because the number of species extinctions in the history of life is almost the same as the number of originations.
>
> (394)

Extinction is balanced by speciation and when both phenomena are in equilibrium, planetary biodiversity remains stable. The examination of fossil records teaches us that the average longevity of a species is of the order of 1–10 million years, at least for marine invertebrates (395). A mean range of 3–20 million years was assessed for corals, molluscs, echinoids and fish (396). This estimate depends on the species involved (Figure 5.17). The lineage of some species may continue for more than 100 million years, but such species are rare.

Palaeoecologists make the distinction between background and mass extinction, although there are no fundamental changes in the processes leading to extinction. It is interesting that marine species, compared to their terrestrial counterparts, tend to live longer (five times longer) (Table 5.2). Gaston and Spicer (397) attributed this difference to the buffer capacity of seawater, although admitting that some inherent properties of marine species may play a role. Indeed, the table also shows that some groups are more resistant than others (e.g. corals versus planktonic foraminifera). The ecological consequences of extinction are also species-dependent. The loss of keystone (or key-structural) species can be dramatic for the whole community.

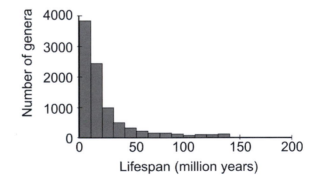

Figure 5.17 Lifespan of 17,500 extinct genera of marine animals, including vertebrates, invertebrates and microfossils.

Source: Redrawn from Raup (394).

Table 5.2 Estimated duration (million years) of some fossil
taxonomic groups.

Marine taxonomic groups	Duration (million years)
Reef corals	25
Bivalves	23
Benthic foraminifera	21
Bryozoa	12
Gastropods	10
Planktonic foraminifera	10
Echinoids	7
Crinoids	6.7

Terrestrial taxonomic groups	Duration (million years)
Monocotyledonous plants	4
Horses	4
Dicotyledonous plants	3
Freshwater fish	3
Birds	2.5
Mammals	1.7
Insects	1.5
Primates	1

Source: From Gaston and Spicer (397).

Table 5.3 Effects of mass extinction on marine species and genera.

Extinction	Age (million years)	Genera (%)	Species (%)
Late Eocene	35.4	15	35 ± 8
End Cretaceous	**65**	**47**	**76 ± 5**
Late Cenomanian	90.4	26	53 ± 7
End Jurassic	145.6	21	45 ± 7.5
Pliensbachian	187	26	53 ± 7
Late Triassic	**208**	**47**	**76 ± 5**
Late Permian	**251***	**84**	**96 ± 2**
Late Devonian	**367**	**55**	**82 ± 3.5**
Late Ordovician	**439**	**61**	**85 ± 3**

Note: Mass extinctions are in bold (species extinction >75%).

Source: From David Jablonski (398). *The age of the Late Permian mass extinction is from Bowring and co-workers (399).

It remains difficult to attribute a cause to past natural extinction. A species becomes extinct when it cannot reproduce either because of extreme conditions or because of competition with a more successful species.

The study of fossil organisms has also revealed mass extinctions where more than 75% of species became extinct (Table 5.3). Five extinctions took place in the Phanerozoic: (1) Late Ordovician; (2) Late Devonian; (3) Late Permian; (4) Late Triassic; and (5) End Cretaceous.

The causes of mass extinction remain in debate, but global climate change seems to have played an important role, even when it was not the primary cause (e.g. end of the Permian or Cretaceous).

5.2.1 The Late Ordovician mass extinction

The Late Ordovician extinction, which occurred 439 million years ago in two phases, eliminated 61% of marine genera and 85% of marine species. At that time, no multicellular organisms had colonised the land. Species particularly affected were brachiopods, bivalves, echinoderms, bryozoans and corals. The cause of the Late Ordovician extinction is thought to be related to a reduction in global temperature, perhaps caused by increased volcanism, triggering glaciation and eustatic regression (400). Using carbonate isotope palaeotherm-ometry, Finnegan and co-workers found that tropical ocean temperatures ranged from 32°C to 37°C, with the exception of short cooling episodes where temperatures dropped by about 5°C at the final Ordovician stage (401). When temperatures dropped, ice sheets increased, reaching a volume comparable to that reached during the LGM. The sea level rose again at the end of the glacial interval about 1 million years later and caused a second burst of extinction (402). Although some brachiopod species (genus *Hirnantia*) successfully evolved during global cooling, they became extinct when the climate warmed. Recently, Rasmussen and Harper provided evidence that changing plate tectonic configurations also modulated biotic extinction and recovery (402).

5.2.2 The Late Devonian mass extinction

The Devonian period is divided into three epochs – Early, Middle and Late – which are subdivided into stages. The mass extinction event occurred around the Frasnian and Famennian stages of the Late Devonian. At that time, about 70% of all species (82% for marine species) disappeared. Marine species such as brachiopods, ammonoids, agnathan and placoderm fish were highly affected. Many coral species disappeared in the Late Devonian, and only 10 out of 157 corals seen in the late Frasnian survived the crisis (403). The causes of the Devonian extinction (or extinctions) are still pending. Many hypotheses have been proposed, ranging from extraterrestrial impacts to **eustatic** sea-level fluctuations and volcanism. The higher rates of extinction for warm-water species suggest that global cooling may be at the origin of this extinction.

Paul Copper hypothesised that the closing of the ocean between Laurasia and Gondwana, which took place at the Frasnian-Famennian boundary about 374 Ma, disrupted the low-latitude circumequatorial flow of warm water (404). This led to the advection of cold waters into equatorial regions on the western margins of the joined continents from high latitudes, creating restricted circulation and anoxic conditions in warm-water basins on the eastern margin. Evidence for a global cooling with strong effects on low-latitude species at this time is confirmed by the disappearance of all genera of the tropical order Atrypida and Pentamerida. Kaiser and co-workers provided evidence for organic carbon burial that followed an episode of global warming (405). Carbon burial was subsequently followed by global cooling at the origin of glaciations and eustatic sea-level changes. As these two scientific reports show, the exact mechanisms for the origin of the Late Devonian extinction remain debated, but climate change seems to be involved.

5.2.3 The Late Triassic mass extinction

About 76% of marine species and 47% of genera became extinct in the Late Triassic. The ammonoid molluscs disappeared. Although scleractinian corals became important during the Triassic (the beginning of the symbiosis between corals and zooxanthellae), this group declined in the Late Triassic. Pitrat (406) estimated that 103 families of marine invertebrates became extinct. He calculated that 20% of ~300 extant families disappeared: cephalopods (loss of 31 families), marine reptiles (loss of 7 families), gastropods (loss of 6 families), bivalves (loss of 6 families) and articulate brachiopods (loss of 5 families). Causes proposed to explain this event range from gradual to catastrophic (407). Gradual explanations invoke sea-level change, anoxia and climate change (e.g. major aridification in the supercontinent Pangaea). Catastrophic hypotheses include intense volcanism and meteorite impact. Volcanism may have injected large quantities of CO_2 into the atmosphere and triggered a runaway greenhouse effect. Both the magnitude of the extinctions among marine and terrestrial groups and the cause(s) of the event remain in debate.

5.2.4 The Late Permian mass extinction

The Late Permian extinction, dated precisely to 251.4 ± 0.3 Ma (399), is the biggest extinction of the Phanerozoic. Trilobites, a taxonomic group that had lived for over 270 million years, were extinguished. Fusulinid foraminifera and Palaeozoic corals (*Rugosa* and *Tabulata*) became extinct (408). Stenolaemate bryozoans, articulate brachiopods and some echinoderms such as Blastoidea were close to extirpation. Because some species were key engineers, some communities collapsed. Between 48.6 and 49% of the marine animal families and 62.9% of the continental animal families disappeared. Using a reverse

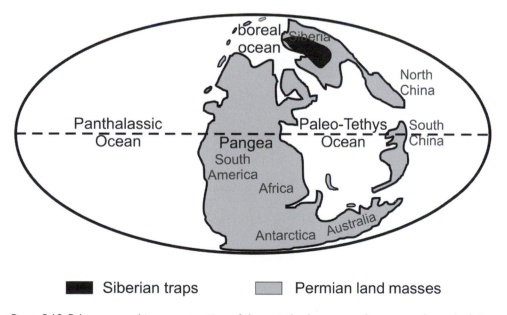

Figure 5.18 Palaeogeographic reconstruction of the main land masses and oceans at the end of the Permian.

Source: Redrawn from Saunders and Reichow (410).

rarefaction technique, Raup (409) estimated that 96% of marine species became extinct. Note, however, that this technique may overestimate the extinction rate by 10–15% so that the actual extinction rate is perhaps closer to 80%. This mass extinction event considerably modified marine biodiversity.

High diversity at that time occurred along the shoreline of the Tethys Ocean (southern China) and along the boundary of the boreal ocean (northern Arctic) (Figure 5.18).

Before the extinction, reef communities were diverse, rich in corals, crinoids, bryozoans, ammonoids and fish (Figure 5.19). Sea urchins, brachiopods, gastropods and foraminifera were abundant. The fauna was, however, richer at the edge of the boreal ocean than in warmer-water ecosystems located along the shore of the Tethys Ocean. The two ecosystems changed drastically after the crisis and became dominated by bivalves of the genus *Claraia* (Figure 5.19).

Many causes have been invoked to explain this mass extinction. A meteorite impact, with the putative identification of a crater in Mexico, has been put forward as a cause. However, this hypothesis has been severely criticised. Volcanism, global warming, acidification and anoxia have all also been proposed. The extinction took many thousands

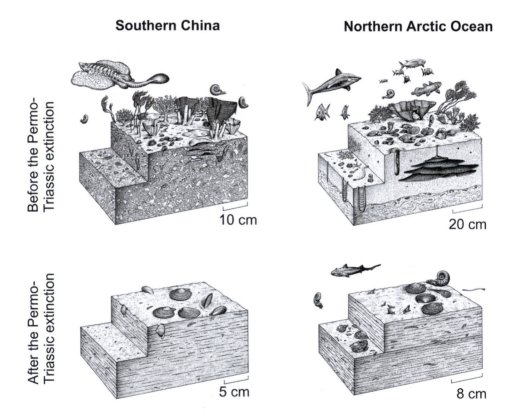

Figure 5.19 Block-diagram reconstructions of the ancient seabed in southern China and the northern Arctic before and after the Permo-Triassic mass extinction. Note the biodiversity change before and after the crisis and the associated changes in burrowing infauna. A marine fauna of 100 or more species was reduced to four or five.

Source: Original artwork from John Sibbick. Courtesy of Michael Benton (*411*).

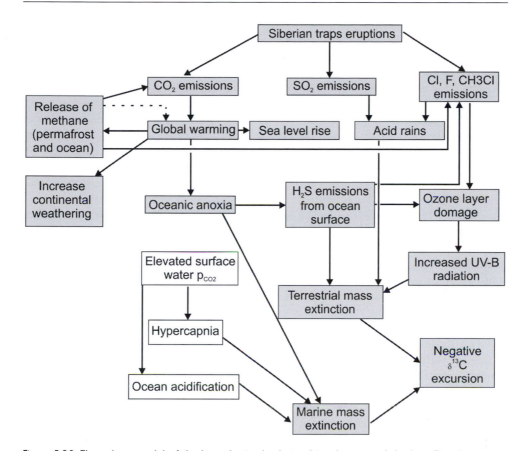

Figure 5.20 Flow chart model of the hypothesised relationships that caused the Late Permian extinction and led to the disappearance of 96% of marine species. White rectangle denotes links that remain in discussion. Dashed line represents a positive feedback. Carbon isotope excursions suggest major disruptions in the global carbon cycle.

Source: Redrawn from Wignall (*413*).

of years, and a new sequence of events has been propounded to explain the mass extinction (*412*). I develop here, in detail, the sequence of events involved in the mass extinction because some processes may also be implicated in the current climatic crisis (Figure 5.20).

At the end of the Permian, the main continental masses formed the supercontinent Pangaea, which was encircled by the Panthalassic Ocean (Figure 5.18). The Palaeotethys Ocean was bordered by Siberia (north), China (east), Australia and Antarctica (south) and Africa (west).

The climate warmed progressively during the 50 million years from icehouse conditions at the beginning of the Permian that were similar to the Quaternary glaciations to hothouse conditions when temperatures were much warmer as the end of the geological period approached. The warming was punctuated by periods of temporary cooling. Over a period of 600,000 years, giant volcanic eruptions took place across Siberia, pouring between 2 and 3 million km³ of basaltic lava over an area of 1.6 million km² (nearly the size of Mexico) to a depth ranging from 400 to 3,000 m (*408*). Prior to these eruptions, the concentration of CO_2 was 5–10 times the pre-industrial level, there were no ice caps over the poles and ocean circulation was probably sluggish. Some researchers have postulated that these 'Siberian Trap' eruptions were caused by a meteorite impact (*414*). The authors calculated that the volume of melt produced by the impact of a 20 km diameter iron meteorite travelling at 10km·s^{-1} would lead to a volume of melt comparable to other large terrestrial igneous provinces ($\sim 10^6$ km^3); however, this hypothesis remains controversial.

The first consequence of these continental flood basalt eruptions was the injection of vast quantities of sulphate aerosol into the lower stratosphere, which probably reduced incident light at the surface and triggered short-duration volcanic winters (Figure 5.20). A major flood basalt eruption of 1,000 km^3 is likely to inject between 3 and 10 gigatonnes (Gt) of SO_2 into the atmosphere (*410*). As a comparison, the Pinatubo eruption in 1991 released 0.02 Gt of SO_2. Photosynthesis reduction, associated with short cooling episodes, has probably led to the extinction of some thermophile species. Sulphate aerosols are, however, rapidly removed in a few years from the lower stratosphere.

A second consequence of these massive basalt eruptions was the injection of large quantities of halogens (e.g. chlorine as HCl) into the atmosphere, which may have reduced stratospheric ozone concentration and enabled more UV-B to reach the surface (Figure 5.20).

The third consequence of the Siberian Trap eruptions was an increase in the concentration of greenhouse gases in the atmosphere and an exacerbation of the greenhouse effect, which was already high in the Late Permian. The eruptions injected between 12,000 (estimate based on 2 million km^3 as basalt lava) and 18,000 Gt carbon (3 million km^3). A release of 18,000 Gt carbon over 1 million years is equivalent to an increase of 0.018 Gt per year, which is much less than the current increase in carbon related to the burning of fossil carbon (7 Gt of carbon per year). This release of carbon may have contributed to an observed change in the ratio of stable isotopes $^{13}C/^{12}C$ and the corresponding negative carbon isotope excursion (CIE), but it is unlikely that it was the only causative agent.

The ratio $\delta^{13}C$ is calculated as follows (expressed as ‰):

$$\delta^{13}C = \left(\frac{\left(\dfrac{^{13}C}{^{12}C} \right)_{\text{sample}}}{\left(\dfrac{^{13}C}{^{12}C} \right)_{\text{standard}}} - 1 \right) \times 1000 \tag{5.1}$$

This ratio varies as a function of global productivity, organic burial and the release of carbon and methane. Methane is constantly produced beneath the ocean floor. Although this chemical is, in part, consumed by archaea and bacteria, it also escapes into the overlying water column as bubbles (*415*). Methane has a very light $\delta^{13}C$ signature. For example, values for biogenic methane and thermogenic methane are close to –60‰ and –40‰, respectively. The release of large amounts of clathrate may alter $\delta^{13}C$ values. Such a carbon isotope excursion (CIE) was observed in the Toarcian 183–175.6 Ma ago and during the Palaeocene-Eocene Thermal Maximum (PETM, 55 million years ago). It has been proposed that the CIE observed at the end of the Permian was also exacerbated by the release of vast quantities of methane from clathrate probably occurring along large submarine slides or warming-induced melting of permafrost (*415, 416*). Nowadays, between 500 and 24,000 Gt may be stored on the oceanic floor. Of this, 400 Gt are estimated to be included in the permafrost. The amount of methane and clathrate on the oceanic floor in the late Permian is not known.

These vast quantities of methane may have exacerbated the greenhouse effect and led to a major increase in global temperature. This scenario is known as the clathrate gun hypothesis or runaway climate change. Based on a CO_2 concentration of 3550 ppmv (parts per million by volume), a methane concentration of 0.7 ppmv, and nitrous oxide of 0.275 ppmv, a solar constant S_0 of 1,338 W·m^{-2}, eccentricity of 0 and obliquity of 23.5°, a simulation showed that the oceans were 4–8°C warmer than the present-day ocean, in agreement with other works (417). As a result, eustatic sea level rose. In the northern hemisphere, winter land temperatures were 10–20°C warmer than in present-day simulations. In the southern hemisphere, surface air temperatures were 10–40°C higher than in the present-day control simulation in the highest southern latitudes (417).

The examination of Permian–Triassic marine sections has provided evidence of oceanic dysoxia (i.e. reduced content of dissolved oxygen), anoxia (i.e. absence of dissolved oxygen) and euxinia (i.e. low oxygen associated with reduced sulphur) (418). Using a three-dimensional global ocean model linked to a biogeochemical model of phosphate and oxygen cycling, Roberta Hotinski and colleagues (419) modelled the oceanic circulation of the Late Permian ocean and reconstructed a low equator to pole temperature gradient, which reduced oceanic circulation and triggered anoxia. Polar warming associated with tropical cooling of sea-surface temperatures caused ocean stagnation and anoxia throughout the deep ocean as a result of both lower dissolved oxygen in bottom source waters and increased nutrient utilisation at the surface.

At the end of the Permian, only 5% of marine species survived. It subsequently took 100 million years for global biodiversity to return to the pre-extinction level, although some reef communities were re-established as soon as 10 million years after the crisis. Bivalves radiated at the expense of brachiopods and gastropods. The runaway model is important because it suggests that current global warming may lead to the breakdown of negative feedback mechanisms and the establishment of positive feedbacks that may trigger a runaway greenhouse. Michael Benton and Richard Twitchett concluded their article 'How to kill (almost) all life: the End Permian extinction event' by stressing that 'some scientists and politicians look to the sky for approaching asteroids that will wipe out humanity. Perhaps we should also consider how much global warming could be sustained, and at what level the runaway greenhouse comes into play' (408).

5.2.5 The End Cretaceous mass extinction

About 76% of marine species and 47% of marine genera went extinct. This major extinction event was characterised by the disappearance of about 70% of the genera and about 30% of the families of scleractinian corals (420). In 1980, Luis Alvarez and co-workers (421) proposed that the Cretaceous–Tertiary extinction was extraterrestrial, related to a massive meteorite impact. They discovered high concentration of Iridium in deposits corresponding to the Cretaceous/Tertiary (K/T) boundary. This chemical element is rare on earth but common in meteorites. Subsequently, shocked quartz that forms exclusively at meteorite impact sites and tidal wave debris in the Caribbean indicative of a massive wave, diamonds with carbon isotope ratios closer to meteorites than terrestrial sources, and the Chicxlub crater (discovered at the northern tip of the Yucatan Peninsula) were all found (422). The Mesozoic was warmer than today, which favoured dinosaurs in comparison to mammals. However, the global cooling that followed the meteorite impact is thought to have unbalanced the interaction between dinosaurs (animals favoured by a high temperature regime) and mammals (animals favoured by a lower temperature regime as nowadays). Even if this extinction was caused by a meteorite impact, global climate change played a major role.

5.3 Natural contemporaneous changes

For decades, oceanographers have recognised the importance of changes in weather and oceanographic conditions for biodiversity (*423–426*). These physical alterations occur on multiple temporal scales. For example, long-term multidecadal changes in eastern North Atlantic sea surface temperatures (SSTs), assessed by **principal component analysis (PCA)** (*427*), encompass signals of different frequencies (Figure 5.21). In this example, the signal with the greatest mean amplitude is the long-term trend (46.8% of the total mean amplitude; Figure 5.21B), followed by the year-to-year (28.6% of the total mean amplitude; Figure 5.21C) and cyclical variability (24.6% of the total mean amplitude; Figure 5.21D).

Here, I separate artificially the influence of meteorological (synoptic) and climatic conditions into intraseasonal, seasonal, year-to-year, ocean–climate anomalies, cyclical or pseudocyclical variability, long-term trends and abrupt shifts. In addition, the examples will show that many factors are at work in the context of global change, revealing the complexity of situations related to many interactive stressors and the difficulty to tease apart anthropogenic and climate forcing (Chapters 7–9).

5.3.1 Synoptic variability

5.3.1.1 Cyclones and coral reefs

The path of tropical cyclones crosses many coral reef regions (Plate 5.3). Coral reef ecosystems have therefore been disturbed frequently by these meteorological events, and may to some extent be adapted (*428*).

On 3–4 February 2010, Cyclone Oli struck western French Polynesia. Both Tahiti and Moorea were hit by the cyclone, which had a mean wind speed of 210 km·h⁻¹. After four days, the laboratory CNRS-CRIOBE (Moorea) carried out a preliminary study on the impacts of Oli. Scientists led by Dr Serge Planes showed that Oli had a marked effect on the physical structure of the outer part of the reef (Plate 5.4). The rugosity index, which is a measure of the linear distance of the reef on the linear distance of a flat reef, fell by 50% down to 30 m because of the effects of waves, broken objects, boulders and increasing sediment concentration. Algal proliferation was observed after the hurricane. The cyclone was not the only phenomenon affecting the reef, however. The ecosystem was already weakened by starfish predation (*Acanthaster*).

In the Caribbean, a meta-analytical study (1980–2001) revealed that coral reef cover decreases by ~17% on average in the year after being hit by a hurricane (*429*). The influence of a hurricane on a reef is a function of the cyclone intensity and the time elapsed from the last disturbance by a cyclone. When human impact on the reef is moderate, the ecosystem can rapidly return to a similar ecological state (*428, 430*). However, in the Caribbean, Gardner and colleagues found no sign of a recovery for at least eight years after the impact. When a coral reef is already affected by anthropogenic pressures, tropical cyclones might precipitate its collapse.

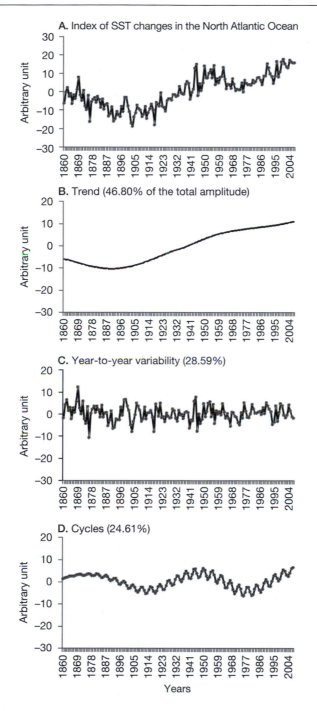

Figure 5.21 Temporal decomposition. (A) An index reflecting multidecadal changes in eastern North Atlantic SST. (B) Long-term trend representing 46.8% of the total mean amplitude. (C) Year-to-year variability representing 28.6% of the total mean amplitude. (D) Cyclical variability (a large cycle between 67 and 100 years and a smaller cycle of 5.7 years). The cycles represented 24.6% of the total mean amplitude. The temporal decomposition was performed by singular spectrum analysis (SSA).

5.3.1.2 Cyclones and mangrove forests

In October 2005, Hurricane Wilma hit the south-western part of the Florida peninsula. The powerful hurricane destroyed about 30% of the mangrove forests (Figure 5.22). This event provided an opportunity for researchers to examine forest recovery (431). As coral reefs, mangroves are present in many regions impacted by cyclones, and are therefore likely to be adapted to these major disturbances. Immediately after Wilma hit the forest, the loss of biomass and leaf area affected Net Ecosystem Production (i.e. total photosynthetic carbon assimilation minus total respiratory losses; NEP) and CO_2 flux significantly increased in 2007 in comparison to 2004 (Figure 5.22).

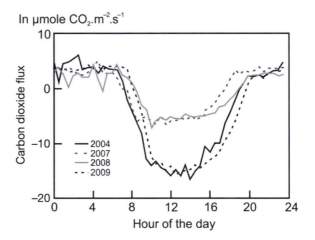

Figure 5.22 Impact of Hurricane Wilma (October 2005) on the mangrove forest in the south-eastern region of the Florida Peninsula. Daily CO_2 fluxes in April for 2004 (before Hurricane Wilma) and 2007–2009 (after Hurricane Wilma).

Source: Adapted from Sarmiento and co-workers (431).

Only daytime fluxes were altered by the hurricane, suggesting that respiration processes were not considerably affected. The major reduction in NEP continued for several years after the impact, only returning to similar conditions to 2009 four years after Hurricane Wilma. However, an integration of NEP throughout the year revealed that annual NEP remained lower in comparison to 2004. Some caution should be made in interpreting these results because the metrics of NEP might not reflect all the subtle changes that may have happened following the Wilma impact.

5.3.1.3 Cyclones, storms and rocky shores

Species inhabiting rocky shores can live a long time. For example, the algae *Ascophyllum nodosum* can live 25 years and the limpet *Patella vulgata* 17 years or more (294). They can withstand the effects of recurrent adverse meteorological conditions. In autumn and winter when storm intensity is maximal, the mussels *Mytilus californianus* are removed from boulders 10 times more frequently than in summer by large waves associated with storms.

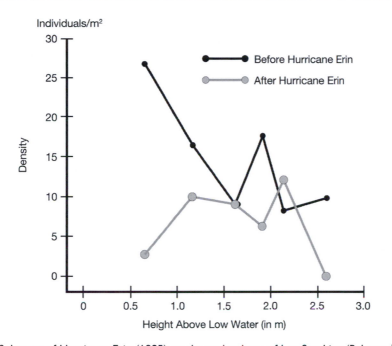

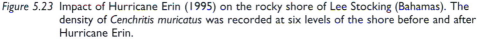

Figure 5.23 Impact of Hurricane Erin (1995) on the rocky shore of Lee Stocking (Bahamas). The density of *Cenchritis muricatus* was recorded at six levels of the shore before and after Hurricane Erin.

Source: Redrawn from Emson and colleagues (432).

Hurricane Erin destroyed part of the limestone shore of the Bahamas (Lee Stocking) in 1995. The heavy wave action associated with the cyclone reduced the abundance of the population of the snail *Cenchritis muricatus* from 20 to 2.8 individuals per m² (432). The number of *C. muricatus* per m² was significantly altered on the lower part of the shore after the hurricane (Figure 5.23). The density of *C. muricatus* decreased by 89% at a height of 64 cm above low water. At 1.05 m, its density diminished by 45%. Below low water, the species was apparently less affected.

Decadal fluctuations in cyclonic storm activities have been observed, although causes are not at present well understood. In the North Atlantic for the period 1944–2000, cyclonic activities have been related to the state of the Atlantic Multidecadal Oscillation (AMO; Chapter 2) (433). The period 1971–1994, corresponding to a cooler AMO phase, was characterised by low hurricane activity, whereas the period 1995–2000 (warm AMO phase) was characterised by a 2.5-fold increase in major hurricanes (i.e. hurricanes with a mean wind speed ≥50m·s⁻¹). Over the Caribbean, the increase was even more pronounced (fivefold increase). The authors hypothesised that the relationship between hurricanes and the AMO took place through an increase in Atlantic SST that promotes hurricane activity and a reduction in **tropospheric vertical shear** in the principal hurricane development region (10–14°N and 20–70°W), both phenomena characterising a warm AMO phase. Tropospheric vertical shear is an inhibitor of the cyclones development because it alters the temperature gradient between the hot oceanic surface and the colder temperature of the upper troposphere. Using 1,400 years of climatic data from the control simulation of the atmosphere-ocean general circulation model HadCM3 (Hadley Centre Coupled Model, version 3), Knight and colleagues showed a positive relationship between the AMO and hurricane activity (*434*).

5.3.1.4 Hot conditions

Rocky shore species are sensitive to time spent out of water. When sunny weather coincides with neap tides, many species can be damaged. On Cumbrae Island (Scotland), individuals of *Fucus spiralis* were severely harmed in the upper part of their vertical distribution on the shore after spending several days out of water in dry and warm weather conditions (*435*). Five events of this type were recorded between May and September 1975. One occurred in August and damaged individuals of *Ascophyllum nodosum* and *Pelvetia canaliculata*. Colin Little and co-authors concluded that only a few days of summer hot conditions might determine the upper vertical limit of fucoids on the shore; this clearly depends on the resonance between the tidal cycle and weather conditions.

5.3.1.5 Heatwaves

The 2003 European heatwave was an extreme meteorological event (*436*). Mean summertime temperatures went beyond the 1961–1990 average by more than four standard deviations over a large part of western and central Europe. This summer was probably the warmest since 1540. The maximum temperature ever recorded in the United Kingdom was broken on 10 August with 38.1°C. Sea surface temperature anomalies relative to 1979–2002 were up to 3°C in August 2003 (Plate 5.5). In the English Channel, temperatures offshore were the warmest ever measured (*437*). As a response to this exceptional meteorological event, many species had either unusual behaviour, or vertical or spatial distributions. The edible crab *Cancer pagurus*, usually caught in shallow waters, was observed as a juvenile stage in large quantities at depths ranging from 150 to 200 m. A melon-headed whale *Peponocephala electra*, a dolphin living in the tropical and subtropical oceans (35°S to 40°N), beached at Oleron Island on 27 August 2003. This was the first record of this species in Europe. Michel Houdart and colleagues from IFREMER also reported a harmful algal bloom of *Karenia mikimotoi* in Saint-Brieux Bay (Brittany, France), which was associated with high mortality of fish such as John Dory (*Zeus faber*) and European seabass (*Dicentrarchus labrax*).

Warm-water diatoms were recorded along the Brittany coasts. The diatoms *Asteromphalus flabellatus* (recorded in 1993, 1994 and 1995) and *Chaetoceros peruvianus* (recorded in 1993 and 2002), previously identified in autumn at concentrations lower than 1,000 cells per litre, were observed at concentration ranging from 1,000 to 10,000 cells per litre. The occurrence of *C. peruvianus* coincided with high trout mortality in Camaret Bay (Brittany). Another rarely observed diatom *Hemiaulus sinensis* reached concentrations of 280,000 cells per litre. Last but not least, two diatoms were identified for the first time in 2003: *Eucampia cornuta* and *Chaetoceros rostratus*. Subsequent monitoring revealed the occurrence of some colonies of *Eucampia cornuta* in September 2005 (*438*).

5.3.2 Seasonal variability

The seasonal climatic influence on biodiversity is prominent in extratropical regions, but also occurs in many equatorial and tropical regions (e.g. monsoons, tropical cyclones,

Madden-Julian oscillation, upwelling; Plate 5.6). The seasonality in eastern equatorial regions and over upwellings is also patent. In contrast, subtropical gyres have low chlorophyll concentration throughout the year.

5.3.2.1 The North Atlantic spring bloom

The North Atlantic spring bloom has been meticulously investigated (*152, 197, 439*). I therefore offer only a brief overview, placing emphasis on the climatic influence. Our understanding of this phenomenon is due to researchers such as Gran and Sverdrup (*159, 440*). According to Sverdrup's model, or the Gran effect, the general seasonal pattern of chlorophyll concentration in the North Atlantic Drift Province (NADR) is a function of several factors and parameters (Figure 5.24). The first factor is mixing, which brings nutrients into surface water. Mixed Layer Depth (MLD) can reach several hundred metres in winter in the North Atlantic (*65, 156*). Phytoplankton cells are mixed inside MLD down to the pycnocline (P; Figure 5.24).

The second factor is **irradiance** (here, mainly the latitudinal seasonal distribution in incoming solar radiation). The depth at which irradiance reaches 1% of its surface level roughly coincides with the **compensation depth** (D_c; Figure 5.24), the depth at which gross photosynthesis balances phytoplankton respiration. At D_c, net photosynthesis equals 0. When both respiration and phytoplankton production are integrated over the whole water column, the depth at which net photosynthesis equals 0 defines Sverdrup's critical depth (D_{cr}; Figure 5.24). According to Sverdrup's model, a bloom is only possible when MLD (or the pycnocline) is shallower than D_{cr}. Climate, through its influence on wind and temperature, influences the MLD.

The third factor is vernal stratification that guarantees that phytoplankton cells remain in the **euphotic zone**.

The spring bloom was intensively studied during the North Atlantic Bloom Experiment (NABE) in 1989. Sampling was carried out along the 20°W meridian at ~47°, 52°, 56° and 60°N (*441, 442*). At these latitudes, spring blooms occurred in mid-April, early May, mid-May and early June, respectively (*443*). This timing difference is commonly associated with increasing water stability and temperature (*439, 444*). In winter, deep convection modulated by wind stress occurs and the pycnocline is deeper than Sverdrup's critical depth (D_{cr}; Figure 5.25).

Deep convection penetrates into the layer of high nutrients concentration, bringing them up to the surface. However, in the absence of stratification, phytoplankton cells spend most of their time at a depth where photosynthesis is not possible because of the absence of light. During this period, sea temperatures decrease. Mann and Lazier (*439*) assessed that in the North Atlantic (40°N, 40°W) in December, solar radiation provides 4,300 kJ·m^{-2} per day (10 hours of sunlight), whereas at the same time the ocean loses heat by evaporation and, to a lesser extent, by sensible heat (19,000 kJ·m^{-2}). Irradiance plays a key role in photosynthesis through photosynthetically active radiation (PAR), but also by its action on sea temperature (Figure 5.26). From Figure 5.26, we see that most loss is related to the latent heat of evaporation.

The bloom initiates when MLD is shallower than the critical depth (Figures 5.24 and 5.25) (*159*). This occurs slightly before density stratification increases as a result of wind stress reduction and temperature rise. Temperature increases as a result of the absorption of short-wave radiations (*439*). Using the example from Isemer and Hasse (*445*), Mann and

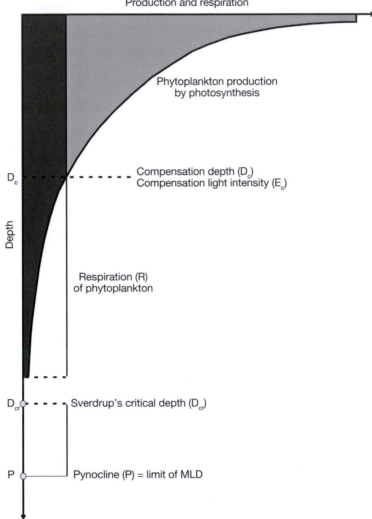

Figure 5.24 Simplified diagram showing the theoretical vertical distribution of phytoplankton production and respiration. The dark grey area shows both phytoplankton respiration and production. The light grey area shows phytoplankton production alone. Respiration (vertical solid black line) is assumed to be constant throughout the MLD.

Source: Adapted from Sverdrup (*159*).

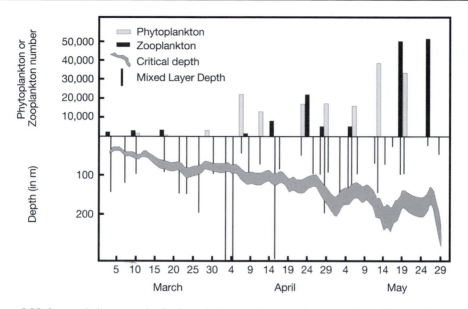

Figure 5.25 Seasonal changes in both phytoplankton and zooplankton abundance (Norwegian Sea; 66°N, 2°E) in 1949 in relation to MLD and Sverdrup's critical depth.

Source: Simplified from Sverdrup (*159*).

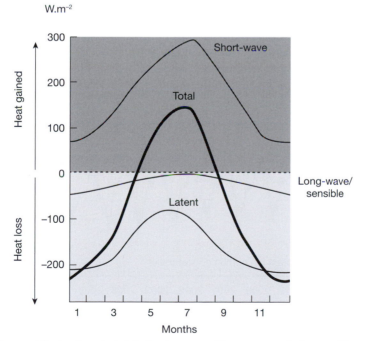

Figure 5.26 Seasonal fluctuations in total, short-wave and long-wave radiations and latent heat at 35°N and 48°W.

Source: Adapted from Mann and Lazier (*439*). Data from Isemer and Hasse (*445*).

Lazier (439) showed that net energy gain (after accounting for evaporation loss) for a day in July in the North Atlantic (40°N, 40°W) was 6,700 kJ·m⁻² in the first 5 m of the ocean. Application of Equation 5.2 showed that this amount of heat enabled an increase in temperature of 0.32°C.

$$\Delta T = \frac{\Delta Q}{mc} \qquad\qquad (5.2)$$

Here ΔT reflects temperature changes, ΔQ the energy gain or loss (kJ), m the mass of the water (one m³ of water ≈ 1,000 kg) and c the specific heat of water (4.18k J·kg⁻¹°C⁻¹).

When the pycnocline becomes shallower than the euphotic zone, increasing irradiance starts to stimulate phytoplankton cells brought to the surface by mesoscale eddy dynamics.

The spring bloom begins to decrease when nutrient concentration becomes limiting. Nutrients are rapidly depleted by phytoplankton uptake and sinking. The bloom, first dominated by diatoms, becomes rapidly controlled by flagellates (lower surface/volume, favouring nutrients uptake) (290). From the beginning of September, a second but less intense increase in chlorophyll concentration can be observed. The timing of the blooms changes latitudinally. While at the southern part of NADR the two blooms are clearly separated in time, they tend to aggregate into a single summer bloom at higher latitudes. Oligotrophic conditions propagate northwards at an average rate of 3° of latitude per month in NADR.

5.3.2.2 The Dilution-Recoupling Hypothesis

Sverdrup's Critical Depth hypothesis has been re-evaluated many times because it has been frequently observed that the bloom can start before stratification can be detected (446, 447). Some authors have even suggested that the concept be abandoned (164). Using data from SeaWIFS (Sea-viewing Wide Field-of-view Sensor), Michael Behrenfeld showed that the phytoplankton bloom initiated prior to the shoaling of the MLD (Figure 5.27).

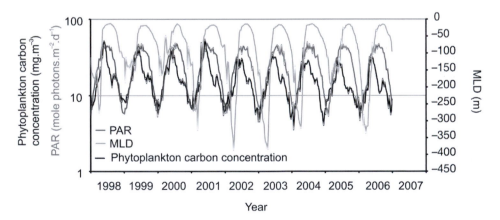

Figure 5.27 Seasonal changes in phytoplankton carbon concentration in relation to MLD (dark grey) and PAR (light grey) in the middle part of the North Atlantic (35°W–25°W and 45°N–50°N).

Source: Data from Behrenfeld (164).

This represents clear evidence against the Sverdrup hypothesis. Although relatively little discussed by the author, the results show a strong relationship between the increase in PAR and chlorophyll concentration.

The author proposed an alternative hypothesis to Sverdrup's hypothesis, which he called the Dilution-Recoupling Hypothesis. This posits that inside the ecosystem, there is usually a strong coupling between phytoplankton production and loss. Net population growth (r) represents the balance between the bulk phytoplankton growth rate μ and losses:

$$r = \mu - g - s - p - f \tag{5.3}$$

Where g is the mortality due to grazing, s the mortality due to sinking, p the mortality due to virus/parasitism and f losses by processes such as vertical mixing. The seasonal mixed layer deepening dilutes the potential grazers, which reduces g and increases r. Behrenfeld made the analogy with dilution experiments that are performed to measure phytoplankton net specific growth rate (r) maintaining grazing at a low level by adding water. In late spring, phytoplankton starts to decline, due to nutrient depletion or the pressure exerted by grazers (g). It should be remembered that early findings found that large diatom cells could be consumed as rapidly as they are produced by photosynthesis (152). In his book, Longhurst reported the study of Harvey and co-workers (448), who found that 98% of phytoplankton production was eaten by zooplankton in several weeks in the English Channel and that the Southern Ocean phytoplankton standing stock only represented 2% of the daily production (449). Increasing grazing pressure can act in synergy with an increasing mortality related to the coagulation of phytoplankton cells (s in Equation 5.3). Behrenfeld also showed that phytoplankton net specific growth rate (r) increased well before the increase in irradiance (his figure 4A) despite the close relationships observed between chlorophyll biomass and PAR (Figure 5.27). When MDL shoals, the coupling between μ and g increases because both phytoplankton and grazers increase.

I have two problems with the Dilution-Recoupling Hypothesis. The first is related to the fact that Behrenfeld compared the phytoplankton net specific growth rate (r) with the level of irradiance. By doing so, he compared a measure of rate of change (r is a slope, a derivative of the chlorophyll concentration) with a measure of state (irradiance). By doing so, he mathematically induced a lag, which explains why the two time series are negatively correlated. I would have put the emphasis on Figure 5.27, which shows a clear relationship between irradiance and chlorophyll concentration while the MLD continues to thicken. The second objection is that seasonal zooplankton patterns estimated from the **Continuous Plankton Recorder (CPR)** seems to match seasonal changes in phytoplankton concentration in the area (NA-5) investigated by Behrenfeld (Figure 5.28). Zooplankton seems to lag slightly both the increase and decrease in phytoplankton.

This above discussion provides evidence against the Dilution-Recoupling Hypothesis. At the initiation phase of the spring bloom, there is no sign of a zooplankton dilution. At the termination phase, there is no sign that grazing might control the bloom because both indicators decreased approximately at the same time.

It is clear that the concept of the spring bloom should be readdressed. Further development should examine the respective influence of nutrient depletion and grazing. We know that primary production is limited in stratified water by vertical nutrient supply. Furthermore, the direct temperature influence on the initiation and development of the bloom should be assessed. Based on a series of transects from the Azores (38°N) to the

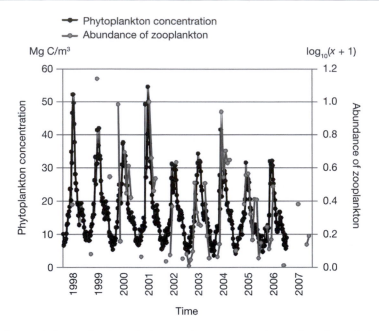

Figure 5.28 Seasonal changes in phytoplankton carbon concentration and total zooplankton abundance ($\log_{10}(x + 1)$).

Source: Phytoplankton carbon concentration data are from Behrenfeld (*164*). Total zooplankton abundance was assessed from the CPR survey (1998–2007).

Subarctic Gyre (54°N), Strass and Woods (*450*) showed that subsurface chlorophyll maximum moved polewards with the 12°C isotherm, suggesting a possible temperature control on the bloom. To conclude, this section illustrates well how climatic variability may modify the timing and intensity of the spring bloom by acting through many physical and biological channels. Behrenfeld's hypothesis explains the inverse correlations found between temperature and chlorophyll concentration in high latitudes (*451*). Climate is clearly implicated in the initiation of the bloom, whereas it is nutrient depletion that is responsible for its termination (both actions of climate and organisms themselves are therefore implicated).

5.3.2.3 Blooms of coccolithophores

Hurlburt (*452*) classifies coccolithophores as stenoecious species (i.e. a species that can only tolerate small environmental fluctuations). The group forms blooms in highly stratified waters where MLD is generally deeper than 30 m (*453*). The blooms modify the optical properties of the surface mixed layer and can therefore be detected from space. Plate 5.7 shows a bloom of *Emiliania huxleyi* occurring in NADR in summer in the English Channel.

High light intensity may be an important condition for coccolithophores, and no photoinhibition at high light intensity has been observed in laboratory experiments (*454*), which may explain why blooms generally forms in late summer. Other explanations invoking mechanisms not related to climate have also been proposed. For example, biogeographic studies suggested that the distribution of *E. huxleyi* coincided in time and space with low silicate concentration.

5.3.2.4 Temperature and diversity

The relationship between pelagic zooplankton biodiversity and climate is patent at a seasonal scale (Plate 5.8). In the extratropical North Atlantic, the seasonal increase in biodiversity parallels the increase in sea surface temperatures (*182*).

However, the rise in copepod biodiversity lags by two months the spring bloom, mainly above 52°N (*182*). Many factors such as summer stratification or food availability may contribute to the delay observed between the timing of the spring bloom and the diversity maximum. The lag identified between the timing of the spring bloom and mesozooplanktonic epipelagic diversity could be explained by the time needed for mesozooplanktonic diversity to rebuild after each winter minimum. Such a lag is compatible with copepod generation time (about one month, although this obviously depends on temperature). Maximum species diversity is observed in late summer when oligotrophic conditions prevail.

5.3.2.5 Deep-sea ecosystems

Recent studies have drawn attention to the strong sensitivity of deep-sea soft-sediment ecosystems to variation in surface food supply (*190*). Deep-sea soft-sediment (medium sands to clays) ecosystems, representing about 54% of the earth's surface (*191*), are influenced positively by the energy that arrives on the seafloor as particulate organic carbon (POC). As we have seen above, this source of endosomatic energy is controlled by climatic and biological processes, as well as surface nutrient concentration. Because of the seasonality in surface primary production (*190*), deep-sea soft sediment ecosystems might be expected to exhibit seasonality, at least in some regions where primary production is seasonal. Seasonality has been shown for isopods. Some bivalves and echinoderms also display recurrent seasonal patterns in ovarian maturation and spawning (*12*). Herring (*12*) reported seasonality in egg production (small and numerous eggs of ~0.1 mm) for one starfish, one urchin and three brittle-stars. He also provided evidence for a reproduction cycle mediated by POC arrival for some sponges, sea anemones and cumacean crustaceans. The activity patterns of urchins and sea cucumbers also tend to be modulated by the arrival of phytodetritus on the seafloor. However, other species do not display seasonality and produce only a few big yolky eggs, in agreement with Thorson's rule (Chapter 3) or Orton's pattern (i.e. deep-sea species with even reproduction).

5.3.3 Year-to-year variability

Year-to-year variability in biodiversity can be large and difficult to explain. In some cases, however, this high-frequency variability results from changes in some key environmental parameters controlled by large-scale hydro-climatic variability. Here, I detail some examples from the North Atlantic, the northern and equatorial Pacific Ocean, and the Southern Ocean (the Antarctic Peninsula and the Scotia Sea).

5.3.3.1 Responses of Calanus finmarchicus to the North Atlantic Oscillation

The use of large-scale indices of climatic variability is thought to better consider the holistic nature of the climatic system than parameters such as temperature, wind and precipitation (*455*) (see also discussion in Chapter 2). Year-to-year variability in the abundance, spatial distribution and **phenology** of species has been correlated to the NAO in both terrestrial and marine systems. In the terrestrial realm, investigations showed that some species of European birds (e.g. skylark *Alauda arvensis*) and amphibians (e.g. *Triturus vulgaris*) bred earlier as a result of frequent positive NAO phases in the 1980s and the beginning of the 1990s (456). In the marine environment, the NAO is thought to influence a large range of processes such as the North Atlantic spring bloom (457, 458) and the alternation observed between herring (*Clupea harengus*) and sardine (*Sardina pilchardus*) in the eastern North Atlantic (459).

It is important to remember (Chapter 2) that the North Atlantic Oscillation has different regional effects on sea surface temperature (SST), implicating contrasting ecosystem responses (Plate 5.9). When the NAO is in its positive phase, it has a negative influence on SST over the Subarctic Gyre (460). As a result, the number of warm-water species diminishes in the south-western part of the gyre (60). In the eastern North Atlantic (including the North Sea), the NAO influence is opposite and the number of warm-water species increases during a positive NAO phase. Because the main NAO influence on species is through temperature, the spatial pattern of correlations between this parameter and the climatic oscillation explains the synchrony in the (opposite) changes observed in the number of warm-water species in the North Atlantic and illustrates well **Moran's theorem** or **Moran's effect**.

Fromentin and Planque (138) discovered that year-to-year changes in the NAO and the abundance of the subarctic species *Calanus finmarchicus* around the British Isles were correlated negatively (Figure 5.29A). They also showed that the warmer-water pseudo-oceanic species *C. helgolandicus* was positively related to the NAO, although the relationship was weaker and with a one-year lag. Many hypotheses were proposed to explain the link between the NAO and these copepods: wind-induced turbulence and prey–predator interaction, the influence of temperature on competition between the two species, bottom-up control, changes in biogeographical boundaries between both species, the volume of Norwegian Sea Deep Water in which *C. finmarchicus* is known to overwinter and the flow of the European shelf edge current (462, 463).

Subsequently, Planque and Reid (465) attempted to forecast North Sea *C. finmarchicus* abundance as a function of the NAO. Such forecasting would have been very helpful, as the species is of great importance for higher trophic levels. However, the relationship between the NAO and *C. finmarchicus* abundance broke down. This was unexpected as the relationship held for more than 30 years. The timing from which the relationships ceased was apparently 1996, a year marked by one of the most extreme negative NAO phases. However, by using sliding correlation analysis, it becomes apparent that the correlation between the two variables ceased at the time of an abrupt ecosystem shift (AES) at the end of the 1980s (Figure 5.29B; see Sections 5.3.6). This result shows that a correlation between a large-scale source of hydro-climatic variability and a key species may vary through time. At a local scale, the correlation may only hold for a limited time period because the response is conditioned by the interaction between the species ecological niche and more direct environmental or climatic factors such as temperature (464). This result will be further discussed in Chapter 11.

5.3.3.2 Year-to-year variability in high-latitude ecosystems

High-latitude ecosystems are very productive, supporting important commercial fisheries (466). The main controlling environmental factors are sea-ice and temperature. The timing

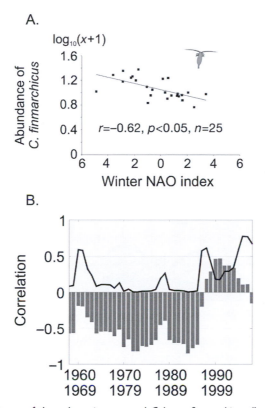

Figure 5.29 (A) Abundance of the subarctic copepod *Calanus finmarchicus* ($\log_{10} (x + 1)$) as a function of the NAO index during the period 1963–1987. (B) Sliding correlation analysis between the winter NAO index and the mean annual abundance of *C. finmarchicus* ($\log_{10}(x + 1)$) for the period 1958–2007 using a time window of 10 years. Grey bars represent correlation values and the black line shows the probability values after accounting for temporal autocorrelation. Upper and lower years indicate the beginning and end of the period for the correlation analysis, respectively.

Source: From Beaugrand (464).

of sea-ice retreat is crucial for the seasonal dynamics of these ecosystems, and especially for the intensity and duration of the phytoplankton bloom on which the whole annual production is based. In the Bering Sea (467), when winter sea-ice is high (cold years), melting occurs in April or May and is immediately followed by an ice-associated bloom (Figure 5.30A).

Large copepods such as *Calanus marshallae* dominate, influencing both carbon exportation and benthic-pelagic coupling (high). The abundance of small copepods is low and a large part of the phytoplankton bloom fuels benthic ecosystems. The survival anomalies of yellowfin sole (*Limanda aspera*), which feeds on benthos, increase (469). In contrast, when winter sea-ice formation is low (warm years), the retreat occurs towards the middle of March (Figure 3.30B). At this time, solar radiation cannot trigger a phytoplankton bloom, which is delayed until May or June when winds reduce, allowing water column stabilisation. Small zooplankton species develop, the phytoplankton bloom is more consumed by zooplankton

A. Cold years

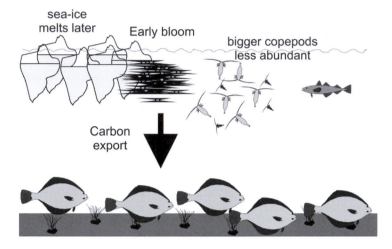

B. Warm years

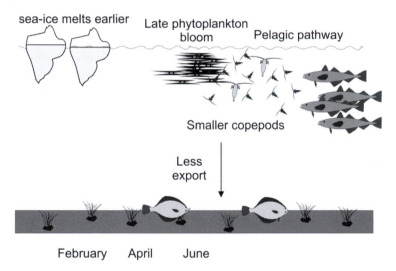

February　April　June

Figure 5.30 Schematic illustration of the influence of sea-ice retreat on polar ecosystems and benthic-pelagic coupling. (A) Cold years: sea-ice retreats later and the energy flow goes mostly towards the benthos. (B) Warm years: sea-ice retreats earlier and the phytoplankton bloom occurs later. Energy flow remains mainly in the pelagic ecosystem.

Source: Redrawn from Drinkwater and colleagues (467) and from Hunt and co-workers (468).

and fuels pelagic ecosystems; benthic-pelagic coupling is reduced. Survival anomalies of walleye pollock (*Theragra chalcogramma*) increase (469).

5.3.3.3 Year-to-year variability in the Southern Ocean

Southern Ocean pelagic ecosystems are often dominated by krill (*Euphausia superba*) or salp (*Salpa thompsoni*). Krill, exploited as a food resource for domestic animals and fish (470), occur mainly in the western South Atlantic (Antarctic Peninsula and Scotia Sea) where chlorophyll concentration is higher. Krill populations can reach hundreds of million tonnes and represent one of the largest protein sources on earth for fish, penguins and whales. Between 58% and 71% of the Antarctic krill are located in these regions (471). In contrast, salps are more widely distributed around the Antarctic continent, being more abundant in regions poorer in phytoplankton. At a year-to-year (and decadal and multidecadal scales; 1926–2003), Atkinson and colleagues (471) showed that western South Atlantic ecosystems were dominated by krill during cold years, whereas they were controlled by salps during warmer years. Salps have a shorter life cycle and quicker growth than krill. They can therefore respond rapidly to climate-caused environmental fluctuations. In this study, krill density was correlated positively with sea-ice extent the previous (austral) winter and winter sea-ice latitude of 15% ice cover (Figure 5.31). Summer food and sea-ice extent were the two key environmental parameters by which climate influences the abundance of this species that can live up to 5–7 years and reach 6cm long.

The intermediate mechanism by which sea-ice extent influences positively the krill density may be through sea-ice algae, which are a critical food resource. In the Antarctic Peninsula and the Southern Scotia Arc, high sea-ice algae concentration stimulates early adult spawning in spring and increases survival of the overwintering larvae the following year.

5.3.3.4 Tuna and the El Niño Southern Oscillation

The El Niño Southern Oscillation (ENSO; see Chapter 2) is the most prominent source of global-scale hydro-climatic variability (110) and may have a large impact on marine biodiversity. For example, it is thought that the super El Niño event of 1997–1998 killed 16% of coral reefs worldwide (472). The warm pool in the western part of the equatorial Pacific provides a suitable habitat for skipjack tuna (*Katsuwonus pelamis*). A convergence zone is detected at the eastern boundary of the warm pool. This zone results from the advection of cold and saline waters from the eastern part of the central equatorial Pacific that encounters warmer and less saline waters in the west due to sporadic bursts of westerly wind (473). This convergence front is indicated by a salinity front where nutrients concentrate and are at the origin of higher plankton abundance. Lehodey and colleagues (473) observed a significant correlation between the ENSO index, the mean location of the convergence zone (isotherm 29°C) and the mean skipjack tuna longitudinal position (Plate 5.10). During the El Niño event of 1991–1992, a zonal movement of more than 40° of longitude eastwards of the warm pool was associated with the zonal displacement of the relative abundance of the skipjack tuna with the same magnitude.

These zonal displacements were consistent with observed movements of tagged skipjack tuna. During the 1991–1992 El Niño, the eastward movements of tagged skip-jack tuna were prominent (473). These results had important implications for the fishing industry because it now enables the mean position of the species to be forecasted several months in advance.

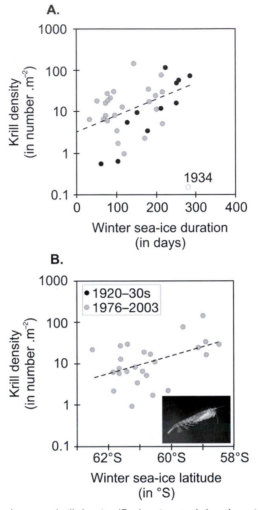

Figure 5.31 Relationships between krill density (*Euphausia superba*) and sea-ice in the western South Atlantic. (A) Annual mean krill density as a function of winter sea-ice duration the previous winter (South Orkneys). (B) Annual mean krill density as a function of winter sea-ice latitude of 15% ice cover. This was assessed along a transect by calculating the mean September latitude at which sea-ice concentration reached 15% (Western Scotia Sea).

Source: Redrawn from Atkinson and co-workers (*471*). Photo courtesy of Professor Uwe Kils.

5.3.4 Ocean–climate anomalies

I use the term 'ocean–climate anomaly' for an anomaly that is observed more than one year but lasting less than five years. Such ocean–climate anomalies have been observed in many ecosystems. Here, I show two examples: (1) Terre Adélie; and (2) the North Sea.

5.3.4.1 Hydro-climatic anomalies in Terre Adélie

In Terre Adélie, using the longest time series available on Antarctic large predators (1952–1999), Barbraud and Weimerskirch (474) found a reduction of 50% in the emperor penguin (*Aptenodytes forsteri*) population near Dumont d'Urville Station (66.7°S, 140°E): a decline from ~6,000 to ~3,000 breeding pairs. The predator forages only at sea and feeds on krill, fish (primarily *Pleuragramma antarcticum*) and squid. The collapse mainly occurred during a warm event (end of the 1970s) that reduced sea-ice extent and adult survival. The warm climate anomaly was attributed to the Antarctic Circumpolar Wave (475), which takes 8–10 years to encircle the continent, propagates eastwards with the circumpolar flow and regulates the climate system within the Southern Ocean. Barbraud and Weimerskirch hypothesised that in years characterised by high sea temperatures, the emperor penguins had more difficulty feeding, which increased mortality. However, processes are quite complex. A large sea-ice extent has two opposite effects. It may increase food availability, as krill increases when sea-ice extends. However, it may also increase the distance between the colony and the sea in winter. Other climatic fluctuations such as prolonged periods of blizzard and early breakout of sea-ice holding up the breeding colony can have devastating impacts.

5.3.4.2 Climate anomalies in the North Sea

Edwards and colleagues highlighted the importance of episodic oceanographic events in North Sea marine ecosystems (476, 477). Such events (1978–1982 and 1989–1991) had strong impacts on marine communities. The first (1978–1982), possibly associated with the Great Salinity Anomaly (64) or a change in the intensity and path of the Icelandic Current (477), initiated a cold hydro-climatic period (478) associated with an increase in boreal species (Figure 5.32). It led to a rapid switch in the ecosystem state with a stepwise increase in the number of subarctic (e.g. *Calanus finmarchicus*, *Pareuchaeta norvegica*) and arctic (e.g. *C. hyperboreus* and *Metridia longa*; not shown in Figure 5.32) species associated with a drastic decrease in the number of warmer-water species such as *C. helgolandicus* (Figure 5.32; mean number of temperate pseudo-oceanic species). As a result, North Sea copepod biodiversity diminished.

The North Sea cold ocean–climate anomaly, also associated with a reduction in decapod larvae and chlorophyll concentration, has often been attributed to the Great Salinity Anomaly (64). However, Turrell (480) highlighted that this event may have been triggered by a decrease in the intensity of the North Atlantic Current and the European Shelf Edge Current, which allowed the East Icelandic Current to carry more cold and low saline water towards the North Sea.

The second event (1989–1991) was characterised by warmer temperatures and higher salinities (481, 482) (Figure 5.32). Chlorophyll concentration and copepod biodiversity increased (483) in most North Sea areas. Subarctic and Arctic (not shown in Figure 5.32) species decreased while warmer-water pseudo-oceanic and continental shelf species augmented (Figure 5.32). Alistair Lindley and colleagues (484, 485) reported the presence of the oceanic species *Doliolum nationalis* and the calanoids *Metridia lucens* and *Candacia armata* in 1989. They concluded that an exceptional oceanic inflow took place into the North Sea.

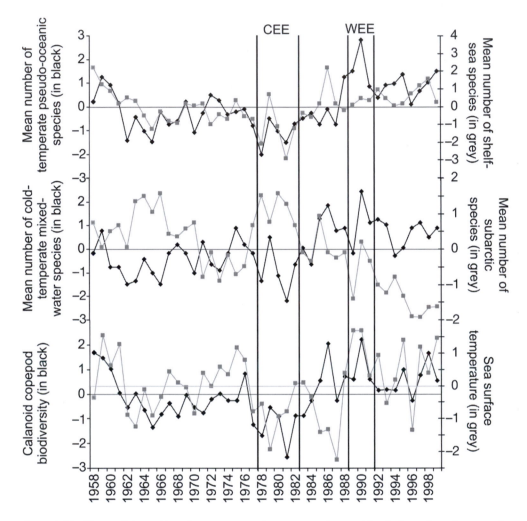

Figure 5.32 Effects of the cold and warm ocean-climate anomalies on North Sea calanoid copepod biodiversity. The cold event diminished copepod biodiversity and warm-water species (e.g. temperate pseudo-oceanic, shelf-sea species). In contrast, the number of cold-water species (e.g. subarctic species) increased. The warm event had approximately the opposite effects on copepod biodiversity. Shelf-sea species did not increase during the warm event. CEE: Cold Episodic Event; WEE: Warm Episodic Event.

Source: Modified from Beaugrand (*479*).

5.3.5 Pseudo-cyclical or cyclical variability

In this section, I show how pseudo-cyclical (or quasi-periodic) or cyclical variability may affect ecosystems and their biodiversity using four examples: (1) the Bohuslän herring periods; (2) sardine/anchovy oscillations in the Pacific; (3) sardine/herring oscillations in the English Channel; and (4) Russell's cycle.

5.3.5.1 Bohuslän herring periods

An example of pseudocyclical (or quasi-periodic) variability is given by the Bohuslän herring fishery (459). The Bohuslän region is located off the Swedish coasts (East Skagerrak). Environmental conditions are highly variable in the region because the area represents a boundary between the North Sea and the Baltic Sea. Periodically, large quantities of herring (*Clupea harengus*) arrive in autumn to overwinter in fjords and over small rocky islands named skerries (Figure 5.33).

These periods are known because of their implication for the regional economy that was normally a farming area. For example, during the second half of the eighteenth century, corresponding to the eighth period of herring abundance, up to 50,000 people came during the fishery season. About 216,000 tonnes of herring were landed during the fishery season 1895–1896 (486). These periods, which lasted a few decades (between 20 and 50 years), were times of economic prosperity. They were interrupted by phases of about 50–70 years

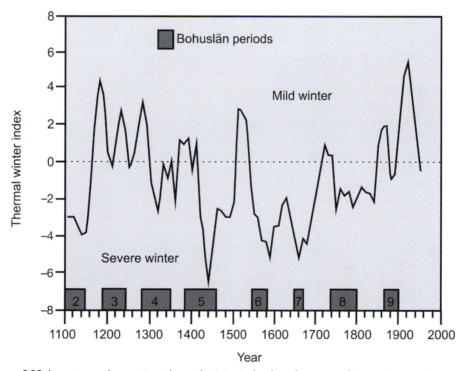

Figure 5.33 Long-term changes in a thermal winter index based on mean January temperature over England in relation to the occurrence of the Bohuslän periods.

Source: Redrawn from Alheit and Hagen (459).

where the species became rare or absent. Since the tenth century, nine periods have been reported.

Alheit and Hagen (459) convincingly explained Bohuslän periods by the North Atlantic Oscillation (NAO). We saw in Chapter 2 that this oscillation is controlled by the strength and position of the Icelandic Low and the Azores High that determine the characteristics of the Belt of Westerlies (BOW). BOW determines the position of a frontal region that separates cold from warm air masses to the south. Seven of the nine Bohuslän periods corresponded to low winter (January) temperature typical of a negative NAO phase. These periods were also characterised by anomalously widespread ice cover around Iceland. The prolonged negative NAO phase can only be explained by sustained negative anomalies in ocean temperature further south.

5.3.5.2 Periodic alternations of sardine and anchovy in the Pacific

Sardines and anchovies show antagonistic fluctuations in the Pacific Ocean. Long-term changes in the landing of sardines tend to be synchronous off California, Peru, Chile and Japan (139). Sardines generally dominate for periods of 20–30 years and are then replaced by anchovies. The dominance of sardines and anchovies occurred during warmer periods (1925–1950 and 1975–mid-1990s) and cooler periods (1900–1925 and 1950–1975), respectively. These shifts in dominance are related to large-scale hydro-climatic variability, even if the exact mechanisms that are in operation remain to be identified. The fact that the two species are not present at the same time could be the reflection of interspecific competition.

The prey-to-predator loop hypothesis is based on the fact that many preys can feed on the eggs or larval stages of their predators (487). When the predator declines because of an external force such as fishing, this can lead to the outbreak of its prey, which may in turn prevent the recovery of the predator because of predation pressure at the egg or larval stage. A key assumption here is the reduction of the predator by fishing. Note that this hypothesis may hold if the predator has been affected by adverse environmental conditions.

The predator pit loop is a hypothesis that is based on the empirical observation that two opposite states seem to occur for forage (prey) fish population, one characterised by a low abundance (refuge) and another characterised by outbreak abundance (487). This hypothesis invokes prey-predator interactions. When the prey becomes low, the predator searches for alternative preys and the specific mortality rate of the prey diminishes. This virtual refuge allows the species to survive and to subsequently recover. When the prey becomes more abundant, the predator will only catch a constant fraction corresponding to the satiation point and the prey is likely to continue to increase. In some circumstances, this can lead to an explosive increase. However, species interactions are probably modulated by climate.

A special type of competition called school trap mechanism has been invoked to explain the opposite cycle of dominance of both pelagic species (488). The conjecture is based on the fact that individuals must rapidly integrate a school to increase their probability of survival. When the species is abundant, it forms large monospecific schools. The less abundant species nevertheless integrate these schools. However, the school is likely to forage environment that corresponds best to the ecological niche (*sensu* Hutchinson) of the dominant species to increase its fitness. As the principle of competitive exclusion of Gause (489) stipulates, two species cannot have the same niche, and it is therefore probable that the rarer species integrated in the school is maintained at a low level by this mechanism. This positive feedback is likely linked to a tipping point.

Fishing pressure or a change in the climate regime may provide a force that might explain the rapid shift from one state to another. However, as mentioned above, to explain the synchrony (or teleconnection) between remote stocks, the cause is likely to be climatic. Cycles of dominance of sardines and anchovies coincide with cycles (or pseudocyclical variability) in the state of many hydro-climatic indices in the Pacific Ocean.

5.3.5.3 Periodic alternations of herring and sardine in the English Channel

The English Channel is an area where both herring (*Clupea harengus*) and sardine (*Sardina pilchardus*) may co-occur. As the sardine thermal preferendum is higher than herring (Arcto-boreal species), sardine is more abundant during warm than cold years. Such opposite cycles were observed along both Devon and Cornish coasts for the last 400 years. Alheit and Hagen (459) asserted that this antagonism was also observed during the **Little Ice Age** during the second half of the seventeenth century where sardine catch was low and herring catch high.

5.3.5.4 Russell's cycle

Monitoring of both chemical and biological conditions was carried out in the western region of the English Channel (Plymouth) from the end of the nineteenth century onwards (490, 491). At the beginning of the 1930s, Russell and colleagues observed a shift from a community dominated by the cold-water chaetognath *Sagitta elegans* to a community dominated by the warm-water species *Sagitta setosa* (490). A decline in large zooplankton was paralleled by a collapse in the number of young herring (*Clupea harengus*) replaced after 1935 by pilchard (*Sardina pilchardus*). These biological shifts were apparently associated with an alteration of the chemical properties of seawater (e.g. reduction in winter nutrient concentration). The change reversed in the middle of the 1960s. The chaetognath *Sagitta elegans* returned and was accompanied by other cold-water species such as the medusa *Aglantha* and the siphonophore *Nanomia* in summer and by the euphausid crustaceans *Meganyctiphanes* and *Thyssanoessa* in winter (492). Copepods of the genus *Calanus*, the pteropod *Spiratella retroversa* and larvae of the starfish *Luidia sarsi* increased. Phytoplankton also shifted after 1967 with pronounced increased in the diatoms *Biddulphia sinensis* and *Coscinodiscus concinnus* (490). The alteration in plankton composition propagated towards higher trophic levels. At times when *Sagitta elegans* was abundant, a high level of herring egg concentration was observed, whereas at times where *Sagitta setosa* dominated, pilchard egg concentration was elevated. Changes in both plankton and fish composition paralleled a shift in sea temperature and nutrient concentrations (e.g. phosphorus) (490).

Edwards and colleagues (136) related Russell's cycle to the Atlantic Multidecadal Oscillation (AMO; Chapter 2). For example, multi-decadal changes in both herring and sardine exhibited very similar trends to the AMO signal (Figure 5.34). The English Channel is close to a biogeographic boundary between the cold waters of the Atlantic Polar Biome and the warm waters of the Atlantic Westerly Winds Biome (152). In the English Channel, warming periods observed during positive AMO phases were associated with sardine dominance, whereas cold periods observed during negative AMO phases were characterised by herring dominance (Figure 5.34A). It is interesting to note that in the Norwegian and Barents Seas, herring stocks are close to their northern distributional range and warming was associated with a high level of herring abundance (Figure 5.34B).

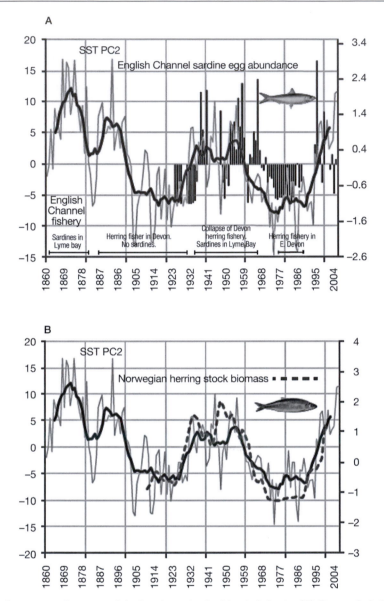

Figure 5.34 Long-term changes in fish abundance in the North Atlantic. (A) Detrended changes in annual SST reflecting long-term changes in the AMO and sardine egg abundance (bars) from the English Channel from 1924 to 2004. Sardine egg abundance values standardised to zero mean and unit variance (bars were used because the time-series was not continuous). Fluctuations in the fisheries of the English Channel are superimposed. The period after 1995 was dominated by sardines in the western English Channel. (B) Detrended changes in annual SST reflecting long-term changes in the AMO and long-term trend in Norwegian herring stock biomass (dashed line) for 1908–2000. Norwegian stock biomass values were standardised to zero mean and unit variance. The correlation between detrended annual SST reflecting the AMO signal and both species were significantly positive (sardine: *r* = 0.49; herring: *r* = 0.91).

Source: Modified from Edwards and co-workers (*136*).

5.3.6 Abrupt ecosystem shifts

Natural climate change can also be at the origin of abrupt ecosystem shifts (AESs), also termed **regime shifts** (59, 427, 483). The term regime shift has been utilised to describe large decadal scale switches in the abundance and composition of plankton and fish. This term reflects an abrupt and substantial shift from one dynamic regime to another (493). Such phenomena, also called critical transitions (494), are characterised by a sudden and substantial change in the community/ecosystem state (493) and involve major biological modifications, often with implications for exploited resources (495).

De Young and colleagues (496) defined regime shift as stepwise changes between contrasting, persistent ecosystem states. Although this term is operationally useful, Beaugrand and co-workers (497) preferred to call this phenomenon an AES because this term does not implicate the existence of stable states. Beaugrand and co-workers associated a shift to an increase in the local variance of the system. Periods of low variance alternate with periods of more pronounced variability (shift), with substantial socio-economic consequences (59, 498). Many abrupt changes (e.g. the North Pacific Ocean (59) and the North Sea (483)) have only been reported years or decades after they had actually occurred (59, 483). It is likely that climate change will intensify the frequency of these phenomena (499).

An AES was observed in the north-east Pacific Ocean in the mid-1970s. The shift was associated with a modification in large-scale hydro-climatic forcing expressed in sea level pressures, temperatures, changes in MLD and Ekman transports (500). Changes in the Aleutian Low increased the strength and frequency of storminess and westerly winds, which allowed a deeper mixing and more nutrients to be brought to the surface. This climatic forcing modified the carrying capacity of the central North Pacific gyre (501). Venrick and colleagues (502) reported an increase in phytoplanktonic biomass after the shift. Prior to the shift, the system was dominated by commercially valuable shrimps and crabs, and after by highly productive gadoids in large areas of the Bering Sea and the Gulf of Alaska (503, 504). Salmon production also increased in Alaskan regions while it diminished in California, Oregon and Washington (504, 505).

5.4 Is climate the primary factor?

In this chapter, we focused on the natural temporal variability in biodiversity on timescales ranging from intraseasonal to million years. Although interpretation of this temporal variability is complex, we have seen that many external forcings influence long-term variability in the marine environment. Palaeoecological changes can be explained by plate tectonics, meteorite impacts, eustatic variations or volcanic activity. However, even at these timescales, climate often plays a key role through changes in solar radiation, greenhouse gas concentrations and global atmospheric and oceanic circulations. Climate also explains frequently contemporaneous biodiversity changes in time and space (see also Chapters 3 and 4). Although many climatic and environmental parameters (wind, photosynthetically active radiation, macro and micronutrients) play a key role, temperature seems to be the most conspicuous factor that explains directly and in a mechanistic way biodiversity changes in time and space. We see in the next chapter by which processes temperature may influence biological systems at all organisational levels.

Chapter 6

Temperature and marine biodiversity

Using a variety of examples from different scientific fields, I show why temperature is probably the most important parameter by which climate affects marine biodiversity.

6.1 Temperature from the origin of the universe to early life

The universe has a thermal history. Since the Big Bang, decreasing temperatures have allowed creation and evolution of matter and enabled quarks, atoms, molecules, stars, planets and the first cells to appear and evolve (323). Many authors such as Lineweaver and Schwartzman have argued that the relationship between temperatures and life is almost deterministic and that connections between them are so fundamental that we can estimate how life is likely to be if temperatures change (323, 333, 335). While Christian de Duve (506) suggested that biogenesis is biochemically unavoidable (the Vital Dust hypothesis), Lineweaver and Schwartzman propounded that life is thermally inevitable. Providing that both the distance of the planet from its star and the size of the planet are compatible with the establishment of an **ecosphere**, both the beginning and the evolution of the biosphere are controlled by temperatures. They proposed that life is a cosmic phenomenon constrained by temperatures. Temperatures on earth may also reflect the long-term interaction between the geosphere and the biosphere (Chapter 5). According to the Gaia theory, life has been controlled by temperatures and has in turn influenced earth's global temperatures (334).

At a palaeoclimatic scale, Günter Wächtershäuser postulated that both biogenesis and microbial evolution advanced in a deterministic way from hyperthermophiles to mesophiles (507). The negative correlation found between the phylogenetic distance of some organisms (archaea, bacteria and eukarya) from the last common ancestor (LCA) and maximum growth temperatures tend to corroborate this statement (Plate 6.1).

Schwartzman and Lineweaver (508) also drew a scatter diagram of the phylogenetic distance from LCA as a function of maximum growth temperature of some microbial organisms (archaea and bacteria; Figure 6.1). A clear negative correlation was detected, indicating that both archaea and bacteria have a thermal regime that corresponds to their emergence time.

According to Schwartzman and Lineweaver, the analysis suggests that surface temperature is a critical factor for microbial evolution (508, 509). Why is temperature so influential for biodiversity? In the next sections, I come back to first principles and review mechanisms at the molecular, cellular and organismal levels because they enable a better understanding of how organisms and communities might react to temperature changes.

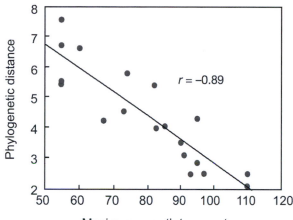

Figure 6.1 Phylogenetic distance (from the LCA) of hyper- and thermophile archaea (correlation coefficient *r* = –0.87) and bacteria (*r* = –0.88) as a function of the maximum growth temperatures (*Tmax*). Correlation with both archaea and bacteria: *r* = –0.89. The phylogenetic distance was assessed using 16S ribosomal RNA.

Source: Simplified from Schwartzman and Lineweaver (*508*).

6.2 Basics and first principles

Species use a mixture of strategies to regulate their body temperature. Most (99.99%) animals present on earth are **ectothermic** and depend on external heat sources (e.g. solar radiation and environmental temperature). Their temperatures are little different from the field. Most ectotherms, but some cavernicole species, are **poikilotherms**; their body temperatures vary. In contrast, birds (9,956 described species) and mammals (5,416 described species) are **endotherms**. They have internal mechanisms that produce heat internally, a process called **thermogenesis**. They also have mechanisms that ensure thermoregulation (or thermal homeostasis, thermostasis). Endotherms are therefore **homeothermic**. When endotherms live in regions colder than their inner temperature, they need to be insulated to minimise heat loss. In contrast, when they live in regions characterised by high temperatures, they need to lose heat excess, especially when they are active. Their **resting metabolic rates** are approximately five times higher than an ectotherm of equivalent body size.

The limit between ectotherms and endotherms is not straightforward. Some big fishes (e.g. some sharks and tuna) can maintain their muscles at a temperature 10–15°C above field temperatures. Their bodies are large and they are said to perform inertial **homeothermy**. Some animals are **heterotherms**. They can switch between **poikilothermy** and **homeothermy**, either in space or in time. Some seabirds or turtles may also perform regional **heterothermy**, keeping the temperature at the periphery of their body close to the temperature of their surrounding environment and their core temperature at more elevated values. The leatherbacks *Dermochelys coriacea* use countercurrent heat exchangers called *rete mirabile* to retain heat generated by their muscular flippers (*510*). Michael and Mrosovsky (*510*) caught some leatherbacks feeding off Nova Scotia (Canada) and measured their body temperature. They found that their inner temperature was, on average, 8.2°C warmer than the sea.

6.2.1 Comparison of the effect of temperature on marine and terrestrial species

Because of the high calorific capacity of water, the aquatic domain has not been favourable to the development of homeothermy. The high conductance of water makes marine ectotherms closer to the temperature of their environment than terrestrial ectotherms. For example, temperature of a lizard can be far from the temperature of its environment because it uses direct solar radiation to warm its body. Direct solar radiation cannot be exploited similarly throughout most of the marine realm. Pough (*511*) hypothesised that poikilotherms are better adapted to low resource levels than homeotherms because the latter requires more resources to sustain their metabolism. **Poikilotherms** are not as constrained, as they do not perform thermoregulation, which increases the basal metabolic rate. As a result, poikilotherms can potentially resist to long periods of limited resource.

Marine organisms are also more **stenothermic**. They have a narrower tolerance range compared to their terrestrial counterparts (*512*). Temperature fluctuations are much more constrained in the marine than in the terrestrial realm because of the high **specific heat capacity** of water, the 'buffer capacity' of water. Daily fluctuations in temperatures are weak and seasonal variations much smaller. Water has a specific heat capacity of 4,185 $J \cdot kg^{-1} \cdot K^{-1}$). About 336 kJ of energy are needed to boil a kettle of one litre of water originally at ambient temperature (20°C), which is equivalent of keeping alight an electric light bulb of 20 W (1 watt = 1 joule per second) during 4 hours and 40 minutes. The **latent heat of vaporisation** (i.e. convert water from a liquid to a gas state) is 2,260 $kJ \cdot kg^{-1}$ at 100°C while the **latent heat of fusion** is 335 $kJ \cdot kg^{-1}$ at 0°C. Extreme conditions may have larger impacts on terrestrial organisms while changes in the mean state may have deeper implications for marine biodiversity. This does not mean, however, that parameters such as minimum and maximum annual temperatures are more relevant for the study of changes in marine species. Because these species are so stenotherm, El Niño events have a large impact on corals, showing that conditions above normal during several weeks can also have a substantial impact on marine biodiversity (Chapters 5 and 7).

Thermal conductivity in water is about 25 times greater than in the air, which explains why we cool more quickly in water. The percentage of water in a marine organism is often higher than in a terrestrial organism making their conductivity even higher (*512*). For these two reasons, **homeothermy** offers no advantage in the sea, explaining why most marine organisms are **ectothermic**. Lipid tissues, which contain between two and three times less water, can protect the organisms against thermal fluctuations. Marine organisms that populate polar regions often have more lipid than their temperate counterparts. For example, the marine copepod *Calanus hyperboreus*, rich in lipid (wax esters) is well adapted to survive in the glacial waters of the Arctic Ocean and Nordic Seas (*513*). Its congeneric species *C. finmarchicus* has much less lipid reserve and is best suited to survive in the less cold waters of the North Atlantic Ocean and the Barents Sea. This adaptation is also necessary to resist periods when environmental conditions are less favourable (e.g. lack of food).

6.2.2 Fundamental laws

In the next sections, I detail some laws that describe how temperature can influence organisms and reveal that species are not equally sensitive to temperature changes. These laws are the Law of Conduction, the Surface Law, the adaptation of the Law of Fick to species and the Law of Van t'Hoff.

6.2.2.1 Law of Conduction

The rate of temperature change between the surrounding environment and a poikilotherm can be calculated as a function of time t (249):

$$T_1 - T_2 = (T_0 - T_2) \exp(-\lambda t) \tag{6.1}$$

Where T_1 is the body temperature, T_2 the ambient temperature, T_0 the initial body temperature and t the time. Heat exchange between the organism and its environment is a function of the initial thermal difference, time and a constant λ. Peters (249) showed that λ was a function of fresh body mass (W, kg):

$$\lambda = aW^{-b} \tag{6.2}$$

With a and b being two empirical coefficients, varying among species. By combining Equations 6.1 and 6.2, we obtain:

$$T_1 - T_2 = (T_0 - T_2) \exp(-taW^{-b}) \tag{6.3}$$

Equation 6.3 provides key information on the potential response of ectotherms to temperature change. It also suggests that large (big) organisms are likely to be less affected by the thermal fluctuations of the environment because of the greater time it takes for temperature changes to propagate through the animal. The loss or gain of temperature by conduction becomes more important when organism size decreases. However, we should not expect big organisms to be less sensitive to climate change, but instead to be more resistant to high-frequency variability in temperature.

6.2.2.2 The Surface Law

Thermal exchange between an organism and its surrounding environment takes place at surfaces. The Surface Law stipulates that the volume of an object increases more quickly than its surface (514). The basis of this law can be clearly understood mathematically and graphically. Mark Blumberg took the example of a growing cube to understand the basis of the law (515). When the cube grows from a one-unit to n-units, the surface-to-volume ratio diminishes because its surface increases slower than volume (Figure 6.2).

Transposed to an animal, after first assimilating a species to a hypothetical sphere, the volume increases quicker than the surface by a factor $R/3$ (Figure 6.3A):

$$S = 4\pi R^2 \tag{6.4}$$

With R the radius of the sphere. The volume V is calculated as follows:

$$V = 4/3\ \pi R^3 \tag{6.5}$$

Combining Equations 6.4 and 6.5, we obtain:

$$V = \frac{R}{3} S \tag{6.6}$$

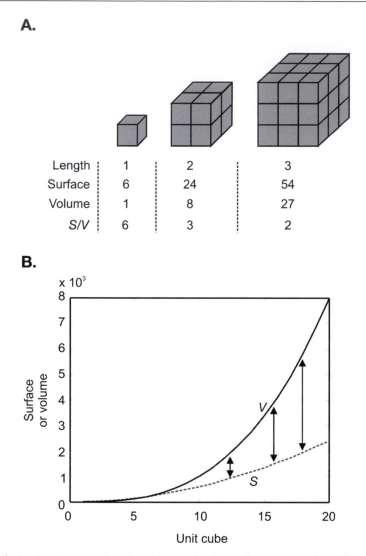

Figure 6.2 Relationships between length, surface and volume of a growing cube. (A) Increase in the surface and volume of a cube from a unit to a three-unit cube. *S/V*: surface-to-volume ratio. Note that the surface-to-volume ratio diminishes when the cube size increases. (B) Increase in the surface (*S*, dashed grey line) and the volume (*V*, black line) of a cube from a one-unit cube to a 20-unit cube. The double arrows highlight the differences.

Figure 6.3 shows that the volume increases more quickly than the surface when the size of an organism increases. However, the spherical shape is rarely met for organisms bigger than a single cell, for example a phytoplankton. So let us assimilate the shape of an organism to a cylinder. The surface S of a cylinder is calculated as follows:

$$S = 2\pi R(R + L) \tag{6.7}$$

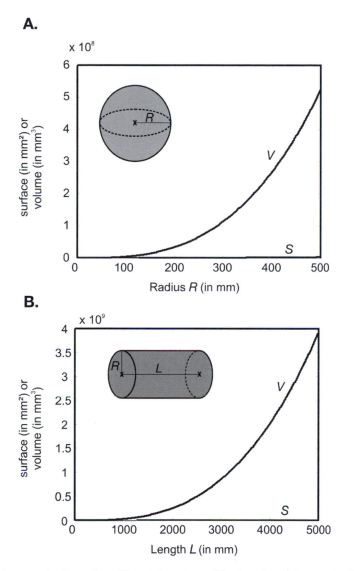

Figure 6.3 (A) Increase in the surface (S) and the volume (V) of a sphere when radius R increases from 1 to 500 mm. (B) Increase in S and V of a cylinder when length L increases from 1 to 5,000 mm. R = L/10.

With R the radius and L the length of the animal. The volume of a cylinder is calculated as follows:

$$V = L\pi R^2 \tag{6.8}$$

For simplification, R = L/10. Assuming that the length of such species varies from 1 (e.g. a small copepod) to 5,000 mm (e.g. a long pelagic fish), we see again that the volume increases quicker than the surface (Figure 6.3B). These theoretical calculations show that a big fish can retain more heat than a copepod simply because the surface increases slower than the volume. Bigger species are more prone to perform inertial **homeothermy** than smaller species. Note that more complex equations to calculate the volume of a copepod or a fish can be found in the literature (516).

6.2.2.3 The adaptation of the Law of Fick to species

This law states that the quantity of heat Q transferred by conduction between the organism and the environment is a function of the temperature difference and the distance L that separates the two points of temperature measurement, the exchange surface S and thermal environmental conductivity k:

$$Q = \frac{k\, S(T_2 - T_1)}{L} \tag{6.9}$$

With T_1 the temperature at a point inside the animal, T_2 the temperature at a point outside the animal and L the distance between the two points T_1 and T_2. From Equation 6.9, we can see that when the internal maximal temperature is reached (this can also be internal maximum temperature at which growth or reproduction remains possible), the increase in Q resulting from an increase in environmental temperature should be minimised. Assuming that organisms cannot rapidly adapt by playing on parameters L and k, the only possibility is to play on the exchanger surface S. When the exchange is through the whole body, the organism can minimise Q by decreasing its surface and therefore its size.

Frontier and Pichod-Viale (11) reported that when the calanoid copepod C. *finmarchicus* is born in water at about 15–16°C (summertime generation), the adult reaches a size of 2.8–2.9 mm. When individuals develop in waters of 8–9°C (wintertime generation), the adult reaches a size of 3.6 mm. This phenomenon has been termed **cyclomorphosis**. This phenomenon is also a manifestation of the **temperature–size rule** at the individual level. This law, also described by the informal expression 'hotter is smaller' (517), stipulates that higher temperatures during ontogenic growth lead to organisms of smaller size (518). Although higher temperatures decrease development time and the age at maturity, they indirectly reduce the surface S of the exchanger.

Atkinson investigated the effects of temperature on the size of diverse groups of protists, plants and ectotherms (208). Of the 109 studies he examined, 90 studies (~83%) found a negative relationship between size and temperature. Positive relationships were observed exclusively in insects. He subsequently reviewed 61 studies that examined the effects of temperature on aquatic **ectotherms** (210). He found that increased rearing temperatures caused a depression in organism size at a given developmental stage in 90.2% (55 studies) of the cases.

6.2.2.4 The Law of Van t'Hoff

Temperatures have a large influence on processes such as respiration, photosynthesis, digestion, excretion, growth and reproduction because most enzymatic reactions are controlled by temperature. The Law of Van t'Hoff states that the speed of chemical reactions increases exponentially with temperature. The coefficient that measures this increase has been termed temperature quotient or Q_{10} (or the Law of Q_{10}).

$$Q_{10} = \left(\frac{P_2}{P_1}\right)^{10/(T_2-T_1)} \qquad (6.10)$$

With P_1 and P_2 the measure of a process at temperature T_1 and T_2, respectively. The temperature quotient measures the thermo-sensitivity of a given physiological rate. For example, a Q_{10} of 3 means that when temperature increases by 10°C, the reaction speed is multiplied by 3. However, inside an organism, Q_{10} cannot remain constant when temperature rises because the speed of metabolic reactions becomes rapidly incompatible with cellular functioning and Q_{10} decreases beyond a critical temperature threshold depending on species and its life cycle. This leads to the notion of lethal temperature (lower and upper), which is classically measured in laboratory by T_{L50}, the temperature above which less than 50% of the individuals survive.

6.3 Effects of temperature at the physiological level

During an annual cycle, many extratropical ectotherms can tolerate large thermal fluctuations by altering their biochemical and physiological properties. Here, I review some key effects of temperature at the physiological level and the physiological adaptations used by organisms to withstand thermal changes.

6.3.1 Speed of enzymatic reactions

The influence of temperature at the physiological level is particularly strong. Temperature alters all aspects of the behaviour and physiology of ectotherms, including speed of enzymatic reactions and protein denaturation (Figure 6.4).

An increase of 10°C often increases the activity of most enzymes by 50–100%. Even thermal fluctuations as subtle as 1–2°C can increase the speed of enzymatic reactions by 10–20%. However, when temperature reaches a certain threshold, the speed of enzymatic reaction collapses (Figure 6.4). This often occurs at 40°C and can be related to protein denaturation, although many proteins cease to function at temperatures lower than the denaturation thermal threshold.

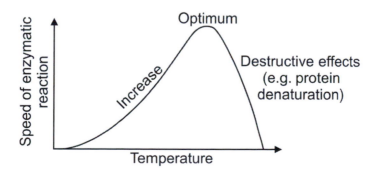

Figure 6.4 Theoretical thermal influence on speed of enzymatic reactions.

6.3.2 Individual performance

Individuals are designed to work within a certain thermal range because the structural and kinetic coordination of molecular and cellular processes are adapted to a specific thermal habitat (Figure 6.5). When temperature is too distant from the species optimum, a loss of **physiological integration** occurs (519). Note the dissymmetry of the response curves shown in Figures 6.4 and 6.5.

Using the terminology defined by Shelford (520) to describe the Law of Tolerance, Frederich and Pörtner described the changing performance of the spider crab *Maja squinado* by examining how haemolymph oxygen tensions, ventilation and heart rate varied from 0°C to 40°C (521). They determined the range of optimum performance by identifying the *pejus* (meaning turning worse) temperature thresholds (T_p) of the animal (Figure 6.6).

Full aerobic scope takes place in the optimum range (between the lower and upper *pejus* temperatures) where haemolymph oxygen tensions are maximal. The interval of temperature inside the optimum range enables both maximum reproduction and growth. In the *pejus* range between the lower/upper *pejus* and critical (T_c) temperatures, haemolymph oxygen tensions diminish, indicating the reduction in aerobic scope. Less energy becomes available for reproduction and growth. When the critical thermal threshold T_c is reached, the species enters the *pessimum* range. At this point, the anaerobic metabolism starts and survival becomes time-limited. The breadth of both optimum and *pejus* ranges reflects both ventilation and oxygen circulation within an organism.

6.3.3 Swimming performance

For many marine species, swimming is a fundamental activity influencing reproduction, settlement and foraging. Swimming is particularly important for coral reef fish, allowing them to survive in areas exposed to waves and tidal forces. Johansen and Jones (522) showed that increasing temperatures altered the **aerobic scope** and swimming performance of damsel-fishes. Aerobic scope is a good proxy of the species' capacity to preserve cellular processes while carrying on physical activity (522). This parameter gives an indication of the energy

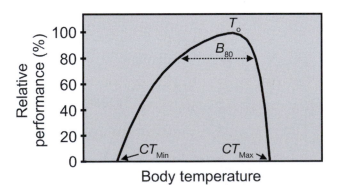

Figure 6.5 Typical response of individual performance to temperature. T_O: thermal optimum that maximises performance; B_{80}: thermal range within which individual performance is equal to 80% of its maximum; CT_{Min}: critical thermal minimum; CT_{Max}: critical thermal maximum.

Source: Modified from Angilletta and co-workers (519).

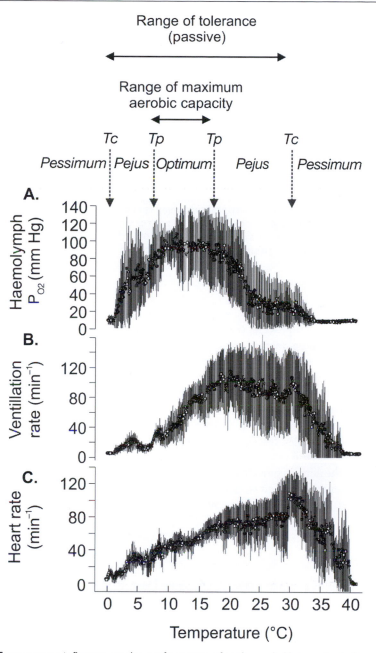

Figure 6.6 Temperature influence on the performance of spider crab *Maja squinado* from 0°C to 40°C. (A) Changes in haemolymph oxygen tensions (P_{O2}) as a function of temperatures. (B) Changes in ventilation rate. (C) Variation in heart rate. The profiles enable the determination of the lower and upper thermal pejus and critical thresholds *Tp* and *Tc*, which determine the optimum, pejus and pessimum range. *Tp* is the threshold at which P_{O2} starts to decrease and *Tc* are thresholds from which anaerobic metabolism starts.

Source: Modified from Frederich and Pörtner (*521*). Courtesy of Dr Pörtner.

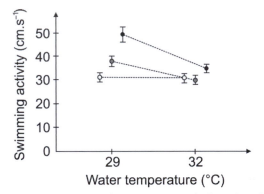

Figure 6.7 Influence of temperature on critical swimming speed of coral reef fishes of the genus *Pomacentrus*. Black circle: *P. coelestis*; grey circle: *P. lepidogenys*; white circle: *P. moluccensis*. Vertical bars are standard deviation. The reduction in critical swimming speed observed in *P. coelestis* and *P. lepidogenys* between 29°C and 32°C was statistically significant.

Source: Simplified from Johansen and Jones (522).

amount an organism can provide, once its basal metabolic rate is considered, to continue activities such as foraging, mating and escaping predators.

On 10 investigated species of Pomacentridae (genera *Chromis*, *Dascyllus*, *Neopomacentrus*, *Pomacentrus*), Johansen and Jones (522) found that a 3°C warming significantly increased the **resting metabolic rate** and significantly affected the critical swimming speed, **maximum aerobic metabolism** and **aerobic scope**. When water temperature increased from 29°C to 32°C, critical swimming speed of *Pomacentrus coelestis* and *P. lepidogenys* significantly declined by 29% and 21%, respectively (Figure 6.7), although no significant change was detected in *P. moluccensis*.

Furthermore, the authors provided evidence for an average 27.4% decline in aerobic scope for the 10 Pomacentridae. Their work therefore demonstrates that a 3°C rise in temperature significantly alters the capacity of damselfishes to carry on normal activities and may reduce survival.

6.3.4 Growth

Many strong relationships between temperature and growth have been reported (523, 524). In microbiology, it is well recognised that temperature is a cardinal factor for bacterial growth (523). Bacterial growth rates r can be modelled as a function of temperature T (K) (525):

$$\sqrt{r} = b\left(T - T_{min}\right)\left(1 - \exp\left(c(T - T_{max})\right)\right) \tag{6.11}$$

With T_{min} and T_{max} the minimum and maximum temperatures at which growth rate = 0; b and c are constant depending on species. For example, the thermophile bacteria *Bacillus stearothermophilus* has an optimum temperature of T_{min} = 303 K, T_{max} = 341 K and T_{op} = 331 K (Figure 6.8).

From an analysis of 181 published estimates of generation time and growth rates for 33 marine copepods at temperatures between −1.7°C and 30.7°C, Huntley and Lopez provided

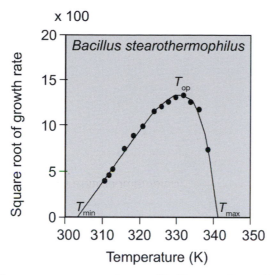

x 100

Figure 6.8 Influence of temperature on growth rate of *Bacillus stearothermophilus*. T_{op}: optimal
temperature; T_{min} and T_{max}: minimum and maximum temperature when growth rate = 0.

Source: Simplified from Ratkowsky and co-workers (525).

evidence that temperature explained more than 90% of the variance in growth rate and
even concluded that food was not limiting (526). The authors modelled instantaneous
copepod growth rates *g* as follows:

$$g = 0.0445 \exp (0.111\ T) \tag{6.12}$$

With *T* the water temperature (°C). The model explained 91% of the variance of the
instantaneous copepod growth rates (Figure 6.9A). They also modelled the generation time
{tau} as a function of temperature (*T*, °C):

$$\tau = 128.8 \exp (0.120\ T) \tag{6.13}$$

Their model explained 91% of the variance of the generation time (Figure 6.9B).

Huntley and Lopez proposed that the adult's weight W_a (µg carbon) was a function of
both the egg's weight W_0 (µg carbon) and temperature *T* (°C):

$$W_a = W_0 \exp (5.7316\ (\exp (-0.009\ T))) \tag{6.14}$$

Although a significant part of the variance remained unexplained, the authors provided a
physiological basis for the **temperature–size rule** and for **Bergmann's pattern** for ectotherms
(Chapter 3). Figure 6.10 shows the effect of an egg's weight and temperature on an adult's
weight.

Huntley and Lopez pursued this by stating that instantaneous growth rate was
independent of body size. This is probably because variation of body size among copepods
is relatively small. The metabolic theory of ecology, which has been tested for species with
body size varying by about 21 orders of magnitude (246), relates individual performance

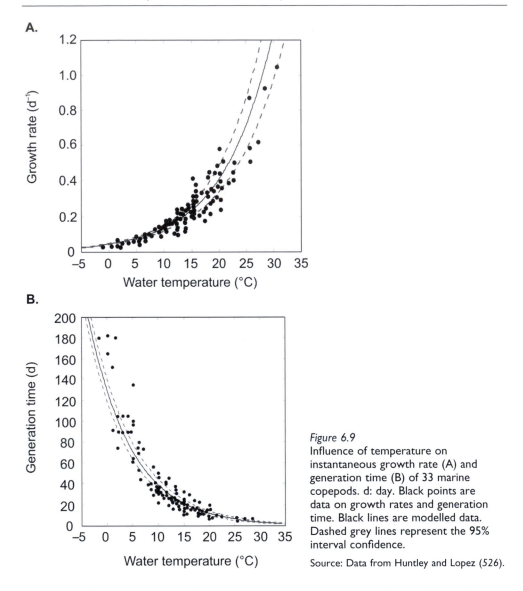

A.

B.

Figure 6.9
Influence of temperature on instantaneous growth rate (A) and generation time (B) of 33 marine copepods. d: day. Black points are data on growth rates and generation time. Black lines are modelled data. Dashed grey lines represent the 95% interval confidence.

Source: Data from Huntley and Lopez (*526*).

and life history traits such as ontogenic growth rate, fecundity, survival and mortality to both temperature and body size (*247*) (Chapter 4). From this theory, providing that a species inside a taxonomic group does not vary substantially with body size, many life history traits and parameters related to individual performance are largely influenced by temperature.

6.3.5 Photosynthesis

Solar radiation influences carbon fixation by photosynthesis. However, this process is also controlled by temperature. Cold lakes have low photosynthetic activity per moles of photons absorbed that some authors attribute to the slowdown of the enzymatic reactions (*275*).

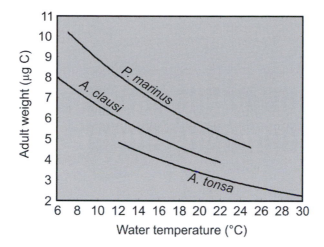

Figure 6.10 Influence of temperature and the egg's weight (µg carbon) on the adult's weight (µg carbon) of the copepods *Acartia tonsa* (W_0 = 0.028 µgC), *Acartia clausi* (W_0 = 0.035 µgC) and *Pseudodiaptomus marinus* (W_0 = 0.047 µgC). The thermal range over which the model was applied differs among species, corresponding to the amplitude they experience in the field.

Source: Data from Huntley and Lopez (526).

6.3.6 Biological macromolecules

Temperature changes also influence the properties of biological macromolecules. Molecular agitation can break the weak interactions (hydrogen bonds, ionic interactions, van der Waals forces, hydrophobic effects) and alter the three-dimensional protein shape; the whole process is called thermal denaturation. By changing protein structure, denaturation affects its ability to function. The protein loses its catalytic activity and becomes biologically inactive, which can cause severe problems for cell functioning. Many different proteins are present and each has its own denaturation threshold. If the temperature rise remains moderate, the process is reversible and renaturation takes place once the thermal regime returns to normal conditions. The sensitivity of these processes to temperature is poorly understood. The upper thermal limit of growth is often determined by the thermolability of molecules such as RNA or DNA, enzymes and membranes. The mitochondria membrane has an upper thermal limit of ~60°C for aerobic eukaryotes.

A hyperthermic stress may perturb the production of intermediate compounds and therefore imply metabolic dysfunction that will have cascading effects at the cellular level. This unbalances coordination between mutually dependent biochemical reactions, leading to a loss of **physiological integration**. When temperature increases, macromolecules are denatured at a higher rate than they are formed. However, many organisms can synthesise heat shock protein. The protein, detected from prokaryotes to mammals, protects macro-molecule three-dimensional structure and favours membrane transport.

6.3.7 Biological membranes

Biological membranes (e.g. cells, mitochondria) are highly sensitive to temperature changes (527). Among important functions they operate, membranes are physical barriers to solute

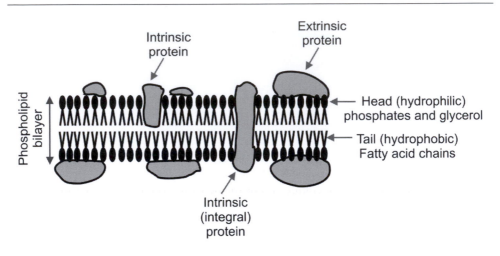

Figure 6.11 Typical structure of the plasma membrane, consisting of a phospholipid bilayer in which proteins are influenced both strong and weak interactions.

diffusion. They control the transmembrane movement of specific solutes and the transmembrane ion gradients that regulate mitochondria energy production. The pronounced thermo-sensitivity of membranes comes from the structure physical stability, which depends on weak chemical bonds. For example, the arrangement of fatty acid chains of phospholipids is due to the **van der Waals force** (Figure 6.11).

The protein-lipid complex develops a number of interactions, some of which are related to: (1) the hydrophobic effect when the protein is entirely embedded within the phospholipid bilayer; (2) the electrostatic (ionic) interaction; or (3) the hydrogen bonds when a protein is located at the surface of the membrane and interacts with the hydrophilic phospholipid heads (Figure 6.11). Contrary to covalent bonds that are strong chemical bonds, the weak interactions listed above form and break continuously in the cell and are controlled by temperature. When temperature rises, the resulting increase in a molecule's movement can destabilise electrostatic interactions or hydrogen bonds to affect membrane structure and function. In contrast, when temperatures decrease, the van der Waals interactions reinforce, which increases membrane viscosity. Because many functions exerted by plasma membranes require a certain level of fluidity (exocytosis and endocytosis), poikilotherms have developed a variety of mechanisms to ensure membrane function stability, a process referred to as homeoviscous adaptation (528). **Ectotherms** make use of the diversity of lipid structures to ensure membranes' physical stability by adjusting the degree of saturation of the fatty acid chains. When the saturation of a fatty acid chain decreases (i.e. increase in the number of double bonds), this increases the fluidity of the plasma membrane because it reduces the number of van der Waals interactions. The ratio of phosphatidylcholine to phosphatidylethanolamine (i.e. a fatty acid that destabilises the phospholipid bilayer) decreases in many ectotherms acclimated to lower temperatures. For example, this ratio declines in the rainbow trout *Salmo gairdneri* from 1.71 to 0.78 in their kidney plasma membranes when acclimated from 20°C to 5°C (529). Obviously, there are limits to this way to acclimatise to temperature change.

6.3.8 Oxygen transport

Temperature has a direct effect on oxygen transport in ectotherms (530). When the thermal regime moves too far away from the species thermal optimum, metabolism, and therefore oxygen demand, increases. The *pejus* temperature T_p, the first threshold beyond which aerobic metabolism starts to diminish, occurs earlier at higher than lower temperature because adaptation to higher temperature is more difficult (Figure 6.12). When this threshold is reached, the aerobic scope diminishes, which in turn affects reproduction and growth. When the thermal stress reaches the critical temperature Tc, anaerobiosis starts and the activity ceases. If the thermal stress continues, torpor takes over the organism, and survival is rapidly threatened; survival is further exacerbated by negative species interactions (e.g. parasitism, disease, predation). When temperature attains the denaturation temperature threshold T_d, the animal loses molecular functions, which lead rapidly to the animal's death. These physiological thresholds constrain species biogeography and their response to climate change.

Working on the eelpout (*Zoarces viviparus*) in the southern part of the North Sea, Pörtner and Knust explained the reduction in eelpout density by the direct influence of temperatures on tissues oxygen supply (530). In laboratory, they showed that the species' growth was highly influenced by temperature (Figure 6.13A).

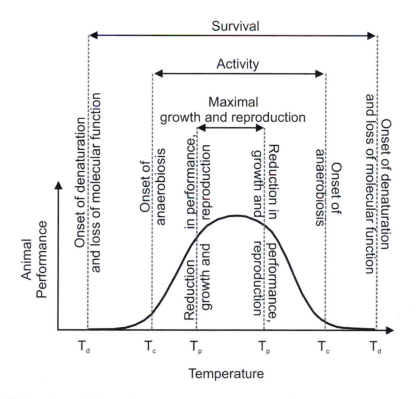

Figure 6.12 Physiological effects of temperature on an ectotherm. T_p: pejus thermal threshold from which aerobic metabolism reduces; T_c: critical thermal threshold from which activity ceases; T_d: denaturation thermal threshold beyond which denaturation initiates.

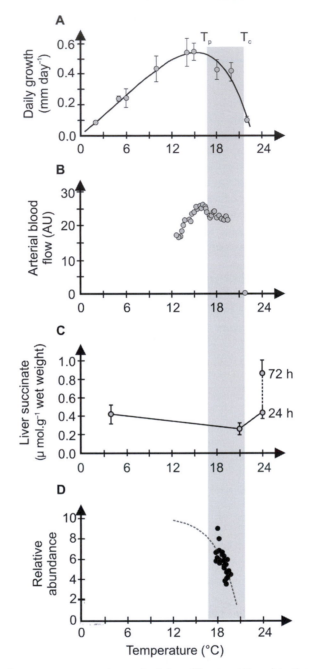

Figure 6.13 Effect of temperature on eelpout physiology (*Zoarces viviparus*) in the North Sea.
(A) Daily growth. Points are observations and the black line represents a model fitted
to the data. (B) Arterial blood flow, measured by nuclear magnetic resonance imaging
techniques. AU: Arbitrary Unit. (C) Accumulation of succinate in the liver as a result of
an increased unbalance between oxygen demand and supply above critical temperature
T_c. (D) Relationships between relative abundance and summer water temperature in
the Wadden Sea. Points are observations and the dashed line represents a model fitted
to the data. The shaded area denotes the pejus range between T_p and T_c.

Source: Simplified from Pörtner and Knust (*530*).

At first, growth increases with temperature. However, rapidly when the *pejus* temperature threshold is reached, the cardiorespiratory system cannot supply all oxygen needed by the animal to satisfy the increasing demand. The **aerobic scope** diminishes, reducing in turn individual performance. This explains the rapid collapse of the growth curve. Above upper *Tp*, the authors found, by using a non-invasive nuclear magnetic resonance technique, that the animal's circulatory capacity became limited (Figure 6.13B). This takes place prior to ventilatory limitations. The mismatch between oxygen demand and supply ultimately leads to the onset of anaerobiosis, as indicated by the accumulation of succinate in the liver (Figure 6.13C). Maximum individual growth rate coincided to the maximum eelpout circulatory capacity and maximum population growth (Figure 6.13A–B). The authors found that eelpout abundance decreased when maximum summer temperature was higher than average (Figure 6.13D).

6.3.9 Reproduction timing

Temperature affects species **phenology**. This has been shown in terrestrial plants by the use of the degree-day model, also called the thermophysiological method. This model stipulates that the sum of mean daily temperatures (sometimes subtracted by the temperature of nil development) is constant for a given development phase (*531*). Although this concept has been well applied in agronomy, it has been difficult to apply it in marine biology to establish a clear link between temperature and the timing of species reproduction. Some studies suggest this may explain phenological shift (*532*) and the mean annual abundance of species (Plate 6.2). In this study, we showed that the number of days needed to reach a given temperature diminished in the North Sea (*533*) and we found a strong correlation between changes in echinoderm larvae number and the number of days above 6°C, the temperature from which gametogenesis in *Echinocardium cordatum* initiates (Plate 6.2). The number of days to reach a given temperature diminished substantially after 1987, which coincided with an increase in both phytoplankton concentration and the abundance of echinoderm larvae.

6.3.10 Sex determination

Temperature can also determine the individual sex in some marine species (*534*). Temperature-dependent sex determination (TSD) has been observed in some marine fish and in all marine turtles. TSD was first observed in the Atlantic silverside *Menidia menidia* (*535*). Conover and Kynard found that the sex of this fish was controlled by both the genotype and temperature during a specific period of the larval development called the thermo-sensitive period. However, some studies have shown that TSD is not widespread among fish (*536*). In general, higher temperatures increase the proportion of males, whereas lower temperatures increase the proportion of females (Figure 6.14).

In contrast, TSD has been observed for all marine turtles (e.g. *Dermochelys coriacea, Chelonia mydias, Eretmochelys imbricata*). Chevalier and colleagues (*534*) investigated TSD of leatherbacks in the French Guiana (Atlantic Ocean) and in Playa Grande (Costa Rica). For these marine turtles, an alteration of less than 2°C had a considerable effect on the sex ratio of the hatchlings (Figure 6.15). Although some differences were observed between populations, a pivotal temperature threshold was detected for this species at about 29.5°C. All eggs developed into males below 28°C, whereas all eggs developed into females above 30°C. Egg thermo-sensitivity from the French Guiana was more prominent.

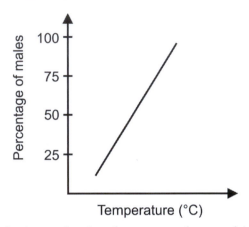

Figure 6.14 Proportion of males as a function of temperature for some fish. Note that the response line is often truncated.

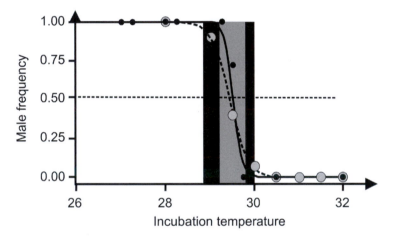

Figure 6.15 Observation of sex ratio in two populations of leatherbacks (*Dermochelys coriacea*) at temperatures ranging from 26°C to 32°C for eggs from the French Guiana (black circles and black line) and from Playa Grande (grey circles and dashed line). The shaded regions indicate the thermal range where between 5% and 95% of males were observed.

Source: Redrawn from Chevalier and co-workers (534).

Similar results were found in the green turtle *Chelonia mydas*, with eggs developing into males and females at temperatures cooler than 28°C and warmer than 29.5°C, respectively. We will examine the potential consequences of global warming for marine turtles in Chapter 7.

6.3.11 Seasonal acclimatisation

All physiological thresholds are difficult to identify precisely because they are dependent on **seasonal acclimatisation** and on the species developmental stage (537). Organisms (e.g. crustaceans and fish) can acclimatise seasonally to seasonal thermal changes (538)

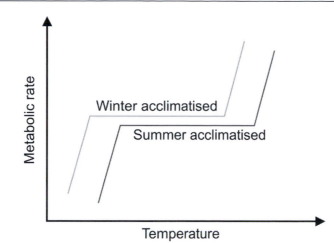

Figure 6.16 Theoretical alterations in the metabolic rate as a function of temperature in winter
(slight grey) and summer (dark grey).

Source: Modified from Levinton (51).

(Figure 6.16). Seasonal acclimatisation enables the species to be better adapted to cold in winter and to warm in summer (Figure 6.16).

As a result, the depression in metabolic rate that occurs at the cold end of the performance curve in winter or the warm end in summer is delayed. For example, when temperatures are close to 3°C, an individual mole crab (*Emerita talpoida*) consumes four times more oxygen than an individual of the same species in summer (51). Acclimatisation also occurs throughout the species latitudinal range and may lead to **physiological races**. At a low temperature, oxygen consumption is higher when the animal is at its polar boundary than at its equatorward limit. Populations of the oyster *Crassostrea virginica* and the sea squirt *Ciona intestinalis* have slightly distinct optima for reproduction throughout their latitudinal range (51). The effects of seasonal acclimatisation on animals are not easy to investigate. Seasonal acclimatisation can be accomplished by: (1) a change in enzyme concentration; (2) a shift in enzyme activities; or (3) the formation of new enzymes, proteins or glycoproteins (e.g. heat shock protein, cryoprotective substance).

6.3.12 Stage-dependent sensivity of species to temperature

Species thermo-sensitivity varies during ontogeny, the organisms being generally less tolerant to environmental fluctuations during the early phases of their biological cycle or at the adult stage (65). For example, both young and adult reproductive stages of fish are more sensitive to temperature than juveniles and young adults (Figure 6.17).

6.3.13 Physiological responses of ectotherms to a thermal stress

6.3.13.1 Resistance of ectotherms to a hypothermic stress

Species have developed physiological adaptations to tolerate large hypothermic stress. An important critical threshold is related to the freezing point of water. When temperatures

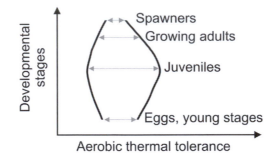

Figure 6.17 Theoretical changes in fish thermo-sensitivity throughout development.
Source: Modified from Pörtner and Farrell (*537*).

fall below zero, organisms migrate or develop adaptations to either avoid freezing or to allow controlled freezing. Many marine fish (e.g. Antarctic notothenioids, some gadoids) live in waters between 0°C and ~−1.8°C. Their blood or haemolymph can be maintained super-cooled providing the fish remove all nucleating agents (e.g. lipoprotein, bacteria or food that remains in the digestive tractus). They prevent freezing by increasing the concentration in some components such as sodium chloride, which increases blood/haemolymph osmolarity. Organisms accumulate cryoprotective substances (e.g. glycerol and sorbitol in insect and glucose in vertebrates).

In the terrestrial realm, some Arctic or alpine insects contain up to 30% glycerol, which enables the insect to decrease their freezing point to −30°C. Some marine organisms (molluscs), living in the intertidal zone, are known to practise controlled freezing. Thanks to nucleating agents (proteins and lipoproteins), they favour freezing in their extracellular space (*512*). The molecules appear in the haemolymph or blood before winter as part of the process of **seasonal acclimatisation**. When ice forms in the extracellular space, its amount must be controlled because the increase in osmotic concentration triggers a movement of water from the cellule to the extracellular space and a concomitant inverse flux of solutes. This, in turn, increases cellular osmolarity, which should remain compatible with the physical pressure at the cell membrane. In most species, the limit is reached when between 60% and 65% of the body water is congealed. Reduction in the photoperiod/temperature triggers the synthesis of antifreeze proteins (e.g. glycopeptids) in some Antarctic notho-tenioids. These proteins fix on ice microcrystals and slow down their progression. Such an adaptation allows Antarctic nothotenioids to live in seawaters close to ~−1.8°C. However, the same species will die if placed in seawater above 6°C. It is also well known that most Arctic fishes die when temperatures are above 10°C.

Some species have developed specific anatomical structures to keep their inner temperatures greater than their environmental temperature. Bluefin tuna (*Thunnus thynnus*) can live in habitats having a wide temperature range. In addition to its lower surface-to-volume ratio that enables the species to practise inertial **homeothermy**, the fish possesses a **rete mirabile**, a specific circulatory system that keeps ~95% of the internal heat generated by the animal's metabolism. Instead of travelling directly towards the centre of the animal, the oxygenated cold blood arriving from the gills travels at the body periphery and passes through a type of counter current heat exchanger. In this network of vessels, cold blood is warmed by vessels that contain warm blood as a result of muscular activity. This anatomical structure allows the fish to maintain a thermal difference between the periphery and the core of the body and explains why muscle temperatures remain between 28°C and 33°C while travelling in waters between 7°C and 30°C. This endothermic lifestyle enables the fish to swim faster in the water, as warm muscles work more efficiently than cold ones, and also allows the animal to improve digestion efficiency.

6.3.13.2 Resistance of ectotherms to a hyperthermic stress

Ectotherms are generally more sensitive to a hyperthermic stress because their capacity to cool is more limited. The resistance of intertidal species to hyperthermia determines their vertical zonation; the higher the resistance to hyperthermia, the higher the species vertical distribution (Figure 6.18).

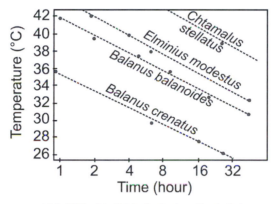

Figure 6.18 Temperature at which 50% of individuals die (median lethal temperature TL50) as a function of time for different barnacles.

Source: Redrawn from Newell (539).

Species with high lethal temperatures tend to be situated towards the supralittoral fringe. The Poli's stellate barnacle *Chthamalus stellatus* is a **thermophile** species inhabiting regions down to the northern part of North Africa. In the English Channel, the species is detected in the upper part of the shore. In contrast, the boreo-Arctic species *Semibalanus balanoides*, detected in the Baltic Sea, is located at the lower part of the shore. Evaluation of the long-term changes in the vertical distribution of such species in a given area could well inform on the impact of current climate change on intertidal species. Of course, temperature is not the only factor that influences species distribution. Animals face other pressures such as desiccation, wave action, type and size of the substratum and salinity (197).

To resist a hyperthermic stress, marine shore organisms can also evaporate water. Evaporation of 1 g of water (with no change in temperature) consumes 584 cal, therefore 2,441 J (512). However, this process involves organism dehydration. At the supra-littoral fringe, the common sea slater *Ligia oceanica* is normally located under stones during day. When temperature is >25°C, evaporation becomes essential but it is rarely achievable under stones where air is often saturated in water vapour. Therefore, the species goes outside where temperatures are higher to evaporate water and diminish its body temperature. Atmospheric humidity conditions the resistance of an intertidal species to a hyperthermic

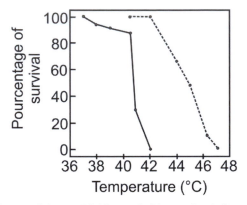

Figure 6.19 Thermoregulation of the sand fiddler crab (*Uca pugilator*). Percentage of survivors as a function of temperature in both dry (dashed line) and saturated atmospheres (full line).

Source: Redrawn from Newell (*539*).

stress. The sand fiddler crab (*Uca pugilator*) survives at higher temperatures in dry atmosphere because it can evaporate water to cool its body (Figure 6.19).

6.3.14 Resistance of endotherms to changes in temperatures

Endotherms maintain their body temperature to a constant temperature to optimise their metabolism. However, they are not less sensitive to temperature changes. I recall the following basic equation:

$$P = C + R + E \qquad (6.15)$$

With P heat production, C loss by conduction, R loss by radiation and E loss by evaporation.

When ambient temperature decreases, endotherms can either increase **thermogenesis** or diminish loss. Tropical species are very sensitive to a hypothermic stress and respond by increasing their metabolism. Another way is to limit heat loss by isolating the body. Organisms have developed special behaviour and some anatomic structures to reduce loss (see **Allen's rule**; Chapter 3). The bird skin of high-latitude anatids (e.g. geese, swans, eider) has duvet and feathers that are waterproofed thanks to an oil substance produced by the uropygial gland (also called preen gland). The preen oil made of diester waxes called uropygiols decreases integument thermal conductivity. Emperor penguins have very tightened feathers that isolate efficiently the bird down to ~−10°C. In marine mammals, the high conductance of water is accelerated by convection when the animal swims. Baleen whales (*Mysticeti*) have a subcutaneous layer of fat, called blubber or panniculus, for extra insulation. For example, the bowhead whale (*Balaena mysticetus*) has a blubber layer ~50 cm thick that allows the mysticeti to live in cold waters. Blubber represents up to 27% of a blue whale's (*Balaenoptera musculus*) body weight, 29% of a grey whale (*Eschrichtius robustus*) and between 36% and 45% of right whales (*Eubalaena australis*, *Eubalaena glacialis*, *Eubalaena japonica*). Some pinnipeds also have blubber ~7 cm thick. The layer maintains the temperature of the skin at a temperature close to the surrounding environment, which limits loss by conduction and keeps an important thermal contrast between peripheral organs

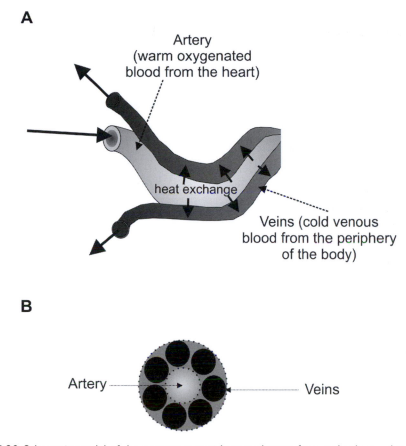

A

Artery
(warm oxygenated
blood from the heart)

heat exchange

Veins (cold venous
blood from the periphery
of the body)

B

Artery —————————————— Veins

Figure 6.20 Schematic model of the countercurrent heat exchanger frequently observed in some marine mammals (e.g. dolphin). (A) Veins are located around the artery. This allows the venous blood to gain heat when returning from the periphery of the body. (B) Transverse plane of the countercurrent heat exchanger.

Source: Simplified from Schmidt-Nielsen (*161*).

and the integument. Sea otters have a hydrophobic lipid substance, called squalene, which keeps their fur waterproof. Whales (*Mysticeti* and *Odontoceti*) also have a lower surface-to-volume ratio.

Three other mechanisms contribute to regulate body temperature. First, tegumental vasoconstriction occurs when the animal's body is in a cold environment. This mechanism cannot be pursued too long because of the risk of cyanosis, which can lead to chilblain. Second, the countercurrent circulation (embedded arteries and veins flowing in opposite direction) is frequently observed in bird and marine mammals (Figure 6.20). Veins have a thin wall. When veins come back to the heart, the blood they contain can be close to 0°C but progressively warms towards the heart. Vasodilatation and vasoconstriction are used antagonistically as a way to ensure thermal homeostasis. During vasodilatation, peripheral vessels in the fins, flukes and flippers in cetaceans and exclusively to the underside of the flippers in pinnipeds dilate to enable the species to lose heat. Third, a mechanism specific to marine mammals is the shunt of tegumental circulation. When individuals want to retain heat, there is a shunt of the circulation and blood alimentation reaches capillaries under the blubber layer. When they want to dissipate heat, the circulation reaches the skin and heat loss is maximal.

6.4 Influence of temperature at the species level

Many studies have provided compelling evidence for the cardinal effect of temperature on marine organisms. For example, Richard Kirby and colleagues (*540*) showed that multidecadal fluctuations in annual sea surface temperature (SST) and decapod larvae in the North Sea were positively correlated (Figure 6.21).

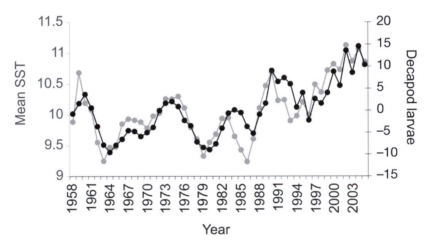

Figure 6.21 Relationships between fluctuations in decapod larvae abundance (black) and annual SST (grey) in the North Sea (1958–2005).

Source: Redrawn from Kirby and colleagues (*540*).

6.4.1 Temperature and the species' bioclimatic envelope

As we saw earlier, each organism has a thermal preferendum and both lower and upper lethal temperatures. These points determine population size at both local and regional scales, and the entire species' spatial range at the biogeographic scale (*541*). To illustrate this point, I take the example of the little auk (*Alle alle*), which has an estimated global population of ~80 million individuals (*542*). This 150 g seabird feeds on zooplankton of the genus *Calanus* (*C. finmarchicus*, *C. glacialis*, *C. hyperboreus*). Fort and colleagues investigated the winter survival of this endotherm species (*543*). They deployed miniature electronic loggers on little auks breeding in Greenland (Plate 6.3A).

This analysis revealed that the bird migrates in winter into a restricted area off Newfoundland, travelling a distance of more than 2,000 km from their summer breeding sites (Plate 6.3B). We investigated why the species selected this area to overwinter. By examining the relationships between the seabird abundance, temperatures and its *Calanus* prey abundance (Plate 6.3C–D), we showed that the species chose this region because it is the warmest place where its prey (mainly *C. finmarchicus*) remains abundant in winter.

The thermal regime ranged from 0°C to 5°C for air temperature and from 2°C to 6°C for sea surface temperature.

Because species are not isolated from each other, lethal temperatures are rarely reached in the field. When a species approaches its lethal temperature, the level of physiological stress is so elevated that species are more prone to negative species interactions such as competition, predation, parasitism or disease.

Each species has its own thermo-sensitivity. Some are **stenotherm** and cannot tolerate large thermal fluctuations, whereas others, the **eurytherms**, can live in environment characterised by large thermal fluctuations. All types of thermal amplitudes (i.e. the breadth of the thermal niche) have been documented. Antarctic notothenioid fishes are extreme stenotherms, inhabiting seawaters between −1.86°C and 4°C. In contrast, the shortjaw mudsucker (Gillichthys seta), an intertidal goby, is an extreme eurytherm with body temperatures varying between 8°C and 40°C (544).

Each species has a different optimum. The marine hyperthermophilic species Aquifex pyrophilus has an optimal growth temperature of 85°C and can live up to 95°C. The unicellular nanoplanktonic cyanobacteria Synechococcus is observed in hot spring at temperatures of ~75°C. The eukaryotic algae Cyanidoschyzon lives in water up to 57°C. The 200 mg freshwater fish Cyprinodon diabolis inhabits a spring called Devil's Hole (temperature regime of 33.9°C). The species has the highest upper lethal temperature (~43°C) known for a fish. Most tropical fishes die at 10°C (512). On the other side, **psychrophiles** (also called **cryophiles**) can inhabit thermal regimes ranging from −15°C to 10°C.

Both strength and sign of correlations observed between species and temperature vary spatially, being determined by the position of the regional thermal regime on the species' **bioclimatic envelope** (Chapter 11). If the thermal regime of a region is close to the centre of the species' bioclimatic envelope, normal temperature fluctuations do not have a strong effect (545).

6.4.2 Temperature and thermotaxis

One important way an **ectotherm** can control its temperature is by altering its behaviour. Behavioural thermoregulation has been largely documented among species on land, and it can be at the origin of **thermotaxis**. Mori and Ohshima have designed an apparatus for studying thermotactic behaviour of the nematode worm Caenorhabditis elegans. The centre of an assay plate was cooled to 17°C, whereas the periphery was warmed to 25°C. The ectotherm moved along concentric isotherms representing their cultivation temperature with thermal variation of ~0.1°C. For example, an individual reared at 20°C circled along the concentric 20°C isotherm (Figure 6.22).

In the marine realm, thermotaxis is probably more widespread than previously envisioned. Many species (e.g. plankton, fish, benthic invertebrates) exploit habitat thermal variability of the marine environment to maximise physiological efficiency. For example, plankton and fish adjust their vertical position in the water column to be as close as possible to their thermal optimum.

6.4.2.1 Copepod thermotaxis

The two copepods Calanus finmarchicus (subarctic species) and C. helgolandicus (temperate **pseudo-oceanic** species) have sometimes clearly distinct vertical distribution. In a study close

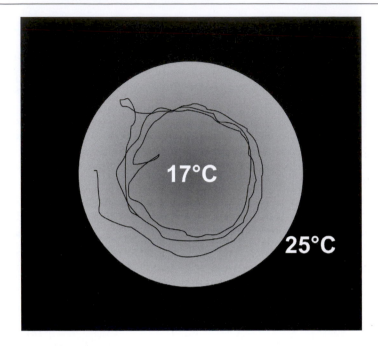

Figure 6.22 Apparatus revealing the behavioural thermoregulation (thermotaxis) of *Caenorhabditis elegans* on a radial thermal gradient ranging from 17°C at the centre to 25°C at the edge of an assay plate.

Source: Simplified from Mori and Ohshima (546).

to Dogger Bank in the North Sea (547), the temperate species was observed in surface, whereas the cold-water species occurred at depths of 30–50 m (Figure 6.23). The vertical separation between the two species is not always so clear, however (Figure 6.23). This study (547) draws attention to the fact that the vertical dimension could represent a way for these pelagic species to control their inner temperature by moving in a microenvironment closer to their bioclimatic envelope. This way to adjust to changes in the thermal regime might rapidly find its limits for herbivore zooplankton, which have to remain in the euphotic zone where phytoplankton develops. Other studies have documented examples of marine organisms migrating offshore in winter to escape colder temperatures that occur frequently in shallow coastal regions.

6.4.2.2 Fish thermotaxis

Experimental studies have clearly revealed thermotaxis in bony fish and elasmobranchs (548). Wallman and Bennett (549) designed a thermal gradient apparatus to investigate the thermal preferendum of Atlantic stingray *Dasyatis sabina* collected with landing nets from shallow waters of St. Joseph's Bay in Florida (Figure 6.24).

First, they showed that pregnant females exhibited statistically higher preferred median temperatures (26.1°C) than non-pregnant females (25.3°C). Although this difference was small (0.8°C), the authors argued it would diminish gestation time by two weeks. Second, Atlantic stingrays were exposed to a thermal gradient between 24°C and 30°C during

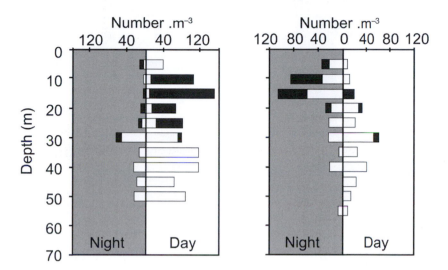

Figure 6.23 Vertical and diurnal changes in the abundance (numbers per m³) of female *Calanus helgolandicus* (dark grey bars) and *C. finmarchicus* (white bars) at day (white) and night (light grey) at two stations close to Dogger Bank in 2005. (A) Station 5 (56°N, 4°E). (B) Station 3 (56.3°N, 3.7°E).

Source: Redrawn from Jónasdóttir and Koski (*547*).

temperature preference trials, which approximate summer field conditions. They showed that unfed stingrays occurred in waters between 24.5°C and 31°C, whereas fed fish were found between 23.5°C and 27.5°C. The authors speculated that moving to cooler waters after feeding enabled stingrays to increase nutrient uptake efficiency by reducing evacuation rates.

In Lough Hyne (south of Ireland), Sims and colleagues proposed that the male of lesser-spotted dogfish *Scyliorhinus canicula* exploits its thermal environment by hunting in warmer waters and resting in cooler waters (*550*). This thermotactic behaviour is achieved by making diel vertical migrations in thermally stratified waters. By thermal choice experiment, the authors showed that male dogfish moved into warm water only for food search. After food consumption, male dogfish returned to rest and digest in cooler waters. By utilising a bioenergetic model, the researchers assessed that this thermotactic behaviour reduces daily energy costs by about 4%.

Thermotaxis has been suggested in a macroecological study on the response of both exploited and non-exploited fishes to climate change in the North Sea (*551*). Perry and colleagues showed a northward movement of 15 out of 36 species, which paralleled an increase in sea temperature. They also showed that some species, which did not move latitudinally, migrated vertically. For example, the plaice (*Pleuronectes platessa*) and the cuckoo ray (*Leucoraja naevus*) were two among six species that responded to sea warming by moving deeper into the water column.

A.

B.

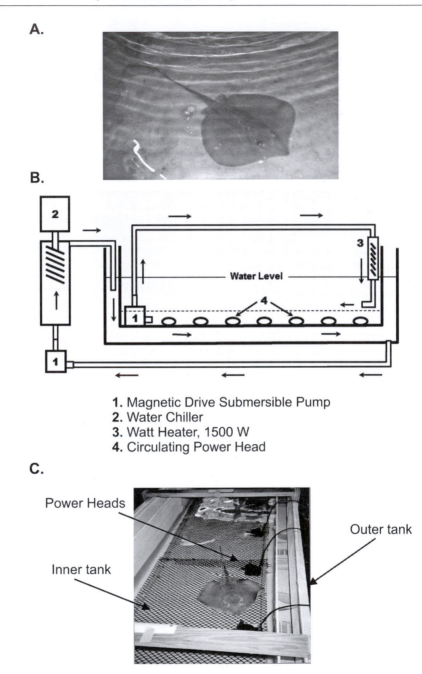

1. Magnetic Drive Submersible Pump
2. Water Chiller
3. Watt Heater, 1500 W
4. Circulating Power Head

C.

Figure 6.24 (A) Photo of Atlantic stingray. (B) Thermal gradient experimental design. The arrows
indicate the circulation of water into the apparatus. (C) Photo of the apparatus.

Source: Courtesy of Wallman-Jordan and Bennett (*549*), University of West Florida.

6.4.2.3 Seabird thermotaxis or kleptothermy

Mammals or birds also develop behavioural adaptation to face a hypothermic stress. The emperor penguin (*Aptenodytes forsteri*) is probably the most spectacular example of behavioural adaptation to extreme cold temperature in endotherms. The most southerly penguin species of the Antarctic is distributed in about 40 colonies and has an estimated global population of about 200,000 breeding pairs. Although its diet is composed of fish (e.g. Antarctic silverfish *Pleuragramma antarcticum*), crustaceans (e.g. gammarid amphipod *Abyssorchomene rossi* and krill *Euphausia crystallorophias*) and to a much lesser extent cephalopods such as squid (552), colonies can be located as far as 100–200 km from the sea.

Both males and females fast 45 days during the pairing period. Subsequently, the males fast an additional period of about 70 days incubing a unique egg at temperatures that can go well below –30°C in the middle of the austral winter (553). Despite the glacial winter, the seabird must keep the egg temperature above 35°C. **Thermogenesis** is activated when ambient temperature is far below their lower critical temperature TL_{cri} = –10°C (554). However, the activation of this physiological process involves a substantial energetic cost. The males can survive because they have lipid reserves that represent 40% of their body mass. However, physiologists have calculated that penguins, without any additional processes to minimise the adverse effects of low temperatures, must oxidise 25 kg of lipid to stay alive. Penguins do not have that amount of lipid available, however. As the mean body mass of a male is 38 kg at the beginning of the fasting period, they can lose ~15kg (555). So, penguins also adopt brief periods of huddling behaviour to help optimise their energetic balance in winter (Figure 6.25A). This behaviour is intermittent, occupying the bird ~38% of the day (556). Physiologists have shown that huddling behaviour among males enables them to reduce their metabolic rate by at least 25% when compared to isolated individuals (554, 557).

A few key results clearly show the energetic benefits of huddling (553). The metabolic rate of little groups having between 5 and 10 individuals, which cannot huddle effectively, is 39% lower than isolated birds, with 32% of the metabolic benefit due to wind protection (Figure 6.25B). Furthermore, metabolic rate of free-ranging emperors, which can move freely and huddle, is on average 21% smaller than little groups of 5–10 individuals.

When huddling, the males experience ambient temperatures that are within their thermoneutral zone, between –10°C and 20°C (554). Furthermore, the reduction in body temperature of 2.2°C observed when the seabird huddles represents an efficient strategy for reducing energy expenditure. A reduction of 1°C enables the bird to diminish its metabolic rate between 7% and 17%. The behavioural adaptation of emperor penguins is vital to successful incubation. Isolated individuals or small groups abandon their egg before the female returns (557). This way to regulate its body temperature by sharing the metabolic thermogenesis of another individual is called **kleptothermy**.

6.4.3 Torpor

An organism must realise its seasonal cycle in habitats characterised by a thermal regime that remains inside its **bioclimatic envelope**. If not, summer of winter torpor during the incompatible period or migration is the only alternative. The copepod species *Calanus finmarchicus*, inhabiting North Atlantic subarctic regions, is epipelagic (0–200 m) during spring and summer but diapauses between 300 m and 1,200 m in winter. This is also the case for *C. hyperboreus* and *C. glacialis*.

A.

B.

Decline in metabolic rate		39%		21%		
Behaviour	Isolated bird	Small groups (5–10 individuals)		Free-ranging seabird		
Rectal temperature	37.7°C ± 0.3	36.6°C ± 0.4		35.5°C ± 0.4		
				Loose grouping	Loose huddling	Tight huddling
Wind speed	4.9 m.s^{-1}	~0 m.s^{-1}		~0 m.s^{-1}	~0 m.s^{-1}	~0 m.s^{-1}
Ambient temperature	−16.6°C	−16.6°C		<−10°C	>−10°C	>−10°C
% of time	100%	100%		10%	41%	49%

Micro-climate (wind)   Micro-climate (temperature)   Reduced surfaces to cold

Figure 6.25 Example of behavioural adaptation of the emperor penguin (*Aptenodytes forsteri*) to cold temperature. (A) Photo of incubating male emperor penguins huddling tightly during a blizzard. (B) Influence of the behaviour of the emperor penguin on microclimatic conditions and consequences on metabolic rate.

Source: Courtesy of Dr Caroline Gilbert. Redrawn from Gilbert and colleagues (553).

6.5 Thermal influence at the community level

6.5.1 Community state

Significant correlations between thermal fluctuations and the abundance of marine species ranging from plankton to fish and seabirds have been reported (558). For example, Kirby and Beaugrand analysed long-term changes in the North Sea community state during the period 1958–2005 (559). The community state was characterised by applying a **principal components analysis (PCA)** on calanoid copepods (*pseudocalanus* spp. and *Calanus finmarchicus*), decapod and bivalve larvae, jellyfish, Atlantic cod (*Gadus morhua*), common sole (*Solea solea*) and plaice (*Pleuronectes platessa*). We observed a very strong correlation between decadal changes in the community state and annual sea surface temperatures (SSTs) on both original and detrended time series (Figure 6.26).

Many examples of such strong correlations between temperature and biological changes have been documented at the species and community levels (*140, 141*). We also saw a few examples in Chapter 5.

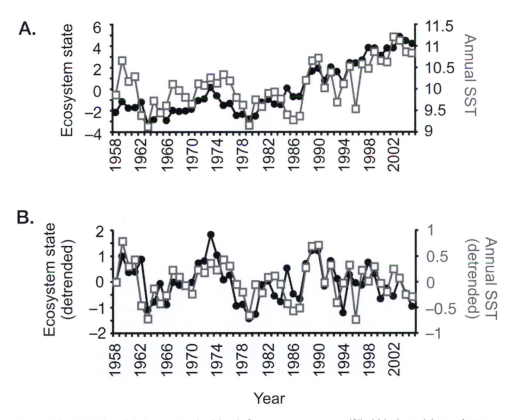

Figure 6.26 Multidecadal changes in the North Sea ecosystem state (filled black circle) in relation to annual SSTs (grey squares) during the period 1958–2005. (A) Original time series. (B) Detrended time series.

Source: Simplified from Kirby and Beaugrand (*559*).

6.5.2 Community body size

Community body size largely determines the types and strengths of the energy and material flows in ecosystems, affecting ecological networks and the way ecosystems are structured and function (560, 561). Biomass, production, predator-prey interactions, cannibalism and carbon export are among the quantities or processes that respond to the community size structure (560–562). Peters (249) recognised well the potential effect of temperature on organisms. However, he emphasised that the effect of temperature was modulated by body size, which could be of great relevance in the context of global climate change. Peters suggested that smaller animals responded quicker to temperature changes than bigger animals. We saw why in the early sections of this chapter. Community reassembly should propagate from smaller to bigger animals (bottom-up effect). Even if, individually, temperature effects are stronger on smaller individuals, bigger individuals depend on smaller for food. I showed a very strong negative relationship between mean copepod community size and annual SST in the North Atlantic (Plate 6.4). Larger copepod community size was prominent in the cold waters of the Subarctic Gyre, whereas in the warmer waters influenced by the Gulf Stream and its extension the North Atlantic Current, mean community size was much smaller.

Spatial changes in calanoid copepods conformed to Bergmann's pattern (Chapter 3). The **temperature–size rule** has been proposed to explain the response of ectotherms to temperature. As we saw, this rule states that most **ectotherms** mature at a larger size when rearing temperatures are lower.

6.5.3 Biodiversity

Changes in community structure also affect species biodiversity. Around the United Kingdom, the stepwise increase in temperature reorganised many components of pelagic and benthic ecosystems. This, in turn, strongly altered calanoid biodiversity. As a result, biodiversity increased substantially between the cold and the warm period (Plate 6.5). This increase in biodiversity was especially pronounced for spring and summer months to the north of the British Isles.

Part 2

Marine biodiversity changes in the Anthropocene

'The population problem has no technical solution; it requires a fundamental extension in morality.'

Garrett Hardin, 'The tragedy of the commons' (564)

Chapter 7

Biodiversity and anthropogenic climate change

We saw that the state of our planet has continuously evolved for 4.56 billion years (Chapter 5) and that impermanence is an ecological law. However, in more recent times, particularly during the last century, anthropogenic activities have exerted huge pressures on living systems (565). Global change reflects all types of human-induced changes on the ecosphere, including global warming.

Wolfram Mauser wrote: 'global change summarises the growing interference of human beings with the Earth's metabolism and its relation to the natural variability of the Earth System' (566). Mankind's pressure is now so often beyond natural variability that the Earth System is said to function in a no-analogue state. The total of the human demand on planet natural resources is thought to exceed the earth's capacity by 20% (567) and humanity uses more than half of all easy-to-get freshwater. The Nobel Prize Winner Paul Crutzen proposed the name Anthropocene for the epoch after the Industrial Revolution in the later part of the eighteenth century. More accurately, he proposed the beginning of this new geological period to be fixed in 1784, the year of the invention of the steam engine by James Watt (5).

Global change is probably the result, at least in part, of the increase in human population. As early as 1798, the economist Thomas Robert Malthus warned that human population exponential growth would lead to major ecological and societal issues. The biologist Paul Ehrlich talked about 'the population bomb' (Bomb P) and asserted that most of the ecological crisis of the 1960s originated from human population increase (568). While it took about 11,850 years for the world human population to grow from 1 million to 1 billion, the population grew from 1 billion to 2 billion in less than 150 years, from 2 billion to 3 billion in less than 50 years, from 3 billion to 4 billion in less than 20 years, from 4 billion to 5 billion and 5 billion to 6 billion in about a decade (Figure 7.1).

The increase in human demography is not the only factor, however. The effect of mankind on nature is also function of the per capita resource utilisation of human populations (569). Ehrlich and Holdren measured the negative impacts of humanity I as a function of human population P and a measure of the per capita effect $F(P)$, also influenced by human population in a non-linear way because of interactive effects:

$$I = P \times F(P) \tag{7.1}$$

The per capita effect also varies as a function of technology and available resources. Global change also results from the humanity socio-economic model and its unappeasable energy appetite (271). Exponential growth in human population paralleled a substantial increase in energy use and the conversion of 30–50% of the planet's land surface (570). Energy under

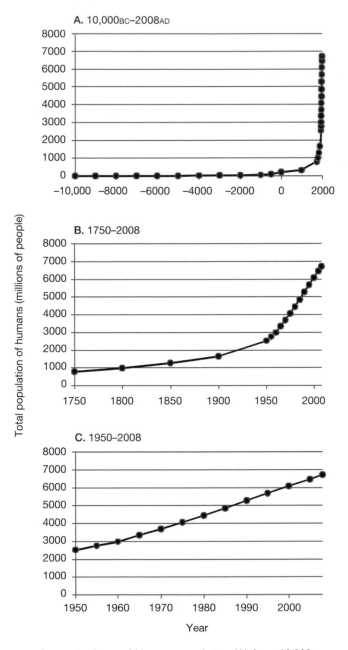

Figure 7.1 Long-term changes in the world human population (A) from 10,000 BC to AD 2008; (B) 1750–2008; and (C) 1950–2008.

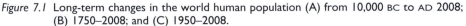

Source: Data from United Nations Population Division, Department of Economic and Social Affairs.

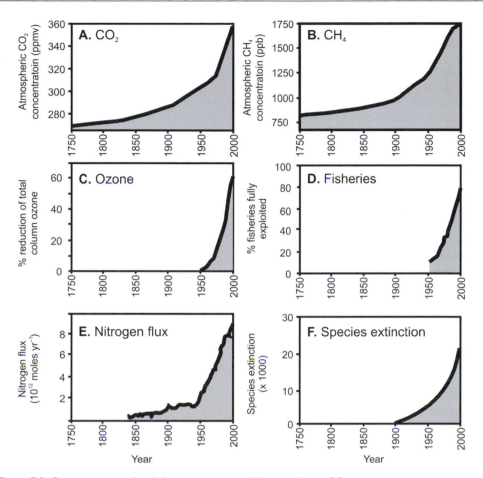

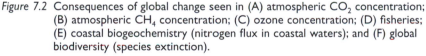

Figure 7.2 Consequences of global change seen in (A) atmospheric CO_2 concentration; (B) atmospheric CH_4 concentration; (C) ozone concentration; (D) fisheries; (E) coastal biogeochemistry (nitrogen flux in coastal waters); and (F) global biodiversity (species extinction).

Source: Redrawn from Steffen and co-workers (*571*).

all its forms is needed to maintain human society organisation. As a result, in many eco-regions, we are assisting to a reckless destruction of the global environment, humanity exerting a strong interference on natural processes and functional units that contribute to the earth's homeostasis (Figure 7.2).

In a few centuries, mankind will have exhausted all sources of fossilised carbon that took hundreds of millions of years to constitute (5) (Figure 7.2A). As a result of human activity, CO_2 concentration has risen from 270 ppm prior to the Industrial Revolution to ~401.3 ppm (part per million) in April 2014 (Mauna Loa). Methane concentration has increased substantially from 700 to ~1,890 ppb (part per billion; December 2009; Figure 7.2B) due to cattle population (~1.4 billion ruminants) growth that paralleled human population increase (*572*). Other methane sources include rice agriculture, the gas and oil industry, biomass burning and coal mining. Increase in greenhouse gas concentration is at the origin of the current warming. Anthropogenic climate change represents a further pressure, which will exacerbate human impact on the ecosphere.

Ozone destruction increased exponentially from 1950 to 2000 (Figure 7.2C). A slowing down in the reduction rate of stratospheric ozone concentration was, however, observed in 2003, 18 years after the Antarctic ozone hole discovery and 16 years after the Montreal Protocol (1987) that started to limit substances that deplete stratospheric ozone layer. The percentage of fisheries fully exploited rose from 10% to 80% (Figure 7.2D), nitrogen flux increased to ~8 × 10^{12} moles per year in 2000 in coastal systems and species extinction was estimated to be ~25,000 in 2000 (Figure 7.2E–F).

In this chapter, I examine the potential influence of anthropogenic climate change on marine ecosystems and their biodiversity. Anthropogenic climate change is defined as a shift in climate that results from increasing atmospheric greenhouse gas concentrations resulting from the combustion of fossil fuel and to a lesser extent land-use change.

7.1 Human alteration of the greenhouse effect and the radiative budget of the planet

Let us consider that the energy emitted by the earth is equal to the energy absorbed (constant global temperature). From Chapter 2, we know that solar radiation reaching the earth's surface can be estimated as follows:

$$E_i = \frac{S_0}{4}(1-\alpha) \tag{7.2}$$

With S_0 = 1,367 W·m^{-2} and α the global albedo. Assimilating the earth to a black body, it is possible to assess energy emitted by the earth by applying Stefan-Boltzmann's law:

$$E_e = \sigma T^4 \tag{7.3}$$

With σ Stefan-Boltzmann's constant (σ = 5.67.10^{-8} W·m^{-2}·K^{-4}) and T the earth's surface temperature (kelvin). Because $E_i = E_e$, we obtain:

$$\sigma T^4 = \frac{S_0}{4}(1-\alpha) \tag{7.4}$$

From the above equation, it is possible to calculate the earth's theoretical temperature:

$$T = \sqrt[4]{\frac{S_0}{4\sigma}(1-\alpha)} \tag{7.5}$$

A calculation with α = 0.3 gives a temperature T = 254.9 K or –18.25°C. This contrasts with the currently observed temperature of ~15°C. The difference of ~33°C is the greenhouse effect. The greenhouse effect is therefore fundamental for the ecosphere. The current problem is the increase in greenhouse gas concentrations, which unbalances the radiative budget of the earth system.

As early as 1827, the French scientist Jean-Baptiste Joseph Fourier proposed that the earth's atmosphere could be compared to a huge garden greenhouse, catching heat to warm the planet. He baptised the process 'un effet de verre', which means 'an effect of glass'. Acknowledging the French mathematician, Swante Arrhenius was among the first to warn on potential relationships between atmospheric greenhouse gas concentrations and climate change (573). The physicist estimated that Arctic temperatures would increase by 8–9°C if CO_2 concentration rose 2.5–3 times the pre-industrial period. This warming estimate was not far from current estimates from the atmosphere–ocean general circulation model (574). Revelle and Suess (575) warned that:

> Human beings are now carrying out a large-scale geophysical experiment of a kind that could not have happened in the past nor be reproduced in the future. Within a few centuries we are returning to the atmosphere and oceans the concentrated organic carbon stored in sedimentary rocks over hundreds of millions years.

Long-term monthly changes in this greenhouse gas are shown in Figure 7.3. A remarkable feature is the seasonal CO_2 concentration variability of ~5–6 ppm, which is attributed to photosynthesis. CO_2 concentration has increased everywhere on the planet, including the South Pole (Figure 7.3).

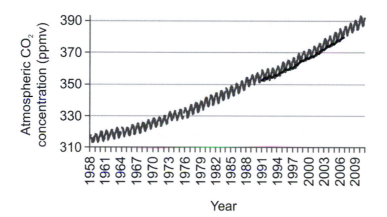

Figure 7.3 Long-term monthly changes in CO_2 concentration in Mauna Loa (Hawaii) from March 1958 to February 2011 (grey) and in the South Pole from April 1991 to December 2006 (black).

Today's concentration has not been exceeded during the past 800,000 years (326, 576). Figure 7.4 shows long-term atmospheric CO_2 changes at Dome C (Antarctica) for the last 800,000 years. CO_2 concentrations oscillated between 170 ppm during glacial periods and up to 280 ppm during interglacial periods. Adding CO_2 concentration measured in Mauna Loa (black), we can see that the current increase represents a major anomaly on a geological timescale. Such a value was observed at the peak warming of the early Pliocene 4–5 million years ago when global temperatures were about 4°C warmer at equilibrium than pre-industrial conditions (577).

The increase in CO_2 concentration associated with other anthropogenic greenhouse gases such as methane and nitrous oxide has perturbed the planet's radiative budget (Figure 7.5).

The radiative forcing related to CO_2 concentration was 1.66 W·m⁻² (1.49–1.83 W·m⁻²) in 2005. Other greenhouse gases contributed to another 1 W·m⁻². Whereas stratospheric ozone exerted a negative forcing of –0.05 W·m⁻², tropospheric ozone increased the radiative forcing by an average 0.35 W·m⁻². Although uncertainties related to aerosols remained high, these compounds contributed to reduce the positive forcing of greenhouse gases by an average $-1.2\ \text{W·m}^{-2}$. Note that Equation 7.5 shows that the earth's energetic balance depends strongly on the albedo. Because anthropogenic climate change could influence the type, height and the nature of cloud or ice coverage (Chapter 2), it may modify these parameters and further unbalance the system. Uncertainties on cloud albedo are currently large.

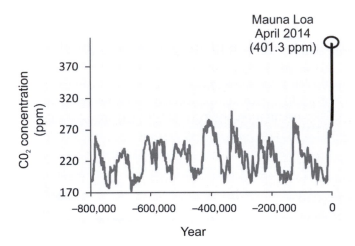

Figure 7.4 Long-term changes in CO_2 concentration from 800,000 years BP to 2014. Data from 798,512 BP to 137 BP (grey) are from EPICA Dome C ice core (*576*). Data from AD 1958 to AD 2014 (black) are from measurements of atmospheric CO_2 in Mauna Loa.

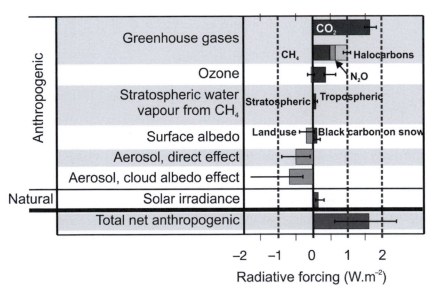

Figure 7.5 Main anthropogenic and natural parameters that influenced average radiating forcing in 2005. Note that uncertainties (black horizontal lines) are not symmetrical. Impact of volcanic eruptions is not reported because of their episodic nature.

Source: Simplified from the IPCC (*55*).

Rises in atmospheric greenhouse gas concentrations have probably contributed to a 0.3–0.5°C temperature increase per century during the period 1906–1996 and represent the main warming cause after 1976 (578). Models predict that atmospheric greenhouse gases will continue to increase during the twenty-first century leading to a global warming ranging between ~1°C and ~5°C. Global warming might radically alter the face of the planet and the current increase in greenhouse gas concentrations might affect the climatic system for the next 50,000 years, leading to an exceptionally long interglacial (579).

7.2 Increase in global air and sea surface temperature

Global warming has become unequivocal, being detected in the atmosphere, the ocean, the cryosphere and the hydrosphere (55). Global temperature rose by 0.78°C between the 1850s and the 2000s (Figure 7.6).

The increase in global temperature has been more prominent in the northern hemisphere, reaching more than 1°C in some years (e.g. 1998 and 2010) (Figure 7.6). The 20 warmest years measured since 1850 occurred after 1982 (Table 7.1). The warmest year of the instrumental period was 1998. The year corresponded to the strongest El Niño event ever recorded. The second warmest year was 2005. It should be noted that no El Niño event occurred during this year and that solar activity was low.

All years of the twenty-first century are among the 15 warmest years since the beginning of the instrumental record. These observations raise concerns about the role of natural climatic variability, which seems surprisingly low. Indeed, we should expect some cold years if natural variability was of the same magnitude as anthropogenic climate change. Although the decade is too short to draw any conclusion, this might indicate that anthropogenic climate change is substantial, perhaps stronger than currently assumed. Stephan Rahmstorf and colleagues recently compared observational CO_2 trends, global temperature and sea level change, and found that projections of changes established by the Intergovernmental Panel on Climate Change (IPCC) in 2001 were too conservative, especially for global temperature and sea level change (580). For example, during the period 1993–2006, satellite data exhibited an increase of 3.3 ± 0.4 mm·yr^{-1} in sea level, whereas this increase was estimated to be <2 mm·yr^{-1} by the IPPC in 2001.

As we showed in Chapter 5, Moberg and co-workers assessed secular changes in northern hemisphere temperature anomalies from AD 1–1979 (327). The curve they drew revealed a much larger temperature variability than the 'hockey stick' graph produced by Mann and colleagues (581). The prospect that the hockey stick underestimated past temperature variability was one of the major arguments advanced by climate sceptics against global warming (387). However, despite the more pronounced variability in the past 2,000 years, the exceptional nature of the period after the 1980s appeared when both proxy data from Moberg's group and northern hemisphere temperature (NHT) anomalies are combined on the same graph (Figure 7.7). Uncertainties remain, as the spatial resolution of NHT anomalies is obviously much higher than the spatial resolution of proxy data. Therefore, it is difficult to give a definitive answer on whether or not the last two decades were warmer than the Medieval Warm Period. However, there is a consensus that the last two decades have been exceptionally warm.

Global warming has altered many natural systems, especially the oceans (55, 87, 582). Ocean temperatures have warmed by ~0.6°C between the period 1900–1920 and the years

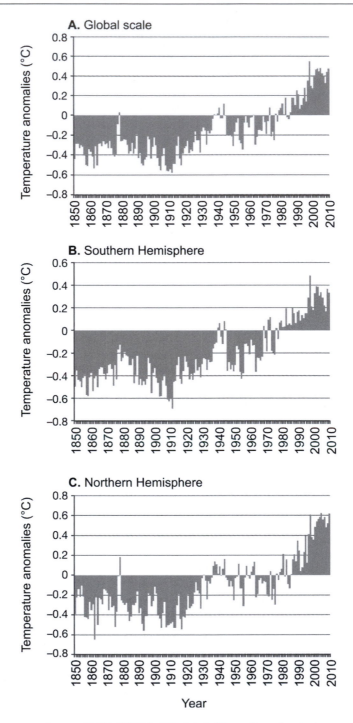

Figure 7.6 Long-term changes (1850–2010) in (A) global, (B) southern hemisphere and (C) northern hemisphere SST anomalies.

Source: Data are from the Hadley Centre for Climate Prediction and Research (HadCRUT3).

Table 7.1 The 20 warmest years during 1850–2010.

Rank	Year	Average global temperature	El Niño/La Niña events	Annual sunspot number
1	1998	0.548	Super El Niño/La Niña	**64.3**
2	2005	0.482	–	29.8
3	2010	0.476	El Niño/La Niña	16.5
4	2003	0.475	El Niño	**63.7**
5	2002	0.465	El Niño	**104**
6	2004	0.447	El Niño	40.4
7	2009	0.443	El Niño	3.1
8	2006	0.425	El Niño	15.2
9	2001	0.408	–	**111**
10	2007	0.402	La Niña	7.5
11	1997	0.352	Super El Niño	21.5
12	2008	0.325	La Niña	2.9
13	1999	0.297	La Niña	**93.3**
14	1995	0.275	El Niño/La Niña	17.5
15	2000	0.271	La Niña	**119.6**
16	1990	0.255	–	**142.6**
17	1991	0.213	El Niño	**145.7**
18	1988	0.18	El Niño/La Niña	**100.2**
19	1987	0.179	El Niño	29.2
20	1983	0.177	Super El Niño	**66.6**

Note: Data are from the Hadley Centre for Climate Prediction and Research (HadCRUT3). Both El Niño and La Niña events are indicated. The oceanic Niño Index (ONI) was used to determine both El Niño and La Niña years. Data originated from the NOAA Climate Prediction Center. Data on annual sunspot number are from the Solar Influences and Data Analysis Center. Values above 60 are in bold.

after 1989. The oceanic hydrosphere has absorbed 84% of the heat added to the climate system (oceans, atmosphere, continents and cryosphere) over the last 40 years (583). This has, by thermal expansion, contributed to at least 25% of sea level rise since the 1950s (584).

As previously noted by Levitus and colleagues (585), warming in the oceans has been more pronounced during two periods: the period 1920–1940 and the period after the end of the 1970s. Warming has been more important in the northern ($\Delta T = 0.62$ between the periods 1900–1920 and 1990–2007) than in the southern hemisphere ($\Delta T = 0.59$ between the periods 1900–1920 and 1990–2007). Temperature rise began earlier in the southern than the northern hemisphere. Analysing temperature data for the period 1955–1998, Levitus and colleagues (583) estimated that ocean heat content increased by 14.5×10^{22} J while Beltrami and colleagues (586) highlighted that the continental lithosphere only gained 0.91×10^{22} J during the period 1950–2000. Levitus and colleagues (87) reported an increase of 0.66×10^{22} J in the global atmosphere during the period 1955–1996, a rise of 0.81×10^{22} J in continental glaciers for the same period, an increase of 0.32×10^{22} J in Antarctic sea ice between the 1950s and the 1970s and an increase of 1.1×10^{22} J in mountain glaciers for the period 1961–1997.

Different scenarios have been considered by the IPCC in their effort to predict the current and future magnitude of global warming. These scenarios are described in the Special Report on Emissions Scenarios (SRES) and were used to establish projections in IPCC AR4

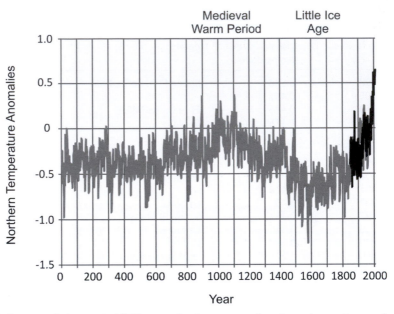

Figure 7.7 Centennial changes in NHT anomalies from proxy data (tree ring, pollens and boreholes) for the period AD 1–1979 (grey) and instrumental temperature data for the period 1850–2010 (black).

Source: Proxy palaeoclimatic temperature data are from Moberg and colleagues (*327*) and NHT anomalies are from the Hadley Centre. All data are anomalies based on mean NHT for the period 1961–1990.

(Assessment Report 4) (*587*). Scenarios are based on socio-economic, technological and environmental projections for the emissions of greenhouse gases and aerosols, and produce a range of predicted effects on global climate. The scenarios are frequently classified as 'optimistic', 'most likely' and 'pessimistic'. For example, Scenario B1 is a scenario of rapid introduction of clean and efficient technologies. In this scenario, the best estimate is an increase of 1.1°C of global temperature. More generally, the best estimates of 'optimistic' scenarios are for the end of this century of ~2°C, the likely range being between 1°C and ~4°C. For most 'pessimistic' scenarios, the best estimates ranged from ~3–4°C and the likely range is between ~2°C and 6.4°C. In a worst-case scenario (ΔT = 6.4°C; Scenario A1FI, fossil-intensive), this might place the earth system just before the Palaeocene-Eocene Thermal Maximum (PETM; 55 million year ago) when CO_2 concentration was ~1,000 ppm (*588*). A strong CO_2 increase from 1,000 to 1,700 ppm happened next in a few millennia and involved a further increase in SST of 4–5°C in the tropics and 8–10°C in high-latitudes (*589*). In Scenario A1B, perhaps the most likely scenario, global temperature increases on average by 2.8°C, but the increase might be as great as 4.4°C.

Using planktonic foraminifera, Pagani and colleagues assessed the climatic sensitivity of the climatic system (i.e. the average global temperature response to a doubling of atmospheric CO_2 concentrations through radiative forcing and associated feedbacks) during the Pliocene and estimated it was 9.6 ± 1.4°C per CO_2 doubling at equilibrium (*577*). The Pliocene was a period when CO_2 concentration was similar to today (401.3 ppm in Mauna Loa in April 2014).

As part of IPCC Assessment Report 5, new climate scenarios, termed Concentration Representation Pathways (RCPs) are used (590). Four possible climate futures are envisioned by their effect on earth's radiative forcing (and not emission scenarios as above): low RCP2.6 (radiative forcing of 2.6 W·m^{-2}), medium-low RCP4.5 (4.5 W·m^{-2}), medium-high RCP6.0 (6.0 W·m^{-2}) and high RCP8.5 (8.5 W·m^{-2}). Changes forecasted by GCMs at the end of this century are greater than changes that happened during the Holocene (e.g. LIA, MWP).

As current greenhouse gas emissions are not controlled yet, it is possible that warming reaches 4°C by the end of this century, which is close to the warming experienced by the planet between the LGM and today. The important difference is the rate of warming: 4°C in a century versus the LGM and the Holocene warming that took place over ~5,000 years. Global biodiversity changed radically since the Last Glacial Maximum (83, 591). Such a global warming rate would be unreached for at least the last 50 million years. Even half of the predicted warming would have a strong influence on regional physical and biological systems. How such a large and rapid climate change will alter current marine biodiversity is investigated in the next section.

7.3 Species responses to anthropogenic climate change

Effects of both CO_2 concentration (Chapter 8) and anthropogenic climate change on marine biodiversity have just started to emerge in the scientific literature and many aspects of these influences remain difficult to understand. Global climate change is expected to constrain species to respond in four different ways: (1) by adjusting their physiology (see also Chapter 6); (2) by triggering adaptive evolution; (3) by altering time/space distribution, a process called species niche tracking; and (4) when none of these three responses are possible, extinction occurs. Note that the first three possibilities are not mutually exclusive. The potential effects of global warming on biological systems are summarised in Figure 7.8.

7.3.1 First type of response to climate change: physiological adjustment

Global warming, by its effects on sea temperatures, will unquestionably affect species physiology (e.g. growth and reproduction; Chapter 6). However, this influence is complex and varies according to the location of a species with respect to its distributional range, the season and warming intensity (Figure 7.9). In summer, a heatwave may trigger a hyperthermic stress at the origin of massive mortality for an extratropical species at the southern part of its distributional range (Figure 7.9). At the northern edge of the distributional range, warming is likely to increase growth and reproduction. In winter, a warm event at the northern edge of the species' distributional range might enhance survival, whereas only small effects are expected to occur at the southern edge. As a result, population variance is likely to increase at the edge of the species' distributional range (and thermal niche), making species abundance more difficult to assess. This picture depends, however, on the **acclimatisation** potential and the degree of stenothermy/eurythermy of species (Chapter 11).

The effect of warming on species' physiology depends on the magnitude, frequency and degree of persistence of extreme warm events (592). A heatwave is likely to trigger acute heat deaths that might occur at timescales ranging from hours to weeks, whereas an increase

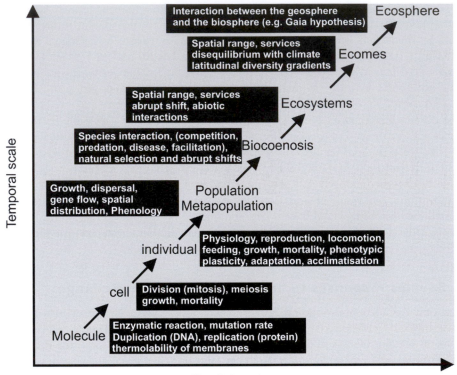

Figure 7.8 Potential effects of global warming on biological systems from the molecular to the ecosphere level.

in mean annual temperature influences more species fitness, progressively affecting population size. It is easier for a species to acclimatise or adapt to this type of warming than to resist acute warming experienced during an extreme event. The possibility that a species faces climate change is also a function of microhabitats available and the other species with which it interacts. Positive interaction such as facilitation might help the species, whereas negative interaction such as competition or predation, parasitism or disease might precipitate the collapse of its population (Chapter 4). To illustrate the complexity of species' physiological responses to climate change, I focus in this section on coral reefs. Other potential influences of global warming through the physiology and behaviour of species can be found in Chapter 6.

7.3.1.1 Global warming and coral bleaching

Coral reefs occur in waters that have average temperature of 27.6°C with maxima ranging from 24°C to 29.5°C (*428*). Coral reefs cover about 600,000 km², approximately 0.16% of the earth's surface. Previous estimates assessed that coral reefs contribute to about 0.05% of the estimated net CO_2 fixation rate of the global oceans (*593*). Many tropical corals and sea anemones live in symbiosis with endosymbiotic dinoflagellates called zooxanthellae (*594*). These zooxanthellae (*Symbiodinium* spp.) are predominantly responsible for the high

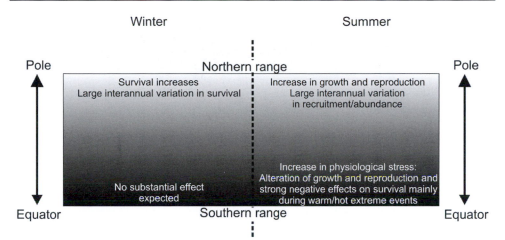

Winter   Summer

Pole   Northern range   Pole

Survival increases   Increase in growth and reproduction
Large interannual variation in survival   Large interannual variation
                                          in recruitment/abundance

                                          Increase in physiological stress:
                                          Alteration of growth and reproduction and
No substantial effect   strong negative effects on survival mainly
expected   during warm/hot extreme events

Equator   Southern range   Equator

Figure 7.9 Potential effects of oceanic warming on a hypothetical extratropical species according to its geographical position and the season.

productivity of coral reefs in tropical waters, which are poor in nutrients. The dinoflagellate absorbs nutrients and uses solar energy to supply >95% of the energy needed to fuel coral metabolism. This symbiosis also allows the polyp to have elevated calcification rates. Although productivity might be as low as 0.01 $gC·m^{-2}·day^{-1}$ in the oceanic areas close to coral reefs, algal turfs and corals can fix 280 $gC·m^{-2}·day^{-1}$ and 40 $gC·m^{-2}·day^{-1}$, respectively (595). Another study estimated that gross CO_2 fixation was close to 700×10^{12} $gC·year^{-1}$, although the majority of this fixed carbon is then recycled within the ecosystem (593).

The sensitivity of corals and their *Symbiodinium* to rising ocean temperatures has been deeply investigated (596). When seawater temperatures surpass summer maxima by 1–2°C for several weeks (>3 weeks), the hyperthermic stress leads to the expelling of the zooxanthellae from the coral, which inhibits photosynthesis. This interference of the symbiotic relationship is referred to as coral bleaching (or whitening) because the polyps lose the colours that normally mask their white calcareous skeleton (Plate 7.1).

The plate shows the consequences of the 2006 bleaching event in the Great Barrier Reef (GBR). The 2006 bleaching event was restricted to the southern part of GBR and affected mainly the inshore reefs. The Keppel Islands had previously bleached in 1998 and 2002, but the 2006 event was by far the worst in this area, killing ~30% of all corals. The ability to recover is then a function of the frequency or the severity of bleaching. Many corals can survive mild bleaching events. Bleaching alters growth (e.g. tissue growth and skeletal accretion), reproduction and disease resistance (597). If the hyperthermic stress continues, this inevitably leads to death. A large-scale bleaching event is generally associated with meteorological forcing, whereas smaller-scale bleaching events can be triggered by pollution, sedimentation, salinity anomaly or bacterial infection.

7.3.1.2 Interactions between global warming and El Niño events

An unprecedented increase in the frequency of coral bleaching events has been documented since the beginning of the 1980s (598). This increase does not seem to be primarily driven by increasing observational pressure as coral bleaching was not observed in

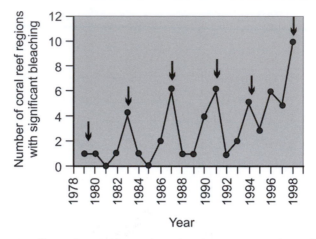

Figure 7.10 Year-to-year fluctuations in the number of reef ecoregions with significant bleaching episodes during the period 1978–1998. Arrows denote significant El Niño events.

Source: Redrawn from Hoegh-Guldberg (599).

well-monitored sites such as Heron Island in Australia or Florida Keys in the USA prior to 1980 (599). Large-scale coral bleaching has been observed in the Pacific and Indian oceans and in the Caribbean Sea since 1982 (600). These large-scale bleaching episodes have often paralleled El Niño events (Figure 7.10).

Widespread bleaching events occurred during the El Niños of 1982–1983, 1987–1988 and 1997–1998. The 1982–1983 Super El Niño event exposed many coral reefs to abnormally high SSTs (601). As we saw in Chapter 2, the 1997–1998 Super El Niño event increased global temperature by ~0.2°C. The phenomenon was responsible for a major episode of coral bleaching (the 1998 global bleaching event) that was observed in >60 countries and led to the mortality of 16% of the world's coral reefs (472). The phenomenon was particularly prominent in the Indian Ocean where a 50% mortality of coral reef was reported. In contrast to previous bleaching episodes that affected reefs shallower than 15 m, the 1998 global bleaching event penetrated to 50 m in some locations. About 70% of the corals died along the coasts of Kenya, the Andamans and the Maldives. However, a pronounced spatial variability was observed, probably because other secondary (both extrinsic and intrinsic) factors modulated the effects of the Super El Niño event.

7.3.1.3 Effect of global warming through an increase in mean temperatures

It is likely that anthropogenic global warming exacerbated El Niño effects on both local and regional temperatures. For example, 1983 (Super El Niño) was the twentieth warmest year on record, whereas 1998 (Super El Niño) was the warmest year ever measured (Table 7.1). Between the 1960s and the 2000s, Beaugrand (60) showed that SSTs increased by about 0.2–0.5°C in most of the regions where coral reefs occur (60). In Jamaica (17.5°N, 76.5°W), the rate of warming was 0.23°C per decade for the period 1981–1999, and in Tahiti it was ~0.14°C per decade for the same time period (599). Using the scenario of global climate change, Donner and colleagues (597) stressed that bleaching could become an annual or biannual event for the vast majority of the world's coral reefs in the next 30–50 years

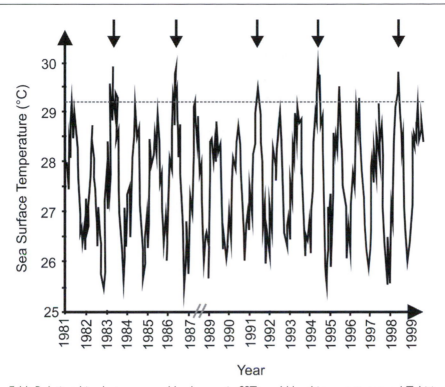

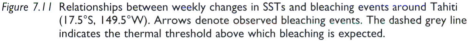

Figure 7.11 Relationships between weekly changes in SSTs and bleaching events around Tahiti (17.5°S, 149.5°W). Arrows denote observed bleaching events. The dashed grey line indicates the thermal threshold above which bleaching is expected.

Source: Simplified from Hoegh-Guldberg (*599*).

without an increase in thermal tolerance of 0.2–1.0°C per decade. Some regions such as Micronesia and Western Polynesia might be vulnerable to global warming. Other studies suggest that the Caribbean Sea and the Great Barrier Reef might be affected annually by bleaching as early as 2020 (599).

In French Polynesia (17.5°S, 149.5°W) bleaching episodes occurred in 1983, 1986, 1991, 1994, 1996 and 1998 when SSTs surpassed the thermal threshold 29.2°C (Figure 7.11). An SST increase of >1°C above maximal temperatures lasting for at least 2–3 days is a useful predictor of subsequent bleaching. A satellite-derived index, called coral bleaching hotspot, is now in operation to highlight areas that have temperature anomalies that exceed by 1°C maximal temperatures during the warmest months in regions where coral reefs occur. Assessed temperatures from which coral reefs bleach vary only slightly between sites; the thermal threshold ranges from 28.3°C in Rarotonga (21.5°S, 159.5°W) and over the southern part of the GBR to 30.2°C in Phuket (7.5°N, 98.5°E). Note, however, that a global thermal threshold for the whole reef is just an approximation as thermal tipping points are species-dependent. In other words, every coral species and their associated *Symbiodinium* spp. have a distinct thermal preferendum. During the 2002 bleaching event in Raiatea (French Polynesia), although some corals such as *Acropora anthocercis* were severely influenced, others (e.g. *Leptastrea purpurea*) were not (Figure 7.12). Intermediate levels of bleaching were observed for species such as *Pocillopora verrucosa*.

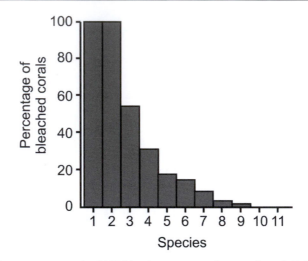

Figure 7.12 Specific responses to the 2002 bleaching event in Raiatea, French Polynesia. (1) *Acropora anthocercis*, (2) *A. retusa*, (3) *Montipora tuberculosa*, (4) *Pocillopora verrucosa*, (5) *M. caliculata*, (6) *Leptastrea transversa*, (7) *P. eydouxi*, (8) *P. meandrina*, (9) *L. bewickensis*, (10) *Porites lobata*, (11) *L. purpurea*.

Source: Redrawn from Hughes and colleagues (602).

Bleaching is also a way scleractinian corals may adapt to global warming (603). Baker and colleagues identified in five Indian Ocean sites symbionts that were present in corals before, during and after the 1997–1998 Super El Niño event by using a molecular technique called restriction-fragment length polymorphisms (large subunit ribosomal DNA). Three clades were detected: Clades A, C and D. *Symbiodinium* D were the most thermally tolerant as chronic photoinhibition arises at higher temperatures (604). In Panama, 43% of corals of the genus *Pocillopora* that contained *Symbiodinium* D in 1995 were not concerned by the 1997 bleaching episode, whereas those that contained *Symbiodinium* C were substantially impacted. In 2001, 63% of *Pocillopora* in the same colony contained *Symbiodinium* D, an increase of 20%. The same result was found in corals of the Persian Gulf. *Symbiodinium* D was more abundant after a severe bleaching event. The authors propounded that changes in *Symbiodinium* type might represent a rapid way to acclimatise to fast warming.

Hughes and colleagues (602) recognised the importance of this mechanism for scleractinian coral acclimatisation to global warming and challenged earlier models that were based on constant temperature thresholds (Figure 7.13A). We saw previously that different coral species are likely to exhibit different thresholds (Figure 7.13B). Hughes and colleagues propounded that thermal thresholds are likely to be altered more or less continuously as climate warms and that thresholds are species-dependent (Figure 7.13C).

7.3.1.4 Effect of global warming through an increase in major hurricanes

Global warming will also act through existing hydro-meteorological channels. For example, global warming is expected to increase the number of major hurricanes (135). The year 2005 was the most active North Atlantic hurricane season on record (a hurricane season

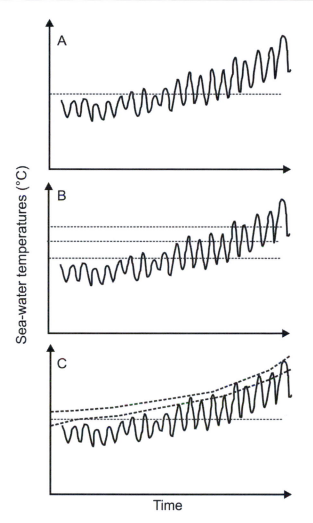

Figure 7.13 Different models of coral reef responses to global warming. (A) Model based on a constant threshold (no possibility of acclimatisation). (B) Models based on different thresholds. (C) Model based on continued thermal threshold alteration due to acclimatisation or adaptation.

Source: Simplified from Hughes and co-workers (*602*).

starts on 1 June and ends on 30 November). This year, SSTs in the tropical North Atlantic where hurricanes form (10–20°N) were 0.9°C above the average for the reference period 1901–1970. We have seen in Chapter 5 that tropical cyclones can have devastating effects on coral reefs at a regional scale. For example, we saw in this chapter the impact of the tropical cyclone Oli on Polynesian coral reefs. We also summarised in this chapter a meta-analytical study conducted in the Caribbean for the period 1980–2001 that revealed that coral reef cover decreased by about 17% on average the year after it had been hit by a hurricane (*429*). However, on a global scale, the effect of global warming on tropical cyclones is likely to be less important than the average expected increase in SSTs.

7.3.1.5 Effect of global warming through changes in mean precipitation

Global warming might also influence coral reefs through precipitation changes. Tropical precipitation has increased by 4% between 1979 and 2003 over the oceans between 25°S and 25°N and the frequency of intense rainfall events is expected to reinforce in the next decades (55). This increase in precipitation may increase sediment discharge and deposition near river mouths, and in turn affect nearby coral reefs. The increase in the frequency and intensity of droughts may alter vegetation cover and land use, which in turn might increase erosion and sediment stress when rains go back. However, many uncertainties remain, and such consequences are likely to vary regionally and on a year-to-year basis.

7.3.1.6 Effect of global warming through disease

The effects of increasing SSTs on corals also take place through disease. Figure 7.14 shows the apparent positive relationships between SSTs and bleaching of *Oculina patagonica* in the Mediterranean Sea.

Major bleaching episodes have been detected when summer SSTs surpassed the thermal threshold of 29–31°C. However, here the bleaching is due to the aetiological bacterium *Vibrio shiloi* (606). The disease, more prominent when SSTs are high, induces a partial necrosis of coral tissues (Plate 7.2). Non-mutually exclusive hypotheses to explain the link between temperatures and disease are related to a thermal modulation of: (1) host sensitivity to the pathogen; (2) pathogen virulence; or (3) transmission frequency via a vector (605).

Other pathogens have been discovered. For example, the pathogenic bacterium *Vibrio coralyticus*, discovered in the Indian Ocean, is responsible for a partial necrosis of coral tissue of *Pocillopora damicornis* (607). Infection of the coral species with *V. coralyticus* takes place when SSTs are ≥27°C because of a rapid lysis of tissues. At temperatures <24°C, no lysis occurs. The bacterium produces an extracellular proteinase, which plays a role in the lysis of tissues. The proteinase is present at higher concentrations at temperatures >27°C.

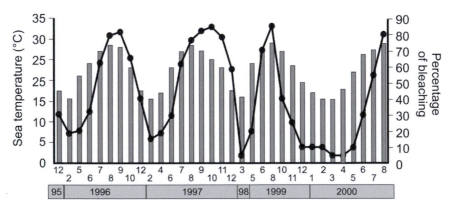

Month and year

Figure 7.14 Relationships between seasonal bleaching of *Oculina patagonica* (bar) and seawater temperatures (grey dot) in the Mediterranean Sea.

Source: Modified from Rosenberg and Ben-Haim (605).

These examples show how temperatures might trigger a disease that affects metabolic performance, growth, reproduction and competitive ability, and eventually lead to the animal death. Although it remains unclear to what extent warming might trigger infection, diseases such as Black Band Disease, White Pox and White Syndromes tend to be more prevalent and virulent when temperatures are elevated during summertime. Ocean warming can therefore indirectly kill corals by affecting their immune system and making them more susceptible to infections. Furthermore, the warming might reinforce both bacterial and fungal pathogens, making them more virulent. However, not all disease can be related to SST rise. For example, the origin of the outbreak of White Band Disease observed in both the Red and the Caribbean Seas remains unclear (Plate 7.3). The disease is identified because of the presence of a white band of bare coral skeleton at the basis of the genus *Acropora*. The disease then progresses towards the tip of the branches at the speed of several mm·day^{-1} (605).

7.3.1.7 Conclusions

Although it is incontrovertible that global warming has played a role in the increasing frequency and magnitude of bleaching events, many local natural processes such as atmospheric circulation, upwelling or cloudiness can alleviate or amplify bleaching conditions (608). For example, weather patterns dominated by clear skies and an absence of wind and wave increase the effects of solar radiation and exacerbate the background influence of warming. Oceanic currents can transport warm or cool waters that persist for months before dispersing, modulating the influence of meteorological conditions. Such conditions eventually lead to stochastic resonance.

The distributional range of reef-building corals is limited by annual minimum temperatures of ~18°C. Global warming might therefore be expected to increase the spatial extent of reef-building corals polewards. However, this increase might not be sufficient to fully alleviate the adverse warming influence on coral reefs, especially at a time when other more direct anthropogenic influences (e.g. pollution, overexploitation) might prevent corals to recolonise. This positive influence of global warming on coral reef spatial distribution remains poorly documented.

7.3.2 Second type of response to climate change: adaptive evolution

The second type of species response to anthropogenic climate change is adaptive evolution. A modification in life history or functional traits is considered to be adaptive if the altered phenotype confers to the individual an advantage in the new environment; in other words, if the new phenotype increases individual **fitness** (609). Adaptation embraces both phenotypic plasticity and evolutionary (genetically based) changes. The limits between the two processes are fuzzy. Via and colleagues (610) defined **phenotypic plasticity** as 'an environmentally based change in the phenotype'. Phenotypic plasticity manifests itself when a species alters one or several life history traits such as behaviour, shape, size or environmental tolerance. The evolution theory teaches us that individuals composing a population are not identical and that they should therefore react differently to temperature change. This is what is called adaptive phenotypic plasticity. Evolution, defined as the shift over time in one or several inherited traits found in species populations, is measured as change in the frequency of some alleles in a population.

7.3.2.1 Plastic versus genetic effects

It is not easy to quantify the relative influence of plastic and genetic effects. Phenotypic plasticity can place a species in a new environment where evolution occurs (611). Phenotypic plasticity can subsequently be assimilated genetically, accelerating the process of adaptive evolution. Such responses to climate change have been mostly documented in the terrestrial domain. By using a multidisciplinary approach, Bearhop and colleagues showed that European blackcap (*Sylvia atricapilla*) changed its migration behaviour and that this plastic trait became heritable (612). Since the 1960s, there have been more blackcaps overwintering in Great Britain. This bird that used to migrate over North Africa and southern Iberia was observed in 31% of British gardens between October 2003 and March 2004. With warmer winters, the advantage to remain on site relies in occupying the best territories. A female in a high-quality territory may improve its body condition because of better food availability, which also increases the probability to have larger clutches. Birds that overwinter in northern Europe (southern Germany and Austria) are 2.5 times more likely to breed assortatively than randomly. **Assortative mating** is favoured by natural selection as hybrids inherit migration directions and distances that are intermediate to those of parents. Birds remaining in breeding sites do not spend energy needed for travelling and thus have more resources to devote to reproduction. This study shows that a plastic trait (migration) can become heritable and be subjected to natural selection.

7.3.2.2 Global warming and plasticity

Nussey and colleagues also showed that plasticity plays an important role in the species' ability to adapt to climate change (613). They investigated long-term changes (32 years) in the phenology of great tits (*Parus major*) in Hoge Veluwe, a Dutch national park. During the 1970s, the bird reproduced at the same time caterpillars were abundant. However, in 1985, the warming triggered a phenological shift of two weeks in the timing of caterpillar prey abundance. Using information on laying dates for 833 females between 1973 and 2004, Nussey and co-workers found that genetic variation relative to the laying date was large and that selection occurred during the last two decades when the trophic mismatch was strong to favour an earlier reproduction to alleviate the trophic mismatch between chicks and their caterpillar preys. It remains unknown whether or not the species will continue to adapt if warming amplifies.

7.3.2.3 Ecological complexity

The two previous studies suggest that species response to climate change might not occur directly, but rather indirectly through an alteration in seasonal events (614). In an experiment, northern mosquitoes (*Wyeomyia smithii*) were transplanted to a simulated southern climate (615). The insect lost 88% of its **fitness** but this decline was not related to higher temperature. Fitness reduction resulted from the incorrect seasonal cue (photic information or day length) experienced by the individuals. In a review on the subject, Bradshaw and Holzapfel stressed that timing was essential, stating that the fitness of species living in temperate and polar biomes is influenced by their aptitude to exploit the growing season and to evade or alleviate the adverse winter influences. Populations that are able to alter the timing of their seasonal activities are likely to remain stable, whereas populations that have not this ability might disappear (616). Selection might favour species that can adjust to latitudinal changes in day length to take full advantage of the growing season and to complete their life cycle.

7.3.2.4 The founder effect

Species individuals are characterised by large variation in ecological traits that can be attributed to genetic recombination and random genomic changes. Phenotypic plasticity is a function of the species 'evolutionary baggage' (i.e. the founder population that creates the species), which has a historical component. The theory of island biogeography (238) teaches us that a species that colonises an island evolves subsequently according to founder genes that limit the contour of genetic adaptation. This is called the founder effect (617). We can illustrate the founder effect by referring to Darwin's finches in the Galapagos Islands. During the expedition aboard the HMS Beagle, the English naturalist investigated the shape of finches' bills. He proposed that the shape of the finches' bill shifts progressively to exploit different resources available. This led to 14 species that formed by evolutionary divergence from a common ancestor, a process that took 3 million years. This study also illustrates that physiological boundaries are phylogenetically limited.

7.3.2.5 Plasticity and the thermal niche

An ecologically important functional trait in the context of climate change is the thermal tolerance that determines species' propensity to respond to climatic variations. Warming (e.g. the increase in both the intensity and frequency of periods of thermal stress) is likely to increase the selection pressure on populations. Some translocation experiments, also called 'common garden' experiments, indicate that some species may adapt quickly to changing environmental conditions. For example, Woodward (618) provided evidence for a rapid adaptation of navelwort (*Umbilicus rupestris*) to low temperature. When placed artificially beyond (north of) its geographical limit, 50% of the surviving population tolerated low temperatures in 1987 that were lethal in 1979. It is not possible from this study to conclude that genetic adaptation was at work, as it could rather be phenotypic plasticity.

7.3.2.6 Plasticity in the multidimensional space of traits

It is intuitive that a single **phenotype** is unlikely to confer to all individuals a high degree of performance in all parts of its distributional range (610). **Phenotypic plasticity** enables a species to adapt to a heterogeneous environment and take less time in contrast to evolutionary response. In a one-dimensional model (when one single trait is used), this can be easy to understand, but it is clear that phenotypic plasticity takes place in the multi-dimensional space of traits, which makes adaptation potential more difficult to appreciate. Phenotypic plasticity can be neutral, maladaptive or adaptive (611). Towards high latitudes, phenotypic changes arise seasonally when SSTs increase. For example, it is well known that pelagic copepods of a same species have smaller size in summer. As we saw in Chapter 6, seasonal size reduction in response to increasing temperature is well known and has been termed **cyclomorphosis**. However, what is less known is whether the reduction in individual size enables the organism to better exploit its environment and therefore whether this modification is adaptive. In the terrestrial realm, the reduction in leaf size that occurs generally when water availability decreases enables the plant to reduce its evapotranspiration.

7.3.2.7 Evidence for rapid evolutionary changes in the terrestrial realm

Phenotypic plasticity is also function of the degree of **genetic polymorphism**. Some studies showed that rapid evolution can occur for some species in response to global climate change. Rapid evolution has been found in the genus *Drosophila* (*619*). Latitudinal clines have been shown in *D. melanogaster* for traits such as body size, developmental time and thermal resistance. Along the eastern Australian coast, there has been a tendency for populations of *D. melanogaster* to acquire a higher frequency of the alcohol dehydrogenase Adh^s allele that characterises more tropical populations. The latitudinal change in the frequency of Adh polymorphism was on average close to 4° (2–6° at confidence limit of 95%) or 400 km within two decades (1979–2004). These genetic alterations were attributed to warmer and drier climate over the past 50 years (*619*). This study provided evidence that allele composition inside a population can be modulated by the climatic regime throughout the species distributional range.

Franks and colleagues (*620*) also showed that the field mustard (*Brassica rapa*) exhibited a rapid adaptive evolutionary response to regional multi-year drought. Prolonged drought was a source of natural selection. Drought abbreviates growing season length and therefore evolution promotes earlier onset of flowering to escape these adverse environmental conditions. The authors demonstrated that post-drought genotypes (2004) had earlier onset of flowering after five consecutive drought years, estimated between 1.9 and 8.6 days, than ancestors that lived during a 'wet' year in 1997. However, this small change might not be sufficient to adjust to rapid climate change. Will the species continue to adapt or has it already reached its genetic limits with respect to this trait? In the past, drought seldom occurs. Is this process a possibility for the species to genetically adjust to past climatic variability or a process to adjust to future rapid and substantial anthropogenic climate change? Huntley (*621*) meticulously dissected Franks's results and concluded that the population was close to its genetic limits for growing season length for two reasons. First, in the short-time treatment (drought conditions), the population from the dry site did not exhibit any difference between descendants (1997) and ancestors (2004). Second, the advance in flowering time in the dry site between descendants and ancestors was only 22% of what was observed on the wet site.

7.3.2.8 Marine dispersal versus potential for adaptive evolution

Back to the marine realm, pelagic groups such as plankton have rapid generation times, which in theory might promote rapid evolutionary change. However, the fecundity of such a group is in general very high, populations are huge and species have large distributional range (*622*). Dispersal, which is often limited on land, can be as far as thousands of km. Perhaps as many as 70% of marine benthic species have a planktonic larval stage prior to settling to their adult habitat (*218*). Although some benthic (aplanic) species can develop directly on sea bottom (e.g. the intertidal gastropods *Littorina obtusata*, *L. saxatilis*), many benthic organisms (e.g. corals, urchins, crabs, worms) have a planktonic phase (meroplankton) that ensures the species' spatial dissemination over periods that may last several weeks. Such larvae are generally separated into three categories (*623, 624*):

- Anchiplanic larvae are short-lived and observed between a few hours and a few days (e.g. the subtidal bivalve *Nucula nitidosa* and the intertidal polychaete *Spirorbis spirorbis*).

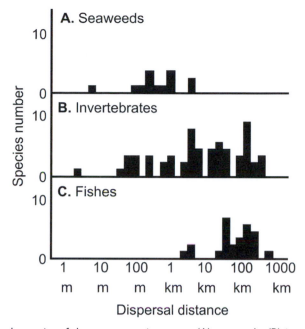

Figure 7.15 Dispersal capacity of three taxonomic groups: (A) seaweeds; (B) invertebrates; (C) fish. Calculations of dispersal distances of propagules (spores and larvae) were based on genetic variability in >100 species.

Source: Redrawn from Kinlan and Gaines (625), Gaines and co-workers (626) and Ayata (627).

- Actaeplanic larvae drift, eat and grow in the plankton between one week and two months (*Semibalanus balanoides*).
- Teleplanic larvae that compose the meroplankton for >2 months. Many demersal or benthic fish (e.g. Atlantic cod, common sole) have planktonic larvae >2 months.

Seaweed propagules can disperse between 10 m and 10 km, invertebrate propagules between 5 m and ~1,000 km and fish larvae between 2 km and ~1,000 km (Figure 7.15).

Offspring disperses much farther in the oceanic than in the terrestrial realm (625). Species with low dispersal (at larval or adult stage) and small population size tend to exhibit a genetic structure and therefore adapt to local conditions. In contrast, species with elevated dispersal have rapid gene flux that slows down adaptation to local conditions and so it is rare to observe genetic structures between populations separated by several thousand kilometres. Palumbi showed by protein electrophoresis and examination of mitochondrial DNA that species with small dispersal capacity tends to have large genetic differences. For example, gastropods of the genera *Nucella* and *Littorina*, catfish of the genera *Arius* and *Bagre*, copepod of the genus *Trigiopus*, the horseshoe crab (*Limulus polyphemus*) have a high difference among **demes** at spatial scales between 1 and 3,000 km. In contrast, urchins of the genus *Strongylocentrotus*, mussels (*Mytilus californianus*), 12 species of reef fish, milkfish of the genus *Chanos* and eels of the genus *Anguilla* had slight or no differences between demes at spatial scales ranging from 1,500 to 10,000 km. Along the western American coast, no genetic differences were detected for sea urchin *Strongylocentrotus purpuratus* over 2,500 km (628). Examining fossil records of marine molluscs, Jablonski showed that development mode influences the spatial extent of larval dispersal, which in turn affects species spatial distribution and genetic variation among populations (629). Late cretaceous gastropods with planktotrophic larvae had larger spatial distribution, extended species durations and both small extinction and speciation rates than gastropods with nonplanktotrophic larvae.

The potential for rapid evolutionary change in the plankton may therefore be limited by their great dispersal and resulting strong mixing within a large gene pool (630). Like many other zooplankton species, the marine copepod C. *finmarchicus* has a large capacity for zoogeographic displacement because of its large dispersal capacity and reproductive strategy (r strategy) (631). This might have lightened pressure for genetic adaptation. Helaouët and Beaugrand (160) did not find any alteration of the thermal niche during the period 1958–2002. Bucklin and colleagues showed in four Norwegian fjords (Lurefjorden, Masfjorden, Sognefjorden and Sorfjorden) that most *Calanus* populations were not genetically distinct (632). However, the authors found that populations of *Acartia clausi* exhibited significantly different genetic structures.

Some estuarine copepod species such as *Eurytemora affinis*, more constrained by geographic factors might be more likely candidates for genetic adaptation (633). Other works suggest that evolutionary rates in marine plankton and fish can be relatively rapid during period of transient isolation (622, 630). Dawson and Hammer investigated populations of jellyfish of the genus *Mastigias*, landlocked in tropical **marine lakes** of Palau (Indo-West Pacific region). These marine lakes formed at the beginning of the Holocene 12,000 years ago from deep topographic depressions and 5,000 years ago for shallower depressions (630). The authors showed that transient isolation, even if short at a geological timescale, may lead to relatively rapid intraspecific differentiation and speciation in jellyfish.

Situations where speciation can happen in the marine realm are more widespread than originally thought. For example, high levels of endemism have been documented for archipelagic fish (634). Although hydro-climatic conditions can promote connectivity between populations, they can create barriers limiting dispersal. At a macroecological scale, I showed previously that the geographical range of some marine copepods (e.g. *Calocalanus* spp. and *Metridia lucens*) was clearly delimited by hydrodynamic barriers (Chapter 4). At a smaller scale, Ayata and co-workers investigated larval dispersal of the honeycomb worm *Sabellaria alveolata* (Plate 7.4A) in the Bay of Mont-Saint-Michel (English Channel) by means of a 3D biophysical model (635). The polychaete is at the origin of large biogenic reefs that constitute biodiversity hot-spots on tidal flats (Plate 7.4B). The authors showed how both local meteorological conditions and hydrodynamics influenced larval dispersal (Plate 7.4C–F). At a relatively high spatial resolution (geographical cell of 800×800 m^2), the model simulated larval dispersal that can typically last between four and eight weeks (actaeplanic larvae). The model that considered advection-diffusion larval transport and gregarious settlement behaviour revealed that oceanographic and meteorological processes might strongly affect the polychaete's recruitment. For example, larvae settlement success in the southern part of the Bay depended on meteorological conditions; more intense west and south-west winds improved the probability of settlement success by a factor of 10 (Plate 7.4C–F). This study illustrates that hydrodynamics and both the geographical and topographic characteristics of a habitat might contribute to species retention and limit dispersal, facilitating genetic differences among populations and setting the stage for speciation.

Even if both physical and biological mechanisms remain difficult to elucidate, many studies have provided evidence that a large part of species recruitment results from local retention.

Working in Saint Croix (north-eastern Caribbean Sea), Swearer and colleagues examined the **otoliths** of the bluehead wrasse, *Thalassoma bifasciatum* (636). Otoliths grow every day, recording the environmental conditions of the water in which an individual evolves. Bluehead wrasse has actaeplanic larvae that can migrate in the **epipelagic realm** during 45 days. Otolithometry therefore enables the reconstruction of the dispersal history of the larvae (i.e. the pelagic pathway undertaken by larvae). The researchers closely examined two study sites: one located on the windward shore of the island and another on the leeward shore. During summer months (June–August), they found that 89% of the leeward-shore recruiting fish had retention signature (otoliths enriched with trace-element) in contrast to windward-shore recruits for which only 32% of the fish had this signature. Despite the possibility for large dispersal, the researchers provided evidence that the recruitment was mainly determined by local retention on the leeward shore of Saint Croix Island. On the northern Great Barrier Reef (Lizard Island), Jones and colleagues marked the otoliths of about 10 million developing embryos of the small coral reef damselfish *Pomacentrus amboinensis* (637). They then sampled 5,000 juveniles recruiting in the same place and found 15 marked individuals. Assuming that they marked about 0.5 ± 2% of embryos, the authors estimated that 15 ± 60% of juveniles may be returning to their natal population.

Although most marine species with large dispersing larvae are expected to be genetically homogeneous, some studies provided evidence for genetic differentiation (638). Taylor and Hellberg (639) investigated genetic variability among populations of the reef-dwelling cleaner goby (*Elacatinus evelynae*) in the Bahamas and Caribbean Seas. This goby has different colour forms: white in the western and blue in the eastern part of the Caribbean Sea and lemon yellow in the northern part of the Bahamas (Plate 7.5). The goby has actaeplanic larvae with pelagic larval duration estimated between 21 and 25 days. The three colour forms rarely coexist in the field. Taylor and Hellberg (639) sampled 246 individuals from 17 populations (Plate 7.5). By polymerase chain reaction (PCR), they amplified and sequenced mitochondrial cytochrome b gene and showed that 78.9% of the genetic variation was partitioned among the three colour forms. Among colour forms, most populations were characterised by a single haplotype, revealing no exchange between most populations at the larval stage. The researchers estimated that between populations of Barbados and Curaçao separated by 1,000 km, the isolation time ranged from 75,000 to 103,000 years.

The study therefore shows that fish larvae retention can arise even in the absence of geographical barrier. Palumbi and Warner, who commented on these results (640), provided two types of explanations. The first explanation is based on the fish that host the cleaner goby. Selection would occur because the large fish would select the goby having the colour the fish expects for cleaning service. However, the explanation does not account for populations of the same colour form that are genetically different. The second explanation is that the fish manages to stay in its native area by mechanisms not yet identified. Such results have also been found for mantis shrimp in Indonesia and crabs, oysters and mussels in different oceanic regions of the North Atlantic and the Mediterranean Sea (640). Strong phylogeographic structures were also detected in western Atlantic reef fish of the genus *Halichoeres* (641). Analysing mitochondrial DNA sequences, Rocha and colleagues (641) showed strong genetic divergence between adjacent but distinct ecologically different **biotopes**, whereas they observed low genetic differences between distant but ecologically similar biotopes. They propounded that **parapatric** speciation, an ecological speciation occurring in adjacent but distinct biotopes, might be responsible for the high level of endemism found in the tropics, inflating tropical biodiversity.

7.3.2.9 Conclusions

Many marine invertebrates have restricted spatial range (e.g. echinoderms, molluscs, crustaceans, annelids), which indicates the probable existence of present and past barriers to gene flow (geographical or allopatric speciation) or alternatively that sympatric speciation (ecological speciation) is common in marine invertebrates (638). What is the implication in terms of genetic adaptation to climate change is at present unknown. Rapid anthropogenic climate change will impose strong directional selection pressures. Bradshaw and McNeilly (642) warned, however, that even if some adaptations are likely, in many cases this might not be sufficient to mitigate the effects of anthropogenic climate change. Thermal acclimatisation, which enables a species to adapt to extreme temperatures, can reduce the **fitness** when the thermal regime comes back to normal conditions (643). Kristensen and co-workers (643) found that cold-acclimated flies (*Drosophila melanogaster*) were 36 times less likely to find food than non-acclimated flies. Huntley (621) also suggested that morphological variation apparent during the Pleistocene was of the same order of magnitude as the spatial variance in the morphological trait observed across the species distributional range.

These examples, added to the few observations of clear adaptive response during the glacial and interglacial periods that punctuated the Pleistocene, suggest that evolutionary adaptation might simply be not achievable for the majority of species in the context of rapid climate change. I recall here the study of Crisp and colleagues (644) who showed that on 11,000 species of plants, evolutionary divergence was barely associated with biome shift in 396 events (3.6%). The biologists Bradshaw and McNeilly (642) reminded that species have commonly responded to climate change by tracking their bioclimatic envelope rather than by evolution. Even if adaptation is possible on a large timescale, current scientific knowledge suggests that the extent to which species might adapt to rapid and substantial global warming is limited.

Furthermore, the existence of 'stable' species borders suggests that rapid adaptation to new environmental change is not generally possible. This is also the case for the contour of species' bioclimatic envelope, which tends to remain constant at a decadal scale. Processes that preclude populations to invade new areas is a subject of active research (645). The most common explanation is that species lacks genes to adapt to new conditions (646). Second, as the growth rate of populations becomes negative towards the edge of the species range, these populations become fragmented and sensitive to processes related to the **Allee effects**. Third, the small populations might lack genetic variation, which limits adaptation. Fourth, the genetic variability might be elevated but rarely expressed because of adverse environmental conditions. As heritability is controlled by both the species genetic pool and the environment, a strong heterogeneous environment is likely to affect negatively potential for evolution. Fifth, there is a further complication that might prescribe adaptive evolution. Adaptation might necessitate the evolution of more than one trait. Sixth, mutations might be counterbalanced by the stirring effect of genes flux coming from the source populations when migration rates are significant. These processes tend to slow down adaptive evolution, which tends to occur at rates corresponding to the geological timescale (Chapter 5).

7.3.3 Third type of response to climate change: species niche tracking

Niche tracking is the time/space movement of some species to track their ecological niche (Chapter 11) (647). Three documented phenomena are probably the result of species niche tracking: (1) thermotactic behaviour; (2) phenological; and (3) biogeographical shifts.

Note that the expression of niche tracking is generally restricted to biogeographical shift. The METAL theory will demonstrate why niche tracking can be extended to phenological shift and thermotactic behaviour in Chapter 11.

7.3.3.1 Underlying mechanisms behind species niche tracking

As we saw, the possibility for a species to adapt to climate change depends on its genetic baggage, which has both a phylogenetic and a historical component. Genetic architecture of traits is probably related to species ecological niche and spatial distribution. Specialist or stenoecious species are likely to be more vulnerable than generalist or euryecious species. Kellermann and co-investigators showed in *Drosophila* that species with narrow ranges of desiccation or cold resistance had less genetic variation in these traits than species with wider ranges of desiccation or cold resistance (648). This study also suggested that stenoecious species, composing much of the earth's biodiversity, may lack genetic variability to extend the amplitude of their ecological niche and might be more vulnerable to extinction. Thus, evolutionary factors might be responsible for the narrow ecological niche of stenoecious species and their restricted spatial range. At the ecological level, positive relationships between the amplitude of species ecological niche and the size of their spatial distribution have been observed (649) (Chapter 11). Generalists may possess life history and demographic traits that enable them to withstand a large range of environmental conditions.

Although adaptive plasticity can lead to ecotypic or niche differentiation, habitat tracking processes tend to have an opposite effect and promote stabilising selection (609). When a species responds to climate change, individuals tend to select climatic conditions experienced prior to the shift, in agreement with their ecological niche. Ackerly (609) stated that this mechanism is likely to lead to niche conservatism through evolutionary time. Helaouët and Beaugrand (160) found strong support for niche conservatism in two key zooplankton species, *Calanus finmarchicus* and *C. helgolandicus*, at a multidecadal scale. Using Continuous Plankton Recorder data (1960–2002), we showed no shift in the species' thermal niches for two contrasting periods: 1960–1987 (cold period) and 1988–2002 (warm period). Despite an increase in average SSTs of ~1°C, the thermal niches were not altered (Figure 7.16).

Similar findings were observed for periods of positive, negative and average North Atlantic Oscillation (NAO) phases (160). The thermal niche remained virtually identical for the three time periods. As the thermal niche remained stable, increasing temperatures modified the dominance of the two species in the North Sea. In 1962, the subarctic species *C. finmarchicus* represented 80% of the *Calanus*. At the beginning of the 2000s, the temperate pseudo-oceanic species *C. helgolandicus* represented 80% of the genus abundance (Figure 7.17).

7.3.3.2 Behavioural adjustment

When climate change is relatively small, climate-caused environmental changes can be overcome by behavioural adjustment. Different possibilities exist. First, a species can search for a microclimate, which depends upon the scale at which it experiences its environment (e.g. size and anatomic structures involved in locomotion). For example, a zooplankton species can adjust its vertical position in the water column to search for cooler waters (Chapter 6). Obviously, this mechanism has limitations, as the species must also find adequate food, which

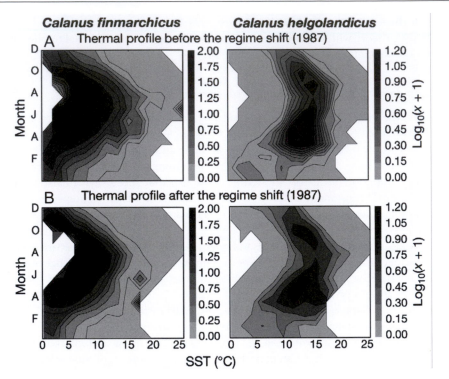

Figure 7.16 Thermal niche of *Calanus finmarchicus* and *C. helgolandicus*. Abundance (decimal logarithm) as a function of months and SSTs for the period 1960–1986 and 1987–2002 (prior to and after the North Sea regime shift).

Source: From Helaouët and Beaugrand (*160*).

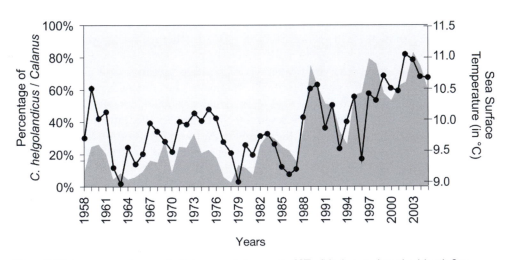

Figure 7.17 Long-term changes in *Calanus* and changes in SSTs (black curve) in the North Sea (1958–2005). Dark grey: percentage of *C. helgolandicus*. Light grey: percentage of *C. finmarchicus*.

Source: Redrawn from Beaugrand (*60*).

is more abundant above the euphotic zone. In Chapter 6, we reviewed some examples of thermotactic behaviour. Second, species can benefit from another (e.g. phoresy) to mitigate the effect of an exceptional warm event, the frequency of which is expected to increase with anthropogenic climate change (Chapters 4 and 6).

Intertidal snails can benefit from thermally heterogeneous environments and may select actively suitable microenvironments (650). Such thermally suitable microhabitats may alleviate the effect of a heatwave on a rocky shore. If global warming is substantial, it is, however, unlikely that microenvironment exploitation will allow individuals to survive in the long term.

7.3.3.3 Phenological shifts

Phenology is the study of periodic biological phenomena. Once microscale behavioural adjustment is not achievable, phenological shift constitutes the first type of species response to anthropogenic climate change (Figure 7.18).

In a given region, when environmental conditions (e.g. temperature) change, the first possibility for a species to adjust to the new conditions is to modify its life cycle in such a way that its critical developmental phases are tuned with seasonal environmental variability. If such changes in timing are possible, the species can stay in the region. If changes are too pronounced, it decreases in abundance or disappears locally, implicating a biogeographical shift (see Section 7.3.3.4). Such phenological changes are species-dependent, reflecting the species' ecological amplitude. Figure 7.18 shows the likely interaction between phenological and biogeographical shifts during a period of climate warming for a theoretical species. In the southern range of its spatial distribution (maximum abundance in spring), a species can respond to a temperature rise by having an earlier occurrence. If phenological shift is not possible because of negative biological interactions or the presence of another limiting factor, the species disappears locally, implicating a biogeographical shift. The possibility for earlier occurrence is limited in extratropical regions by the photoperiod. At the northern range of the species' spatial distribution, the possibility of a phenological shift is generally higher as the seasonal maximum is located in late summer (Figure 7.18). When environmental conditions become more favourable at its northern edge, the species extends its range northwards (phenological dilation). However, in higher latitudes, the photoperiod becomes more and more important and can become limiting for some species, acting as a geographical barrier and triggering species range contraction.

Population geneticists tend to attribute a phenological shift to phenotypic plasticity. For example, Nussey (Section 7.3.2.2) wrote that 'after a spring warming, female passerines often breed earlier than they do after a cold spring. This is the result of phenotypic plasticity' (613). I think that potential phenotypic plasticity can be estimated from the species ecological niche, when the niche is based on the whole species' spatial distribution (Chapter 11). The species ecological niche (*sensu* Hutchinson) may therefore enable potential phenotypic variability to be anticipated. Many works in both the terrestrial and the marine realms have shown that climatic variability can influence species phenology (456, 651–655).

Processes behind phenological shifts are complex, but temperature is often advocated in both the terrestrial and the marine realms (656). For example, phenological shifts in many planktonic species (e.g. dinoflagellates and copepods) have been reported in the North Sea (657) and attributed to regional warming. Mackas and colleagues (658), in a review of phenological shifts in different oceanic systems, also proposed temperature as the main driver,

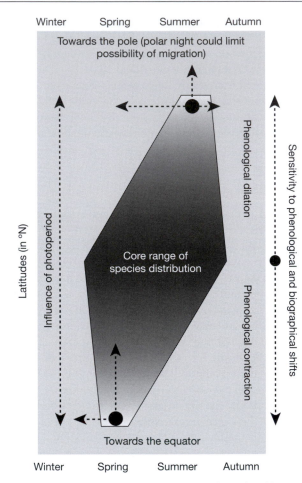

Figure 7.18 Hypothetical relationships between species phenological and biogeographical shifts in the northern hemisphere during a period of climate warming. Levels of grey reflect species abundance. Space/time range is limited by a black line. Species generally have a seasonal maximum in spring in their southern part of spatial distribution while they have a seasonal maximum in late summer in their northern part. In the core range, species have a larger seasonal extent. The impact of photoperiod increases towards high latitudes. During a warming, the northern part of the spatial distribution dilates while the southern part contracts. Bold dashed arrows indicate expected direction and magnitude (proportional to arrow length) of both phenological and biogeographical shifts. Black circle indicates the movement origin. The dotted arrow shows where the photoperiod is likely to play an important role. Sensitivity of both phenological and biogeographical shifts is indicated by dashed arrows.

Source: Modified from Beaugrand (60).

although it was apparent that other factors (e.g. day length, mixed-layer depth) may play a significant role (657). Phenological changes have been identified in both plankton and fish (587).

A recent study on the plankton showed that phenological responses are species-dependent (657). For example, the study suggested that diatoms did not react to warmer spring temperature as their increase is more triggered by seasonal increase in solar radiation (length of a day and solar intensity) while dinoflagellates peaked 23 days earlier. Copepods also showed a less coherent pattern of change despite having, on average, a seasonal maximum 10 days earlier. It has therefore been stressed that these phenological shifts could implicate decoupling of trophic interactions between species. Such community reassembly in time and in space can alter ecosystem functioning and it is thought to be one of the most worrisome consequences of climate change on ecosystems (659).

Plate 7.6 shows changes in the peak of maximum abundance of *Calanus helgolandicus* calculated using the centre of gravity index (660, 661). Associated with SST increase, the seasonal maximum of the species advanced by nearly one month between the 1960s and the beginning of the twenty-first century (2000–2005).

7.3.3.4 Biogeographical shifts

When changes are too substantial, species may spend too much time outside their ecological niche and diminish in abundance or disappear from a region. Contemporaneous biogeographical movements are generally regarded as latitudinal species shifts in response to warming (427, 551, 662–664). Such shifts have become more frequently documented in both the terrestrial and the marine realms (427, 551, 662, 663). Hickling and colleagues investigated the response of freshwater and terrestrial species (e.g. millipedes, spiders, herptiles, birds, mammals) to climate change in Britain (665). On the 329 species they examined, 275 (84%) species exhibited a northward shift in agreement with global warming expectation. Thomas (664) further examined these results and concluded that most freshwater and terrestrial range boundaries were climatically controlled. For example, the endangered leatherback turtle (*Dermochelys coriacea*) has its northerly distribution limited by the location of the 15°C isotherm in summer in the North Atlantic (289). Coccolitho-phores have also extended their range northwards with reports of unprecedented blooms in the Bering Sea (666, 667).

An important aspect of biogeographical shifts is that they should be envisioned in a two-dimensional (biogeographic) space and not only a one-dimensional (latitudinal) space. Recently, I modelled past (Last Glacial Maximum; LGM), present (1960–1969) and future (2090–2099) spatial distributions of *C. finmarchicus* in the North Atlantic (Plate 7.7). Although large changes took place on the eastern side of the North Atlantic, limited shifts occurred on the western side. In their meta-analytic study, Parmesan and Yohe (663) estimated that the mean latitudinal shift of ~1,700 species was 6.1 km per decade polewards. From Plate 7.7, we can see that an average latitudinal rate for a species shift can be misleading because the latitudinal shift varies at different longitudes. Furthermore, the shift is also structured in two dimensions (i.e. zonal and meridional component). Therefore, Parmesan and Yohe probably underestimated the true magnitude of the biogeographical movements in some locations.

7.3.4 Fourth type of response: extinction

As we saw earlier at the geological scale (Chapter 5), periods of global warming have been associated with major extinction events. We have also seen that taxonomic richness is strongly correlated to temperature changes (350, 668). Therefore, anthropogenic climate change may cause species extinction in the next decades. An estimate of the magnitude of such climate-induced extinctions is, however, difficult. Based on the examination of 1,103 species of animals and plants, Thomas and colleagues (669) provided a rough assessment of the potential effects of anthropogenic climate change on terrestrial species extinction by 2050. Estimates were based on bioclimatic models and species-area relationships for three levels of warming and for scenarios of no or full (universal) dispersal. For a high level of warming, they found that climate-caused species extinction might range between 21% and 32% when there is no limitation in species dispersal and between 38% and 52% when there is no possibility of dispersal. In case of a medium climate change, the percentage of species committed to extinction ranged between 15% and 20% (universal dispersal) and 26% and 37% (no dispersal). Using optimistic scenarios of climate change, the percentages were 9–13% (universal dispersal) and 22–31% (no dispersal). Rates of species committed to extinction varied among taxonomic groups and ecoregions.

No climate-induced species extinction has been documented yet in the marine realm. Although some marine species such as the Mediterranean mysid *Hemimysis speluncola* (marine cave species) may become at risk (670), it is currently rarely the case for most species. Climate-caused species extinction depends on a multitude of biological parameters (e.g. fecundity, dispersal and species ecological amplitude). Species vulnerability to climate change is difficult to assess, and both bioclimatologists and ecologists will have to work rapidly on this issue.

7.3.5 Species vulnerability to anthropogenic climate change

Vulnerability is defined as the degree to which a species or a population can be altered by global warming. Vulnerability depends upon both intrinsic and extrinsic factors (671). Vulnerability has three components: (1) adaptive capacity; (2) sensitivity; and (3) exposure.

7.3.5.1 Adaptive capacity

Adaptive capacity (intrinsic factor) is the species' capacity to adapt to climate change. Adaptive capacity is a function of phenotypic plasticity and the species' genetic pool, which determines species' evolutionary potential (see Section 7.3.2). It also depends on the species' evolutionary rate, which is a function of temperature, species size, generation time and dispersal. Adaptive capacity influences species' behaviour and its possibility to shift to more suitable habitats (e.g. plankton shift in vertical distribution, microhabitat refugee for intertidal species). Adaptive capacity therefore determines the species ecological niche (*sensu* Hutchinson).

7.3.5.2 Sensitivity

Sensitivity (intrinsic factor) is the extent to which a species can be altered by global climate change. Sensitivity depends on intrinsic properties such as fecundity rate, demographic strategy (e.g. r and K strategists), life history traits such as dispersal, size and the characteristics of the ecological niche that reflects, in part, species' ecophysiological properties. A specialist species, characterised by low ecological amplitude, is likely to be more sensitive to climate (or more generally environmental) change. In contrast, a generalist species is likely to be less sensitive to climate change. **Stenotherms** will be more sensitive than **eurytherms**. Many Antarctic species live within a narrow range of temperatures, their habitat having temperatures fluctuating between −1.9°C and 1.8°C (Figure 7.19).

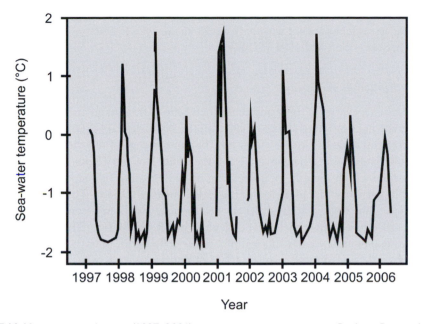

Figure 7.19 Year-to-year changes (1997–2006) in seawater temperatures at Rothera Research Station at 15 m depth.

Source: Redrawn from Peck and co-workers (672).

7.3.5.3 Exposure

Exposure (extrinsic factor) is an indication of the magnitude of climate change, varying in time (season) and space (regionally). Exposure can be approximated using past climatic observations or projections from general circulation models. Species exposure varies inside its spatial range.

7.3.5.4 High sensitivity to climate change

As we have seen earlier, species are highly sensitive to climate change. Peck and colleagues (672) investigated **acclimation** potential of two separate assemblages of brittle stars (*Ophionotus victoriae*). Individuals were progressively acclimated from 0°C to 2°C or 3°C. Individuals were unable to acclimate, with failure taking place from day 19 and day 24 when acclimated at 3°C and 2°C, respectively. All stars died after 32 and 68 days at the +3°C and +2°C experiments, respectively. This acclimation experiment suggests that invertebrates are highly sensitive to warming and have thereby low capability to acclimate to warmer temperatures. This result has been observed for many other invertebrates. Although many psychrophiles can live up to 10°C, critical biological functions are rapidly altered for increase in temperatures of only 1–2°C above current summer maximum seawater temperatures. In addition to the lower number of eggs produced by Antarctic species compared to temperate invertebrates, this result implies that Antarctic invertebrates might be more sensitive to global warming than temperate species (673). Tropical species might also be sensitive because they are generally stenotherm (674). We saw earlier that coral reefs are also very sensitive to climate warming. Acute episodes of coral reef bleaching and high mortality were associated with positive anomalies of ~1°C above long-term monthly averages between 1983 and 1991 (675).

7.3.5.5 Variation in sensitivity

Species sensitivity varies in space and time. For example, Beaugrand and co-workers (2008) examined the implication of climate change on the Atlantic cod *Gadus morhua* throughout its distributional range (Plate 7.8). Towards the warmer edge of the thermal niche (North Sea), small temperature changes have a large effect on cod (Plate 7.8). The thermal influence occurs both directly by influencing physiology and indirectly through the food web, and in particular the plankton at the larval stage of the fish. Towards the centre of the thermal niche around Iceland, where the thermal profile represents a broad plateau, similar temperature changes will have less of an effect on the fish. The Atlantic cod has evolved to exploit a particular thermal niche that reflects not only the effect of temperature on physiology (537), but also the effect of temperature on other components of the environment (e.g. food supply, type and timing) (497, 676). Consequently, at the edge of their thermal niche, cod will be especially sensitive to both the direct and indirect effects of temperature. Sensitivity might be aggravated by other pressure. In the case of the Atlantic cod, overfishing has weakened cod stocks throughout its whole distributional range and has increased its sensitivity to climate change (677) (Chapter 9). When projected changes in SSTs are added (SRES Scenario A2; period 1990–2100), the species might disappear inexorably from the North Sea and might become more sensitive around Iceland. This analysis provides an indication on long-term changes in species sensitivity. Exposure is expected to be larger in the North Sea (magnitude of SST change = 4.9°C for the period 1990–2100) than around Iceland (magnitude of change = 3.4°C for the period 1990–2100).

7.3.5.6 Ontogenic variation in sensitivity

Most marine species have complex life histories and many have stages that have different sensitivities to climate change. (e.g. reproduction, spawning, fertilisation, hatching, embryonic development, larval stage, metamorphosis, settlement, recruitment). This makes it difficult to identify critical stages. Generally, early reproductive and developmental stages are more sensitive than older stages (218). In his seminal paper on marine invertebrate larval ecology, Thorson stated: 'In analysing the ecological conditions of an animal population we have above all to focus our attention upon the most sensitive stages within the life cycle of the animal, that is, the period of breeding and larval development'. Studies suggested that marine invertebrate mortality during early life stages exceeded 90% in their biotope (678), and Hjort (679) proposed that the abundance of year-class of marine fish populations was determined through the early life history. The biologist proposed that slight environmental fluctuations occurring during this period strongly influenced recruitment. This has been clearly shown for North Sea cod (676) and haddock (680). Figure 7.20 summarises the main potential environmental influences and biotic interactions on fish from the egg to the juvenile stage. It also recalls that the sensitivity of adults or spawners tends to be higher than juveniles (Figure 6.17).

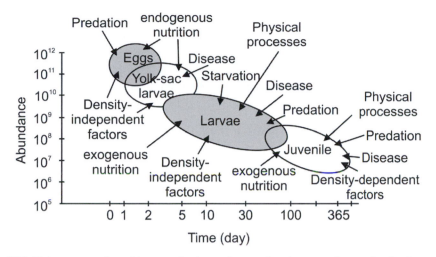

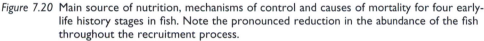

Figure 7.20 Main source of nutrition, mechanisms of control and causes of mortality for four early-life history stages in fish. Note the pronounced reduction in the abundance of the fish throughout the recruitment process.

Source: Redrawn from Houde (*681*).

7.3.5.7 Species sensitivity and extreme events

Extreme events are often accompanied by mass mortalities. By affecting organism physiology, extreme events can make them more vulnerable to predation, parasitism or disease (coral reefs; Chapter 6 and Section 7.3.1). Spicer and Gaston (682) suggested that the rate of environmental change is important: the greater the magnitude of an event, the less plausible the rate of physiological compensation. This study suggests that rapid climate change may enable less physiological compensation. In the North Sea, some cold events had major consequences. Horwood and Millner (683) reported a mortality of 60% of sole (*Solea solea*) in the North Sea after the 1962–1963 cold winters. We saw other examples for intertidal species in Chapter 6 and coral reefs in the present chapter (warm events).

7.3.5.8 Further complexity

Some environmental stimuli may have long-lasting effects if they occur during a critical phase of the species life cycle (e.g. the impact of temperature on sex determination in turtles; Chapter 6). Timing of a physiological event may also be modified by the environment (e.g. temperature), a process termed by Spicer and Gaston (682) as physiological heterochromy.

The impact of climate change is expected to influence deeper K-selected than r-selected species. K-selected species inhabit places where only few random fluctuations typically occur while r-selected species are characteristics of regions that are less predictable. Their capability to adapt is different. Capacity adaptation, the aptitude of an individual to adjust its physiological functions when an environmental fluctuation occurs, is probably higher for r-strategists. Resistance adaptation, the capacity of an individual to resist against the lethal effects of extreme environmental conditions, is more difficult to evaluate and is probably more related to the degree of stenoecy/euryecy for a given environmental parameter.

7.4 Community/ecosystem response to climate change

We can envision different types of climatic effects at the community/ecosystem level: (1) biogeographical community shifts; (2) alterations in the ecosystem trophodynamics; (3) trophic mismatch and changes in species interactions; and (4) long-term community/ecosystem shifts, including abrupt community/ecosystem shifts.

7.4.1 Biogeographical community shift

Based on their spatial distribution and both diel and seasonal variability, Beaugrand and colleagues divided (178) North Atlantic copepod biodiversity (109 species or taxa) into nine species assemblages. They subsequently examined long-term changes in these assemblages and found major biogeographical shifts in warm-water species associated with a reduction in the number of colder-water species (Plate 7.9). All assemblages showed consistent long-term changes that appear to reflect the response of marine ecosystems towards a warmer dynamic regime. In the same region, northward movements in both exploited and unexploited fish were also detected (551, 685, 686).

Comparison of shift rates in the marine and terrestrial realms suggests that marine shifts can be more pronounced (684). Table 7.2 summarises the physical and biological processes or characteristics that may explain differences in shift magnitude observed between the terrestrial and the marine pelagic ecosphere. In their study, Beaugrand and colleagues (684) showed apparent movements of some plankton assemblages (e.g. warm-water pseudo-oceanic species) of ~10 degrees of latitudes, a shift of 23.16 km·yr⁻¹ in 48 years. The rate of change they provided was a coarse indication of the shift, as these biogeographical modifications had a zonal and a meridional component (see Plate 7.7). Although observed latitudinal shifts were more moderate for other species assemblages, they remained high in comparison to the much smaller terrestrial movements reported in the literature. Indeed, Thuiller (687) suggests that a temperature rise of 1°C, as approximately observed in the North Sea, moves northwards the location of ecological zones by 160 km. Examination of fossil pollen suggests that tree shifts during the post glacial warming in North America and Europe ranged between 100 and 1,000 m·yr⁻¹ (688), although recent works suggested such rates could have even been overestimated (689). Thomas and Lennon (1999), investigating 59 breeding bird species over the United Kingdom, detected a northward shift of 18.9 km in 20 years, an average shift of 945 m·yr⁻¹. A meta-analysis conducted on 1,700 species or taxonomic groups at a global scale reported a poleward shift of 610 m·yr⁻¹ (663). Parmesan and colleagues (690) detected a maximum range shift of 200 km for butterfly over a period of 40 years, which represents a latitudinal shift of 5 km·yr⁻¹.

Beaugrand and colleagues (684) proposed five explanations to explicate why biogeographical shifts in the pelagic realm were more rapid than in the terrestrial realm. First, marine organisms are, in general, stenothermic in comparison to their terrestrial counterparts (Table 7.2; Chapter 6). Second, copepods are free-floating species and can track rapidly their bioclimatic envelope contrary to terrestrial species where barriers (e.g. different substrate, presence of a lake or a town) can prevent a species to move, where dispersal is restricted to specific stages and where other climatic parameters become more important (e.g. precipitation, microclimate). Third, regions where the greatest changes were observed are generally located at the boundary between the Atlantic Westerlies Wind Biome and the Atlantic Polar Biome (143, 178, 182). Many authors have drawn attention that species

Plate 2.1 Picture taken by space shuttle a few weeks (8 August 1991) after the volcanic eruption of Mount Pinatubo.

Source: Courtesy of the National Aeronautics and Space Administration (NASA).

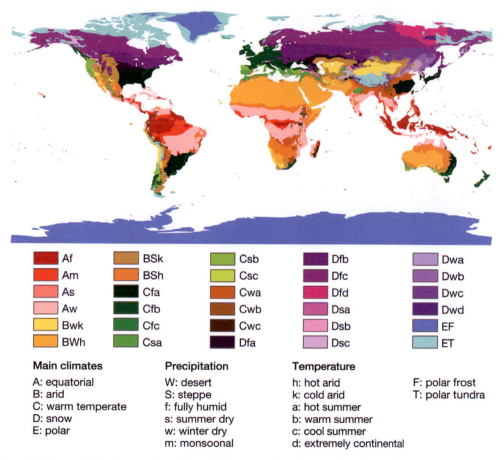

Af	BSk	Csb	Dfb	Dwa
Am	BSh	Csc	Dfc	Dwb
As	Cfa	Cwa	Dfd	Dwc
Aw	Cfb	Cwb	Dsa	Dwd
Bwk	Cfc	Cwc	Dsb	EF
BWh	Csa	Dfa	Dsc	ET

Main climates
A: equatorial
B: arid
C: warm temperate
D: snow
E: polar

Precipitation
W: desert
S: steppe
f: fully humid
s: summer dry
w: winter dry
m: monsoonal

Temperature
h: hot arid
k: cold arid
a: hot summer
b: warm summer
c: cool summer
d: extremely continental

F: polar frost
T: polar tundra

Plate 2.2 Köppen classification of climate. The classification was based on precipitation and temperature data of the period 1951–2000. (A) Regions with an average temperature of coolest month greater or equal to 18°C. (B) Regions with 70% or more of annual rainfall that occur during the warmer six months. (C) Regions with an average temperature of warmest month >10°C and of coldest month ranging between 0°C and 18°C. (D) Regions with average temperature of warmest month >10°C and of coldest month ≤0°C. (E) Regions with average temperature of warmest month <0°C. *f* means absence of drought, *w* and *s* means a dry winter and summer, respectively.

Source: Data from Kottek and colleagues (*63*).

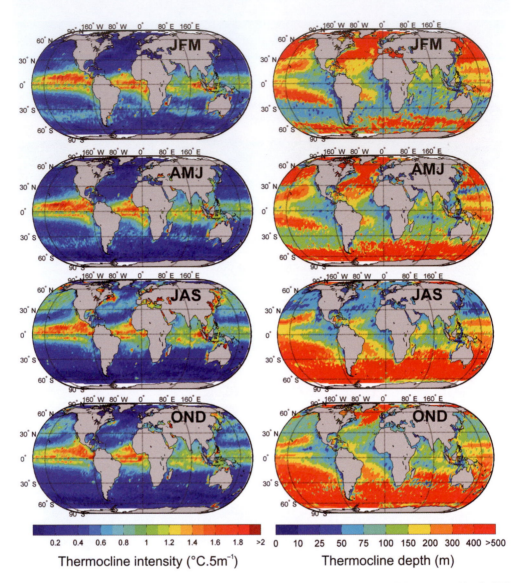

Thermocline intensity (°C.5m⁻¹)

Thermocline depth (m)

Plate 2.3 Seasonal changes in both intensity (left) and depth (right) of the thermocline. JFM: January to March. AMJ: April to June. JAS: July to September. OND: October to December.

Source: Modified from Reygondeau and Beaugrand (*65*).

FACING PAGE

Plate 2.4 (Top left) Mean surface currents in the world oceans and mean sea level pressure. (A) Main surface currents. Currents strictly lower than 0.1 m·s⁻¹ are in white. These low-current areas indicate the centre of the main gyres and high-latitude oceanic regions. The name of the main surface currents is indicated. (B) Mean sea level pressure. The name of the semi-permanent highs and lows are superimposed.

Source: Current data are from the Global Drifter Programme (NOAA, Physical Oceanography Division) and atmospheric (HadSLP2) data are from the Hadley Centre.

Plate 2.5 (Top right) Global patterns of mean sea surface height for the period 1958–2007.

Source: Data are from CARTON-GIESE Simple Ocean Data Assimilation (v2p0p2-4).

Plate 2.6 (Bottom) Relationships between temperature, oceanic and atmospheric circulation. (A) Mean sea surface temperature (SST) zonal anomalies (1979–2012). The main oceanic surface currents are superimposed. (B) Mean air temperature (AT) zonal anomalies and atmospheric circulation (1979–2012). The arrows indicate both the direction and the intensity of wind. A zonal anomaly is the difference between the temperature of a given node minus the mean temperature at the corresponding latitude. The size of the arrows is proportional to wind intensity. NAC: North Atlantic Current. IC: Irminger Current.

Source: Temperature data are from the ERA INTERIM programme (European Centre for Medium-Range Weather Forecasts). Wind data are from the NCEP/DOE AMIP-II Reanalysis (Reanalysis-2) Model.

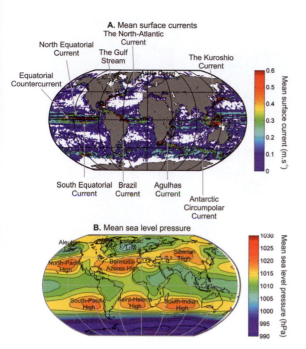

A. Mean surface currents

The North-Atlantic Current

North Equatorial Current

The Gulf Stream

The Kuroshio Current

Equatorial Countercurrent

Arctic Ocean

Mean surface current (m.s⁻¹)

South Equatorial Current

Brazil Current

Agulhas Current

Antarctic Circumpolar Current

B. Mean sea level pressure

Aleutian Low

Icelandic Low

Siberian High

North-Pacific High

Bermuda Azores High

South-Pacific High

Saint-Helena High

South-Indian High

Mean sea level pressure (hPa)

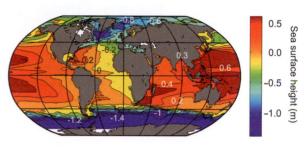

Sea surface height (m)

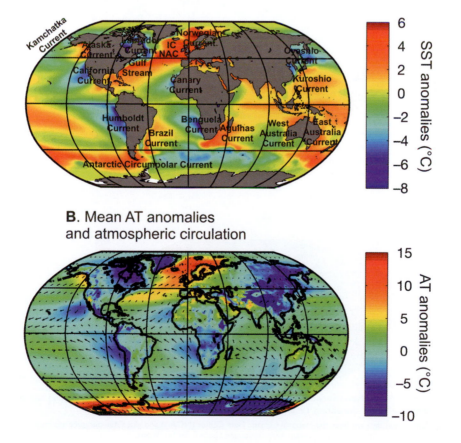

A. Mean SST anomalies

Kamchatka Current

Alaska Current

Labrador Current

Norwegian Current

IC NAC

Oyashio Current

California Current

Gulf Stream

Canary Current

Kuroshio Current

Humboldt Current

Brazil Current

Benguela Current

Agulhas Current

West Australia Current

East Australia Current

Antarctic Circumpolar Current

SST anomalies (°C)

B. Mean AT anomalies and atmospheric circulation

AT anomalies (°C)

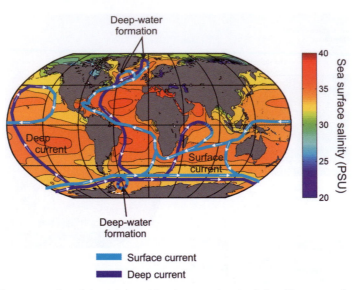

Plate 2.7 Idealised representation of the global meridional overturning circulation. Mean sea surface salinity is indicated. Salinity data are from the World Ocean Database.

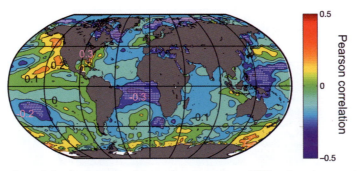

Plate 2.8 Patterns of correlation between annual sea surface temperature (SST) and yearly sunspots for the period 1948–2012. White circles indicate significant correlations. It should be noted that no correction to account for temporal autocorrelation was applied here. Correlations are weak and explain a low percentage of variance.

Source: SST (ERSST V3) data are from the National Climatic Data Center (NOAA) and sunspot data are from the Solar influences Data Analysis Center.

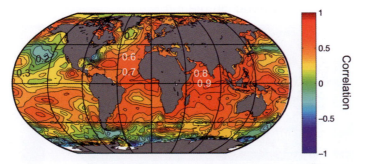

Plate 2.9 Patterns of correlation between annual sea surface temperature (SST) and the global (land and sea) temperature anomalies for the period 1950–2012.

Source: SST (ERSST V3) data are from the National Climatic Data Center (NOAA) and global temperature anomaly data are from the Goddard Institute for Space Studies (National Aeronautics and Space Administration).

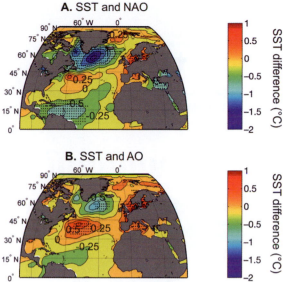

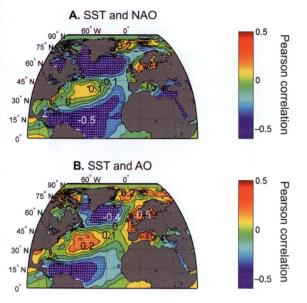

A. SST and NAO

B. SST and AO

A. SST and NAO

B. SST and AO

Plate 2.10 Patterns of correlations between annual sea surface temperature (SST) and both the North Atlantic and the Arctic Oscillations for the period 1948–2012. White circles indicate significant correlations. It should be noted that no correction to account for temporal autocorrelation was applied here.

Source: SST (ERSST V3) data are from the National Climatic Data Center. NAO and AO data are from the Climate Prediction Center.

Plate 2.11 Differences in annual sea surface temperatures (SST) between a positive and a negative phase of the North Atlantic Oscillation (NAO) and the Arctic Oscillation (AO) for the period 1950–2012. Thresholds used here to determine a positive and a negative phase are 0.5 and –0.5, respectively. Black circles indicate significant differences ($p < 0.05$) after using a Kruskal-Wallis test.

Source: SST (ERSST V3) data are from the National Climatic Data Center. NAO and AO data are from the Climate Prediction Center.

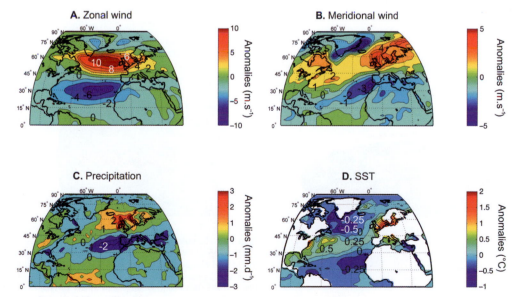

A. Zonal wind

B. Meridional wind

C. Precipitation

D. SST

Plate 2.12 Differences in some winter (December to March) climatic variables between a positive and a negative phase of the North Atlantic Oscillation (NAO) during the period 1979–2012. Thresholds used here to determine a positive and a negative phase are 1 and –1, respectively. (A) Winter zonal wind anomalies between a positive and a negative phase of the NAO. (B) Winter meridional wind anomalies between a positive and a negative phase of the NAO. (C) Winter precipitation anomalies between a positive and a negative phase of the NAO. (D) Winter SST anomalies between a positive and a negative phase of the NAO.

Source: Winter precipitation (GPCP Version 2.2 Combined Precipitation Dataset) data are from the Polar Satellite Data Centre. Winter SST (ERSST V3) data are from the National Climatic Data Center. Both zonal and meridional wind data (NCEP/DOE AMIP-II Reanalysis-2 Monthly Averages) are from the Earth System Research Laboratory. Winter NAO data are from the Climate Prediction Center.

A.

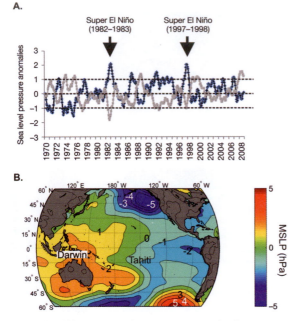

Super El Niño (1982–1983)

Super El Niño (1997–1998)

B.

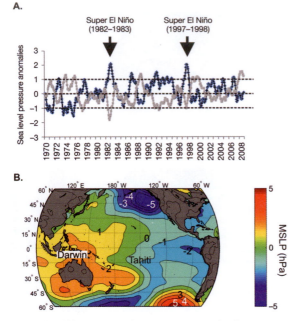

Plate 2.13 (A) Year-to-year changes in mean sea level pressure in Darwin (low pressure centre, in blue) and Tahiti (high pressure centre, in grey) after application of an order-6 simple moving average. (B) Mean sea level pressure anomalies (MSLPA) during the 1997–1998 Super El Niño event.

Source: Data from the Climate Prediction Center of the National Oceanic and Atmospheric Administration. MSLP ERA-INTERIM data (1979–2012) are from the European Centre for Medium-Range Weather Forecasts.

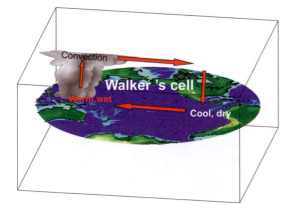

Plate 2.14 The Walker's cell in the Pacific Ocean during normal conditions.

A. Meridional wind anomaly

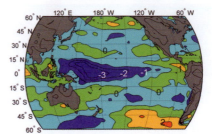

B. Sea Level Height anomaly

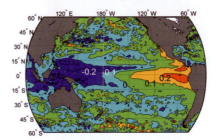

C. SST anomaly

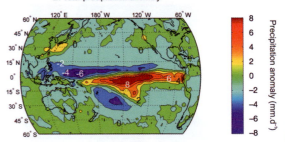

D. Total precipitation anomaly

Plate 2.15 Hydrologic and meteorological changes associated with the El Niño of 1997–1998. (A) Monthly anomalies in the meridional component of the wind in 1997. (B) Monthly anomalies in sea level height anomalies between July 1997 and March 1998. (C) Monthly anomalies in sea surface temperature (SST) anomalies between July 1997 and March 1998. (D) Total precipitation monthly anomalies between January and June 1998.

Source: U-wind data (1979–2012) are from the Earth System Research Laboratory (NCEP/DOE AMIP-II Reanalysis-2). Sea level height data (1958–2007) are from CARTON-GIESE Simple Ocean Data Assimilation (v2p0p2-4). SST data (1979–2012) are from the ERA-INTERIM dataset. Total precipitation data are from the GPCP Version 2.2 Combined Precipitation Dataset.

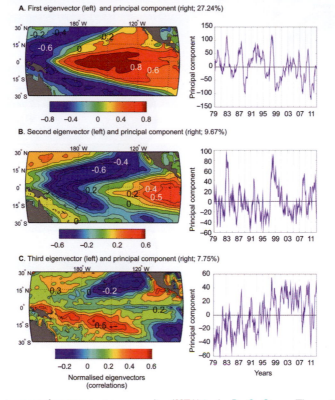

A. First eigenvector (left) and principal component (right; 27.24%)

B. Second eigenvector (left) and principal component (right; 9.67%)

C. Third eigenvector (left) and principal component (right; 7.75%)

Normalised eigenvectors
(correlations)

Plate 2.16 Changes in sea surface temperature anomalies (SSTA) in the Pacific Ocean. The spatial patterns are normalized eigenvectors and time series are principal components originated from a principal component analysis performed on sea surface temperature anomalies (SSTA) for the period 1979–2012. (A) First normalised (left) and principal component. (B) Second normalised (left) and principal component. (C) Third normalised (left) and principal component.

Source: Monthly SSTA data are from the ERA-INTERIM dataset. Anomalies are deviations from the average of each month calculated for the whole time period.

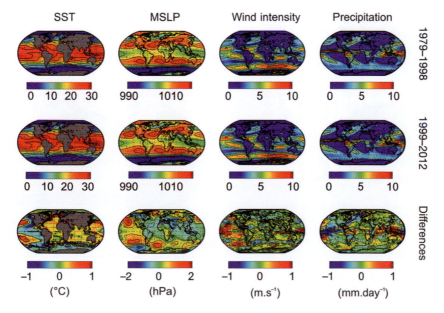

| SST | MSLP | Wind intensity | Precipitation |

1979–1998

1999–2012

Differences

−1 0 1   −2 0 2   −1 0 1   −1 0 1
(°C)   (hPa)   (m.s⁻¹)   (mm.day⁻¹)

Plate 2.17 Changes in some hydro-climatic parameters between 1999–2012 and 1979–1998: sea surface temperatures (SST), mean sea level pressure (MSLP), wind intensity and precipitation. A positive change means an increase from 1979–1998 to 1999–2012 and inversely.

Source: SST, MSLP and wind intensity are from the ERA-INTERIM dataset and total precipitation data are from the GPCP Version 2.2 Combined Precipitation Dataset.

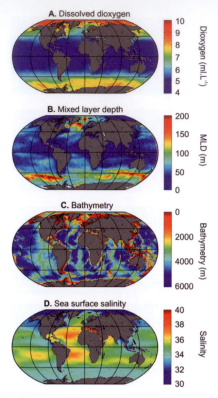

Plate 3.1 Concentration in (A) dissolved dioxygen, (B) mixed layer depth (MLD; density), (C) bathymetry and (D) sea surface salinity.

Source: Dissolved dioxygen and sea surface salinity are from the World Ocean Atlas. Bathymetry data is from the General Bathymetric Chart of the Oceans (GEBCO). MDL (density) data originated from de Boyer Montégut (156).

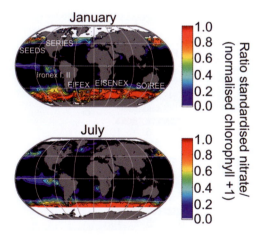

Plate 3.3 High Nutrient Low Chlorophyll areas as revealed by the ratio of standardised nitrates (between 0 and 1) on [standardized chlorophyll (between 0 and 1) +1] in January (top) and July (bottom). Values of chlorophyll greater or equal to 2 mg·m^{-3} and of nitrates greater or equal to 20 µmoles·L^{-1} were fixed to 1 in the standardised values of nitrates and chlorophyll, respectively. Black areas are regions where the ratio is 0.

Source: Data from SeaWIFS and the World Ocean Database. The location of some ocean iron experiments is indicated on the map. IRONEX I and II: IRON fertilization EXperiment (1993 and 1995); SOIREE: Southern Ocean Iron RElease Experiment (1999); SEEDS: The Subarctic Pacific Iron Experiment for Ecosystem Dynamics Study (2001); SERIES: Subarctic Ecosystem Response to Iron Enrichment (2002); EIFEX: European Iron Fertilization Experiment (2004).

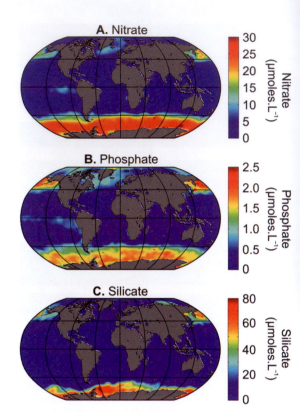

Plate 3.2 Concentration in (A) nitrate, (B) phosphate and (C) silicate in seawater.

Source: Data from the World Ocean Atlas.

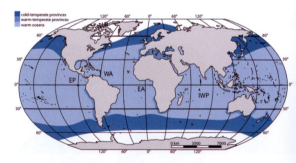

Plate 3.4 The biogeographic division of the marine ecosphere proposed by Briggs and Bowen and based mainly on fish. Cold regions (Arctic and Antarctic) are in white, cold-temperate regions are in medium blue and warm-temperate provinces are in medium blue. WA: Western Atlantic. EA: Lusitania, Black Sea, Caspian, Aral and Benguela provinces in the East Atlantic. IWP: Mediterranean, Sino-Japanese, Auckland, Kermadec, south-eastern Australian and south-western Australian provinces in the Indo-Pacific. EP: California and Peru-Chilean provinces in the East Pacific.

Source: Courtesy of Briggs and Bowen (172).

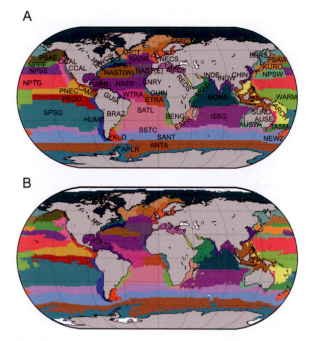

Plate 3.5 Biogeography of the global ocean (i.e. biogeochemical provinces) proposed by (A) Longhurst (2007) and (B) recalculated by a numerical procedure from January 1998 to December 2007.

Source: From Reygondeau and co-workers (*179*).

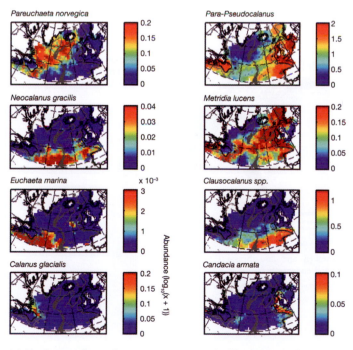

Plate 3.6 Mean spatial distribution of some key species or taxa in the North Atlantic Ocean.

Source: Modified from Beaugrand *et al.* (*181*).

A. Lower bathyal provinces

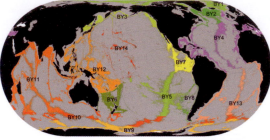

BY1: Arctic
BY2: Northern Atlantic Boreal
BY3: Northern Pacific Boreal
BY4: North Atlantic
BY5: Southeast Pacific Ridges

BY6: New Zealand-Kermadec
BY7: Cocos Plate
BY8: Nazca Plate
BY9: Antarctic
BY10: Subantarctic

BY11: Indian
BY12: West Pacific
BY13: South Atlantic
BY14: North Pacific

B. Abyssal provinces

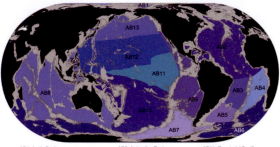

AB1: Arctic Basin
AB2: North Atlantic
AB3: Brazil Basin
AB4: Angola, Guinea, Sierra Leone Basins
AB5: Argentine Basin

AB6: Antarctica East
AB7: Antarctica West
AB8: Indian
AB9: Chile, Peru, Guatemala Basins
AB10: South Pacific

AB11: Equatorial Pacific
AB12: North Central Pacific
AB13: North Pacific
AB14: West Pacific Basins

C. Ultra-abyssal provinces

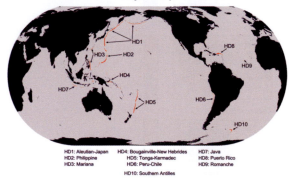

HD1: Aleutian-Japan
HD2: Philippine
HD3: Mariana

HD4: Bougainville-New Hebrides
HD5: Tonga-Kermadec
HD6: Peru-Chile

HD7: Java
HD8: Puerto Rico
HD9: Romanche

HD10: Southern Antilles

Plate 3.7 Lower bathyal (A), abyssal (B) and ultra-abyssal (C) provinces proposed by Watling and colleagues (*146*).

Source: Courtesy of Les Watling, department of Biology, University of Hawaii.

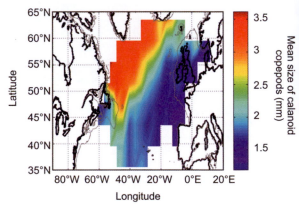

Plate 3.8 Mean spatial distribution of the mean size of calanoid copepods (minimum size of female). The grey line denotes the isobaths 200 m.

Source: Redrawn from Beaugrand and colleagues (*207*).

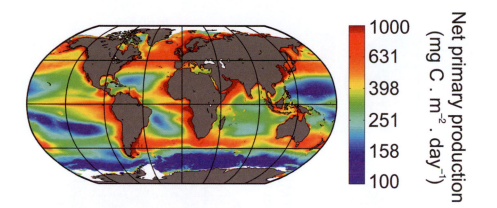

Plate 4.1 Mean spatial distribution (1997–2007) of net primary production estimated from SeaWIFS.

Source: Methods of calculation of net primary production are from Behrenfeld and Falkowski (278). Data from Oregon State University.

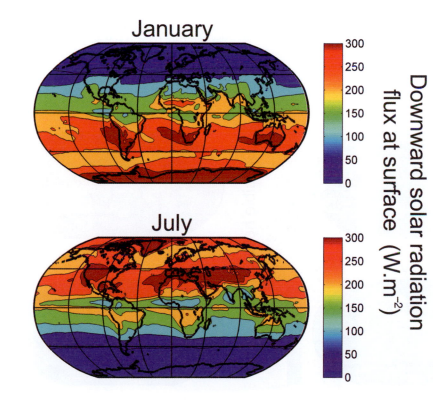

Plate 4.2 Spatial distribution of the downward solar radiation flux at surface (W·m^{-2}) in January and July (1979–2012).

Source: Data from NCEP/DOE AMIP-II Reanalysis-2 Monthly Averages.

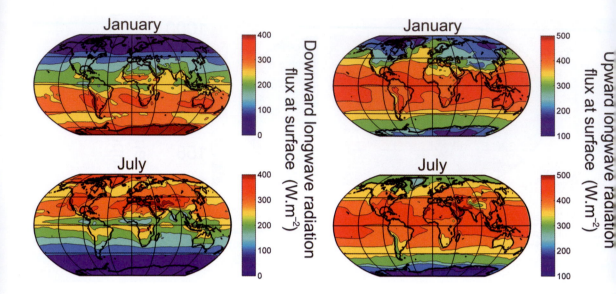

Plate 4.3 Spatial distribution of the downward long-wave radiation flux at surface (W·m⁻²) in January and July (1979–2012).

Source: Data from NCEP/DOE AMIP-II Reanalysis-2 Monthly Averages.

Plate 4.4 Spatial distribution of the upward long-wave radiation flux at surface (W·m⁻²) in January and July (1979–2012).

Source: Data from NCEP/DOE AMIP-II Reanalysis-2 Monthly Averages.

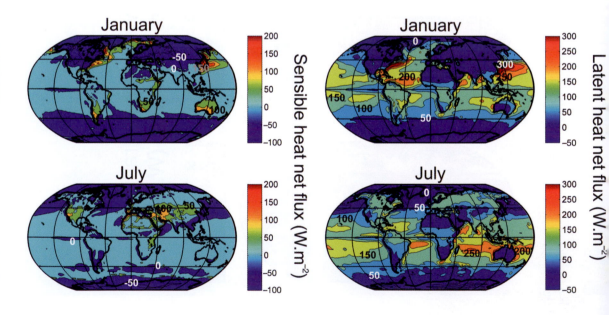

Plate 4.5 Mean surface sensible heat net flux (W·m⁻²) in January and July during the period 1979–2012.

Source: Data from NCEP/DOE AMIP-II Reanalysis-2, downloaded from the Earth System Research Laboratory (NOAA).

Plate 4.6 Mean surface latent heat net flux (W·m⁻²) in January and July during the period 1979–2012.

Source: Data from NCEP/DOE AMIP-II Reanalysis-2, downloaded from the Earth System Research Laboratory (NOAA).

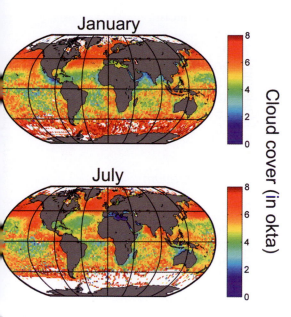

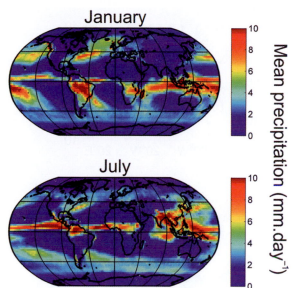

Plate 4.7 Global cloud cover in January and July during the period 1991–2011. Note that cloud cover is expressed in oktas (or eighths). An okta is an estimate of how many heights cover the sky. The okta scale ranges generally between 0 (sky completely clear) to 8 (sky fully cloudy). An okta of 4 means a sky half cloudy. An okta of 9 means a sky obstructed from view. Such a situation takes place in case of dense fog or heavy snow.

Source: Data from the Earth System Research Laboratory (NOAA; NCEP reanalysis).

Plate 4.8 Mean precipitation (mm·day⁻¹) in January and July during the period 1979–2010. Note the seasonal movement of the Intertropical Convergence Zone around the equator.

Source: Data from the GPCP Polar Satellite Precipitation Data Centre and downloaded from the Earth System Research Laboratory (NOAA).

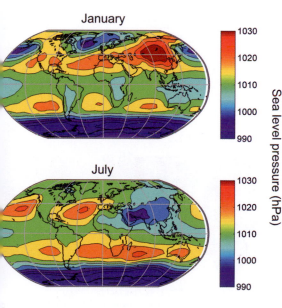

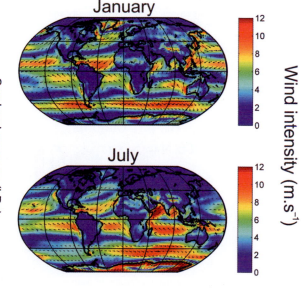

Plate 4.9 Mean atmospheric surface pressure (hPa) in January and July.

Plate 4.10 Mean wind intensity and direction (m·s⁻¹) in January and July during the period 1979–2012.

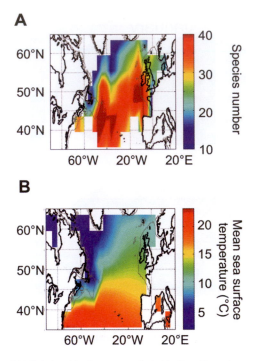

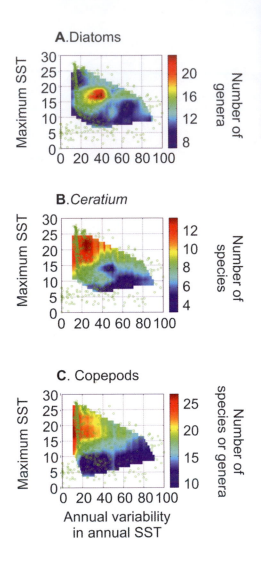

Plate 4.11 Relationships between pelagic biodiversity and mean sea surface temperature in the North Atlantic Ocean. (A) Mean spatial patterns in North Atlantic calanoid biodiversity using data from the CPR survey (1958–2007). Biodiversity was assessed by applying a first-order jackknife procedure on species number. (B) Mean sea surface temperature (1960–2005).

Source: Modified from Beaugrand and co-workers (*207*).

Plate 4.13 Relationships between plankton biodiversity and temperature. Plankton biodiversity as a function of maximum annual SST and an index of annual variability in SST for (A) diatoms, (B) Ceratium and (C) calanoid copepods.

Source: Modified from Beaugrand and co-workers (*207*).

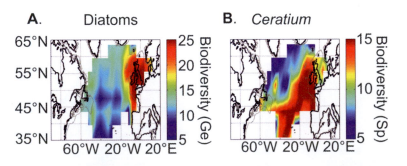

Plate 4.12 Mean spatial distribution in diatom (35 genera) and Ceratium (dinoflagellates; 47 species) biodiversity in the North Atlantic Ocean. (A) Mean spatial patterns in diatom biodiversity. (B) Mean spatial patterns in Ceratium biodiversity. The biodiversity was assessed by applying a first-order jackknife procedure on species number.

Source: Data from the CPR survey (1958–2007). Ge: number of genera. Sp: number of species. Modified from Beaugrand and co-workers (*207*).

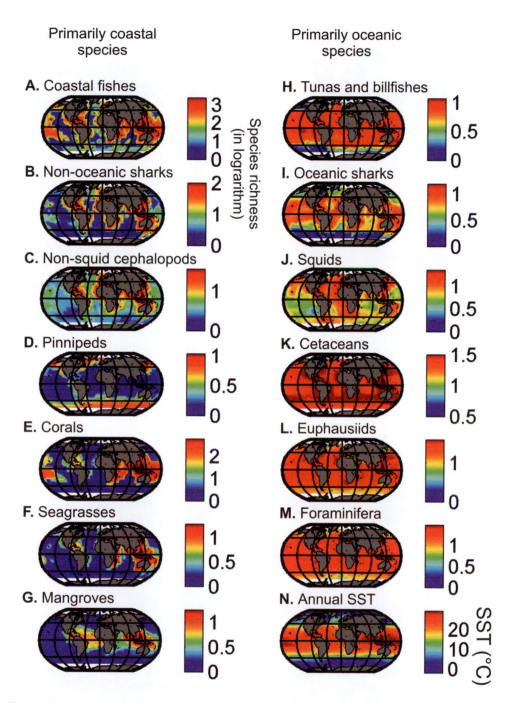

Primarily coastal species

A. Coastal fishes

B. Non-oceanic sharks

C. Non-squid cephalopods

D. Pinnipeds

E. Corals

F. Seagrasses

G. Mangroves

Species richness (in logarithm)

Primarily oceanic species

H. Tunas and billfishes

I. Oceanic sharks

J. Squids

K. Cetaceans

L. Euphausiids

M. Foraminifera

N. Annual SST

SST (°C)

Plate 4.14 Spatial patterns in the diversity of 13 taxonomic groups and annual sea surface temperature. Left: spatial patterns in the diversity (number of species) of (A) coastal fishes, (B) non-oceanic sharks, (C) non-squid cephalopods, (D) pinnipeds, (E) corals, (F) seagrasses and (G) mangroves. Right: spatial patterns in the diversity (number of species) of (H) tunas and billfishes, (I) oceanic sharks, (J) squids, (K) cetaceans, (L) euphausiids and (M) foraminifera. (N) spatial patterns in annual sea surface temperature.

Source: Biodiversity data are from Tittensor and colleagues (292). Temperature data are from the Extended Reconstructed Sea Surface Temperature data set (Version 3).

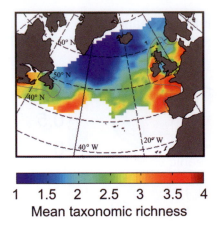

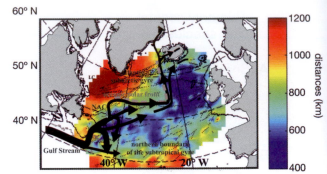

Mean taxonomic richness

| 1 | 1.5 | 2 | 2.5 | 3 | 3.5 | 4 |

Plate 4.15 Copepod biodiversity in the North Atlantic Ocean.

Source: From Beaugrand and colleagues (*288*).

Plate 4.16 Spatial variance in North Atlantic copepod biodiversity. The quantification of the spatial variance was performed by using Point Cumulative Semi-Variograms. In blue: small-scale spatial variance. In red: large-scale spatial variance. The path of the main currents is indicated.

Source: From Beaugrand and colleagues (*224*).

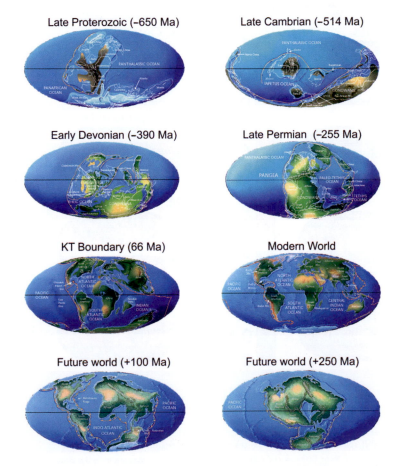

Plate 5.1 Continental drift and paleotopography of earth's land masses from the Precambrian to the present-day and future (Paleomap Project).

Source: Modified from Dr Christopher Scotese (www.scotese.com/).

A. Last Glacial Maximum

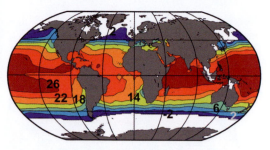

B. 2000–2008

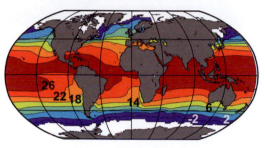

C. Difference

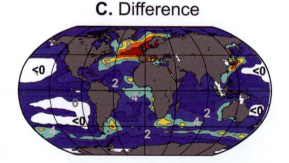

A. Global distribution of coral reefs

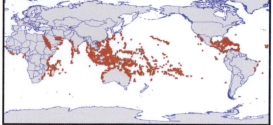

B. Global cyclone tracks for the period 1985–2005

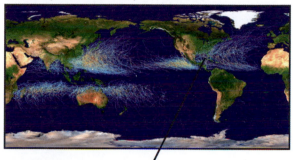

C. Hurricane Katrina

Plate 5.2 Difference in SST between the Last Glacial Maximum (LGM) and the period 2000–2008. (A) Mean annual SST for LGM. Data from Paul & Schäfer-Neth (*355*). (B) Mean annual SST for the period 2000–2008. Sea surface temperature (SST) data was obtained from the Extended Reconstructed Sea Surface Temperature (ERSST_V3; version 3) data set. ERSST_V3 data were provided by the NOAA/OAR/ESRL PSD, Boulder, Colorado, USA (*356*). (C) Difference between the thermal regime of the period 2000–2008 and the thermal regime of LGM.

Plate 5.3 (A) Global distribution of coral reefs. The map resulted from an examination of 7,000 Landsat images for evidence of reefs; it does not include deep-water coral. (B) Global tropical cyclone tracks for the period 1985–2005. As explained in Chapter 2, there are no cyclones near the equator. Over the south-western part of the South Atlantic, the track of cyclone Catarina is visible. The cyclone struck the state of Santa Catarina the night of 27 March 2004. Catarina was the first cyclone known to reach the coasts of Brazil. (C) Hurricane Katrina (August 2005).

Source: Courtesy of the National Aeronautics and Space Administration (NASA).

A. The Vaipahu site before the arrival of the cyclone

B. The Vaipahu site after Cyclone Oli

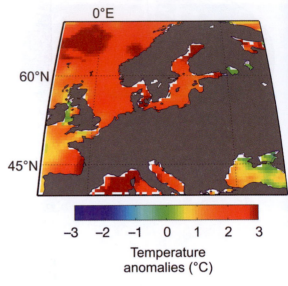

Plate 5.4 The site Vaipahu (North Moorea; 6 m depth) before (A) and after (B) Cyclone Oli arrived.

Source: Courtesy of Mohsen Kamal, University of Perpignan.

Plate 5.5 Map of temperature anomalies in August 2003 relative to the climatological base period 1979–2002.

Source: SST data are from ERA INTERIM.

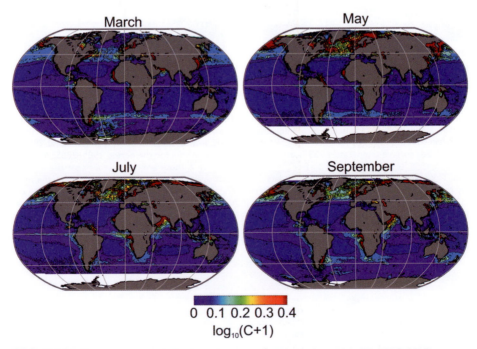

Plate 5.6 Seasonal changes in chlorophyll concentration ($mg \cdot m^{-3}$) inferred from SeaWIFS (1998–2005).

Plate 5.7 Satellite image of a bloom of *Emiliania huxleyi* in the English Channel (24 July 1999).

Source: Landsat false colour image for 24 July 1999 provided by the NERC Earth Observation Data Acquisition and Analysis Service, Plymouth, UK.

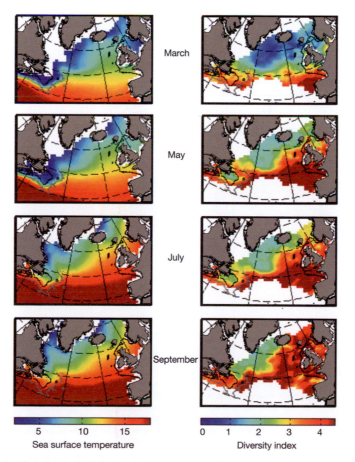

Plate 5.8 Relationships between epipelagic copepod biodiversity and sea surface temperature at a seasonal scale (March, May, July, September) in the extratropical North Atlantic Ocean. The monthly climatology was based on the period 1958–1997. The diversity index used is the mean number of species per sample.

Source: Biological data come from the Continuous Plankton Reorder survey. Temperature data are from the database International Comprehensive Ocean-Atmosphere Data Set (ICOADS). Redrawn from Beaugrand and colleagues (*182*).

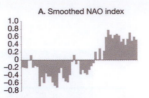

A. Smoothed NAO index

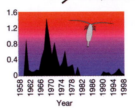

Plate 5.9 Impact of the North Atlantic
Oscillation (NAO) on SST changes and both
warm-water and cold-water species
(1958–1999). (A) Smoothed NAO (weighted
moving average). (B) Correlation between
NAO and SST (untransformed data). (C)
Proportion of warm-water species on colder-
water species in the eastern and western part
of the North Atlantic.

Source: Modified from Beaugrand (461).

B. Correlations between SST and the NAO index

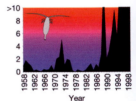

C. Proportion of warm-water species on colder-water species

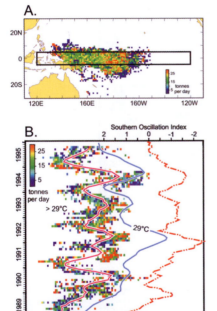

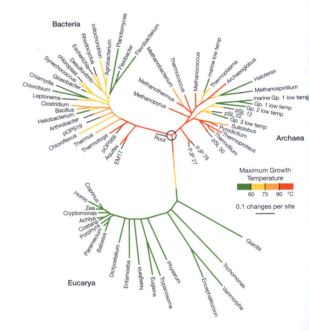

Plate 5.10 Impact of ENSO on skipjack tuna mean location in the
Equatorial Pacific. (A) Relative abundance (catch per unit of
effort; CPUE) of the skipjack tuna. The black rectangle indicates
the region where the mean longitudinal position of the skipjack
tuna was calculated (see panel B). (B) The ENSO index (broken
red line), the isotherm 29°C as an indicator of the convergence
zone (blue line), the mean CPUE of skipjack tuna (coloured
rectangle) and the longitudinal mean position of skipjack tuna
relative abundance (purple line). All indicators were smoothed by
a simple five-month moving average.

Source: Modified from Lehodey and colleagues (473).

Plate 6.1 Phylogenetic tree based on 16 S ribosomal RNA
sequences. The LCA is indicated by the black circle. Maximal
growth temperatures were determined to colour the branch
of the tree. The length of the bar denotes a 0.1 change per
nucleotide.

Source: From Lineweaver and Schwartzman (323). Courtesy of
Dr Lineweaver.

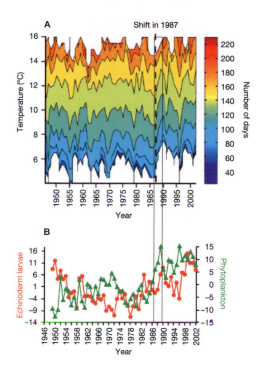

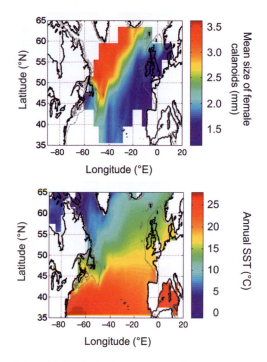

Plate 6.2 Changes in the number of days needed to reach a given temperature in the North Sea in relation to annual phytoplankton concentration and annual abundance in echinoderm larvae.

Source: Modified from Kirby and co-workers (533).

Plate 6.4 Spatial distribution in mean copepod community size (mm; 109 species or genera) in relation to annual SST. (A) Mean community size. (B) Spatial distribution in annual SST.

Source: Modified from Beaugrand and colleagues (207).

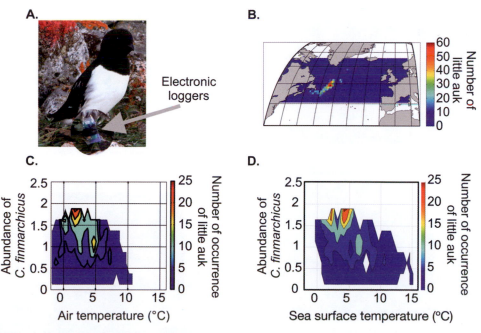

Plate 6.3 Influence of temperature on the winter spatial distribution of little auks in the North Atlantic Ocean. (A) Photo of a little auk. Courtesy of Dr Jérôme Fort. (B) Density of little auk in December. (C) Little auk abundance as a function of air temperature and *C. finmarchicus* abundance. (D) Little auk abundance as a function of sea temperature and *C. finmarchicus* abundance. In winter, *C. glacialis* and *C. hyperboreus* are not generally present in surface waters.

Source: Redrawn from Fort and co-workers (543).

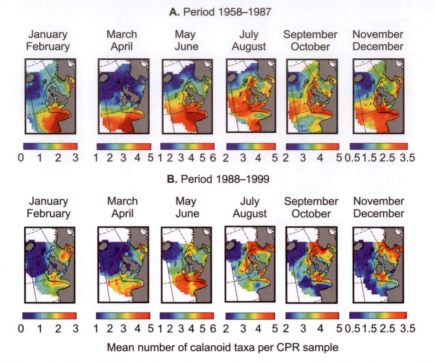

A. Period 1958–1987

| January February | March April | May June | July August | September October | November December |

0 1 2 3 1 2 3 4 5 1 2 3 4 5 6 2 3 4 5 2 3 4 5 0.5 1.5 2.5 3.5

B. Period 1988–1999

| January February | March April | May June | July August | September October | November December |

0 1 2 3 1 2 3 4 5 1 2 3 4 5 6 2 3 4 5 2 3 4 5 0.5 1.5 2.5 3.5

Mean number of calanoid taxa per CPR sample

Plate 6.5 Changes in calanoid biodiversity for each two-month period between the cold and warm period. (A) Cold period (1958–1987). (B) Warm period (1988–1999). Only night samples were used.

Source: Modified from Beaugrand (*563*).

A. Middle Island in the Keppel Island Group

Plate 7.2 Picture of the Mediterranean coral species *Oculina patagonica* showing bleached and healthy regions. Bleaching is caused by *Vibrio shiloi*. An increase in SSTs augments pathogen virulence.

Source: From Rosenberg and Ben-Haim (*605*). Courtesy of Pr Rosenberg, Tel Aviv University.

B. North Keppel Island

Plate 7.1 Coral bleaching of the staghorn coral *Acropora formosa* in the Great Barrier Reef (February 2006). (A) Middle Island in the Keppel Island Group (23 February 2006). (B) North Keppel Island (26 February 2006).

Source: Courtesy of Dr Berkelmans, Australian Institute of Marine Science.

Plate 7.3 The White Band Disease on *Acropora* in the Red Sea.

Source: Courtesy of Pr Eugene Rosenberg, Tel Aviv University.

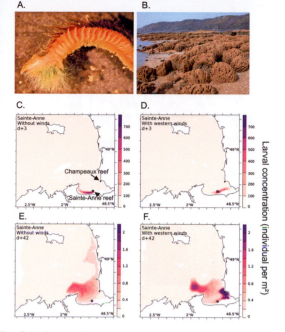

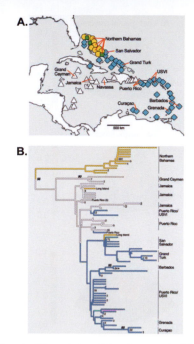

Plate 7.4 Influence of meteorological conditions on the dispersal of the honeycomb worm *Sabellaria alveolata* in the Bay of Mont-Saint-Michel (English Channel). (A) Photo of an adult honeycomb. (B) Biogenic reefs of *Sabellaria alveolata*. Spatial distribution of larvae at day 3 (C and D) and 42 (E and F) without (C and E) and with (D and F) western wind forcing.

Source: From Ayata and co-workers (635). Courtesy of Dr Ayata.

Plate 7.5 Genetic differentiation of the cleaner goby *Elacatinus evelynae* in the Bahamas and the Caribbean Seas. (A) Spatial distribution of the different colour forms. White triangle: white form. Blue diamonds: blue form. Yellow circle: yellow form. Green square: regions where blue and yellow forms co-occur. (B). Neighbour-joining tree of 79 haplotypes (mitochondrial cytochrome b) sampled from 246 individuals. The tree indicates genetic differentiation among populations and colour forms. The colour on the branches indicates the different colour form of the individuals.

Source: Simplified from Taylor and Hellberg (639).

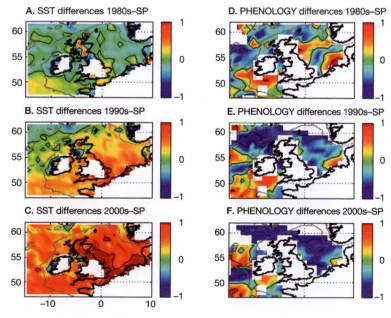

Plate 7.6 Changes in the phenology of *Calanus helgolandicus* (right panels; numbers are months) in relation to SSTs (left panels) in the North-East Atlantic. SP: period 1960–1979. 1980s: period 1980–1989. 1990s: period 1990–1999. 2000s: period 2000–2005. For temperature, differences are in °C. For phenology, differences are in month.

Source: From Beaugrand (60).

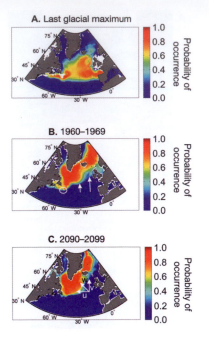

A. Last glacial maximum

B. 1960–1969

C. 2090–2099

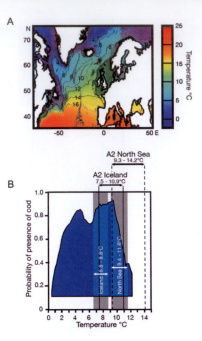

A

B

Plate 7.7 Modelled probability of occurrence of *Calanus finmarchicus* for the LGM (A), the 1960s (B) and the end of the 2090s (C). The model NPPEN (Non-Parametric Probabilistic Ecological Niche), based on both SST and bathymetry, was applied. Future spatial distribution was modelled using a pessimistic scenario (pre-industrial $CO_2 \times 4$). The white circle indicates an absence of northward movement between the LGM and the 1960s and white arrows indicate the magnitude of the northward movement. *u* and *v* is the zonal and meridional component, respectively. *w* is the vectorial sum of *u* and *v*.

Plate 7.8 Dependence of the sensitivity of the Atlantic cod (*Gadus morhua*) to the position along its thermal niche. (A) Spatial distribution of the Atlantic cod (shaded area) and mean annual sea surface temperature (SST). (B) Thermal niche based on mean annual sea surface temperature (1960–2005) and probability of cod occurrence. Both observed (1960–2005, shaded bars, white text and arrows) and projected (scenario A2, 1990–2100, black text and arrows) ranges in annual SST are indicated for Iceland (solid vertical lines) and the North Sea (dashed vertical lines).

Source: Modified from Beaugrand and Kirby (*545*).

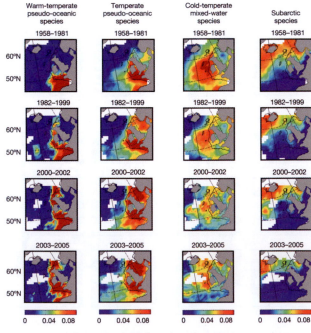

Plate 7.9 Long-term biogeographical shifts in copepods in the North Atlantic. Mean number of species per sample is small because sampling is based on ~3 m³ of seawater filtered and includes day/night surface sampling.

Source: Modified from Beaugrand and co-workers (*684*).

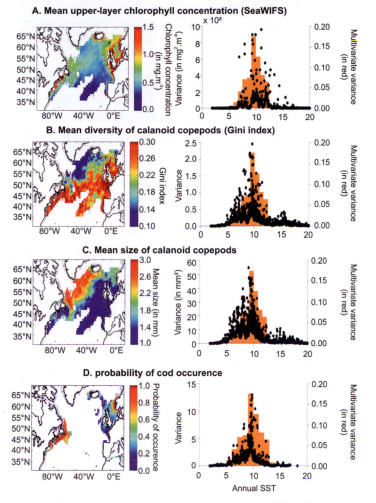

Plate 7.10 Annual mean distribution of upper-ocean chlorophyll concentration (A, left; 1997–2006), diversity (Gini index; 1958–2005) of calanoid copepods (B, left), mean size of female calanoid copepods (C, left; 1958–2005), mean probability of cod occurrence (D, left) and the local variance of these biological parameters as a function of sea surface temperature (right). Each point denotes a geographical pixel on the map. High values of local biological variance are mainly detected between 9°C and 12°C with a maximum between 9°C and 10°C, indicating substantial variability in these functional attributes in regions with a temperature regime of 9–10°C. Red bars show the (multivariate) local variance when all four indicators were combined. Grey lines denote the isobath 200 m.

Source: Modified from Beaugrand and co-workers (497).

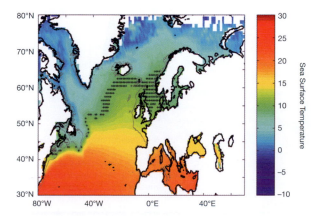

Plate 7.11 Location of the critical thermal boundary (9–10°C; indicated by '+') for the period 1960–1969 in the North Atlantic Ocean and SST.

Source: Modified from Beaugrand and co-workers (497).

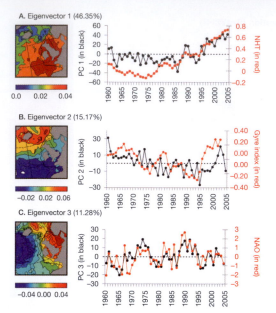

A. Eigenvector 1 (46.35%)

PC 1 (in black) / NHT (in red)

0.0   0.02   0.04

B. Eigenvector 2 (15.17%)

PC 2 (in black) / Gyre index (in red)

−0.02 0.02 0.06

C. Eigenvector 3 (11.28%)

PC 3 (in black) / NAO (in red)

−0.04 0.00 0.04

Plate 7.12 Long-term SST changes in the North-East Atlantic (1960–2005). (A) First eigenvector and principal component accounting for 46.35% of the total variance. NHT anomalies are superimposed in red. (B) Second eigenvector and principal component accounting for 15.17% of the total variance. The subarctic gyre index (98) is superimposed in red. (C) Third eigenvector and principal component accounting for 11.28% of the total variance. The NAO index is indicated in red.

Source: Modified from Beaugrand and co-workers (684).

Plate 9.1 Detection of microplastics in the North Atlantic Ocean and its adjacent seas in samples collected by the Continuous Plankton Recorder survey.

Source: From Edwards and colleagues (818).

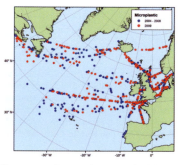

Plate 9.2 Satellite image of a bloom of filamentous nitrogen-fixing cyanobacteria *Cyanobacterium nodularia* in the Baltic Sea.

Source: SMHI, EOS 8211; MODIS 2005-07-11, NASA, processed by SMHI8217 oceanography unit.

A. Foraminifera (normalised eigenvectors and first principal component)

B. Coccolithophores

C. *Limacina* spp.

D. *Clione limacina*

E. Echinoderm larvae

F. Bivalve larvae

−0.8 −0.4 0 0.4 0.8
Correlations

Plate 8.1 Decadal changes (1960–2009) in North Atlantic calcifying plankton inferred from standardised principal component analysis. First normalised eigenvectors (left), showing the correlation with the first principal component (right) for (A) foraminifers, (B) coccolithophores, (C) *Limacina* spp., (D) *Clione limacina*, (E) echinoderm and (F) bivalve larvae. A pronounced shift is detected in the mid- to late 1990s.

Source: From Beaugrand and colleagues (710).

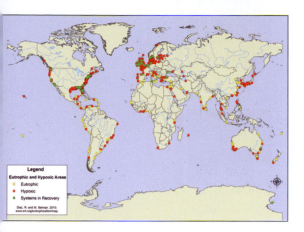

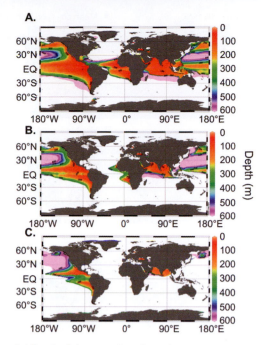

Plate 9.3 World hypoxic and eutrophic coastal areas.

Source: From Diaz and co-workers (*873*).

Plate 9.4 Depth of the upper boundary of oxygen minimum zones (OMZ). (A) Category A (PO2 = 106 matm). (B) Category B (PO2 = 60 matm). (C) Category A (PO2 = 22 matm).

Source: Modified, from Hofmann and colleagues (*881*).

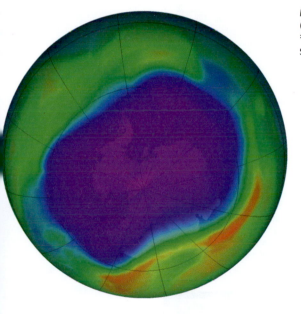

Plate 9.5 The ozone hole (in blue) over Antarctica in September 2006. The maximum extent in the ozone hole occurred on 24 September 2006.

Source: Courtesy of NASA Ozone Watch: http://ozonewatch.gsfc.nasa.gov/.

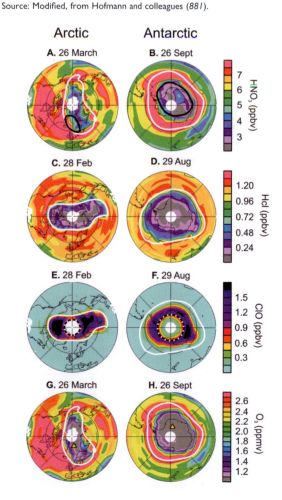

Plate 9.6 Chemical composition of the lower stratosphere over the Arctic Ocean and Antarctica (20 km). (A, B) HNO3 concentration. (C, D) HCl concentration. (E, F) ClO concentration. (G, H) O3 concentration. White and black lines should not be interpreted in this context.

Source: Modified from Manney and colleagues (*903*).

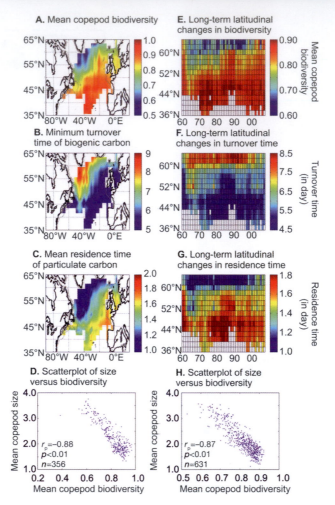

Plate 10.1 Relationships between biodiversity and two size-derived functional characteristics of copepods in the extratropical North Atlantic. Biodiversity was measured by the Gini coefficient. Mean spatial distributions (1960–2007) of: (A) copepod biodiversity, (B) minimum turnover time of carbon incorporated in copepods (days) and (C) mean residence time above 50 m of sinking copepod particles (days). (D) Relationship between mean copepod size and biodiversity based on mean spatial distributions (1960–2007). Long-term latitudinal changes in: (E) copepod biodiversity, (F) minimum turnover time of carbon incorporated in copepods and (G) mean residence time above 50 m of sinking copepod particles. (H) Relationship between long-term latitudinal changes in size and biodiversity. Linear correlation (r_p), probability (p) and degrees of freedom (n) are indicated in (D) and (H).

Source: Modified from Beaugrand and colleagues (207).

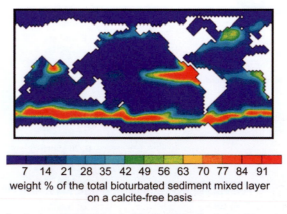

7 14 21 28 35 42 49 56 63 70 77 84 91
weight % of the total bioturbated sediment mixed layer
on a calcite-free basis

Plate 10.2 Opal sediment distribution modelled for the late-Holocene pre-industrial ocean (in weight % of the total bioturbated sediment mixed layer, on a calcite-free basis).

Source: From the Hambourg Ocean Carbon Cycle model. Courtesy of Dr Olivier Ragueneau.

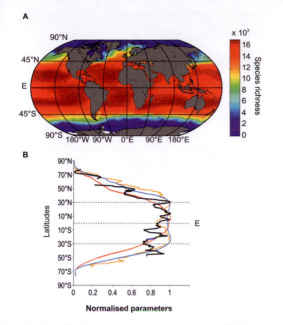

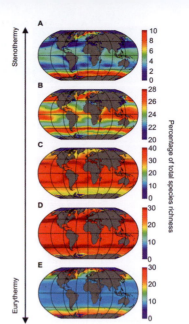

Plate 11.1 Modelled biodiversity patterns from the METAL theory and observed biodiversity patterns in foraminifers and copepods. (A) Modelled spatial diversity patterns from the METAL theory. (B) Relationships between latitudinal changes predicted by the theory (blue line) and observed biodiversity patterns for foraminifers (orange line) and calanoid copepods (black line). Normalised SST (red line) is superimposed. Horizontal dashed lines indicate the equator (E) and the latitude 30°N and 30°S.

Source: Modified from Beaugrand and co-workers (217).

Plate 11.2 Modelled spatial biodiversity patterns from highly stenothermic to highly eurythermic species. (A) Species with a thermal niche breadth ≤6°C. (B) Species with a thermal niche breadth between 7°C and 16°C. (C) Species with a thermal niche breadth between 17°C and 26°C. (D) Species with a thermal niche breadth between 27°C and 36°C. (E) Species with a thermal niche breadth >36°C. Biodiversity is here expressed as a percentage of the total species richness per geographical cell.

Source: Modified from Beaugrand and co-workers (217).

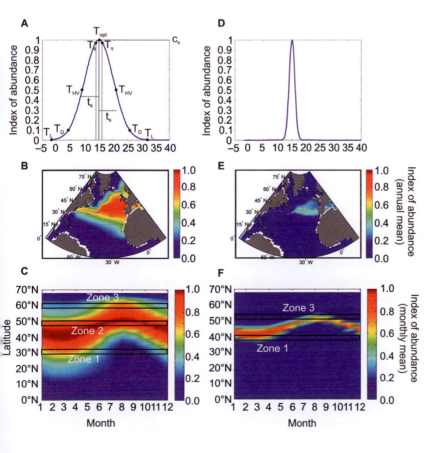

Plate 11.3 Theoretical relationships between species distribution, latitudinal range and phenology of two cold-water species: one eurytherm and one stenotherm. Theoretical thermal niche of a (A) eurytherm and (D) sternotherm. For meaning of Topt, Ts, THV, TD and TL, see Figure 11.8 and text. Theoretical mean annual spatial distribution of a hypothetical (B) eurythermic and (E) stenothermic species. Theoretical changes in abundance as a function of latitudes and months for a hypothetical (C) eurythermic and (F) stenothermic species. Zone 1: Part of the distribution where seasonal maximum occurs in spring/winter. Zone 2: Part of the distribution where seasonal extent is highest. Zone 3: Part of the distribution where seasonal maximum is located at the end of summer.

Source: From Beaugrand and co-workers (1088).

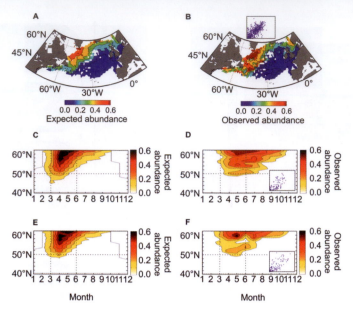

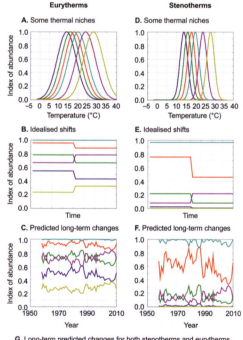

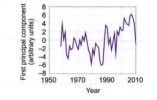

Plate 11.4 Relationships between spatial distribution, latitudinal ranges and phenological shifts of both observed and theoretical abundance of *Calanus finmarchicus*. (A) Expected and (B) observed spatial distribution of *C. finmarchicus* in the North Atlantic. Latitudinal and seasonal changes in both expected and observed abundance of *C. finmarchicus* based on the periods 1960–1979 (C and D, respectively) and 1990–2009 (E and F, respectively). Both expected and observed *C. finmarchicus* abundance were calculated for a meridional band between 30°W and 10°W (North-East Atlantic). Scaled between 0 and 1, scatter plots in B, D and F exhibit expected abundance versus observed abundance for A–B, C–D and E–F, respectively. Both vertical and horizontal dashed lines (C–F) are superimposed to better reveal phenological and biogeographical shifts.

Source: Modified from Beaugrand and co-workers (1088).

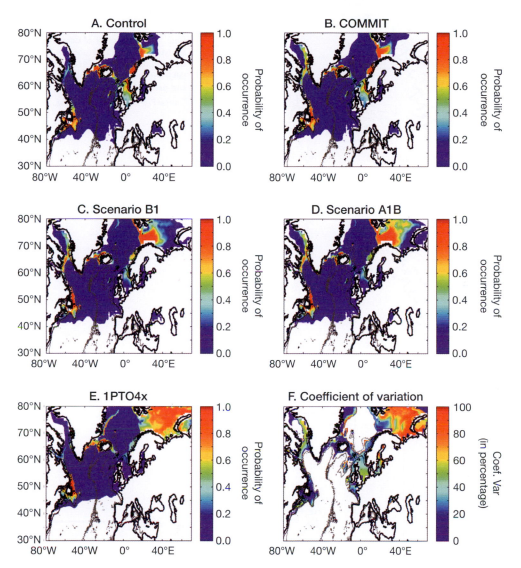

Plate 11.6 Projected long-term decadal changes in the probability of cod occurrence for the period 2090–2099. (A) Scenario 'Control'. (B) Scenario COMMIT. (C) SRES scenario B1. (D) SRES scenario A1B. (D) 1PTO4x. (D) Coefficient of variation (Coef. Var) based on all scenarios and general circulation models. Warming intensity increased from A to E.

Source: From Beaugrand and colleagues (*180*).

Plate 11.5 *(facing page)* Theoretical examples showing how the thermal niche of eurytherms and stenotherms may interact with climate-induced temperature changes to produce community shifts. (A) Thermal niche of six eurytherms (thermal optimum us = 14°C, 16°C, 18°C, 20°C, 24°C, 28°C and a thermal tolerance ts = 5°C). (B) Expected changes associated with an increase of 1°C. (C) Expected changes based on a hypothetical time series in SSTs (mean of 20°C and year-to-year variability corresponding to measured North Sea SSTs between 1958 and 2010). (D) Thermal niche of six stenotherms (us = 14°C, 16°C, 18°C, 20°C, 24°C, 28°C and ts = 2°C). (E) Expected changes associated with an increase of 1°C. (F) Expected changes based on a hypothetical time series in SSTs (mean of 20°C and year-to-year variability corresponding to measured North Sea SSTs between 1958 and 2010). (G) Theoretical long-term ecosystem changes as indicated by the first principal component after using a standardised PCA on the table years (1958–2010) × 12 pseudo-species (six eurytherms and six stenotherms).

Source: From Beaugrand and co-workers (*1088*).

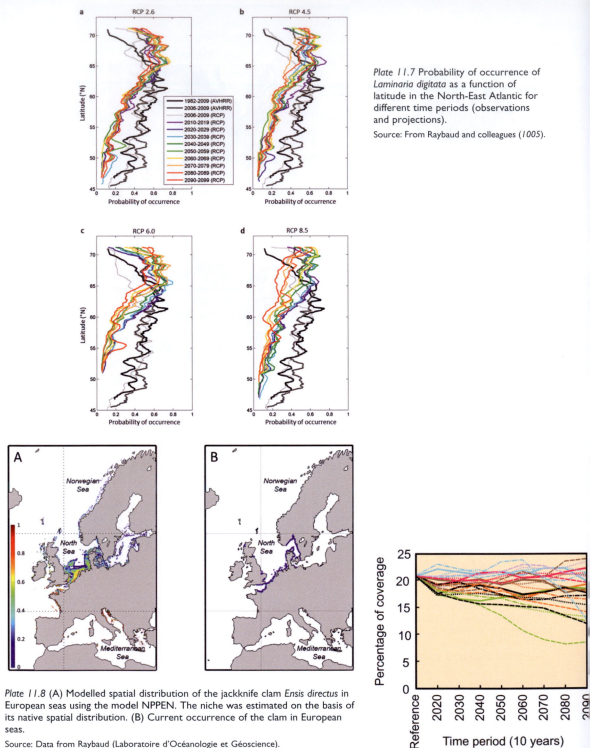

Plate 11.7 Probability of occurrence of *Laminaria digitata* as a function of latitude in the North-East Atlantic for different time periods (observations and projections).

Source: From Raybaud and colleagues (1005).

Plate 11.8 (A) Modelled spatial distribution of the jackknife clam *Ensis directus* in European seas using the model NPPEN. The niche was estimated on the basis of its native spatial distribution. (B) Current occurrence of the clam in European seas.

Source: Data from Raybaud (Laboratoire d'Océanologie et Géoscience).

Plate 11.9 Long-term projected changes in the spatial extent (as percentage of occurrence) of *Castanea sativa* for each decade of the twenty-first century for different warming intensities and using seven general circulation model outputs. The reference period is 1950–2000. Percentages of occurrence were calculated for all climate scenarios (the low RCP2.6, the medium-low RCP4.5, the medium-high RCP6.0 and the high RCP8.5) and seven GCMs: CNRM-CM5 (violet), CSIRO-Mk3.6.0 (orange), IPSL-CM5-LR (blue), HadGEM2-ES (green), MPI-ESM-LR (pink), GISS-ES-R (brown) and CCSM4 (black). The line style denotes RCP climate scenarios: full line: 2.6 W·m⁻²; dotted line: 4·5 W.m⁻²; dashed dot line: 6 W·m⁻²; dashed line: 8.5 W·m⁻².

Source: Simplified from Goberville and co-workers (1138).

Table 7.2 Comparisons of physical and biological processes or characteristics between the marine pelagic and the terrestrial realm.

Processes	Terrestrial realm	Pelagic realm (plankton)
Habitat		
Main climatic limiting factor	Temperature, precipitation, atmospheric circulation	Temperature, atmospheric circulation
Secondary factors	Sunlight, topography, geology, soil type, microclimate	Current, sunlight, stratification, ocean, bathymetry
Physical barriers	High	Smaller (oceanic currents and frontal structures such as the Oceanic Polar Front)
Predictability	Less predictable	More predictable
Interactions with the climate system		
Type of interaction	More complex (e.g. microrefugia)	More simple, although indirect effects of climate through oceanic circulation might make projections more difficult (no microrefugia)
Variation in climate	Higher short-term (daily or monthly) variance making parameters such as minimum or maximum temperature very important	Buffered domain; extremes in climate are less important
Synergistic effect		
Anthropogenic interaction	High to very high (habitat fragmentation, exploitation, human infrastructure)	Much smaller in general at a large scale (existence of more indirect relationships through fishing)
Capacity to track climate change		
Sensitivity to temperature change	Lower (eurytherm)	Higher (stenotherm)
Reproductive characteristics		
Larger size of offspring	Lower	Higher
Low age at maturity	Older	Younger
r/K selection	Stronger interaction between r- and K-selected species	Lesser interaction between r- and K-selected species
Reproductive rate (advantage to high reproductive rates)	Slower (e.g. months to several decades for plants; weeks to several months for insects)	Quicker (e.g. phytoplankton, days; zooplankton, weeks to several months)
Dispersal (advantage to efficient dispersers)	Low/medium (often constrained by the interspecific relationships with plants)	Higher as biotic interactions are smaller; domain more constrained by the physical environment
Interspecific interactions (advantage to community with smaller biotic interactions)	Greater; many species of insects are specific to a type of vegetal community; co-shifts	Smaller; species shift

Source: Modified from Beaugrand and co-workers (684).

and ecosystems are more influenced by climate over transitional systems (*691*). Fourth, the increase in the European shelf-edge current may have triggered the northward advection of plankton polewards (*692*). Finally, the apparent shift of some species groups might be considered as another example of Reid's paradox. Reid's paradox was termed after the difficulties of botanist Clement Reid to understand the relatively rapid spread of oaks across the United Kingdom during the warming of the post-glacial period (*689*). While several hypotheses have been proposed to explain this anomaly, a theory stipulates that colonisation may propagate from local population, persisting in a microclimate (microhabitats) (*693*). The relatively low density could explain why palynological studies were unable to detect pollen occurrence. Transposing this hypothesis to the marine realm, migration rates could appear very high, as colonisation may have propagated from several local points already above the observed limited boundary but not detected because of the coarse sampling of the pelagic realm.

7.4.2 Changes in the ecosystem trophodynamics

Different hypotheses can be proposed to explain how climate may influence a community composed of several trophic levels. In a first hypothesis, we assume an absence of trophic interactions between investigated species. In such a case, climate independently influences

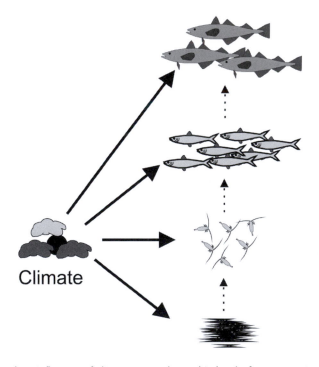

Figure 7.21 Independent influence of climate on each trophic level of a community. In this hypothesis, the propagation of the climatic signal throughout the food web is not detected. Black arrows: direct climatic effect. Dashed arrows: no detectable climatic effect. Four trophic levels are represented: phytoplankton, zooplankton, planktivorous and piscivorous fish.

each trophic level and there is no climatic propagation throughout the food web (Figure 7.21). This hypothesis is rarely tested by researchers, although statistical techniques (e.g. path analysis, causal modelling) can be applied to determine its likelihood (694). In Figure 7.21, the strength of the arrows is the same but it is likely that climate has species and time/space-dependent effects, especially if the climatic signal interacts with the species climatic niche (Plate 7.8; Chapter 11).

Because species are interrelated with each other, it is unlikely that species responses to climate change do not ramify at the community level. The second hypothesis is an impact of the climatic signal at the bottom (phytoplankton), top (predator) or an intermediate trophic level that subsequently propagates throughout the food web (Figure 7.22). Such responses have been termed trophic cascades (695). The prerequisite is the presence of trophic interactions. Bottom-up control is by far the most documented ecosystem response (696), although it is difficult to distinguish statistically this climatic effect from a common response of all trophic levels to climate (Figure 7.21). In bottom-up climatic propagation, all correlations with climate should have the same sign (495). In a top-down propagation, the sign of correlations should alternate from one trophic level to another. In a wasp-waist propagation, the sign of correlations should remain the same towards higher trophic levels and alternate towards lower trophic levels.

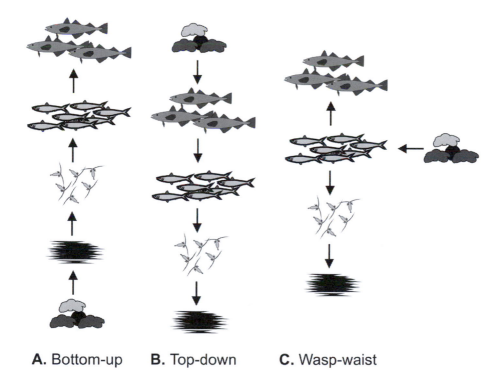

A. Bottom-up   **B. Top-down**   **C. Wasp-waist**

Figure 7.22 Climate-induced trophic cascades. (A) Bottom-up climatic propagation. (B) Top-down climatic propagation. (C) Wasp-waist climatic propagation. Four trophic levels are represented as above.

A. Trophic amplication

B. Independent climatic effect and trophic amplification

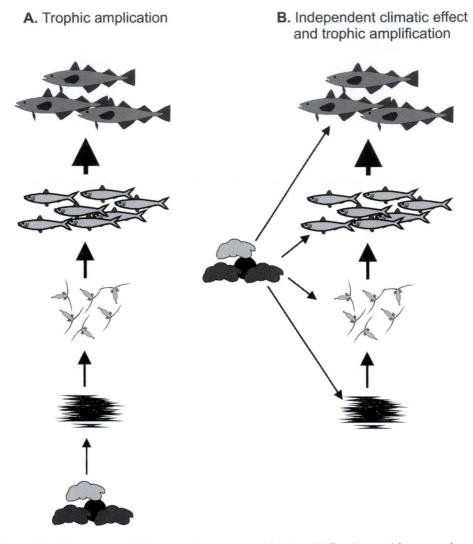

Figure 7.23 Climate-induced (bottom-up) trophic amplification. (A) Trophic amplification and no detectable independent influence on each trophic level. (B) Trophic amplification and detectable independent climatic influence on each trophic level. The size of the arrows is proportional to the climatic influence. Four trophic levels are represented as above.

The third response type is trophic amplification (559). Trophic amplification is the amplification of the climatic signal throughout the food web. Only bottom-up control trophic amplification has been documented. In this response type, the correlation should have the same sign but should become stronger towards higher trophic levels (Figure 7.23). Trophic amplification can be observed without (Figure 7.23A) or with (Figure 7.23B) an independent climatic influence on each trophic level.

7.4.3 Trophic mismatch and changes in species interaction

Because species reactions to climate change are in large part determined by their ecological niche (Plate 7.8) and that each species has a unique niche according to the principle of competitive exclusion (489), species are likely to exhibit distinct responses. Such responses are expected to trigger community reassembly in time or space, which are thought to be the most worrisome consequences of climate change on ecosystems because they may unbalance ecosystem trophodynamics, having the potential to involve trophic mismatch or to perturb prey-predator relationships (659).

The match/mismatch hypothesis originally proposed that interannual variability in fish recruitment was a function of the timing of the production of their food (679, 697). This hypothesis has been subsequently extended to a large number of ecological situations (predator–prey interactions) by Durant and colleagues (698, 699). We reviewed mechanisms in detail for a marine (cod/zooplankton) system, a marine–terrestrial (puffin/herring) system and a terrestrial (sheep/vegetation) ecosystem (698).

Here, I illustrate potential perturbations of the predator-prey interactions by climate change, using fictive examples. The first example is from a system composed of a predator and a prey (Figure 7.24). In this system, the predator and the prey initially have the same phenology (Figure 7.24A), and we assume that the predator has a fixed phenology (i.e. adaptation or species niche tracking is not possible). Warming triggers a phenological shift towards an earlier prey abundance (Figure 7.24B), which leads to a mismatch between the predator and its prey, diminishing predator abundance. Subsequent warming may reduce prey abundance because the phenological shift of the prey may not be possible due to the existence of another limiting factor (photoperiod). This reduction in prey abundance leads to a further decrease in predator abundance.

The second example is more complex and is characterised by a predator and two preys (Figure 7.25). In the first situation, there is a match between the predator and prey 1 (Figure 7.25A). Subsequent climate change reduces prey 1 abundance but increases prey 2 abundance. The result is a mismatch between the predator and prey 1. The increase in prey 2 cannot compensate for the diminution of prey 1 because of the phenological lag between

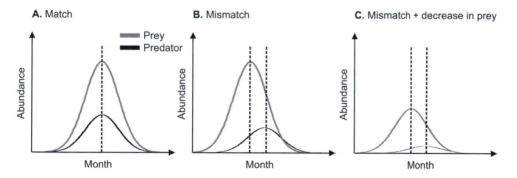

Figure 7.24 Effects of climate change on a hypothetical system with prey-predator interaction. (A) Match between the predator and the prey. (B) Climate-induced phenological shift (earlier prey phenology) and mismatch, diminishing predator abundance. (C) Climate-induced phenological shift and reduction in prey abundance, implicating a further decrease in predator abundance. These effects are explained in detail by the METAL theory (Chapter 11).

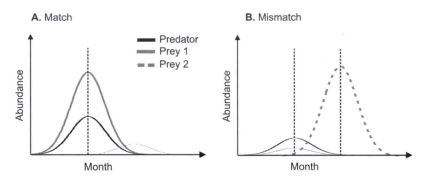

Figure 7.25 Effects of climate change on a hypothetical system with one predator and two preys. (A) Match between the predator and prey 1. (B) Climate-induced decrease in prey-1 abundance and increase in prey 2 abundance, which diminishes predator abundance. Note that prey phenological shifts are not possible here.

prey 2 and the predator. Such a situation may have been observed in the North Sea for the Atlantic cod (*Gadus morhua*) (676). Prior to the North Sea warming, *C. finmarchicus* was highly abundant (Figure 7.17) and its seasonal maximum coincided with larval cod occurrence in spring. After the warming, *C. finmarchicus* decreased substantially and was replaced by the warm-water pseudo-oceanic species *C. helgolandicus*, which had a seasonal maximum in late summer at a time when larval cods start to have a demersal life. As always, the situation was more complicated (700) and other adverse changes were also observed (e.g. decrease in prey biomass and euphausiids) (676, 694).

The third example makes the emphasis on the prey (Figure 7.26). When the predator and the prey have the same seasonal timing, the predator controls its prey abundance (Figure 7.26A). Climate-induced phenological shift in the predator towards an earlier seasonal maximum relaxes pressure on the prey, which increases its abundance (Figure 7.26B). The abundance of the predator may collapse if no alternative preys are available. We showed that an increase in asynchrony between predator and prey peak abundance can lead to increased survival and potentially increased recruitment of the prey in some systems (699).

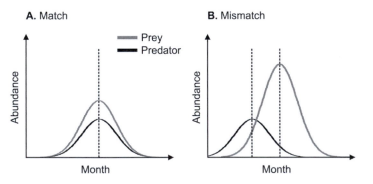

Figure 7.26 Effects of climate change on a hypothetical system with one predator and one prey. (A) Match between the predator and the prey; the predator controls the prey abundance. (B) Climate-induced predator phenological shift relaxing pressure on the prey. These effects are explained in detail by the METAL theory (Chapter 11).

All types of species interactions (Chapter 4) are likely to be altered in the context of climate change. Kirby and Beaugrand showed that trophic interactions were not constant in the North Sea during the period 1958–2005 (559). Some trophic interactions strongly changed after the late 1980s pronounced warming. Trophic interaction impermanence occurs because of coupling/decoupling processes related to mechanisms such as those documented above, making it more complex community change anticipation.

7.4.4 Coupling/decoupling between systems

7.4.4.1 Benthic–pelagic coupling

Global warming may reduce mean size of animals inhabiting pelagic ecosystems as a consequence of the temperature–size rule and Bergmann's pattern (207). Community body size largely determines the types and strengths of flows of energy and materials in ecosystems, altering ecological networks and ecosystem structure and functioning (560, 561). Biomass, production and predator–prey interactions are among quantities or processes that respond to ecosystem size structure (562). Changes in size may therefore have strong consequences for ecosystem functioning and carbon export, as we will see in Chapter 10. They may also have strong effects on benthic–pelagic coupling. In Chapter 5, we saw a clear example of benthic–pelagic coupling that may be unbalanced by global warming. Benthic–pelagic coupling in high-latitude ecosystems is highly dependent on the timing of sea-ice melting (Figure 5.30). Earlier sea-ice melting may benefit to smaller pelagic species and reduce substantially (endosomatic) energy fluxes sinking to the bottom, which alters species dominance.

Richard Kirby and colleagues (540) suggested that recent modifications in the balance between North Sea merozooplankton (increase) and holozooplankton (decrease) reflect a shift in energy partitioning between the benthos and the pelagos. Both decapod and echinoderm larvae increased during the period 1958–2005, and especially after the 1980s North Sea warming. As a result, the burrowing psammivorous spatangoid *Echinocardium cordatum* (larvae and adult) augmented. Complex mechanisms are at work but warming increases gametogenesis and egg production (533). *Echinocardium cordatum* larvae might have also benefited from phytoplankton increase (653). The increase in decapods was also strongly related positively to temperature change (Figure 6.21). As decapods are predator for juvenile cod and bivalve, these changes may have also contributed to maintain low-level bivalves and cod in the North Sea. Decapod/cod dominance oscillations have also been observed in the large eastern Scotian Shelf ecosystem off Nova Scotia (701). Decapods may also control holozooplankton species and changes in their abundance may alter benthic–pelagic coupling.

7.4.4.2 Land–sea coupling

Increase in North Sea decapods was also associated with an outburst in swimming crabs (*Necora puber*, *Liocarcinus depurator* and *Polybius henslowi*) (702). As swimming crabs are a significant food source for lesser black-backed gulls during the breeding season, we suggested that this increase might be at the origin of an increase in some North Sea colonies (702). Inhabiting the land, but feeding mainly at sea, these gulls provide a link between marine and terrestrial ecosystems, since the bottom-up influence of allochthonous nutrient input from seabirds to coastal soils can structure terrestrial trophodynamics. In this way, climate-caused marine ecosystem changes may have some consequences for terrestrial ecosystems.

7.4.5 Abrupt ecosystem shifts

The North Sea is one of the most biologically productive ecosystems in the world. This system supports important fisheries leading to the catch of 5% of the world's total fish and also contributes significantly to biogeochemical cycles (703). Long-term community changes have been observed in the North Sea and frequently associated with changes in SSTs (Figure 7.27; see also Figure 6.26 and section 5.3.6). Rates of changes concentrated around two periods: the mid- to late 1980s and 1990s (704). Such abrupt ecosystem shifts (AESs), detected in both pelagic and benthic realms (705–707), involved an increase in phytoplankton biomass, changes in biodiversity and organism phenology (499, 653, 708). Concomitant modifications occurred in both global and regional temperatures, changes in three trophic levels (phytoplankton, zooplankton and the Atlantic cod) and in both holozooplanktonic (*Calanus finmarchicus*) and merozooplanktonic (decapods) components (Figure 7.28). Copepod biodiversity increased and was paralleled by an increase in the number of warm-water species associated with a decrease in the number of colder-water species (Figure 7.29). All assemblages showed coherent patterns of changes with NHT anomalies.

While original hypotheses invoked the NAO and its potential effects on oceanic circulation (708, 709), subsequent studies provided evidence that temperature was the primary driver of the shift (60). We saw in Chapter 5 that climatic variability can be at the origin of AESs. AESs have become more regularly identified, and it is likely that anthropogenic climate change will increase both the magnitude and the frequency of such phenomena (59, 483, 695, 710).

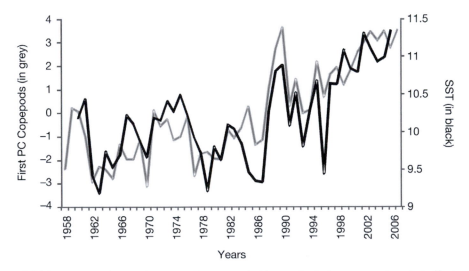

Figure 7.27 Long-term changes in copepods as revealed by a principal component analysis (first principal component) and changes in SSTs.

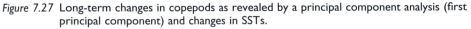

Source: Modified from Beaugrand and colleagues (704).

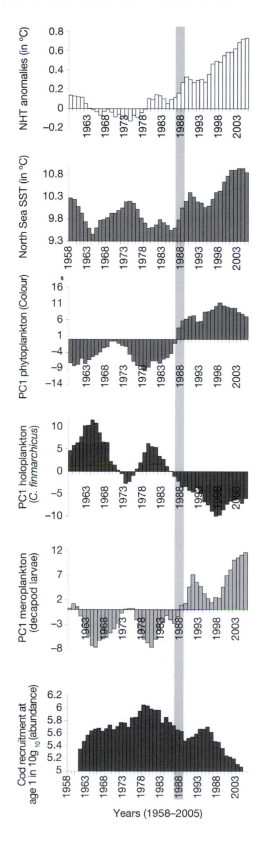

Figure 7.28
Long-term changes in northern hemisphere temperature (NHT) anomalies, North Sea surface temperature, phytoplankton concentration (PC1: first principal component), *C. finmarchicus*, decapod larvae and cod recruitment at age 1.

Source: From Beaugrand (*60*).

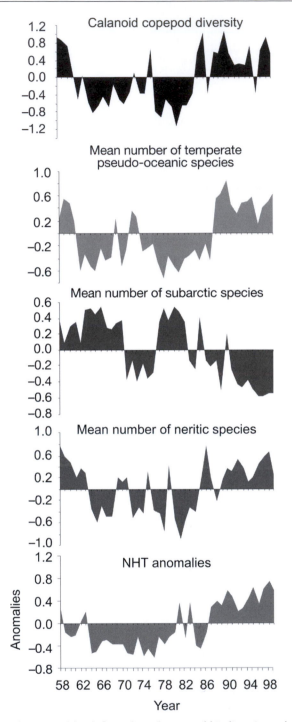

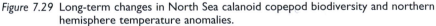

Figure 7.29 Long-term changes in North Sea calanoid copepod biodiversity and northern hemisphere temperature anomalies.

Source: From Beaugrand (*60*).

Beaugrand and co-workers (497) showed that North Sea isotherm movements northwards triggered a poleward shift of the boundary between the temperate and the Arctic ecomes/biomes. Using a statistical technique based on the multiscale (multivariate) variance, the authors first showed that the average spatial distribution of upper-ocean chlorophyll concentration (as measured by SeaWIFS), both diversity and mean size of calanoid copepods (as measured by the Continuous Plankton Recorder survey) and the Atlantic cod (modelled occurrence) exhibited a pronounced non-linearity in their responses to temperature change between 9°C and 10°C (Plate 7.10).

All these key functional attributes showed a maximum in local biological variance at temperatures (i.e. annual mean of sea surface temperature) between 9°C and 10°C. This result suggests that a temperature rise for a system with that mean annual temperature regime will experience pronounced ecosystem effects while an identical increase in another temperature regime (e.g. 12–13°C) will have a more moderate impact. Therefore, all systems are not equally sensitive to temperature rise and 'variance hot spots' (i.e. regional discontinuities characterized by high biological variance) exist. These results also indicate that AESs are likely to be concentrated around critical thermal boundaries. The critical thermal boundary of 9–10°C coincided with the transitional region between the Atlantic Polar and the Atlantic Westerly Winds biome/ecome (Plate 7.11). Our analysis revealed that ecome/biome boundaries are highly sensitive to climate change and that climate-induced modifications in their geographical locations may be at the origin of pronounced ecosystem shifts. Other regions spatially embedded deeply within a major ecome can remain relatively ecologically stable over long periods.

As we saw in Chapter 3, a major ecological compartment (ecome) is by definition in equilibrium with a climatic regime. Therefore, climate change is expected to unbalance these mega-ecosystems (10). These systems are biologically distinguished by their diversity, organism size and carrying capacity (10, 143). In the case of cod, we suggested that the transitional waters between the polar and temperate biome may rapidly change through climate warming, exacerbating other anthropogenic pressures, and having a dramatic impact on cod populations. The methodology proposed in this study, and the conclusion that transitional regions between major biomes are particularly vulnerable to abrupt climate change, might well be applicable to other oceanic ecomes and possibly even in the terrestrial realm. However, exact mechanisms by which biogeographical movements implicates AESs were not proposed. Trophic cascades and amplifications can also be at the origin of AESs (see above). In Chapter 11, we will see that the METAL theory allows a clear explanation of bioclimatic mechanisms at the origin of climate-caused AESs.

7.4.6 Anthropogenic climate change and tipping points

Stochastic resonance is an important mechanism by which anthropogenic climate change may interact with climatic variability to trigger AESs or acute species mortality at the species level. I illustrate this point with an example (Figure 7.30). In this fictive example, neither the stochastic (year-to-year) variability nor the long-term trend can reach a bifurcation point (also called **tipping point**) (Figure 7.30A–B). However, when both signals are combined, they can sometimes reach the critical threshold and trigger a major phase transition (Figure 7.30C). It is extremely difficult to forecast the time at which a tipping point is crossed. The best we can perhaps do is to evaluate the probability of reaching this

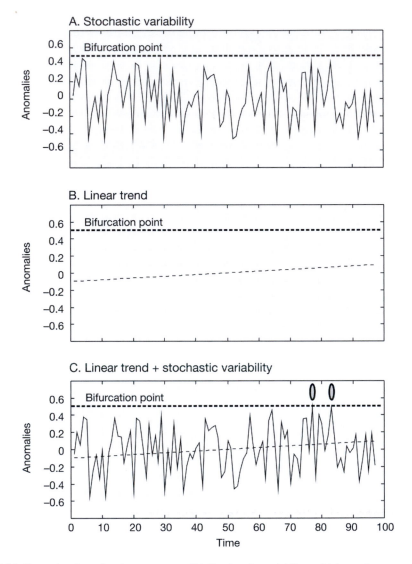

Figure 7.30 Example of stochastic resonance. (A) Stochastic variability, which can be compared to inherent climatic variability. Alone, this source of variability cannot reach a bifurcation point (or critical threshold). (B) Long-term (low-frequency) linear trend. (C) Stochastic variability and trend. Grey ovals indicate that the bifurcation point is reached. The dashed line indicates the hypothetical bifurcation point.

point, as the bifurcation point itself is rarely known with accuracy. Once the tipping point has been crossed, the system can go in quite surprising directions (see Younger Dryas; Section 5.1.4.4) and changes are difficult to reverse.

7.5 Anthropogenic climate change and natural hydro-climatic variability

7.5.1 Comparison of anthropogenic climate change with natural source of climatic variability

Although temperature increase between the periods 1850–1899 and 2001–2005 was 0.76°C, natural forcing can mitigate or reinforce the current global warming. For example, Lean and Rind (108) showed that a major El Niño event can increase the global temperature by ~0.2°C. The 1997–1998 Super El Niño event increased global temperatures by 0.23 ± 0.01 from June to November 1998. In contrast, a major volcanic eruption can decrease global temperature by ~0.3°C. For example, the eruption of the Pinatubo decreased the global temperature by 0.25 ± 0.02°C from November 1991 to September 1992 mainly in regions located between 40°S and 70°N. Solar variability can involve a 0.1°C temperature increase. For example, solar forcing increased the temperature by 0.17 ± 0.01°C from the solar minimum of April 1996 to the solar maximum of February 2002. Lean and Rind also stressed that to best quantify both natural and anthropogenic forcing, it is important to consider them in the same analysis to take into account the effect of other factors when one is quantified. When all these forcings were considered, the empirical models of Lean and Rind explained 76% of the variance of monthly global surface temperature recorded during the period 1889–2006. However, ENSO, volcanic forcing and solar irradiance accounted for only 0.002°C: –0.001°C and 0.007°C per decade, respectively. It is relatively lower than anthropogenic warming, which represented 0.05°C per decade (108). In this analysis, the contribution of solar forcing equalled ~10% of the total warming, far less than the value of 69% reported by Scafetta and West (711). When only the period 1979–2005 was considered (total increase in temperature: 0.177 ± 0.052°C), the ENSO contribution was negative (–0.007 ± 0.005°C), the contribution of volcanic activity positive (0.018 ± 0.004°C) and solar activity negative (–0.004 ± 0.005°C). All natural sources of variability were smaller than the contribution brought by anthropogenic forcing (0.199 ± 0.005°C).

7.5.2 Anthropogenic climate change versus natural hydro-climatic variability in the eastern North Atlantic

In the north-eastern part of the North Atlantic, I recall that long-term changes in plankton and fish have been attributed to climatic variability (e.g. Atlantic Multidecadal Oscillation, North Atlantic Oscillation and the subarctic gyre index; Chapter 5) and anthropogenic climate change. A principal component analysis performed on annual sea surface temperatures (SSTs) showed that long-term changes in SSTs were mainly related to northern hemisphere temperature (NHT) anomalies (first eigenvector and principal component or PC; Plate 7.12A). Many ecosystem changes (e.g. North Sea, Baltic Sea, eastern North Atlantic and Bay of Biscay) have paralleled changes in NHT anomalies (Figures 7.28 and 7.29) (497, 712).

However, long-term changes in PC2 (Plate 7.12B) were related to the Subarctic Gyre Index (an index of the intensity of the gyre circulation in the North Atlantic) between Iceland and Norway. Long-term changes in phytoplankton, zooplankton, fish and pilot whales have been associated with the state of the subarctic gyre and changes in sea surface temperatures in this region (*140*). Long-term changes in PC3 were correlated to the NAO. Many ecosystem changes have also been associated with the effect of the NAO (*713*). Although the variability associated with NHT anomalies represented 46% of the total variance, the variability associated with the gyre index and the NAO represented 26% of the total variance (Plate 7.12).

Anthropogenic climate change is likely to affect organisms through hydro-meteorological and environmental channels. Global warming will also alter other ecological factors such as ocean stratification, nutrients, oxygen concentration, both the path and the intensity of oceanic circulation and upwellings. Mechanisms by which they may affect ecosystems and their biodiversity have also been reviewed in Chapters 2–6.

Chapter 8

Marine biodiversity and ocean acidification

8.1 Introduction

The oceans contain ~38,000 Gt (gigatonnes) of carbon, which is 50 times more than the atmosphere and 20 times more than the carbon stored by the terrestrial biosphere and soils (Figure 8.1).

It is not only a crucial reservoir, but also the most long-term carbon sink. Although atmospheric CO_2 is chemically neutral, it reacts with oceanic water so that aqueous CO_2 only represents 1% of the dissolved inorganic compounds (DIC) occurring in the ocean. The remaining DIC are bicarbonate (HCO_3^-; 91%) and carbonate (CO_3^{2-}; 8%). The relationships between these chemicals are given by these equilibrium equations:

$$CO_{2(atmos)} \longleftrightarrow CO_{2(aq)} \tag{8.1}$$

$$CO_2 + H_2O \rightarrow H_2CO_3 \rightarrow H^+ + HCO_3^- \rightarrow 2H^+ + CO_3^{2-} \tag{8.2}$$

$$CO_2 + H_2O + CO_3^{2-} \rightarrow HCO_3^- + H^+ + CO_3^{2-} \rightarrow 2HCO_3^- \tag{8.3}$$

$$Ca^{2+} + CO_3^{2-} \rightarrow CaCO_{3(s)} \tag{8.4}$$

As other gases, CO_2 follows **Henry's law**, which means that a rise in atmospheric CO_2 increases oceanic CO_2 concentration. The equilibrium between atmospheric and aqueous CO_2 occurs at a monthly scale (Reaction 8.1). Dissolved CO_2 reacts with water to form a carbonic acid, which is then transformed into a bicarbonate ion (Reaction 8.2). Another bicarbonate ion is then formed from a carbonate ion, aqueous CO_2 and water (Reaction 8.3). Therefore, two bicarbonate ions are formed, which increase hydrogen ion (proton) concentration and reduce carbonate ion concentration. The increase in atmospheric CO_2 therefore implicates a drop in pH and a reduction in the oceanic buffer capacity with respect to acidification (714). Carbonate reduction diminishes the saturation state of calcium carbonate ($CaCO_3$), which has a negative effect on shell formation or calcifying organism skeleton (Reaction 8.4). The saturation state Ω is calculated as follows (715):

$$\Omega = \frac{\left[Ca^{2+} \right]\left[CO_3^{2-} \right]}{\left[K_{sp} \right]} \tag{8.5}$$

With [Ca^{2+}] calcium concentration, [CO_3^{2-}] carbonate concentration and K_{sp} the solubility constant of each mineral phase. On short timescales, calcium concentration does not vary

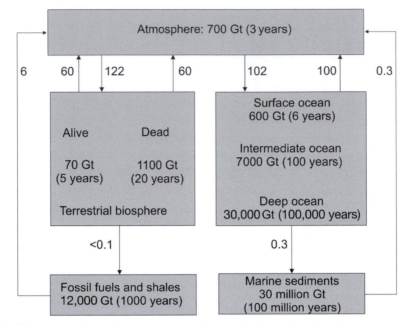

Figure 8.1 Schematic of the global carbon cycle showing both the size of carbon reservoirs (Gt) and fluxes between reservoirs (Gt per year). About 2 Gt of carbon is currently taken up by the marine ecosphere. Another 2 Gt is absorbed by the terrestrial biosphere. About 2 Gt of carbon is accumulating in the atmosphere every year. Note that the current estimation of carbon emission by the Global Carbon Project (www.globalcarbon project.org/) was close to 10 Gt carbon for 2013, which represents a substantial increase in comparison of the 6 Gt of carbon in the figure.

Source: Redrawn from Raven and colleagues (714).

substantially, and thereby this is carbonate concentration that influences the calcium carbonate saturation state. When $\Omega > 1$ (supersaturation), precipitation of calcium carbonate is thermodynamically favourable. When $\Omega < 1$ (undersaturation), it is unfavourable. This state is, however, dependent on species, and more particularly on the type of calcium carbonate they synthesise.

Three forms of calcium carbonate are used by marine species: calcite (e.g. coccolithophores, foraminifers), aragonite (e.g. pteropods, corals) and high-magnesium calcite (e.g. crustose coralline algae). Calcite is more resistant to dissolution than aragonite because of its structure (trigonal versus orthorhombic). In the ocean, the saturation state of calcium carbonate varies for aragonite between 2.41 ± 0.3 in the Arctic Ocean (north of 65°N) and 3.94 ± 0.2 in the North Indian Ocean and for calcite between 3.82 ± 0.4 and 5.93 ± 0.3 in the same oceanic basins (716). All areas are therefore supersaturated with respect to aragonite or calcite, which means that calcification is favoured more than dissolution. However, most species have optimum rate of calcification in the supersaturation range for calcite and aragonite. For example, coral reef occurs in oceanic areas where $\Omega > 3.3$ because calcification is greater than bioerosion (715). Calcium carbonate saturation state decreases with depth, this being influenced by respiration, water temperature and pressure. Below a **saturation horizon** (i.e. the depth below which water becomes undersaturated), the calcium carbonate starts to dissolve.

The biological activity alters oceanic pH. For example, it is well known that productive areas increase oceanic pH. This effect can be understood by the simplified equilibrium reaction of photosynthesis (Reaction 8.6) and respiration (Reaction 8.7):

$$6CO_2 + 6H_2O \ (+ \text{ light energy}) \rightarrow C_6H_{12}O_6 + 6O_2 \qquad (8.6)$$

$$C_6H_{12}O_6 + 6O_2 \rightarrow 6CO_2 + 6H_2O + \text{energy} \qquad (8.7)$$

When photosynthesis (Reaction 8.6) is greater than respiration (Reaction 8.7), dissolved CO_2 decreases and increases pH. The opposite occurs when respiration becomes dominant. During most of the Holocene, average oceanic pH was ~8.2, but this value has varied regionally and seasonally by ±0.3 units. PH fluctuates spatially because cold waters contain more CO_2 than warm waters. Cold waters therefore have lower pH and saturation state of calcium carbonate. This also varies in time because primary production tends to increase pH. Therefore, natural large-scale pH patterns are correlated to SST and are also modulated regionally by upwellings. Other factors influencing pH are salinity, river run-off and sea-ice melting.

8.2 Anthropogenic acidification

Atmospheric CO_2 concentration has changed between ~200 ppm and ~8,000 ppm for 550 million years. Periods of elevated atmospheric CO_2 concentration occurred between the early and middle Palaeozoic (543–400 million years), late Triassic and early Jurassic (200 million years) and Cretaceous (125 million years). After the Cretaceous, atmospheric CO_2 concentration is thought to have only been rarely greater than today's concentration, with some exceptions (e.g. Palaeocene–Eocene Thermal Maximum, or PETM; 55 million years ago). We definitively know that atmospheric CO_2 concentrations fluctuated between ~170 ppm and ~280 ppm in the past 800,000 years (326) (Figure 7.4).

Since 1750, atmospheric CO_2 concentration has risen from 280 ppm to ~400 ppm due to the combustion of fossil fuels, cement production, agriculture and deforestation. Both biogeochemical models and observations suggest that ~140 GtC of the 500 GtC emitted in the atmosphere by human activities since pre-industrial time has been absorbed by the ocean. About 2 GtC·year^{-1} has been taken up by the ocean, which represents 30% of anthropogenic carbon emissions and 50% if only combustion of fossil fuels and cement production are considered (717). Through fossil fuel use, 25 Gt of CO_2 (1 Gt C = 3.67 Gt CO_2) are emitted into the atmosphere each year (718). This is ~100 times greater than natural CO_2 emissions from volcanoes and other natural geologic sources.

Increase in CO_2 has been observed mainly in the first 1,000m of the oceans, but in some regions CO_2 penetration has been deeper (717). This has been the case in the North Atlantic where CO_2 has been observed down to 3,000 m due to profound convection. As a result, the aragonite saturation horizon has shoaled to 50–200m below the surface in all oceanic basins (162). Global pH has been reduced from 8.2 to 8.1 since industrialisation began. In logarithmic pH units, this 0.1 unit reduction seems negligible, but in absolute terms this translates into a 30% increase in surface-ocean acidity. This pH drop has been estimated to be 0.017–0.020 units per decade at ALOHA station (subtropical North Pacific Ocean), BATS station (North Atlantic Ocean near the Bermuda) and ESTOC station (north-eastern part of the North Atlantic subtropical gyre) from 1984 to 2007 (715). It is more important

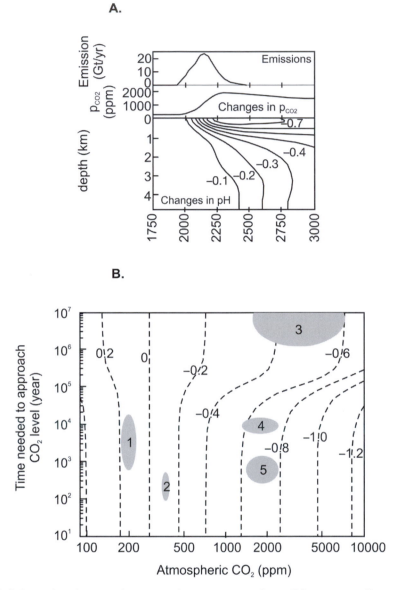

Figure 8.2 Relationships between long-term changes in atmospheric CO_2 emissions P_{CO2} and pH. (A) Long-term observed and projected changes in CO_2 emissions, P_{CO2} and surface ocean pH from 1750 to 3000. (B) Estimated maximum change in surface oceanic pH as a function of final atmospheric partial pressure P_{CO2} and the transition time needed for the system to reach these values from 280 ppm. 1: Glacial-Interglacial P_{CO2} changes. 2: Historical P_{CO2} changes. 3: Slow P_{CO2} changes over the past 300 Ma (million years). 4: Changes in P_{CO2} during PETM. 5: Changes in P_{CO2} related to unabated fossil fuel burning over the next few centuries. Note that P_{CO2} changes during PETM might have been smaller than those seen in the figure because CO_2 concentration was higher than 280 ppm at the beginning of the time period and because the system was less sensitive to carbon perturbations (722).

Source: Modified from Caldeira and Wickett (721) and Zeebe (722).

in higher latitudes such as Tatoosh Island (48.32°N, 124.74°W; 2000–2007) (*719*). As with global warming, this is the rate of change that is worrying. Zachos (*720*) said that the release rate was at least 10 times faster than at the PETM. The researcher stressed that 'whereas nature took a few thousand years to spout out thousands of gigatonnes of carbon, humans could be doing it in a few centuries' (*720*). At the PETM, global temperatures increased by 5°C in a period of less than 10,000 years. A subsequent massive CH_4 and CO_2 release, which may have originated from marine sediments, triggered a major perturbation in the carbon cycle that lasted 170,000 years (*715*). Contemporaneous decadal changes in pH may be 100 times faster than changes that took place during glacial terminations in the Pleistocene.

Caldeira and Wickett (*721*) investigated expected pH changes as a function of atmospheric CO_2 increase (Figure 8.2). They forced the Lawrence Livermore National Laboratory ocean general circulation model with atmospheric CO_2 pressure corresponding to the period 1975–2000 and 2000–2010 using a business-as-usual IPCC scenario (Scenario IS92a). They estimated an atmospheric CO_2 concentration of 1,900 ppm by 2300 and a corresponding maximum pH reduction of 0.77 units. They then compared both present and projected pH modifications to past changes at timescales ranging from 10 to 10^7 years. (Figure 8.2). They found strong oceanic pH sensitivity to atmospheric CO_2 at timescales lower than 10^4 years and a reduced sensitivity at timescales greater than 10^5 years because of the carbonate-silicate geochemical cycle. Over the past 300 million years, they did not detect any pH reduction greater than 0.6 units. Shorter time events (e.g. PETM) might be an analogue for future changes in ocean carbon chemistry even if the system was at that time probably less sensitive to carbon perturbations (*722*). Zeebe (*722*) also warned that the Cretaceous, a time period where a large quantity of chalk formed (e.g. white cliffs of Dover), is not an analogue for the future because the period was a long-term steady state interval, which involved other regulation mechanisms.

Feely and colleagues (*716*) used a biogeochemical model (the National Center for Atmospheric Research Community Climate System Model 3.1) forced by a business-as-usual emission scenario (IPCC A2 scenario) in which atmospheric CO_2 levels reach 800 ppm towards the end of the century. In this scenario, oceanic surface pH drops from 8.2 to ~7.8, increasing the ocean's acidity by about 150% relative to the start of industrialisation. In most oceanic regions, aragonite undersaturation takes place when carbonate concentration is <66 μmol·kg^{-1}. In this work, aragonite undersaturation becomes widespread in the Arctic Ocean by ~2050, although other studies indicate that it may even start earlier (~2020) (*723*). The Southern Ocean (south of 60°S) and some regions of the North Pacific become undersaturated by ~2095. Aragonite undersaturation occurs earlier in the Arctic Ocean than in the Southern Ocean because of ice melting increasing low-alkalinity freshwater inflow. Using a business-as-usual scenario (Scenario IS92a), Orr and co-workers (*162*) showed that the Southern Ocean is expected to become undersaturated with respect to aragonite by ~2050 and that undersaturation may extend to the whole ocean by the end of this century.

Calcite undersaturation takes place when carbonate concentration is <42 μmol·kg^{-1}. Calcite undersaturation may only occur at the end of this century in some parts of the Arctic Ocean, as well as the Bering and the Chukchi Seas (*716*). In the North Atlantic (north of 50°N), the model used by Feely and co-workers suggests that aragonite saturation states drop from 1.47 to 1.08 and 0.76 for a CO_2 **fugacity** of 387, 560 and 840 μatm, respectively (*716*). Calcite saturation drops from 2.34 to 1.72 and 1.21 at the same fugacity values.

8.3 Effects of acidification on biodiversity

Research on the impacts of ocean acidification on marine organisms focuses on calcifying taxa as they are likely to be the most vulnerable to pH modifications and their effects on calcification and other physiological processes (*724*).

8.3.1 Early results

Early warning on the potential adverse effects of ocean acidification on calcifying species has been mainly based on predicted shifts in ocean chemistry with limited experimental supports (*725*). Experimental and field studies have been conducted on some taxonomic groups (*163, 724–727*).

8.3.1.1 Corals

Researchers showed that organisms need a high level of supersaturation. Gattuso and colleagues (*728*) experimentally examined the effects of the alteration in aragonite saturation on calcification rates of the Scleractinian coral species *Stylophora pistilla* (Figure 8.3).

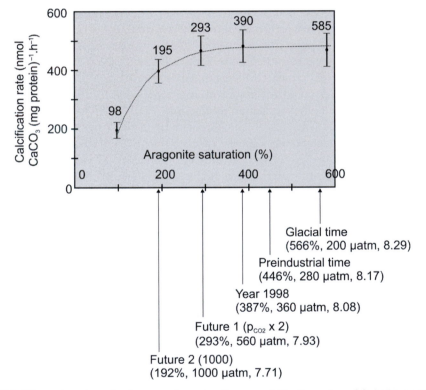

Figure 8.3 Effect of an alteration in aragonite saturation on calcification rates of *Stylophora pistilla*. Values of saturation used in the experiments are indicated above the points. Each arrow indicates a specific time period, characterised by a level of aragonite saturation, CO_2 partial pressure and pH, respectively.

Source: Adapted from Gattuso and colleagues (*728*).

They varied aragonite saturation Ω between 98% and 585% (undersaturation when $\Omega < 100$ and supersaturation when $\Omega > 100$, Ω being expressed as percentage), this last value being close to the current saturation of the tropical oceans. The calcification rate, measured by the alkalinity anomaly technique, fluctuated by ~threefold when aragonite saturation ranged from 98% to 585%. These rates were then compared to both past and projected changes in pH and aragonite saturation (vertical arrows in Figure 8.3). Changes in aragonite saturation between a glacial and an interglacial period were not sufficient to trigger any changes in calcification rate, the level of supersaturation being elevated. A potential strong negative influence occurred when atmospheric CO_2 concentration doubled. At 1,000 µatm, this level would have a strong negative effect on *Stylophora pistilla*. However, it should be noted that current atmospheric CO_2 concentrations are not sufficiently high to strongly affect the coral species and that a doubling of the pre-industrial control would not be high enough to trigger major changes in calcification rates. Only major changes in atmospheric CO_2 concentration (1,000 ppm) would trigger substantial reductions in calcification rate.

8.3.1.2 Coccolithophores

Plankton are a major contributor to carbon cycling and can generate important flux of calcium carbonate from the surface to the bottom (729). Most (>80%) of the biogenic carbonate precipitation is accomplished by plankton (730). Among calcifying plankton, coccolithophores are unicellular marine autotroph algae belonging to the class of *Prymnesiophyceae*. Their cell is composed of a carbonate exoskeleton called **coccosphere** itself resulting from the aggregation of oval calcitic platelets named **coccoliths**. This phytoplanktonic group (280 species) form blooms that extend over vast ecoregions and blooms of *Emiliania huxleyi* cover ~$1.4 \times 10^6 km^2 \cdot year^{-1}$ (731). Brown (731) estimated that coccolithophore blooms produce ~0.4–1.3 million tonnes of $CaCO_3$ annually between 40° and 60°.

Riebesell and colleagues (729) investigated the effects of ocean acidification on coccolithophore calcification. They chose *E. huxleyi* and *Gephyrocapsa oceanica*; the first species forms large blooms in both temperate and subpolar systems, whereas the second occurs over tropical neritic waters. They were reared under environments ranging from 280 (pre-industrial level) to 750 ppm. At high CO_2 level, both coccolithophores showed an increase in photosynthetic carbon fixation (8.5% for *E. huxleyi* and 18.6% for *G. oceanica*) but exhibited a stronger reduction in calcification rate (15.7% for *E. huxleyi* and 44.7% for *G. oceanica*). The calcite/organic carbon ratio was negatively correlated to P_{CO2}. This negative correlation held in different light conditions (photon flux densities of 30, 80 and 150 $\mu mol \cdot m^{-2} \cdot s^{°1}$). Malformed coccoliths and incomplete coccospheres occur commonly in the field. The authors, however, found that the frequency of malformed coccoliths and incomplete coccospheres increased in experiments characterised by elevated P_{CO2}. For example, Figure 8.4 shows the effect of high P_{CO2} on the coccolithophore *Calcidiscus leptoporus* reared in experiments with normal (A, $P_{CO2} = 320$ µatm) and elevated (B, $P_{CO2} = 780–850$ µatm) P_{CO2}.

These laboratory experiments were consistent with results obtained from incubation of natural plankton assemblages sampled in the Subarctic North Pacific (729). The calcification rate was reduced by 36–86% after incubing the coccolithophores during 1.5–9 days in environments (four independent experiments were conducted) characterised by elevated P_{CO2} (800 ppm) when compared to experiments performed in low P_{CO2} (250 ppm).

A.

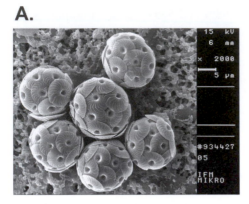

B.

Figure 8.4
Scanning electron microscopy (SEM) photographs of the cosmopolitan coccolithophore species *Calcidiscus leptoporus* reared in experiments with (A) normal (P_{CO2} = 320 µatm) and (B) elevated (P_{CO2} = 780–850 µatm) CO_2 partial pressure. Post-formation calcite dissolution was not considered because water remained supersaturated.

Source: Courtesy of Ulf Riebesell (GEOMAR, Kiel).

The experiment showed that a high level of atmospheric CO_2 concentration, as expected for the end of this century, might alter calcification rates of this taxonomic group.

8.3.1.3 Hypercapnia

Pörtner (*732*) and Fabry and colleagues (*724*) reviewed biological processes and physiological mechanisms involved in marine species sensitivity to ocean acidification. Exposures of marine organisms to increasing P_{CO2} (called by Pörtner ocean hypercapnia) may have strong physiological consequences on ion transport and acid–base regulation. If pH compensation is not possible, this can lead to metabolic depression, which is an adaptive strategy for species to survive during short periods of hypoxia or hypercapnia. Chronic hypercapnia may reduce protein synthesis and alter growth and reproduction.

8.3.2 Complexity

Studies on the effects of ocean acidification on calcareous marine species include bottle experiments, micro- and mesocosms, field experiment and in situ observations. A recent meta-analytical work (*163*) indicated that 139 studies quantified biological effects of ocean acidification and the activity of this research field is probably accelerating under the

	Number of species	Response to rising CO_2			
		A	B	C	D
Calcification					
Coccolithophores	4	2	1	1	1
Planktonic foraminifera	2	2			
Molluscs	6	5		1	
Echinoderms	3	2	1		
Tropical corals	11	11			
Coralline red algae	1	1	1		
Photosynthesis					
Coccolithophores	2		2	2	
Prokaryotes	2		1	1	
Seagrasses	5		5		
Nitrogen fixation					
Cyanobacteria	4		3	1	
Reproduction					
Molluscs	4	4			
Echinoderms	1	1			

Figure 8.5 Responses of major taxonomic groups to ocean acidification from experimental studies. (A) Linear negative response. (B) Linear positive response. (C) No response. (D) Non-linear parabolic response.

Source: Simplified from Doney and co-workers (734).

impulsion of international programmes (733). Some recent contributions begin to show that results might be more complex than previously assumed (734). Of 45 taxonomic groups investigated, 28 exhibited a negative response to ocean acidification, 14 displayed a positive response, six did not show any response and one expressed a parabolic response (Figure 8.5).

8.3.3 Acidification in some marine ecosystems

Hofmann and colleagues (735) reviewed the effect of ocean acidification on five ecosystems: (1) coral reef; (2) open-ocean; (3) coastal; (4) deep-sea; and (5) high-latitude ecosystems.

8.3.3.1 Coral reefs

Coral reef ecosystems may cease to grow and undertake a net dissolution worldwide if atmospheric CO_2 concentration reaches 560 ppm (736). However, many aspects of the responses of coral reefs to ocean acidification remain elusive. Coral calcification takes place inside the animal between the calicoblastic ectoderm and the skeletal surface, a site termed the subcalicoblastic space. In this space, chemical properties are likely to be distinct from the surrounding seawater. In the field, the community experiences diurnal pH fluctuations between 7.84 and 8.56 or in aragonite saturation rate Ω_a between 1.83 and 5.54. Interdecadal pH variability also exists in the range of ~0.3 units (between 7.9 and 8.17), a time period for which no relationship was found between pH and growth of a massive *Porites* on a centurial scale (737). Pelejero and colleagues (737) found a positive relationship between fluctuations in aragonite saturation and the Interdecadal Pacific Oscillation (Chapter 2). The oscillation may therefore attenuate or aggravate ocean acidification. Acclimatisation in coral may occur, but this is a complex issue because corals host diverse *Symbiodinium* communities. At present, it is unknown if a phenomenon comparable to adaptive bleaching (Chapter 7) might occur to override the effects of ocean acidification.

8.3.3.2 Open-ocean ecosystems

Open-ocean ecosystems might also undertake rapid changes as a result of ocean acidification. Although cyanobacteria (e.g. *Trichodesmium* and *Crocosphaera*) might be stimulated by P_{CO2} increase, pteropods, coccolithophores and foraminifers might be affected by ocean acidification. However, conflicting results exist on whether or not acidification will have a negative effect on these three important taxonomic groups.

8.3.3.3 Coastal ecosystems

Coastal ecosystems may be strongly altered by acidification (735). Feely and colleagues (738) made a number of transect lines over the western part of North America from the northern part of Mexico to Canada (25–52°N). The authors identified what they called corrosive upwelling that brought undersaturated waters with respect to aragonite in surface over large continental-shelf regions. One transect off northern California was composed entirely of undersaturated waters. Although this phenomenon is natural along these coasts, ocean uptake of anthropogenic CO_2 may reinforce the phenomenon. Acidification might affect many marine coastal invertebrates and early stages are likely to be the most vulnerable to acidification.

8.3.3.4 Deep-sea ecosystems

Contemporary spatial distribution of deep-sea scleractinian corals seems to be strongly related to aragonite saturation horizon. Therefore, shoaling of aragonite saturation is expected to influence coral-dominated bioherms and seamounts, which support diverse communities of invertebrates and fish. In the North Atlantic, the saturation boundary might rise by 1–2 km by 2100 (735).

8.3.3.5 High-latitude ecosystems

High-latitude ecosystems might be among the first ecosystems to face the adverse effects of ocean acidification. Undersaturation may strike the Southern Ocean by 2050 and the Arctic Ocean by 2020. In this latter system, undersaturation with respect to aragonite is predicted to occur for at least one month per year in 10% of the Arctic Ocean by 2020. The Arctic pteropod *Limacina helicina* reduces its calcification rate when aragonite saturation state diminishes (739). However, as in coastal ecosystems, organisms might be more resistant to ocean acidification. A series of experiments were carried out to investigate the effects of pH reduction (pH between 6 and 8.2) on the pluteus larvae of tropical (*Tripneustes gratilla*), temperate (*Pseudechinus huttoni*, *Evechinus chloroticus*) and polar (*Sterechinus neumayeri*) species. Although a significant reduction in calcification was found for both tropical and temperate species, the calcification of pluteus larvae of the Antarctic sea urchin was not significantly affected by pH reduction (740).

8.3.4 Acidification and life history stages

Byrne (741) and Kurihara (742) have reviewed the potential influence of ocean acidification on marine invertebrate life history stages. Many studies have provided evidence that conditions during early development can have a substantial influence on the subsequent individuals and cohorts performance (743). Benthic organisms have a complex life-history cycle and many stages can be affected by ocean acidification (Figure 8.6).

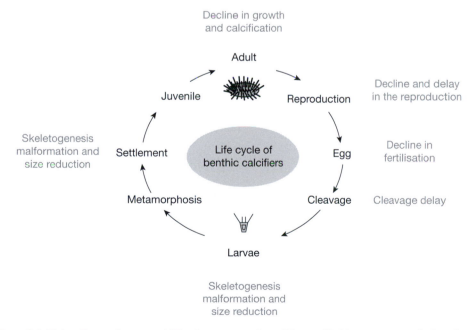

Figure 8.6 Main effects of ocean acidification expected on different life-history stages of a benthic calcifier. These effects are species-dependent and are a function of ocean acidification.

Source: Redrawn from Kurihara (742).

Ocean acidification is expected to delay reproduction, reduce fertilisation, delay cleavage, diminish the size of larvae and induce skeletogenesis malformation. Reduction in fertilisation and reproduction will adversely affect population size, whereas skeletogenesis malformation will alter survival. However, these effects are species-dependent. Corals and shrimps start skeletogenesis at the settlement stage. In contrast, sea urchins and bivalves will be more affected at the larval stage because skeletogenesis occurs during this period.

Dupont and colleagues (744) showed experimentally that an acidification of 0.2 units led to 100% of larval mortality of the brittle star *Ophiothrix fragilis* within eight days compared to the 70% of larval survival that occurred in the control experiment. The authors also showed that abnormal development and skeletogenesis were more frequent when the larvae were exposed to low pH.

8.4 Limitations of past studies on ocean acidification

Recent scientific advances have shown the limits of pessimistic predictions made by earlier studies. I review here some limitations on studies of the effects of ocean acidification on ecosystems and their biodiversity.

8.4.1 Few species have been investigated

A first shortcoming of past studies on ocean acidification impacts has been to extrapolate results obtained from one particular species to a whole taxonomic group or sometimes to all calcifying organisms (162, 729). Kroeker and colleagues (163) applied meta-analytical techniques to explore the biological effects of ocean acidification from 73 studies and a total

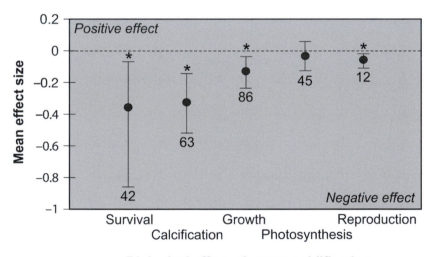

Biological effect of ocean acidification

Figure 8.7 Meta-analysis on expected effects of ocean acidification on species by 2100. The number of experiments on which each effect was quantified is indicated. An asterisk means a significant effect. The mean and 95% confidence interval are shown for each biological variable.

Source: Simplified from Kroeker and colleagues (163).

of 251 experiments. The authors found that ocean acidification had an adverse effect on survival, calcification rate, growth and reproduction (Figure 8.7). The effect was especially prominent for survival and calcification. No effect was found on photosynthesis.

However, when taxonomic groups were separated, very few negative effects of ocean acidification were detected. Whereas a negative effect on calcification was found for corals, no significant effect was detected for calcareous algae, coccolithophores, molluscs and echinoderms. This was even positive for crustaceans. No significant effect on survival was detected (molluscs, echinoderms and crustaceans). For growth, negative effects were found for calcareous algae and corals, whereas positive effects were found for non-calcifiers such as fish and fleshy algae. No effect was found in coccolithophores, molluscs, echinoderms and crustaceans.

8.4.2 Effects of acidification are species-dependent

Ries and colleagues (726) demonstrated that generalisation of the effects of ocean acidification on species based on the examination of one single organisms is not realistic. In 60-day laboratory experiments, the researchers investigated 18 benthic species that belong to various taxonomic groups (chlorophyta, rhodophyta, cnidaria, annelida, crustacea, echinoida, gastropoda, bivalvia). They reared the 18 species under four different conditions corresponding to four CO_2 concentrations: 409 ± 6, 606 ± 7, 903 ± 12 and $2,856 \pm 54$ ppm (or converted as aragonite saturation $\Omega_{ar} \approx 2.5$, 2, 1.5 and 0.7). The authors showed that although some calcifiers were adversely affected by ocean acidification, some did not respond, and perhaps more importantly others were positively influenced by increasing atmospheric CO_2 concentrations (Figure 8.8).

More accurately, six response types of the calcification rate to ocean acidification were highlighted. The first type concerned species that were linearly adversely affected by increasing atmospheric CO_2 concentration (Figure 8.8M–R). This concerned the serpulid annelid *Hydroides crucigera*, two gastropods (periwinkle *Littorina littorea* and whelk *Urosalpinx cinerea*) and three bivalves (bay scallop *Argopecten irradians*, oyster *Crassostrea virginica* and soft clam *Mya arenaria*). For all these species except the Serpulid worm, ocean acidification may have a rapid negative effect on net calcification rate.

The second response type concerned species for which no effect was detected up to ~900 ppm, followed by a pronounced reduction in net calcification for extreme elevated CO_2 concentration (~2,850 ppm) (Figure 8.8I–L). This type of response concerned: one temperate coral (*Oculina arbuscula*), one echinoderm (pencil urchin *Eucidaris tribuloides*), one bivalve (hard clam *Mercenaria mercenaria*) and one conch (*Strombus alatus*). Note that the whelk (first type of response) could also belong to this category.

The third response type was an absence of any effect of ocean acidification on net calcification (Figure 8.8H). This only concerned blue mussel (*Mytilus edulis*).

The fourth response type concerned species for which a maximum in calcification was found between ~400 and ~2,850 ppm (Figure 8.8D–G). A positive effect of ocean acidification was found up to ~900 ppm, followed by a pronounced negative effect. This type was found in four species: limpet (*Crepidula fornicata*), purple urchin (*Arbacia punctulata*), coralline red algae (*Neogoniolithon* sp.) and a calcareous green macroalga (*Halimeda incrassata*).

The fifth response type was a slight positive effect between ~400 and ~900 ppm and a pronounced positive effect at ~2,850 ppm (Figure 8.8C). This effect was detected for the American lobster (*Homarus americanus*).

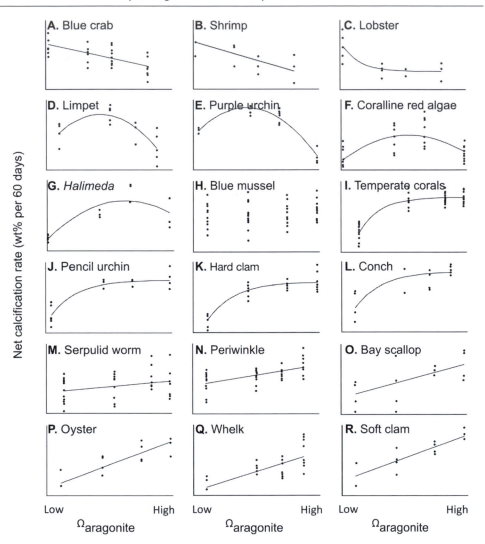

Figure 8.8 Effect of ocean acidification on the calcification rate of 18 benthic species after 60 days at four different atmospheric CO_2 concentrations: ~400 (right), ~600, ~900 and ~2,850 ppm (left). Relationships were assessed using the least squares method. In grey: 95% confidence interval. Points represent replicates. Aragonite saturation Ω_{ar} decreases when atmospheric CO_2 concentration rises.

Source: Simplified from Ries and co-workers (726).

The sixth response type was a linear positive effect of acidification on net calcification (Figure 8.8A–B). This effect was observed for two crustaceans: blue crab (*Callinectes sapidus*) and shrimp (*Penaeus plebejus*). This study therefore revealed the positive influence of ocean acidification on the calcification rate of three crustaceans (American lobster).

When reared during 60 days at a normal and elevated P_{CO2}, the blue crab and the American lobster developed heavier exoskeletons (Figure 8.9). The two species calcified quicker at elevated CO_2 concentration. Crustaceans have existed for a long time with fossil record (Cambrian). They experienced very different conditions with respect to atmospheric

A.

B.

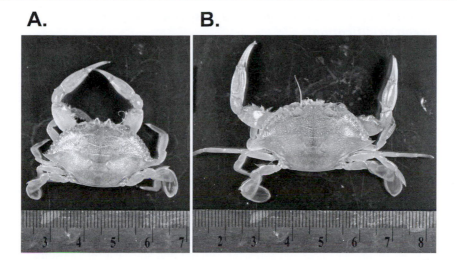

C.

D.

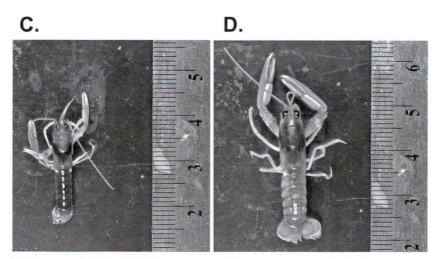

Figure 8.9 Pictures of a blue crab (*Callinectes sapidus*) and an American lobster (*Homarus americanus*) reared for 60 days under CO_2 concentrations of 400 ppm (A and C) and 2,850 ppm (B and D). The two species had heavier exoskeleton when reared at a CO_2 concentration 10 times the pre-industrial level.

Source: From Ries and co-workers (726). Courtesy of Dr Ries, University of North Carolina.

CO_2 concentration, from 10 to 20 times the pre-industrial level (726). If these results can be generalised to important crustacean subgroups, this might have strong consequences on ocean carbon cycling.

As we saw above, Kroeker and colleagues (163) found a significant positive effect of ocean acidification on calcification. An increase in calcification under high P_{CO2} was also observed for the brittle star *Amphiura filiformis* (745). However, the increase in calcification was accompanied by a rise in metabolic rate and a loss in arm muscle mass. An increase in calcification under high P_{CO2} was also observed for the predatory sea star *Pisaster ochraceus* (746).

8.4.3 Diversity of mechanisms involved in species' responses to ocean acidification

Recent studies also showed that mechanisms by which ocean acidification impacts marine calcifiers are likely to be complex and diverse. Here, I cite two examples.

8.4.3.1 Some species are protected by an external organic layer

Some calcifying organisms have an external organic layer that isolates the shell or the exoskeleton against acidic waters. In Ries's paper, organisms with a dense protective organic layer (crustaceans, algae and blue mussel) were not affected or were positively influenced by ocean acidification. Organisms (e.g. clams, oysters, scallops and conchs) with a thinner protective organic layer were adversely affected. Under elevated CO_2 concentration, the aragonite shell of conchs was altered by the resulting acidic waters. In particular, the knobs along the upper lip of the shell, which prevent the animal from being destabilised on the seafloor, exhibited clear signs of dissolution. The presence of an external protective organic layer perhaps explains why Ries and colleagues did not find any clear relationships between the level of magnesium in calcium carbonate and species sensitivity to ocean acidification, except for a high level of CO_2 concentrations (~2,850 ppm).

Rodolfo-Metalpa and colleagues (747) showed by transplantation experiments along a natural gradient of increasing P_{CO2} concentration in volcanic vents off Ischia in the Tyrrhenian Sea (Italy) that external organic protective layers play a crucial role in isolating the calcium carbonate from surrounding acidic waters, enabling species growth to continue. They compared two bivalves, *Mytilus galloprovincialis* with a periostracum protecting the shell and *Patella caerulea* that has no protective organic layer. At pH = 6.8, dead limpet shells dissolved nine times quicker than mussel shells. Dead mussel shells transplanted during three months at pH = 7.2 showed clear sign of dissolution, whereas live mussels that continuously replace their periostracum continued to grow, even after five months at pH = 7.2. They performed a similar experiment, comparing two zooxanthellate corals, *Balanophyllia europaea* with a skeleton entirely covered by organic tissues and *Cladocora caespitosa*, which has parts of its skeleton with no protective tissues. They reached the same conclusions: skeletons protected by tissues were not altered by surrounding corrosive waters ($\Omega_{ar} < 1$), whereas unprotected tissues showed some signs of dissolution. The authors proposed that the protective role exerted by organic layers might explain why the thickness of these tissues increases when seawater pH is reduced. The authors made the distinction between gross calcification and gross dissolution. By doing so, they showed that all species were able to continue to calcify, even in very corrosive waters, demonstrating that: (1) dissolution is the main mechanism by which acidification influences the four species; and (2) protective organic layers are critical for the survival of these species with respect to acidification.

8.4.3.2 Transformation of bicarbonate to carbonate prior to calcification

Although no species can constitute their shell or skeleton from bicarbonate, some species such as crustaceans are suspected to be able to transform bicarbonate into carbonate prior to calcification. Photosynthetic calcifiers could benefit from the increase in P_{CO2}, which favours photosynthesis up to 1,000 µatm, and increases the energy needed to locally elevate the pH needed for bicarbonate to be transformed into carbonate prior to calcification (748). This and the thicker protective organic layer that protect the exoskeleton may explain why crustaceans performed well in Ries's experiments.

8.4.3.3 Cold species may be less sensitive to ocean acidification

Cold species could be less sensitive to ocean acidification because cold waters are more acidic than warm water (726). The temperate urchin (*Arbacia punctulata*) was less affected by acidification than the tropical urchin (*Eucidaris tribuloides*) in the study of Ries and colleagues (726).

8.4.4 Information gained from in situ studies and long-term experiments

Many studies on the impact of ocean acidification have been performed in vitro and have been limited in time. Hall-Spencer and colleagues (749) had the good idea to investigate the effect of ocean acidification on benthic biodiversity along a natural gradient of pH ranging from 8.2 to 6.57 (Figure 8.10A).

This natural gradient occurs in the cold vent areas off Ischia (Italy) where seawater is constantly acidified by natural CO_2 sources. The rocky-shore stations with an average pH between 7.8 and 7.9 (804 µatm < mean P_{CO2} < 957 µatm) were characterised by a drop in 30% in species richness, calcifying organisms being the most affected species (e.g. algae *Halimeda* and corals *Caryophyllia*, *Cladocora* and *Balanophyllia*) (Figure 8.10B). In contrast, some species such as *Caulerpa* and *Sargassum* (invasive species) were resistant to high CO_2 concentrations (Figure 8.10B). In subsequent stations, measured pH values were as low as 7.4–7.6 with an aragonite saturation Ω_{ar} as little as 0.8–1.2 (values not expected in 2100). The anemone *Anemonia viridis* was the only species detected in these undersaturated waters. Corallinaceae, which have high-magnesium calcite skeleton, decreased from more than 60% in coverage to 0 at the core of the vent (Figure 8.10). This result contrasts with a meta-analytical study, which showed that more soluble forms of calcium carbonate such as high-magnesium calcite was more resistant to ocean acidification than calcite or aragonite (163). Sea urchins (e.g. *Paracentrotus lividus*, *Arbacia lixula*), which have high-magnesium calcite skeletons, had virtually nil abundance, whereas the gastropod *Osilinus turbinata* had maximum abundance at mean pH of 7.83 (station 2) (Figure 8.10B). Limpets and barnacles showed a reduction in abundance at Station 2 but were only absent in areas characterised by extreme pH values (Station 3; pH = 6.57; Figure 8.10B). At pH = 8.2, leaves of *Posidonia oceanica* were largely covered by calcified epiphytes (>75%), whereas at pH = 7.6, these epiphytes only covered 2% of the leaves. When transplanted in acidic waters, complete dissolution of the epiphytic corallinaceae occurred in two weeks. A striking feature of these results (Figure 8.10) is that all types of gradients seem to exist, in agreement with the concept of ecological niche or gradient analyses (750). These results confirm that some organisms will benefit from ocean acidification and others will be adversely affected (see Section 8.4.2). Therefore, calcareous organisms will persist even in a world characterised by a high level of atmospheric CO_2 concentration. Biodiversity will reorganise and will affect the dynamic regime of marine ecosystems, ecosystem services and some biogeochemical cycles. Ocean acidification is unlikely to lead to the disappearance of all calcareous organisms, even in cold regions where both aragonite and calcite saturation rates are low.

A recent study investigated both the short- and long-term responses of *Lophelia pertusa* to increasing ocean acidification (751). The cold-water coral species is cosmopolitan and lives in regions with a thermal regime varying between 4°C and 12°C. Two types of experiments were conducted. The first type of experiments examined the short-term (eight days) species response. A diminution in calcification of 26–29% was measured when

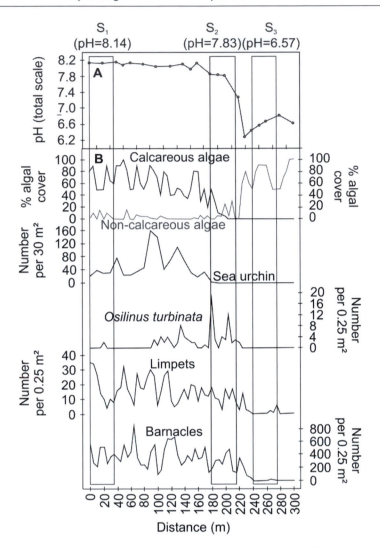

Figure 8.10 Changes in pH, algae coverage and species abundance along a CO_2 gradient in the vent areas off Ischia (Italy). Samples were collected from 18 April to 9 May 2007. (A) Mean pH and standard deviation. (B) Percentage of coverage of calcifying and non-calcifying algae, abundance of sea urchin, *Osilinus turbinata*, limpets and barnacles. S_1, S_2 and S_3 are stations 1, 2 and 3, respectively.

Source: Redrawn from Hall-Spencer and co-workers (749).

pH declines by 0.1 units. At a CO_2 partial pressure of 509 µatm (ambient condition; pH = 8.02), the measured growth was $6.8.10^{-3}$ %·day^{-1} (percentage of skeleton increase per day). It starts to decrease to $1.78.10^{-3}$ %·day^{-1} at P_{CO2} = 605 µatm to reach $1.27.10^{-3}$ %·day^{-1} at P_{CO2} = 981 µatm. In the second type of experiments, the species was exposed to high concentration of P_{CO2} for six months (178 days). In contrast to the short-term experiments, there was no significant relationship between growth and P_{CO2}. At P_{CO2} = 604 µatm, the measured growth was higher ($8.70.10^{-3}$% versus $6.80.10^{-3}$ %·day^{-1} at ambient conditions).

Rates were even higher at more elevated P_{CO2}: $1.14.10^{-2}\%\cdot day^{-1}$ at $P_{CO2} = 778$ µatm and $1.88.10^{-2}\%\cdot day^{-1}$ at $P_{CO2} = 982$ µatm. These experiences illustrate the capacity of cold corals to adapt to ocean acidification. Casareto and colleagues (752) used a natural bloom of the coccolithophore *Pleurochrysis carterae* to investigate the short-term (<39 hours) and long-term effect (seven days) of high P_{CO2} levels on calcification. Net calcification rates were negative in short incubations irrespective of P_{CO2} levels, indicative of dissolution of calcium carbonate. In contrast, the rate of calcification augmented with P_{CO2} in the long-term experiment. The authors concluded that *P. carterae* might have the potential to adapt to increased P_{CO2} levels (~1,200 ppm) with time.

8.4.5 Absence of consideration of the effects of temperature

Another shortcoming has been to neglect the effects of other potentially important stressors (753). Ocean warming might strongly affect marine calcareous organisms. I recall here that warming magnitude may be as pronounced as 4–5°C in worst-case scenarios (old scenario A1FI, new scenario RCP8.5, scenario $[CO_2] \times 4$) by 2100 (55). Such changes would represent the difference between a glacial and an interglacial period (83). Both aquatic and terrestrial systems changed radically since the Last Glacial Maximum (83, 591, 754). Even half of the predicted warming (e.g. Scenarios A1B, A1T, B2; about 2.5°C) (55) may have a strong influence on marine biodiversity because species and ecosystems are very sensitive to temperature (Chapters 3–7).

Studies combining the effects of temperature and acidification suggest that temperature affects in both positive and negative ways the effects of ocean acidification on marine species. Sometimes the thermal effect is higher than the influence of acidification. For example, the development of the abalone *Haliotis coccoradiata* is more sensitive to temperature than P_{CO2} (Figure 8.11). At 20°C, the pH affected the percentage of normal cleaving embryo in contrast to higher temperatures for which the pH had only a small effect. Temperatures reduced by ~60% the percentage of normal cleaving embryo from 20°C to 24°C. Coral larvae were

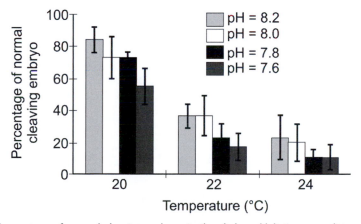

Figure 8.11 Percentage of normal cleaving embryo in the abalone *Haliotis coccoradiata* as a function of temperature and pH, using pH and temperatures values expected under the pessimistic IPCC scenario A1FI for 2100.

Source: Redrawn from Byrne and colleagues (741).

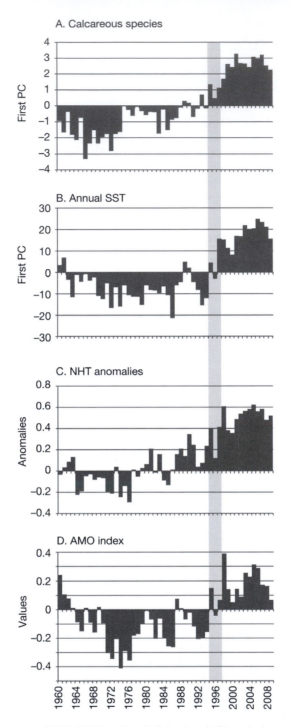

Figure 8.12 Decadal changes (1960–2009) in North Atlantic calcifying plankton in relation to
climate change. (A) Long-term changes in an index reflecting the abundance or
frequency of calcifying plankton. (B) Long-term changes in annual North Atlantic SST.
(C) Long-term changes in northern hemisphere temperature (NHT) anomalies.
(D) The Atlantic Multidecadal Oscillation (AMO) index. The index of the abundance or
the frequency of calcifying plankton was calculated using a standardised PCA on first
principal components reflecting long-term changes in foraminifers, coccolithophores,
Limacina spp., *Clione limacina*, echinoderm larvae and bivalves (see Plate 8.1).

Source: From Beaugrand and colleagues (710).

also sensitive to temperature and changed their settlement preferences when temperatures increased by 3°C (755).

With colleagues, I recently examined long-term changes in the spatial distribution of various calcifying planktonic species or taxa recorded by the **Continuous Plankton Recorder** survey (667). In the eastern North Atlantic Ocean, we examined whether calcifying plankton changes were related to climate, pH and P_{CO2} changes. All calcifying plankton organisms exhibited an abrupt shift in the mid- to late 1990s (Plate 8.1). As all normalised eigenvectors were positively correlated to the first principal component, this indicated that the first principal component directly reflected changes in occurrence frequency or abundance of species. Whereas foraminifers, coccolithophores and echinoderms increased after the shift, both pteropod and non-pteropod molluscs decreased substantially (Plate 8.1). These biological modifications were not correlated to pH or P_{CO2} changes. Rather, these changes were explained by temperature, and pronounced changes in the abundance of calcifying organisms detected in the mid-1990s were explained by large-scale temperature shifts and the state of the Atlantic Multidecadal Oscillation (Figure 8.12).

8.4.6 CO₂ concentration

Some experimental studies have used very high CO_2 concentrations that are unlikely to be reached at the end of the century. For example, Ries and colleagues (726) used atmospheric concentrations of ~2,800 ppm and Anthony and colleagues (756) concentrations of ~1,300 ppm. At more realistic atmospheric CO_2 levels (e.g. 600 ppm), Ries and colleagues found that only 6 of 18 investigated species were adversely affected by ocean acidification (Figure 8.8). Can present acidification really alter marine species in the field? I highlight two reports (757, 758).

The first study is from De'ath and co-workers (757) who investigated 328 colonies of *Porites* corals from 69 reefs in the Great Barrier Reef (11.5–23°S). The authors analysed composite data that contained 16,472 annual records of skeleton density, annual extension and calcification rates obtained by applying X-ray and gamma densitometry techniques. Coral age varied from 10 to 436 years and data were collected during two periods: 1983–1992 and 2002–2005. The authors estimated that *Porites* calcification rate was 1.67 g·cm^{-2}·year^{-1} during 1900–1930 and 1.76 g·cm^{-2}·year^{-1} in 1970. After 1990, the calcification rate dropped from 1.76 to 1.51 g·cm^{-2}·year^{-1} (Figure 8.13). Their results suggest that calcification overall declined by 14.2% after 1990.

The authors extended their records back to 1572 using data from seven reefs and 10 colonies and suggested that the post-1990 decline was unprecedented for ~400 years. Although they proposed that increase in frequency and intensity of recent heat-stress episodes may have contributed to the calcification decline, they also suggested that the 14.2% reduction in calcification observed after 1990 roughly corresponded to the 16% reduction observed in aragonite saturation since industrialisation (162). The authors also referred to a paper that suggested calcification reduction in scleractinian corals between 9% (*Turbinaria reniformis*) and 56% (*Galaxea fascicularis*) for CO_2 concentration doubling to 560 ppm (759). Among the genus *Porites*, calcification reduction ranged from 16% (*Porites porites*) to 38% (*Porites lutea*). Assuming a linear relationship between calcification and atmospheric CO_2 concentration (CO_2 concentration was ~390 ppm and not 560 ppm), they estimated a calcification reduction for the genus *Porites* between 6.28% and 14.92% for a period of >200 years. This estimate was not far from the 14.2% observed decline. The authors recognised,

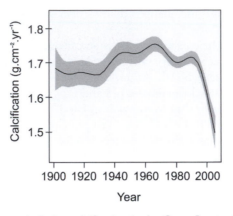

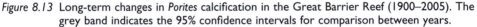

Figure 8.13 Long-term changes in *Porites* calcification in the Great Barrier Reef (1900–2005). The grey band indicates the 95% confidence intervals for comparison between years.

Source: Simplified from De'ath and colleagues (757).

however, the scale difference between aragonite saturation decline (200 years) and calcification (16 years). Changes in aragonite saturation did not decrease sharply after 1990. I think that temperature influence through increase in both frequency and intensity of heat stress episodes is a more likely explanation for the decline in *Porites* calcification. Reid and I investigated long-term monthly SST changes over continental shelves worldwide and found an accelerating warming from the mid- to late 1990s in many regions of the world, including East Australian continental shelves (760).

The second paper reports on the calcification decline of the planktonic foraminifera *Globigerina bulloides* in the Southern Ocean (central subantarctic zone; 45–50°S and 144–151°E). Moy and colleagues (758) investigated shell weights collected from sediment traps as a proxy for calcification. They used G. *bulloides* because the species is uninterruptedly detected back to the late Pleistocene. The researchers documented a pronounced reduction in the average shell weights of G. *bulloides* relative to Holocene shells. A decline of 35% was observed for size range 300–355 μm and 30% for size range 355–425 μm. As De'ath and colleagues, they compared this diminution to reduced calcification observed in experiments for corals and foraminifers under high CO_2 concentrations. They then examined the relationships between long-term changes in shell weights and atmospheric P_{CO2} assessed from ice core record. They found a strong correspondence between the two time series over the past 50,000 years (Figure 8.14).

Shell weights were high during the LGM and lighter during the Holocene. The authors estimated a 20% reduction of shell weights for an 80 ppm increase in atmospheric P_{CO2}. However, the correspondence is as strong/close as with temperature because Antarctic temperature anomalies are strongly correlated positively to CO_2 concentrations (Figure 8.15).

The authors assumed that temperatures did not vary substantially in the Southern Ocean between pre-industrial and modern times, and dismissed the parameter as a potential explanation for observed shell weight reduction. However, Gille (763) used historic shipboard measurements made since 1930 and Autonomous Lagrangian Circulation Explorer (ALACE) float measurements collected in the 1990s to show that the Southern Ocean warmed rapidly between 700 and 1,100 m. The warming rate was $0.004 \pm 0.001°C \cdot year^{-1}$ between 35°S and 65°S for the period 1930–1989 and was more pronounced between 45°S

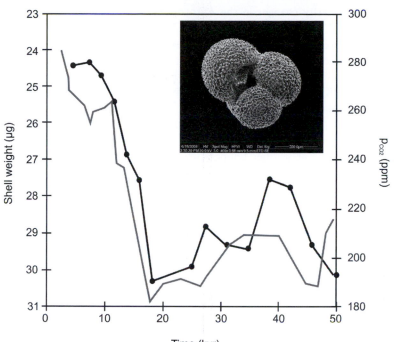

Figure 8.14 Long-term changes in shell weights of *Globigerina bulloides* in relation to changes in atmospheric P_{CO_2} for the past 50,000 years. Atmospheric P_{CO_2} were taken from Vostok. Shell weights were heaviest in the Last Glacial Maximum (18,000–24,000 years) and lowest during the Holocene (0–10,000 years).

Source: Simplified from Moy and colleagues (758).

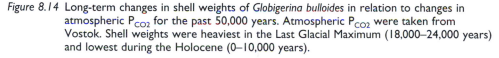

- ■ Taylor CO₂
- • Byrd CO₂
- ▲ Vostok CO₂
- — Dome C Temperature anomalies

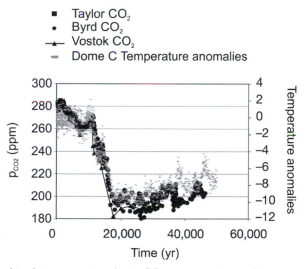

Figure 8.15 Relationships between atmospheric CO_2 concentration and temperature anomalies over the Antarctic.

Source: Byrd CO_2 data are from the National Snow and Ice Data Center (761). Taylor Dome data are from Indermühle and colleagues (762). Vostok CO_2 data are from Petit and colleagues (325). Data from Dr Will Howard and used in Moy and colleagues (758). Temperature anomaly (temperature difference from the average of the last 1,000 years) data are from EPICA Dome Concordia (326).

and 60°S, an area corresponding to the region covered by Moy and colleagues. More recently, Fyfe (*764*) provided evidence for a highly statistically significant mean warming of 0.17 ± 0.06°C between the 1950s and the 1980s. The author estimated an average warming of 0.42°C between 1880 and 2000 (his figure 4). This warming, and the one that took place between the Last Glacial Maximum and the Holocene, might have affected foraminifera mean size as a consequence of the temperature-size rule (Section 3.3.1.2). I recall that this phenomenon is the tendency of individuals reared at lower temperatures to be larger as adults than individuals reared at warmer temperatures. Therefore, we should be careful in the attribution of calcification changes to ocean acidification, as temperature might have the same effect.

8.5 Conclusions

Many studies have shown the potential influence of oceanic acidification on marine biodiversity. Ocean acidification will affect organism physiology (calcification, dissolution), biology (reproduction, skeletogenesis) with strong potential consequences at the community level for ecosystem structure and functioning (resistance to disease, unbalance of predator-prey interaction) and global carbon cycle. Although most studies to date have relied on acute CO_2 exposure (*727*), expected at best for the end of this century, the effects are more difficult to observe at CO_2 concentrations expected in the next decades. Hendricks and colleagues (*725*) also question if marine functional biodiversity will be affected by the level of acidification forecasted by models for the end of the twenty-first century. In addition, recent results showed that some species (e.g. crustaceans) might benefit from higher atmospheric CO_2 concentration. It seems obvious that more research is needed to better understand the potential effects of ocean acidification on marine biodiversity.

Chapter 9

Biodiversity and direct anthropogenic effects

As we saw already, human population has substantially increased for the past 50 years and has increased the pressure exerted on many marine and terrestrial ecosystems. The anthropogenic influence may continue as current demographic projections indicate that human population could reach 9 billion in 2045 and 9.5 billion in 2050. For the end of this century, projections vary from 6.2 (reduction in fertility rate) to 10.1 (reduction in fertility rate in countries with current rates above the replacement level) and 27 billion (765). The last projection, based on a constant fertility rate with respect to the current level, is thought to be unsupportable for the biosphere. Nowadays, 43% of the terrestrial ecosystems have been converted, which corresponds to 0.92 ha for each human currently living on the planet. The meaning of this rate can be better appreciated when it is compared to the rate of change that took place between the last glacial to the current interglacial when 30% of the ice-covered land became ice-free. Assuming a constant rate of per capita use, Barnosky and colleagues (765) estimated that 50% of terrestrial ecosystems might be converted by 2025, corresponding to a human population of 8.2 billion. This may reach 70% by 2060 at a population level of 11.5 billion. These authors postulated that the earth's biosphere may undertake a global phase shift if both demography and human ecological fingerprint are not alleviated (Figure 9.1).

About 70% of the world population now lives within 60 km of the shoreline (14) and the cumulative activities of these people are now strongly altering marine biodiversity (766). Marine ecosystems are being impacted by habitat fragmentation or destruction (173, 767), species invasions (768) and marine resource overexploitation (769). Pollutant transfers from continents provide 80% of all marine pollution and contamination (766). Human activities (e.g. agriculture, sewage and urban development) contribute to alter nutrient concentrations resulting in dramatic disruptions of coastal and estuarine systems (767). An increase in nutrients (i.e. overfertilisation) can have profound effects on biological productivity and ecosystem health, leading to toxic or harmful algal blooms, hypoxia/anoxia and shifts in species dominance (770).

Although it remains challenging to estimate the implications of human activities on the oceans, many results suggest that they are deeply altering marine biodiversity (771). Human impacts are of three types: (1) physical; (2) chemical; (3) biological (Figure 9.2). In the present chapter, I propose an overview of anthropogenic effects on marine biodiversity: (1) exploitation; (2) pollution; (3) nutrient enrichment and eutrophication; (4) oxygen depletion; (5) species introduction and invasion; (6) reduction in stratospheric ozone; (7) tourism and (8) extinction. Some examples of interactive effects between anthropogenic stressors are also examined at the end of the chapter.

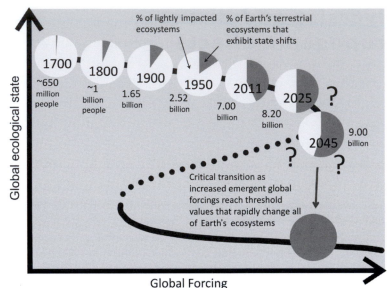

Figure 9.1 Hypothetical effect of land use on the global ecosystem state. In each circle, the dark section indicates the proportion of land that has already been modified by human activities (agricultural and urban land). The size of the global human population is indicated for each time period. The dark line denotes a fold bifurcation with hysteresis.

Source: Modified from Barnosky and co-workers (765). Courtesy of Dr Anthony Barnosky, University of California.

9.1 Exploitation of marine biodiversity

Not long ago, the oceans were viewed as containing an endless amount of resources. However, recent studies have revealed that all groups of large marine species have declined in number, biomass and spatial extent. Humanity now uses 8% of the aquatic primary production. This fraction increases to 25.1% for upwellings and 35.3% for non-tropical continental-shelf ecosystems (772). Overexploitation has both direct effects by reducing stock size and indirect effects through trophic cascades (773, 774).

9.1.1 Direct effects of exploitation

Although fishing has begun for several millennia, it has become a significant driving force since the start of industrialised fishing in the early nineteenth century. After the First World War, fishers started to use stream trawlers, and after the Second World War freezer trawlers, radar and acoustic fish finders (775). Since then, exploited fish stocks have declined steadily in abundance (773, 776). Of the 600 marine fish stocks monitored by the Food and Agriculture Organization (FAO), only 23% of stocks are exploited in a sustainable way, the others being fully exploited (52%), overexploited (17%) or depleted (7%).

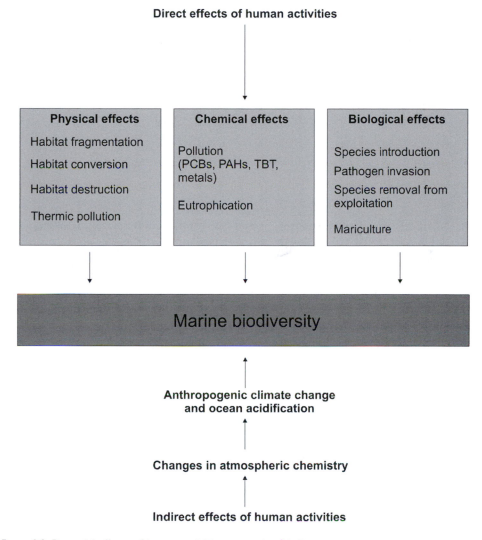

Direct effects of human activities

Physical effects	Chemical effects	Biological effects
Habitat fragmentation	Pollution (PCBs, PAHs, TBT, metals)	Species introduction
Habitat conversion		Pathogen invasion
Habitat destruction	Eutrophication	Species removal from exploitation
Thermic pollution		Mariculture

Marine biodiversity

Anthropogenic climate change and ocean acidification

Changes in atmospheric chemistry

Indirect effects of human activities

Figure 9.2 Potential effects of human activities on marine biodiversity.

9.1.1.1 Reduction in stock size

Ransom Myers and Boris Worm (*777*) provided evidence that industrialised fisheries diminished community biomass by 80% within 15 years of exploitation and estimated that large predatory fish biomass may only represent 10% of pre-industrial levels. Figure 9.3 shows the rapid reduction in the biomass of a tropical open-ocean community after a few years of exploitation.

Fishing reduces the abundance of target species populations and has caused in some cases, regional stock collapse (e.g. Canadian cod fishery). Local extirpation has been observed for the giant sea bass *Stereolepis gigas* in California and the skate *Raja batis* in the Irish Sea.

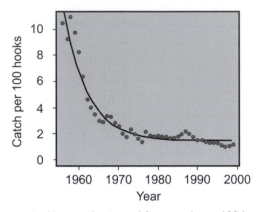

Figure 9.3 Trends in community biomass (estimated from catch per 100 hooks) of a tropical Atlantic open-ocean ecosystem. Points are estimates of relative biomass from the beginning of industrialised fishing. The trend (black) is superimposed. These results show rapid depletion of a community after industrialised fishing initiation.

Source: Simplified from Myers and Worm (777).

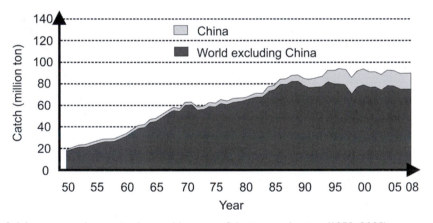

Figure 9.4 Long-term changes in the world capture fisheries production (1950–2008).

Source: Redrawn from the FAO Fisheries and Aquaculture Department (778).

Since the late 1980s, world fisheries landings have declined by ~0.7 million tonnes each year (775), reaching 79.5 million tonnes in 2009 if Chinese reports are excluded (Figure 9.4). When they are included, more than 90 million tonnes of fish are caught every year.

Five species out of the 15,716 that populate the marine ecosphere account for about 15% of the landings. Most captured marine species are anchoveta and, to a lesser extent, gadoids, herring, tuna and mackerel (Figure 9.5).

The increase in tuna fisheries stopped after a maximum catch of 6.5 million tonnes reached in 2007. Gadoid catches were <8 million tonnes in 2008, a minimum value for the period 1967–2008 after an annual catch of 14 million tonnes in 1987. North Atlantic cod stocks have halved from 1.6 million tonnes in 1980 to 0.8 million tonnes in 2000 (685). After 15 years of exploitation, it is not rare to observe a decline of 90% of the stock size. This has been shown for some stocks of Clupeidae (e.g. herring and sprat), Gadidae

Catch (million tonnes)

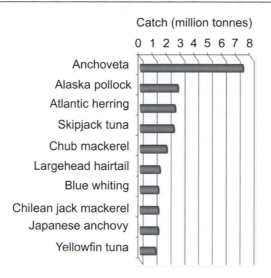

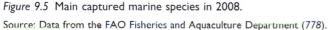

Figure 9.5 Main captured marine species in 2008.

Source: Data from the FAO Fisheries and Aquaculture Department (*778*).

(e.g. cod and haddock), Scombridae (e.g. mackerel and tuna), Sparidae (snapper), Pleuronectidae (plaice and halibut) and Scorpaenidae (e.g. redfish) (*779*). Among the 90 marine fish stocks examined by Hutchings, the decline ranged from 13% to 99%.

Using the largest data set available for the western North Atlantic, Baum and colleagues (*780*) provided evidence for rapid and substantial declines in large coastal and oceanic shark populations. Hammerhead sharks such as scalloped hammerheads (*Sphyrna lewini*) declined by 89% since 1986, white sharks by 79% and tiger shark by 65%. *Carcharhinus* species exhibited a diminution ranging from 49% to 83%. Oceanic sharks also declined. Thresher sharks (e.g. common thresher *Alopias vulpinus* and bigeye thresher A. *superciliousus*) diminished by 80%. Many sharks are at risk of local extirpation.

Marine mammals have also been substantially affected. Between the 1890s and the 1990s, populations of blue whale (*Balaenoptera musculus*) decreased by 99.75% in the Southern Ocean and populations of fin whale (*Balaenoptera physalus*) by 97%. Polar bears were hunted at very large scale for their fur and their meat so that their populations declined to less than 10,000 in the early 1970s. In 1973, the five nations that have their north territories falling into the distributional range of the marine species signed an international agreement for their protection and the population increased to ~20,000–40,000 individuals at the beginning of the twenty-first century.

9.1.1.2 Reduction in resilience

Fishing also has other effects. Many teleosts have an extended longevity, which enables species to persist in some areas during unfavourable time periods. This high longevity represents a storage effect (also called buffering capacity) similar to the seed bank of plants (*781*). However, size-selective fishing leads to the rapid depletion of the oldest and largest individuals in a population. As a result, the number of size class diminishes at a spread and magnitude proportional to fishing intensity. The age distribution becomes skewed towards

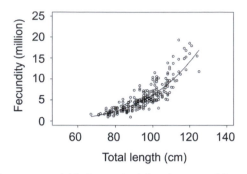

Figure 9.6 Relationships between cod (*Gadus morhua*) female size and fecundity around Iceland in 1995.

Source: Modified from Marteinsdottir and Begg (*785*). Courtesy of Gudrun Marteinsdottir, University of Iceland.

younger ages and earlier maturity at a smaller size (*782, 783*). Because the number of age class declines, truncation of population age structure reduces breeding population size (i.e. spawning stock biomass) and exacerbates the stock sensitivity to environmental variability. This occurs for two main reasons. First, a population of smaller individuals has a shorter spawning season, which may reduce the probability that larvae hatch when oceanographic conditions favour growth and survival (*784*). Second, larger (and so older) individuals produce more eggs, which apparently tend to have a better survival rates than eggs released by younger fish. Figure 9.6 shows that cod females of more than 1 m release 20 times more eggs than females less than 50 cm.

Another example illustrates well the potential effect of fishing on egg production rates. A female red snapper (*Lutjanus campechanus*) of 61 cm (12.5 kg) contains a similar egg number (~9 million) than 212 females of 42 cm (1.1 kg each) (*775*). It is therefore clear that fishing might substantially reduce egg number produced by female.

9.1.1.3 The maternal effect

The non-genetic influence of the female fish on offspring performance has been termed the **maternal effect** (*786*). For example, Berkeley and colleagues (*781*) showed that older black rockfish (*Sebastes melanops*) females produce larvae with larger oil globules and have a better growth and survival than those produced by younger females. Such larvae are less sensitive to mismatch situations in the sense of the match/mismatch hypothesis (*697*) (Chapter 7). Fisheries based on one or two year-class thereby become highly dependent on successful recruitment, and years of poor recruitment due to sustained adverse environmental conditions may cause stock collapse.

9.1.1.4 Migration patterns

Another fishing effect is to affect migration patterns. Because fish tend to reproduce every year in the same place, the targeted removal of older fish that is said to diffuse the knowledge on migration patterns to other individuals may have severe consequences on spawning migration, at least for some species such as herring. This phenomenon has been termed conservatism in migration (*787*).

9.1.1.5 Genetic structure

Fishing changes the genetic structure of populations, selecting individuals that grow quicker and leading to lower age and size at maturity. When fishing becomes very selective, it can alter stock genetic characteristics and favours fish of less economic values. Selective fishing can cause heritable differences in yield and life history traits.

9.1.1.6 Population interactions

Planque and colleagues (788) also stressed that fishing alters interaction between populations of the targeted species. Metapopulations are composed of subunits called populations, which are relatively isolated from each other. Despite their relative isolation, they interact. Random demographic processes and environmental forcing affect these populations differently. If adverse environmental conditions lead to the diminution of a population, then it can rebuild from others. However, fishing makes population reconstitution more difficult because it affects several populations. Fishing therefore erodes or contracts species distributional range. In this way, it exacerbates stocks' sensitivity to environmental fluctuations (677).

9.1.1.7 Serial depletion

Everywhere, the sequence of events is the same. In general, fishing concentrates on large marine fish. These fishes can be apex predators because such fish, at the top of the food chain, have generally high commercial values. For example, in January 2010, a bluefin tuna of 232 kg caught off Aomori north of Honshu Island was sold for €122,000 at a Tokyo fish auction. This was the second most expensive tuna of history. In 2001, a tuna of similar size was sold for €155,000. Exploiters fish down through the food web. This way of exploiting fish, also called 'fishing down the food web' by Pauly and colleagues (773), is termed serial depletion. Pauly and co-workers used a simple procedure to show how fishing can alter ecosystem trophodynamics. They calculated the mean trophic level of fish exploited in a given ecosystem or area. For each exploited fish, species trophic level (TL_s) was determined as a function of the trophic level of its prey (TL_p):

$$TL_s = TL_p + 1 \tag{9.1}$$

Phytoplankton has a trophic level of 1, herbivorous zooplankton 2 and carnivorous zooplankton 3. Large zooplankton or small fish have a trophic level of 3. Cod or tunas have a trophic level ranging from 3.5 to 4.5 (773, 775). Using two datasets of the FAO on 220 species, they applied the technique and revealed that mean trophic level of fisheries landings declined from 1950 to 1994 (773). The mean trophic level decreased from 3.3 at the beginning of the 1950s to 3.1 in 1994 (Figure 9.7). The pronounced reduction in the mean trophic level between the 1960s and the early 1970s was related to substantial catches (i.e. >12 million tonnes per year) of Peruvian anchoveta (lower trophic level). The fisheries collapsed in 1972–1973 at the time of an El Niño event. Fishing simplifies the food web and removes many interactions that contribute to stabilise the ecosystem. As a result, it has been suggested that many ecosystems exhibit more year-to-year fluctuations. Fishing alters ecosystem trophodynamics because the decline of a target species can propagate through the whole food web (495).

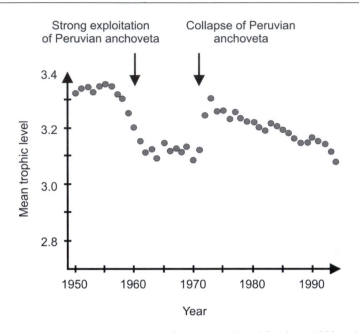

Figure 9.7 Global trends in mean trophic level of marine exploited fish from 1950 to 1994.

Source: Redrawn from Pauly and co-workers (773).

9.1.1.8 Coral reefs

Although coral reefs only occupy 0.1% of the ocean surface, they are in serious decline worldwide. About 58% of coral reefs are threatened (429) and a report on the status of world coral reefs recently estimated that 19% of areas originally covered by tropical reefs have already been lost (789). Projections using a 'business as usual' scenario indicate that 15% of reef ecosystems might be lost in the next 10–20 years and that 20% might disappear in 20–40 years. An estimated 46% of coral reefs remained in a healthy state in 2008 (789). The impact on coral reefs is even more pronounced across the Caribbean and an 80% reduction in coral reef cover has been reported. Over-harvesting is among the factors that have been held responsible for the collapse of these ecosystems. The total global annual yield is estimated to be 1.4–4.2 million tonnes, representing between 2% and 5% of global fisheries catches (472).

The humphead Maori wrasse (*Cheilinus undulates*) is the largest living member of the family Labridae and is among the largest fish associated with coral reefs in the ocean. The fish can exceed 2 m and a weight of 190 kg. It is highly prized, and individuals can be sold up to €110·kg⁻¹ in Hong Kong restaurants (789). As a result, the fish has been heavily fished. Sadory and co-workers (790) showed that in areas where fishing is high, the species density is lower (Figure 9.8).

9.1.1.9 Fish in deep environments

Fishing is expected to affect fish in different ways, depending on their life history traits as well as their individual and social behaviours. In general, fishing affects more strongly species

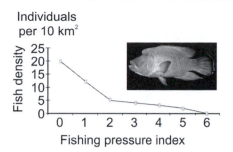

Individuals
per 10 km²

Figure 9.8 Negative relationship between the density of humphead Maori wrasse and an index of fishing pressure.

Source: Adapted from Sadory and colleagues (790). Courtesy of Dr John Randall, Honolulu University.

that have slow growth rates, therefore low maturity at age. Species that live in deep environments have long generation times. They are very sensitive to exploitation. Since the beginning of orange roughy (*Hoplostethus atlanticus*) exploitation, the species, which inhabit continental slopes at depths between 700 and 1,500 m (*791*), has declined by 70% in many areas. The species has a high longevity and can live up to 130 years. Off New Zealand, the fish reaches maturity after 20 years, corresponding to a length of 25–40 cm. Such fish are vulnerable to exploitation because fishing generally uses trawl gear with 10 cm size (sometimes less). In Porcupine Bank (eastern North Atlantic), macroscopic analyses revealed that 50% of female *Hoplostethus atlanticus* reached maturity at ~27 years (37 cm) (*792*). eastern North Atlantic orange roughy matures at a larger size and generally have a higher mean fecundity than those found in the southern hemisphere. This result may be attributed to different environmental forcing or fishing pressure. Species with rapid population growth and maturity are less vulnerable to fishing. However, in all cases, overexploitation is likely to affect species.

9.1.1.10 Destructive fishing practices

Some fishing practices are especially destructive at a local scale, causing acute damage. In coral reef ecosystems, the use of chemicals impacts fish as well as invertebrates (corals). This type of exploitation has led to the substantial decline of species such as the humphead wrasse (*Cheilinus undulatus*). McAllister (*793*) calculated that ~150 tonnes of sodium cyanide are utilised each year on Philippine reefs to capture aquarium fish. Nowadays, cyanide fishing is illegal in many Indo-Pacific countries.

Blast (or dynamite) fishing occurs in many reef ecosystems of the Atlantic, Pacific and Indian Oceans. Like cyanide fishing, blast fishing is illegal but remains, however, widespread in Southeast Asia where it is estimated to threaten ~50% of the reefs (*794*). Although it remains difficult to quantify its impacts on reef species and ecosystems, a classic charge eradicates most marine species within a radius of 77 m (*795*). The ecological effects of blast fishing depend on the frequency of explosions. Fox and Caldwell (*794*) examined the recovery of reefs from two different types of blast fishing. In the first type of blast fishing (i.e. a single acute blast), they showed that the reef partially recovered after five years. In the second type (i.e. chronic blasting over large temporal and spatial scales), they showed that the reefs did not recuperate over the six years of the study and that originally diverse ecosystems became rubble fields.

Drive netting is a technique that captures fish leaving in reef ecosystems. This technique needs people to scare and direct fish towards an encircling net (795). Weighted scarelines, used to frighten fish, damage the reefs. During a single *muro-ami* (a type of drive netting technique) operation, involving 50 fishermen who each struck the bottom 50 times with a 4 kg weighted scareline, ~6% of a one-hectare reef ecosystem was damaged (796).

Shark finning refers to the amputation of shark fins and the discard at sea of the remaining body. After the fin's removal, the shark cannot swim and sinks slowly, being eaten by invertebrates and fish. Researchers estimate that ~100 million sharks are killed annually for their fins. Shark fins are seen as a symbol of wealth, and are eaten in soups at Chinese wedding celebrations.

9.1.2 Indirect consequences of fishing

9.1.2.1 Bycatch

Fishing practices also have strong indirect consequences for non-targeted species. Bycatch is defined as 'the incidental take of undesirable size or age classes of target species (e.g. juveniles or large females), or to the incidental take of other nontarget species. Individuals caught as bycatch can be unharmed, released with injuries, or killed' (797). Global

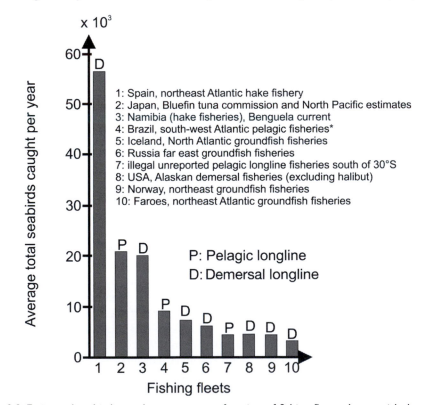

Figure 9.9 Estimated seabirds caught per year as a function of fishing fleets. An asterisk denotes the maximum number of seabirds killed annually.

Source: Redrawn from Anderson and co-workers (798).

bycatch has been estimated to be ~27 million tonnes per year. This represents ~33% of global annual marine captures (Figure 9.4). Many species or taxonomic groups (sharks, turtles, seabirds and marine mammals) are at risk of extinction because of fisheries bycatch. Anderson and co-workers (798) estimated that at least 160,000–320,000 seabirds are killed each year. Albatrosses, petrels and shearwaters are the most frequently captured seabirds. Fishing fleets that killed the highest number of seabirds are the eastern North Atlantic hake fisheries (gran sol area), North-Pacific tuna fisheries and the hake fisheries in the Benguela Current (Figure 9.9).

Bycatch of Pacific loggerhead (*Caretta caretta*), green (*Chelonia mydas*) and leatherback (*Dermochelys coriacea*) turtles has been well documented. Lewison and colleagues (799) estimated that >200,000 loggerheads and 50,000 leatherbacks were captured in 2000 as a result of pelagic longline bycatch (Figure 9.10).

The decline of some small cetaceans has been related to gillnet, driftnet, purse seine and trawl fisheries (797). Read and colleagues published a first estimate of the magnitude of US fisheries bycatch for marine mammals between 1990 and 1999 (800). The authors concluded that 6,215 ± 448 marine mammals were killed annually. Bycatch of most cetaceans (84%) and pinnipeds occurred in gill-net fisheries. They then estimated marine mammal global bycatch by extrapolating their first estimate from US statistics to data on fleet composition from the FAO. They assessed that 649,154 marine mammals (466,392 for cetaceans and 182,763 for pinnipeds) were caught by bycatch fisheries in 1990.

9.1.2.2 Benthic ecosystems

Fishing activities also have strong indirect consequences on benthic ecosystems through destruction by bottom trawling. Pauly and colleagues (775) made the analogy between the goal of bottom trawling and 'cutting forests in the course of hunting deer'. Some fishing practices destroy the entire ecosystem on which many species depend on for reproduction, food and survival. Destruction by bottom trawling can be considerable in some regions.

Both dredging and trawling substantially modify benthic biodiversity, exerting different adverse influences. The first effect is to kill most community members. Colonisation starts soon and a new succession initiates, mainly triggered by opportunistic r-strategy organisms. If both the frequency and the magnitude of the disturbance are low, biodiversity impacts may remain limited and species richness may even increase (the intermediate-disturbance hypothesis; see Chapter 4). However, if the disturbance is frequent or intense, the phenomenon leads to a complete alteration of benthic biodiversity. Large organisms are replaced by smaller ones and trophic guilds are altered; carnivores diminish and scavengers increase. Organisms are also expected to become smaller and both structural complexity and biodiversity of the community decrease.

Substantial trawling of ecosystems dominated by *Posidonia oceanica* in the Mediterranean Sea is suspected to reduce primary production (795). Trawling effects on benthic ecosystems depend on the mass and the type of fishing gears as well as the degree of contact the trawling has with the seafloor. In the North Sea in some areas, beam trawling can be as intense as 25,000–58,200 hours per year and otter trawling as high as 25,000–45,300 hours per year (Figure 9.11).

Thrush and Dayton (802) reviewed the effect of trawling and dredging for benthic biodiversity. Trawling and dredging alter both biogenic and physical structures of the habitat and reduce its biocomplexity. Trawling also diminished the density of echinoderms,

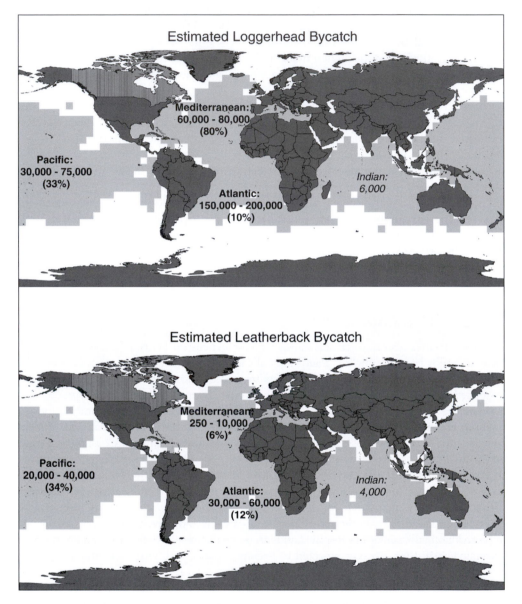

Figure 9.10 Global estimates of loggerhead and leatherback turtles captured as pelagic longline bycatch in 2000. The number in brackets is a measure of the observer coverage for a given oceanic region. In the Indian Ocean, numbers are in italics because no bycatch data exist. The asterisk highlights the lack of zero recorded for leatherbacks that may be at the origin of the small observer coverage in the Mediterranean Sea.

Source: From Lewison and co-workers (799).

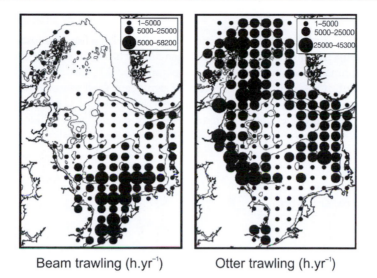

Beam trawling (h.yr^{-1})   Otter trawling (h.yr^{-1})

Figure 9.11 North Sea bottom trawling in 1998. Data from Germany, Netherlands, Denmark, Norway, England and Scotland were merged.

Source: Modified from Callaway and co-workers (*801*). Courtesy of Dr Callaway, School of Biosciences, Swansea University.

polychaetes and molluscs. Regular trawling strongly reduces benthic biodiversity. Cryer and colleagues (803) documented a reduction in the diversity of large benthic invertebrates along a continental slope (200–600 m) off New Zealand, which was attributed to a scampi (*Metanephrops challengeri*) fishery.

Trawling leads to sediment resuspension, modifying type and particle sizes and altering infauna and epifauna, and biogeochemical flux rates. Large, fragile and slow-growth organisms are impacted first. When the pressure increases, K-strategy are replaced by r-strategy organisms and more opportunistic species appear. These alterations are likely to reduce density of both benthic and demersal fish. These fishing effects can also affect the biodiversity of poorly known benthic ecosystems (e.g. deep-water corals and seamounts). Deep-water corals are especially vulnerable to fishing because they are fragile and have a slow growth and reproductive rates. *Lophelia pertusa* is a deep-water coral that reaches between hundreds of metres and several kilometres in diameter and 45 m high. High biodiversity is often associated with these biogenic structures.

Seamounts are also subjected to intensive fishing (e.g. orange roughy). On seamounts off Tasmania, Koslow and colleagues (23) suggested that tonnes of coralline material are destroyed when trawling occurs in new regions. On seamounts where fishing was rare, the authors found high marine invertebrate biodiversity (262 species), which was dominated by the cold coral *Solenosmilia variabilis*, sponges, ophiuroids and sea stars. Benthic biomass of unfished seamounts was 106% greater than seamount biomass where intensive fishing occurred and species richness per sample was 46% greater in pristine than in heavily fished seamounts.

9.1.3 Aquaculture

Aquaculture refers to the raising of aquatic organisms such as fish, molluscs, crustaceans and aquatic plants in controlled conditions. Aquaculture has been practiced for thousands of years (e.g. China 4,000 years ago and Mesopotamia 3,500 years ago). Although aquaculture is ancient, it has only become a globally significant food source for a few decades. Both aquaculture and fishing provided ~103 Mt (65 + 1 + 29 + 8 Mt) of food in 2000 (Figure 9.12). This estimate includes aquaculture of freshwater species, but excludes the 22 Mt of fishmeal that serve as food for pigs, chickens and other reared animals.

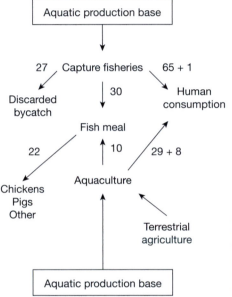

Figure 9.12
Simplified flow chart of both capture and farmed fisheries in 2000. Numbers are in megatonnes (million metric tonnes). When a '+' is indicated, it refers to the additional exploitation of seaweeds.

Source: Simplified from Naylor and colleagues (804).

Table 9.1 Global production and commercial value of the main aquaculture fish in 1997.

Species	Production (kilotonnes)	Commercial value (US$ million)
Common carp	2,237	2,709
Grass carp	2,662	2,444
Silver carp	3,146	2,917
Nile tilapia	742	885
Channel catfish	238	372
Atlantic salmon	639	2,113
Milkfish	393	697
Giant tiger prawn	490	3,501
Pacific cupped oyster	2,968	3,164

Note: There are both freshwater and marine fish. Weight includes shell.

Source: Simplified from Naylor and co-workers (804).

In 2000, fish produced by aquaculture accounted for ~25% of all fish directly consumed by humans (804). Table 9.1 shows the production and commercial value of the main fish produced by aquaculture in 2000.

In 2009, the total world fisheries (aquaculture and capture of both inland and marine species) increased to 145.1 million tonnes (778). In 2007, both capture and aquaculture provided 1.5 and 3 billion people with 20% and 15% of their per capita consumption of animal protein, respectively.

9.1.3.1 Aquaculture needs a lot of endosomatic energy

On the total world fisheries (145.1 million tonnes) in 2009 for both inland and marine realms (778), aquaculture accounted for 55.1 million tonnes. In the marine realm, capture equalled to 80 million tonnes (Chinese reports excluded) and aquaculture to 20.1 million tonnes in 2009. It has been proposed that aquaculture could counterbalance the decline observed in global catches. Although freshwater fish such as carp and filter feeders such as shellfish necessitate low (carp) or no (shellfish) wild fish, carnivorous fish and marine shrimp need a large quantity of fishmeal and oil (Table 9.2). These fish products are sources of essential amino acids (e.g. lysine and methionine) and fatty acid (e.g. eicosapentanoic acid, termed EPA), which are not found in or are deficient in plants.

It follows that current practices of marine fish aquaculture cannot counterbalance the reduction in global fisheries production because it needs between 2.46 and 5.16 times more wild fish to produce 1 kg of farmed fish (see, however, ratio < 1 for milkfish; Table 9.2). Current practices of marine aquaculture therefore increase the pressure on marine resources and compromise the growth of this industry.

9.1.3.2 Aquaculture and habitat alteration

Marine aquaculture development has been at the origin of habitat modification (804). Mangroves and coastal wetlands have been transformed into milkfish and shrimp ponds. In the Philippines, mangroves have suffered from severe degradation and only 35% of the original area occupied by these ecosystems subsists today. In Thailand (2000), ~65,000 ha of mangroves were converted to shrimp ponds. Naylor and colleagues quantified wild fish loss

Table 9.2 Wild fish needed to produce farmed fish in 1997.

Farmed species	Production of farmed fish (kilotonnes)	Wild fish needed (kilotonnes)	Ratio wild fish/ farmed fish
Marine finfish	754	1,944	5.16
Eel	233	546	4.69
Marine shrimp	942	2,040	2.81
Salmon	737	2,332	3.16
Milkfish	392	74	0.94
Molluscs	7,321	0	–

Note: Finfish excludes salmon and includes flounder, halibut, sole, cod, hake, haddock, redfish, seabass, congers, tuna, bonito and billfish. Note that the category molluscs include both freshwater and marine species.

Source: Simplified from Naylor and co-workers (804).

resulting from this habitat conversion and estimated that 400 g of captured fish/shrimp was lost per kilogram of shrimp farmed.

Aquaculture of milkfish, shrimp and tuna needs larvae caught in the field and not reared in a hatchery. Milkfish larvae used in aquaculture represent only 15% of a total field catch, 85% being discarded (*189, 804*). Assuming that 1.7 billion wild milkfish are produced annually in the Philippines, the authors estimated that this bycatch represents a loss of ~10 billion other fish.

9.1.3.3 Aquaculture and substances release

The growth of aquaculture has been accompanied by increasing wastes in the marine environment (*189*). Aquaculture is generating two types of substances: (1) organic; and (2) chemical. Organic wastes originate from an excess of food and faeces that sink to the bottom and may lead to eutrophication (nitrogen and phosphorus) and sediment hypoxia or anoxia. Eutrophication and hypoxia/anoxia may not only involve a stress in caged fish, but it may also contribute to reduce growth and weaken their immune system. In 2002, total annual milkfish production in Southeast Asia was ~350,000 tonnes, and half of this was produced in the Philippines alone (*805*). This industry generates solid wastes (e.g. uneaten feed and faeces) that settle down to the sediment, increasing oxygen consumption and altering benthic ecosystems. Holmer and colleagues (*805*) found clear effects of fish pens in the Bolinao area (Philippines). Sediments were enriched with organic matter, which released a large quantity of nutrients and generated large oxygen consumption. These alterations were accompanied by a shift from large oxygen-sensitive species (e.g. mud-lobster) to small sulphide tolerant worm of the genus *Capitella*.

Many chemicals are used in aquaculture (e.g. disinfectants, chemotherapeutants, antifouling agents, algaecides, pesticides feed additives). Salmonids are treated by pesticide (e.g. pyrethrin) against parasite copepods that cause sea-lice infestations. Disinfectants are also utilised to prevent virus spread (*806*). In salmon aquaculture, recent sea lice epizootics of *Lepeophtheirus salmonis* and *Caligus elongatus* instigated large losses to the industry. To prevent or treat those infestations, aquaculture uses a large spectrum of chemotherapeutants such as hydrogen peroxide, pyrethrin, cypremethrin, dichlorvos or azamethiphos (*806*). Haya and colleagues (*806*) also mentioned medicated feed that may contain ivermectin, emamectin benzoate, diflubenzuron or teflubenzuron. Large release of such chemicals into the environment may adversely influence marine species.

Haya and colleagues (*806*) investigated the effects of the azamethiphos on some species of molluscs and crustaceans. The organophosphate insecticide is the active molecule contained in the Salmosan used to prevent or treat fish against sea lice. The researchers exposed some species to the same concentration of azamethiphos used for salmon (100 $\mu g \cdot L^{-1}$) and estimated the time needed for these species to show a mortality of 50% (Table 9.3). Although green crab and molluscs were not affected by the different concentrations of azamethiphos during the experience time, the chemical had a strong effect on lobsters, and to a lesser extent on shrimps. They also exposed different larval stages of lobsters to azamethiphos, pyrethrin and cypermethrin at concentrations corresponding to treatment in bath against sea lice infestations: 100, 10 and 5 $\mu g \cdot L^{-1}$, respectively (Table 9.4). Results show that sublethal concentrations are far below concentrations used to treat salmon against sea lice. Although no significant difference was detected for both azamethiphos and cypermethrin, young stages were more vulnerable to pyrethrin than older stages. Adults were

Table 9.3 Time (hour) from which some marine invertebrates exhibit 50% mortality (LT50) when exposed to different concentrations of azamethiphos (Salmosan).

Marine species	Concentration of azamethiphos ($\mu g \cdot L^{-1}$)		
	50	100	500
Green crab	NM	NM	NM
Shrimp	8.50	8.50	1.58
Clams	NM	NM	NM
Scallops	NM	NM	NM
Lobster	1.20	0.71	0.17

Note: NM: No Mortality.

Source: Modified from Haya and colleagues (806).

Table 9.4 Lethal concentrations (in $\mu g \cdot L^{-1}$) corresponding to 50% mortality (LT50) of different types of chemicals on larval stages and the adult stage of the American lobster *Homarus americanus* after 48-hour exposure.

Lobster larvae	Type of chemicals		
	Azamethiphos ($100 \ \mu g \cdot L^{-1}$)	Pyrethrin ($10 \ \mu g \cdot L^{-1}$)	Cypermethrin ($5 \ \mu g \cdot L^{-1}$)
Stage I	3.57	4.42*	0.18
Stage II	1.03	2.72*	0.12
Stage III	2.29	1.39	0.06
Stage IV	2.12	1.02	0.12
Adult	1.39		0.04*

Note: Concentrations used for treatments against sea-lice infestations are indicated into brackets. Asterisks denote significant differences between a given stage and the others.

Source: Modified from Haya and colleagues (806).

more sensitive to cypermethrin than larval stages. This study therefore demonstrates that chemicals used in salmon aquaculture may represent a potential hazard to indigeneous species.

9.1.3.4 Antibiotics

The widespread utilisation of antibiotics has led to the development of resistance in bacterial populations (807). High rates of antibiotic resistance to tetracycline, oxytetracycline, oxolinic acid, furazolidone and chloramphenicol were observed in bacteria from fish ponds. In Thailand, most shrimp farmers used antibiotics on a prophylactic basis. The collapse of shrimp farming in Taiwan in 1988 has been attributed to the unselective utilisation of antibiotics, which increases short-term survival but reduces long-term resistance of the crustacean to pathogens (808).

9.1.3.5 Escape of farmed species in the wild

Farmed salmon can also escape and mix with wild salmon. It has been estimated that about 40% of salmon caught in the North Atlantic north of the Faroe Islands are from aquaculture (809). Assuming between 146 and 172 million adult salmon in 2000 and an escapement

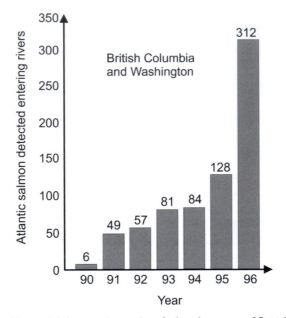

Figure 9.13 Number of farmed Atlantic salmon identified in the rivers of British Columbia and Washington State from 1990 to 1996.

Source: Simplified from Gross (*810*).

rate between 1% and 2%, Gross (*810*) assessed that between 1.5 and 3.4 million adults of Atlantic salmon escape into the wild. Escaped Atlantic salmon have been detected in 50 rivers flowing to the North Pacific, and this number increased from 1990 to 1996 (Figure 9.13).

9.1.3.6 Indirect biological effects of aquaculture: pathogen invasions

More than 220 species of shellfish and fish are currently farmed (*804*). Millions of salmon (Atlantic salmon *Salmo salar* and to a lesser extent the genus *Oncorhynchus*) are reared annually in the world to such an extent that nowadays, 50 farmed fish correspond to one wild fish in Europe (*811*). This led to a substantial increase of ectoparasitic sea lice (mostly *Lepeophtheirus salmonis*) in the world's coastal marine waters. The sea louse has two free-living planktonic nauplii stages (Figure 9.14).

The next stage, copepodite, becomes infective. The copepodite moults of the chalimus stage use a special frontal filament to attach to salmon (*812*). Two immature pre-adult stages can migrate freely on the fish. Adults can overwinter on salmon. They are frequently located on the head and behind the fins. Sea lice graze on the host, removing mucus, skin and the underlying tissue and involving bleeding and tissue necrosis (Figure 9.15).

The ectoparasite impairs growth and immunocompetence. They are frequently located on the head and behind the fins. Although these parasitic copepods develop in farmed salmon, there are increasing concerns they may also contaminate wild salmon that often interact with farmed salmon during their migration. Within a few years, farmed Atlantic salmon have been at the origin of many sea-louse epizootics on young wild salmon. Researchers

Lakselus: *Lepeophtheirus salmonis*

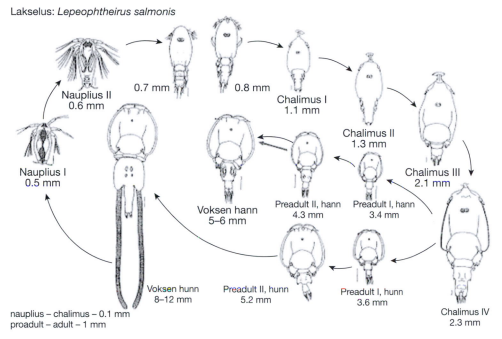

Nauplius II
0.6 mm

0.7 mm

0.8 mm

Chalimus I
1.1 mm

Chalimus II
1.3 mm

Nauplius I
0.5 mm

Chalimus III
2.1 mm

Voksen hann
5–6 mm

Preadult II, hann
4.3 mm

Preadult I, hann
3.4 mm

Voksen hunn
8–12 mm

Preadult II, hunn
5.2 mm

Preadult I, hunn
3.6 mm

Chalimus IV
2.3 mm

nauplius – chalimus – 0.1 mm
proadult – adult – 1 mm

Figure 9.14 The life cycle of the sea louse *Lepeophtheirus salmonis*. Nauplius I and II and copepodites are free-living planktonic stages.

Source: Courtesy of Professor Ken Whelan, Atlantic Salmon Trust (*812*).

Figure 9.15 Picture of a female sea louse on a young salmon.

Source: Courtesy of Stan Proboszcz, Watershed Watch Salmon Society.

calculated that Norway's farmed Atlantic salmon produced about 1.45×10^{11} sea-louse eggs in 1990 during the critical two-month (April–May) spring migration of juvenile salmon (813). Global warming may also exacerbate the frequency of outbreaks, but this remains to be investigated.

9.2 Pollution

Pollution is an important problem. Of the 50,000 km of main rivers in China, 80% have no fish due to pollution (814). The word pollution, which originates from the Latin word *pollutionem* (meaning make dirty) has been defined by the United Nations Convention on the Law of the Sea as:

> The introduction by man, directly or indirectly, of substances or energy into the marine environment, including estuaries, which results or is likely to result in such deleterious effects as harm to living resources and marine life, hazards to human health, hindrance to marine activities, including fishing and other legitimate uses of the sea, impairment of quality for use of the sea water and reduction of amenities.

Pollution is the environmental damage induced by the existence of wastes in oceans and seas (815). Pollution scientists talk about inputs to designate wastes, contamination to indicate their accumulation in the sea and pollution to emphasise their effects on organisms and ecosystems. Marine pollution is so pervasive that many marine organisms, from plankton to whales and polar bears, are currently being contaminated with anthropogenic chemicals such as pesticides, polychlorinated biphenyls (PCB) and polycyclic (or polynuclear) aromatic hydrocarbons (PAHs). Pollution affecting marine biodiversity originates from the atmosphere, freshwaters, seas and lands.

9.2.1 Plastic debris

Plastic debris is comprised of cheap synthetic organic polymers. They have been made for a century but mass production began after the Second World War. Global plastic production accounted for 265 million tonnes in 2010 (816). They decompose very slowly (817). Pollution by plastic debris and microplastics has now become so pervasive that they globally affect marine pelagic ecosystems (Plate 9.1).

Law and colleagues (819) reported that 60% of 6,136 surface plankton net tows caught buoyant plastic debris in the North Atlantic subtropical gyre. Plastic debris occur everywhere on beaches, shorelines and harbours. For example, plastic debris on the beach of South Wales (United Kingdom) and St Lucia (Caribbean Sea) represent 63% and 51% of collected items, respectively (817). Plastic debris accumulate towards the centre of the oceanic gyres in the Atlantic, Pacific and Indian Oceans.

Plastic debris may also sink to the seabed; they represent between 80% and 85% of debris that sink on the seabed in Tokyo Bay (820). Because most plastics float, they are often ingested by marine pelagic species such as seabirds, marine mammals, most turtles and fish. Large species such as seals and dolphins can be entangled in plastic debris. Entanglement can be at the origin of skin lesions and deformation, and may eventually lead to the animal's death. Six of the seven marine turtles, 32 out of 115 marine mammals and 51 out of 312 seabirds have reports of entanglement (816). Ghost fishing can also be considered as a form of entanglement.

Of 1,033 birds examined off the coast of Carolina (USA), 55% had plastic debris in their guts (*821*). These plastic particles were mistaken for prey. The researchers found that plastic debris with a certain form and colour were selected by the birds. Plastic debris in the birds' stomach diminish its storage volume and cause fitness reduction. Plastic debris ingestion can eventually lead to a bird's death. A dead white-faced storm-petrel (*Pelagodroma marina*) was found with plastic pellets in its gizzard at Chatham Islands in New Zealand (*822*). Some marine mammals also died after ingesting plastic debris. A young male pygmy sperm whale (*Kogia breviceps*) and a West Indian manatee (*Trichechus manatus*) died after plastic debris such as corn chip bags occluded part of their digestive tracts (*817*). A recent study conducted in the North Pacific Central Gyre (NPCG) found that 35% of the 670 sampled fish (five mesopelagic species such as *Symbolophorus californiensis* and *Myctophum aurolanternatum* and the epipelagic species *Calolabis saira*) had ingested plastic debris in their gut (*823*).

9.2.2 Chemical pollution

Chemical pollution strongly affects marine biodiversity but it is difficult to quantify its large-scale effects.

9.2.2.1 Oil pollution

Hydrocarbon release has been estimated to range between 600,000 and 1 million tonnes each year (*824*). The effect of oil pollution depends on oil type. Very light oils such as jet fuel or gasoline evaporate rapidly. Their effects are acute for species living in the neuston (i.e. the first few centimetres of the water column). However, because this oil type disappears quickly, there are no long-term effects at the ecosystem level. When oil weight increases from light (e.g. diesel) to heavy oils (most crude oils), the pollution starts to have more long-lasting effects on taxonomic groups such as seabirds and marine mammals. Oil pollution can be accidental or chronic.

9.2.2.1.1 ACCIDENTAL POLLUTION

Although accidental pollution only represents a tiny fraction (~10%) of oil products entering the sea, their consequences can be locally devastating for species and ecosystems (e.g. Exxon Valdez in 1989, Erika in 1999, Prestige in 2002). Oil spill effects vary among species and early developmental stages are more sensitive. Surface eggs are strongly affected, whereas eggs close to the bottom are less impacted. Laboratory experiments showed that a low concentration of petroleum hydrocarbons (i.e. 50 ng·g^{-1}) stimulates phytoplankton photosynthesis because such levels of petroleum hydrocarbons have a nutritive effect, whereas higher concentrations (i.e. >50 ng·g^{-1}) tend to depress photosynthesis (*815*). Concentrations higher than 250 ng·g^{-1} alter copepod (e.g. *Acartia*) feeding. Fish can be affected by oil spill when they ingest oil or oiled prey or when their gills are altered.

Accidental pollution has dramatic direct and rapid consequences on seabirds. Seabirds are isolated from the environment thanks to their feathers. When they are oiled, the animals cannot keep their inner temperature and die from hypothermia. Seabirds such as razorbill (*Alca torda*), guillemot (*Uria aalge*) and puffin (*Fratercula arctica*) are very sensitive to oil pollution. On 24 March 1989, the supertanker Exxon Valdez spilled 41 million litres of crude oil over 30,000 km^2 at Prince William Sound (Alaska). More than 30,000 seabirds,

Table 9.5 Estimated recovery time of various ecosystems after oil damage.

Ecosystems	Recovery time
Plankton	weeks/months
Sand beach	1–2 years
Exposed rocky shore	1–3 years
Sheltered rocky shore	1–5 years
Saltmarsh	3–5 years
Mangrove	>10 years

Note: These estimates vary from one author to another. Some authors showed that some saltmarshes can have a recovery time of ~10 years.

Source: From ITOPF (*828*).

which belonged to 90 species, died in the months following the catastrophe (*825*). Most affected birds were murres (88%), other alcids (7%) and sea ducks (5.3%). The total number of birds that died was estimated at between 100,000 and 300,000. However, after two years, the effects of the oil spill diminished rapidly and although some seabirds remained impacted in late 1991, most species recovered (*826*). These results suggest that bird populations may be resilient to short-term anthropogenic perturbations. I caution here that lethal (acute) effects are easier to detect than sublethal (chronic) effects. In a review on the effects of the Exxon Valdez oil spill, Peterson and colleagues (*827*) stressed that 'unexpected persistence of toxic subsurface oil and chronic exposures, even at sublethal levels, have continued to affect wildlife' and that 'delayed population reductions and cascades of indirect effects postponed recovery'. They continued by stating that 'development of ecosystem-based toxicology is required to understand and ultimately predict chronic, delayed, and indirect long-term risks and impacts'. More than 300 harbor seals were killed from inhalation of toxic fumes that trigger brain lesions and disorientation (*827*).

   Oil spill impact also varies as a function of regional characteristics (e.g. bathymetry, topography, high or low-energy environment, oil damage sensitivity) and recovery after an oil spill varies from one system to another (Table 9.5). Intertidal regions are typically highly affected by oil spills. Oil pollution effects also depend upon regional thermal regimes. In warm areas, the immediate effects are enhanced but also diminished quicker because of a more rapid degradation of oil products by bacteria. Rocky shores are high-energy systems, and when they are not sheltered oil is rapidly eliminated. Exposure to fresh oil is lethal for some taxa such as green and red algae, but as oil rapidly loses its toxic constituents it quickly becomes less toxic (*815*). Seaweeds may secrete mucins to protect themselves against oil. In low-energy areas such as sandy beaches, oil can penetrate deeper into the sediments and remain there for a long time, as bacterial degradation is low in hypoxic sediments. In saltmarsh, oil pollution effects are seasonally dependent; oiled plants in bud are unlikely to develop flowers, oiled plants in flowers are unexpected to produce seeds and oiled seeds do not germinate. Mangroves are very sensitive because the coating of tree pneumatophores (e.g. *Avicennia*, *Rhizophora*) that constitute the basis of these ecosystems impairs gases exchange and may lead to asphyxiation. The recovery of these ecosystems takes time.

   The Amoco Cadiz was among the worst oil spills ever. The spill released 220,000 tonnes of oil off Brittany (France) in April 1978. An estimated quantity ranging from 10,000 to

92,000 tonnes accumulated in subtidal sediments (48). The polluted site was, by chance, monitored one year before the catastrophe, which enables some researchers to have a relative reference state to better understand the ecological effects of this spill. They found a clear decline in the abundance, biomass and production of the fine-sand community of Pierre Noire (Brittany). In particular, *Ampelisca* species, which have no pelagic larvae and tend to maintain insular populations, were deeply affected by the oil spill. Only one species persisted on the site, although in low density. The sediments were subsequently quickly depolluted and the researchers observed the recovery of the community. Recolonisation by *Ampelisca* took at least 10–15 years because of the amphipod demographic strategies (e.g. low fecundity and dispersal). Dauvin (48) provided evidence that the low abundance of *Ampelisca* during this time may have had some consequences for higher trophic levels. Some fish feed on macrobenthic fauna, especially crustacean amphipods, polychaetes and bivalves, and he estimated that ~50% of the benthic biomass is available for demersal fish. For a 2.5 km² area (Pierre Noire), community alteration after the spill involved a decline in fish biomass of ~0.65 tonnes of fresh weight during the 11-year period. Assuming a commercial value (in French francs) of about 10 FF (1998) per kilogram, he assessed that the deficit of the fishery was 650 × 10 FF × 11 annual cycles = 71,500 FF (US$12,000) for the 11-year period following the pollution event.

9.2.2.1.2 CHRONIC POLLUTION

Chronic pollution is continuous and hard to detect (829, 830). Chronic pollution has multiple origins such as leaks in marine terminals, atmospheric fallout from partial combustion of oil in vehicles, urban and industrial wastes. For example, ~80% of urban sewage released into the Mediterranean Sea is not treated. Although such chemical pollution gets far less attention than accidental pollution, it is absorbed by species and generally concentrated from lower to higher trophic levels. As a result, many exploited and unexploited fish, as well as mammals, may concentrate high levels of pollutants with strong metabolic and biological effects (e.g. reproduction, locomotion, foraging). Oil damage occurs in plants, invertebrates, fish and vertebrates through coating, asphyxiation and poisoning.

Huang and colleagues (831) investigated the effects of chronic pollution on phytoplankton in the subtropical Bay of Yueqing (China). The researchers exposed eight phytoplanktonic groups to crude oil **water accommodated fraction (WAF)** for 15 days under laboratory conditions. Crude oil WAF is now used to standardise the experiments and make them more comparable. Chlorophyll a and cell density were estimated, and phytoplankton species were identified every day. They found that elevated concentrations of oil pollution (≥ 2.28 mg·L^{-1}) diminished phytoplankton growth and cell density. In contrast, low pollution (≤ 1.21 mg·L^{-1}) increased growth. At the community level, species diversity declined in all seasons and species seasonal succession was altered because species had different oil pollution sensitivity.

9.2.2.2 Persistent organic pollutants

Persistent organic pollutants (POPs) are characterised by high persistence, bioaccumulation and toxicity. They are also abbreviated PBTs (persistent, bioaccumulative and toxic) or TOMPs (toxic organic micro pollutants). POPs include carcinogen PAHs (polycyclic

A. Naphthalene

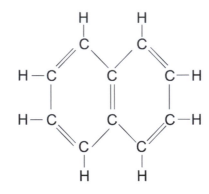

B. Benzo[a]anthracene (carcinogen)

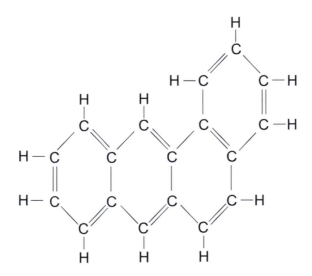

Figure 9.16 The chemical structure of two polycyclic aromatic hydrocarbons (PAHs): (A) naphthalene; (B) benzo[a]anthracene.

aromatic hydrocarbons), halogenated hydrocarbons and organo-metallic substances such as tributyltin (TBT). The toxicity of these chemicals is enhanced by the process of bioconcentration (i.e. accumulation of a POP though the organism life) and biomagnification (i.e. accumulation of a POP in the food chain from phytoplankton to top predator).

   PAHs are organic compounds that, when present in large concentration, are toxic or carcinogenic to plants and animals including humans. Typically, PAHs are by-products of petroleum processing or combustion. PAHs have two or more aromatic (benzene) rings (Figure 9.16). Naphthalene ($C_{10}H_8$) has two aromatic rings and benzo[a]anthracene ($C_{18}H_{12}$) has four rings.

A. DDT

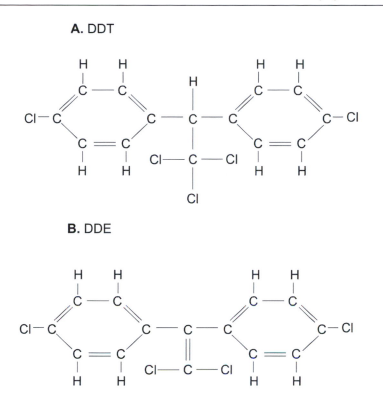

B. DDE

Figure 9.17 Chemical structure of halogenated hydrocarbons: (A) DDT (dichlorodiphenyltrichloro-ethane); (B) DDE (dichlorodiphenylethane).

Two groups exist: (1) low-molecular weight PAHs with two or three rings (e.g. naphthalene and fluorene), which are toxic to aquatic species; and (2) high-molecular weight PAHs with four to seven rings (e.g. chrysene and coronene), which are harmless. Biomass burning and volcanic activities are natural sources of PAHs, whereas partial burning of organic matter, automobiles and some industrial wastes are anthropogenic sources. Because of their lipophilic nature, PAHs can penetrate biological membranes and accumulate in organisms. Fish are particularly vulnerable to PAHs, but more studies are needed to determine the consequences of PAHs on marine biodiversity.

Halogenated hydrocarbons are chemicals comprising halogens such as chlorine (organochlorines or chlorinated hydrocarbons), bromine, fluorine or iodine. Pesticides such as DDT belong to this category (Figure 9.17A). As metals, they are conservative pollutants, being resistant to chemical oxidation or bacterial action (*815*). DDT can then be transformed to DDE and 80% of derivative of DDT occur in the marine environment as DDE (Figure 9.17B).

Insecticides such as dieldrin and endrin are halogenated hydrocarbons. Polychlorinated biphenyls (PCBs) are another example of halogenated hydrocarbons (Figure 9.18). They are used in plastics and electrical equipment. Plastic debris ingestion by some species leads to PCB absorption. These **xenobiotic** substances are stored in the fatty tissues and start to have an effect on a species when the fatty tissues are metabolised.

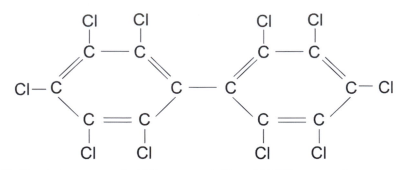

Figure 9.18 Chemical structure of a PCB compound. About 200 PCB compunds exist.

Accumulation rates of POPs depend on species, the level of fat it contains and its position into the food web. Top predators or carnivores typically contain more POPs than herbivores because of biomagnification. For example, common sole (*Solea solea*) had the higher rates of halogenated hydrocarbons (PCBs and DDT) in a study made in the Weser Estuary (Germany; Figure 9.19). This fish contained a high level of fat and is a carnivore. The shrimp *Crangon crangon* accumulated high level of PCBs in contrast to other halogenated hydrocarbons (Figure 9.19). The molluscs *Mya arenaria* and *Arenicola marina* accumulated high levels of PCBs and DDT. The cockle *Cerastoderma edule* accumulated only a small fraction of halogenated hydrocarbons. Accumulation of chlorinated hydrocarbons varies among organs. In the common sole, both the brain and the liver are the organs that accumulated the highest level of DDT. Accumulation also differs between male and female of the same species.

Some PCBs, pesticides and organo-metallic compounds are similar to oestrogens, which are sex hormones that control sex organ development (*815*). In the 1960s–1970s, cuprous oxide (Cu_2O), a biocide used in antifouling paint, was substituted by tributyltin (TBT). Rapidly, some studies provided evidence that TBT affected non-target species such as the commercially cultivated Pacific oyster *Crassostrea gigas* in Arcachon (France), female dogwhelks (*Nucella lapillus*) in the United Kingdom, sting winkles (*Ocenebra erinacea*) in France, emarginate dogwinkles (*Nucella emarginata*), channel dogwinkles (*Nucella canaliculata*) in California and eastern mudsnail (*Nassarius obsoletus*) in Connecticut (*832*). The organo-metallic compound triggered masculinisation of many female gastropods (*833*). Even with a low level of TBT (undetectable by analytic chemistry), all females developed a 'penis-like' structure, a phenomenon termed '**imposex**' or **pseudohermaphroditism**. Laboratory experiments later demonstrated that TBT was the causative agent. The TBT has also been found responsible for intersexuality or '**intersex**' (i.e. the gradual transformation of a female into a sterile male) (*833*). This phenomenon has been found in the periwinkle *Littorina littorea* for high TBT concentrations. The antifouling agent also acts as a growth inhibitor. TBT was partly banned (vessels <25 m) from French waters in 1982 and Irish waters in April 1987. In marine mammals, sex is determined genetically, but in other invertebrates the same phenomenon can arise when concentration of some halogenated hydrocarbons is high.

Intersex has also been documented in wild populations of gonochorist teleosts and in reptiles living in polluted environments (*834*). For example, this phenomenon has been detected in the Mediterranean swordfish *Xiphias gladius*. On the 162 individuals tested by de Metrio and colleagues, 40 (~25%) showed evidence of intersex. The authors

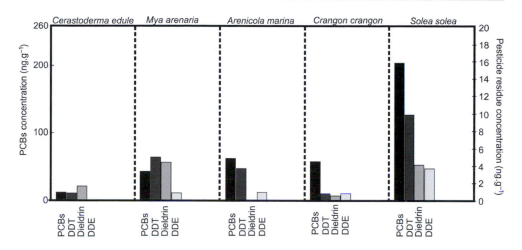

Figure 9.19 Average concentration of halogenated hydrocarbons (PCBs, DDT, dieldrin, DDE) in some animals in the Weser Estuary.

Source: Simplified from Clark and colleagues (*815*).

hypothesised that this was related to the presence of oestrogen-mimicking substances in the environment (PCBs, pesticides, PAHs). TBT is also very toxic for coral zooxanthellae. The organo-metallic compound inhibits Photosystem II at concentration as low as $50ng·L^{-1}$ or 50 parts per trillion (*428*).

9.2.2.3 Pollution by toxic metals

Metals are conservative pollutants and many species have small capacity to excrete these pollutants from their organism. As a result, they tend to accumulate (bioaccumulation) and concentrate from low to high trophic levels (biomagnification) (*815*). Islam and Tanaka (*824*) listed the 10 most toxic metals, by order of decreasing toxicity: mercury, cadmium, silver, nickel, selenium, lead, copper, chromium, arsenic and zinc. Most (e.g. mercury, silver, copper, cadmium and zinc) are classified as heavy metals, which are defined as metals having a specific density of more than $5g·cm^{-3}$. Although being a metalloid, arsenic is also classified as a heavy metal. Because heavy metals are a poorly defined category, I prefer to use here the term toxic metals. Contrary to POPs that tend to concentrate in tissues rich in lipid, toxic metals tend to concentrate in tissues rich in proteins such as muscle and liver. Therefore, they may be more important in the gonads of female fish than males.

Plankton are sensitive to a variety of pollutants, and both plankton abundance and diversity have been used as water quality indicators. Because they often concentrate pollutants (high surface-to-volume ratio), the chemical analysis of plankton provides information on the level of some heavy metals that may be difficult to measure directly in seawater.

Primary production has been shown to be inhibited by toxic metals. Kaladharan and colleagues (*835*) investigated the influence of five metals (cadmium, nickel, lead, chromium and mercury) on primary production off Cochin (Indian Ocean). They found that all metals inhibited rapidly primary production, even a low concentration (Figure 9.20).

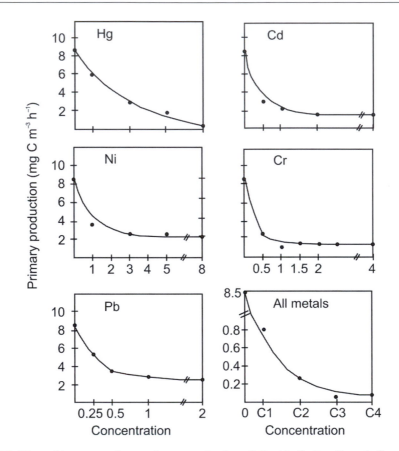

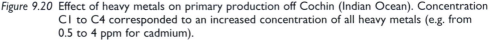

Figure 9.20 Effect of heavy metals on primary production off Cochin (Indian Ocean). Concentration C1 to C4 corresponded to an increased concentration of all heavy metals (e.g. from 0.5 to 4 ppm for cadmium).

Source: Redrawn from Kaladharan and colleagues (*835*).

During the in situ experiment, the main phytoplankton species were *Rhizosolenia calcaravis, Biddulphia mobiliensis, Ceratium* sp. and *Asterionella japonica*. When all metals were considered together, primary production diminished quickly with increasing metal concentrations.

In general, toxic metals concentrate throughout the species life cycle. For example, in the Adriatic Sea, Storelli and colleagues (*836*) showed that mercury concentration and the weight of the blackmouth catshark *Galeus melastomus* were significantly positively correlated (Figure 9.21).

Some species such as tuna (*Thunnus* spp.), swordfish (*Xiphias gladius*) and marlin (*Makaira indica*) contain high mercury concentration (*815*). Halibut (*Hippoglossus hippoglossus*), which may weigh 300 kg and live over 50 years, contains sometimes high methylmercury concentrations that prevent its human consumption.

Marine invertebrates, and especially bivalves, concentrate toxic metals. For example, oysters accumulate copper by a factor of 7,500. The flesh can even become green when copper reaches high levels. Among filter-feeders, scallops can concentrate high metal levels in their

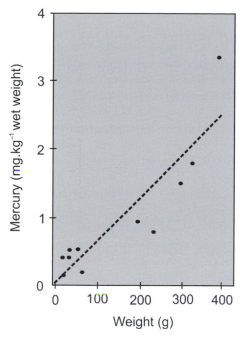

Figure 9.21 Mercury concentration in dorsal muscle as a function of the weight of the blackmouth catshark *Galeus melastomus* in the Adriatic Sea.

Source: Redrawn from Storelli and colleagues (*836*).

Table 9.6 Concentration factor (mean ± standard deviation) in different compartments of the tropical scallop *Comptopallium radula* after four days of exposure.

Compartments	Cadmium	Cobalt	Manganese	Zinc
Digestive gland	107 ± 29	27 ± 4.1	29 ± 11	549 ± 92
Kidneys	391 ± 69	603 ± 298	33 ± 12	184 ± 52
Gonad	76 ± 75	3.1 ± 1.8	14 ± 6	55 ± 14
Muscle	7 ± 1.9	0.6 ± 0.3	2 ± 0.4	18 ± 5.3
Gills	142 ± 29	3.2 ± 0.5	24 ± 5.3	96 ± 25
Remaining tissues	3.2 ± 0.6	0.2 ± 0.1	0.6 ± 0.2	4.2 ± 0.9

Note: The bivalves originated from Noumea City (New Caledonia).

Source: From Metian and colleagues (*838*).

tissues (*837*). King scallops (*Pecten maximus*) concentrate nickel in their kidneys (22.9 ppm), their digestive gland (3.54 ppm) and to a lesser extent in their muscle tissue (0.04 ppm). Similar results were found in the tropical scallop *Comptopallium radula* off New Caledonia (*838*). The metals mainly accumulated in the kidneys, the gills, and the digestive gland, and to a lesser extent in the gonad and muscles (Table 9.6).

Both gonad and kidneys are, in general, the compartments that accumulate lead in king scallop (*837*). This same species accumulates cadmium mainly in the kidneys (concentration factor FC = 928 ± 547), the digestive gland (FC = 322 ± 175), the gills (FC = 277 ± 102) and the foot (FC = 265 ± 74). The gonad and the adductor muscle do not concentrate cadmium with $FC = 45 \pm 65$ and $FC = 21 \pm 6$, respectively (839). Similar results were found for the pectinidae *Chlamys varia*.

When herring (*Clupea harengus*) eggs were exposed to mercury, the metal was at the origin of ultrastructural changes (e.g. deformation of the sarcoplasmic reticulum and mitochondria and size reduction of myofibrils in muscle cells of fish larvae) (840). Similar findings have also been found when *C. harengus* larvae were exposed to zinc, aluminium and copper.

Toxic metals may alter the function of immunocompetent cells by a diversity of physiological pathways and lead to immunosuppression (841). Studies have shown that toxic metals affect phagocytosis, killer cell activity and transformation of lymphocytes in some marine mammals such as the harbour seal (*Phoca vitulina*). For example, lymphocyte division is inhibited by lead, cadmium, mercury and beryllium in newborn pups of harbour seals (842). In contrast, metal pollutants may be at the origin of immunoenhancement that triggers hypersensitivity and autoimmunity.

Experimental intoxication of harp seals (*Pagophilus groenlandicus*) with methylmercury (25 mg·kg^{-1}) triggered lethargy, weight loss and led to animal death (843). Shlosberg and colleagues (844) showed that an excess of lead was responsible for liver damage and the death of a bottlenose dolphin (*Tursiops truncatus*). Further investigation revealed that death was caused by renal failure, uraemia and toxic hepatitis.

Since the early 1980s, dozens of new diseases appeared on corals. Causes of such infection are not clear, but may reflect ecosystem health. In the 1980s, a bacteria close to Cholera killed many corals of the genus *Acropora* in the Caribbean Sea.

Pollutants alter fish behaviour through an effect on cognition (e.g. learning and awareness), brain function (e.g. dysfunction of neurotransmitters), senses (e.g. olfaction, vision, taste) and prey-predator relationships, and may have transgenerational effects (845). A study showed that one-month larvae of mummichog (*Fundulus heteroclitus*) bred from adults exposed to contaminants (Piles Creek, PC, New Jersey) were more prone to predation than larvae originated from unpolluted sites (846).

9.2.3 Radioactive substances

Radioactive substances occur naturally in the oceans and seas (847), and 1 m^3 of seawater classically contains 1 **becquerel** (Bq; 1 Bq = 1 disintegration per second) of tritium (^{3}H), 4 Bq of carbon-14 (^{14}C), 40 Bq of uranium-238 (^{238}U), 4 Bq of radium-226 (^{226}Ra), 4 Bq of polonium-210 (^{210}Po) and 12,000 Bq of potassium-40 (^{40}K) (848). However, since the end of the Second World War, human activities related to nuclear weapons, nuclear-powered ships and nuclear reactors have produced, by nuclear fission, radioactive substances such as strontium-90 (^{90}Sr) and caesium-137 (^{137}Cs). These radionuclides are chemically similar to calcium and can thereby be assimilated by species. Their half-life (i.e. the time needed for a half disintegration of the radionuclide) is ~30 years. As heavy metals, they can bioaccumulate in species and biomagnify in the food web and cause potential hazards. Other examples of anthropogenic radionuclides are plutonium-239 (^{239}Pu) and technetium-99 (^{99}Tc), which have half-lives of 24,000 and 210,000 years, respectively (847).

Many accidents have triggered radionuclide pollution in the marine environment (815): nuclear-powered satellite in 1964, crash of a US B-52 bomber off Spain and loss of US and Russian nuclear-powered submarines. The Chernobyl accident in April 1986 had considerable consequences over Europe. The Baltic Sea and, to a lesser extent, the North

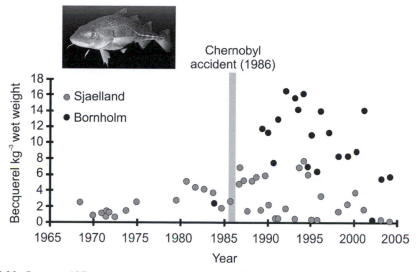

Figure 9.22 Caesium-137 concentrations in Atlantic cod *Gadus morhua* sampled at Bornholm and Sjaelland (Baltic Sea) before and after the Chernobyl accident.

Source: Redrawn from Dahllöf and Andersen (*847*). Photo of the Atlantic cod from FishBase (www.fishbase.org).

and Irish Seas were affected. Caesium-137 concentration increased from 0.018 Bq·L^{-1} prior to the accident to 0.18 Bq·L^{-1} in October–November 1986 in the Baltic Sea, an increase of a factor of 10. Concentrations of ^{137}Cs in the Atlantic cod *Gadus morhua* increased substantially after the accident in two sites of the Baltic Sea (Figure 9.22).

Similar results were also found for herring and flatfish (*847*). Although levels of radiocesium were not apparently dangerous for humans (*815*), even low exposure levels may constitute an issue. Zhao and colleagues (*849*) also showed that the mangrove snapper *Lutjanus argentimaculatus* accumulated ^{137}Cs from ingested preys. Biomagnification took place rapidly, and ^{137}Cs accumulated in the muscles.

The Fukushima Daiichi nuclear accident on 11 March 2011, which was the consequence of an earthquake and a resulting tsunami, discharged large quantities of anthropogenic radionuclides in the north-western Pacific Ocean. In July 2011, levels of ^{137}Cs were 10,000 times more elevated than levels measured in 2010 in the coastal waters off Japan (*850*). The amount of ^{137}Cs detected in coastal regions off Fukushima was well above previous concentrations ever recorded in the marine environment (Figure 9.23). For example, radiocesium was 1,000 times higher than ^{137}Cs found in the Baltic Sea after the Chernobyl accident in 1986.

Before this nuclear accident, concentrations of ^{137}Cs off Japan ranged between 1 and 2 Bq·m^{-3}. Although it remains too early to draw any firm conclusions on this accident, recent studies showed that ^{137}Cs off Japan has started to bioaccumulate and biomagnify (*851, 852*). Buesseler and colleagues (*852*) showed that radionuclides were dispersed from Fukushima and exhibited a spatial distribution that followed hydro-dynamics. In particular, the Kuroshio Current formed a barrier against radionuclide dispersal southwards. The authors quantified the concentration of radionuclides (i.e. the degree of radionuclide enrichment in species relative to seawater) to examine the potential implications for seafood

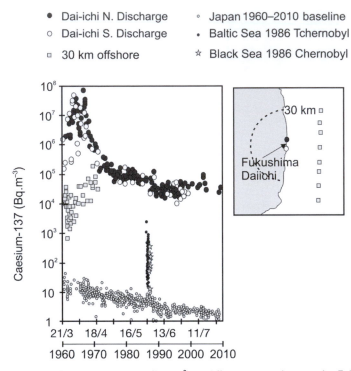

Figure 9.23 Caesium-137 concentrations (Bq·m^{-3}) in different sites close to the Fukushima Daiichi nuclear power plant measured between 21 March and 31 July 2011. These measurements are compared to historical records off the coast of Japan (in grey) and concentration data of ^{137}Cs measured in the Black and Baltic Seas after the Chernobyl accident.

Source: Redrawn from Buesseler and colleagues (*850*).

consumption. They found a median concentration factor of 44 for ^{137}Cs for zooplankton, a value corresponding to the recommended International Atomic Energy Agency (IAEA) value. However, concentration factors ranged between 10 and 285 depending on species composition and location. For example, concentration factors were 19 for gelatinous zooplankton (*Chrysaora* sp.), between 9 and 39 for *Euphausia pacifica* and reached 285 for a sample of copepods.

Along the Californian coast, Pacific bluefin tuna (*Thunnus orientalis*) contaminated by ^{137}Cs from the Fukushima Daiichi nuclear accident were found (*852*). Madegan and colleagues (*852*) measured high ^{137}Cs concentration (6.3 Bq·kg^{-1} dry weight) in 15 Pacific bluefin tuna sampled in August 2011 along the Californian coasts (Table 9.7). No caesium-137 was detected in tuna prior to the Fukushima Daiichi nuclear accident.

These results suggest that transportation of radionuclide pollution by some marine organisms such as the Pacific bluefin tuna is likely (Figure 9.24). Such transportation may be expected for other species such as fish (salmon sharks), turtles (loggerhead turtle) and seabirds (sooty shearwater) (Figure 9.24).

Table 9.7 Measured concentrations of caesium and potassium in Pacific bluefin tuna in 2008 and 2011.

Year	Statistics	Length (cm)	Body mass (kg dry weight)	Age (years)	^{134}Cs	^{137}Cs	^{40}K
2008	Mean	66.2	1.5	1.4	0	1.4	266
	SD	1.2	0.09	0.05	0	0.2	43
2011	Mean	66.2	1.5	1.5	4	6.3	347
	SD	3.6	0.2	0.1	1.4	1.5	49

Note: SD: standard deviation. Radionuclide concentrations are in Bq·kg^{-1} dry weight. Potassium-40 is a natural radionuclide.

Source: Simplified from Madigan and co-workers (*852*).

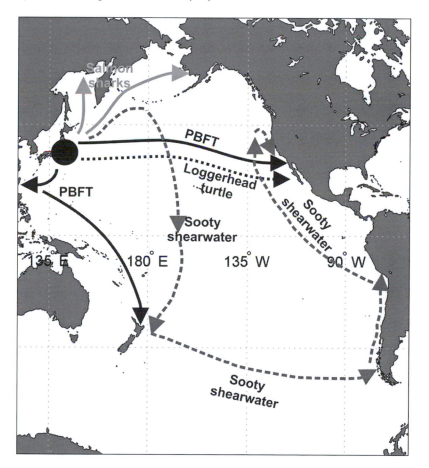

Figure 9.24 Simplified migration patterns of some marine species that may transport radionuclide from Japanese coasts to other ecoregions of the Pacific Ocean. Migration path of salmon sharks (grey arrow), sooty shearwaters (grey dashed arrows), Pacific bluefin tuna (PBFT; black arrows) and loggerhead turtles (black dotted arrows). The black circle is the main area contaminated after the Fukushima Daiichi nuclear accident.

Source: Redrawn from Madigan and co-workers (*852*).

9.2.4 Noise pollution

Infrastructures such as ship engines and sonars make the oceans and seas so noisy that they may threaten marine biodiversity. Noise pollution may have strong consequences on seabirds and marine mammals. Noise from aircraft can disrupt seabird colonies in different ways. Parents can abandon their eggs or chicks for a while long enough to increase predation vulnerability (766). Noise such as naval sonar or seismic surveys can lead to cetacean death. Strandings of beaked whales (family Ziphiidae), one of the deepest-diving cetaceans, coincided temporally and geographically with naval sonar operations (853). These marine mammals died within hours by stranding or died directly at sea from ephemeral exposure to moderate levels of mid-frequency military sonar. Fernandez and colleagues investigated lesions of beaked whales after a mass stranding that took place in the Canary Islands during military naval exercises on 24 September 2002. Strandings started only four hours after the onset of mid-frequency sonar activity. Post-mortems were carried out on eight Cuvier's beaked whales (*Ziphius cavirostris*), one Blainville's beaked whale (*Mesoplodon densirostris*) and one Gervais's beaked whale (*Mesoplodon europaeus*). Whereas no pathogen was detected, the whales showed severe and diffuse congestion (i.e. excessive accumulation of blood in vessels) and haemorrhage, mainly around the acoustic jaw fat, ears, brain and kidneys. Gas bubble-associated lesions and fat embolism were detected in the vessels and the parenchyma of vital organs. Fernandez and colleagues termed this pathology the 'gas and fat embolic syndrome'.

9.2.5 Thermal pollution

Thermal pollution occurs when waters used for cooling industrial plants (e.g. nuclear energy stations and electric thermal stations) are discharged in the sea. The efficiency of power stations is comprised between 40% and 45%, which suggests that the amount of water discharged by power plants is substantial (854). A 1,800–2,000 MW power station utilises ~60 m³·s⁻¹ of cooling water. As a result, temperatures may rise as high as 6–12°C and may alter coastal biodiversity. If the new thermal regime becomes outside species' thermal niches, growth and reproduction may rapidly impair. On the other hand, thermal pollution may allow some species usually limited by a cold thermal regime to extend their seasonal length and their growth and reproduction. By this way, thermal pollution may facilitate the establishment of new species. Thermal pollution affects both **meroplankton** and **holozooplankton** species.

The consequences of a power plant on copepods and some meroplanktonic species were examined in the Bahía Blanca Estuary (Argentina) (855). Mortality rates were calculated for the copepods *Acartia tonsa* and *Eurytemora Americana*, the crab larvae *Chasmagnathus granulata* and the invading cirriped *Balanus glandula*. Values of mean total mortality were up to four times more elevated at the site where water discharge takes place than at the control site. In the Indian Ocean, the addition of warm water along the coast increased the proliferation of some species such as blue-green algae, which have thermal preferendum ranging between 35°C and 40°C (856). Thermal pollution also reduces oxygen concentration (i.e. an increase of 10°C reduces by 20% oxygen concentrations in the water column). Thermal pollution is, in general, a local issue and has sometimes no major effect on organisms, especially on benthic communities (857), but also on copepods (858).

9.3 Nutrient enrichment and eutrophication

To synthetise organic compounds, phytoplankton, seaweeds and plants require nitrogen and phosphorus. The below reaction summarises the relationships between carbon, hydrogen, oxygen, nitrogen and phosphorus in the marine environment for a photosynthetic species.

$$106CO_2 + 16NO_3^- + HPO_4^{-2} + 139H_2O \longleftrightarrow C_{106}H_{261}O_{109}N_{16}P +$$

$$138O_2 + 18OH^- \tag{9.1}$$

This reaction shows why nitrogen is limiting for photosynthetic organisms and confirms the theory of the German agricultural chemist Justus von Liebig (859). Photosynthetic organisms also need four other essential nutrients: potassium, calcium, magnesium and sulphur. They further require seven micronutrients: iron, chlorine, copper, manganese, zinc, molybdenum and boron. Some taxonomic groups also need other chemicals to build their skeleton; diatoms use silica to build their **frustule**.

9.3.1 Causes of eutrophication

Eutrophication is caused by nutrients accumulation (phosphorus and nitrogen), which stimulates algal production in estuaries or coastal systems. Figure 9.25 shows that primary production increases with dissolved inorganic nitrogen.

Eutrophication has been reported in many estuaries and coastal systems, from the Baltic, Adriatic, and Black Seas, to the estuaries and coastal waters of Europe, China, Japan and Australia, as well as in North America (861). Nitrogen and phosphorus loading has increased dramatically in both estuarine and coastal systems (Figure 9.26).

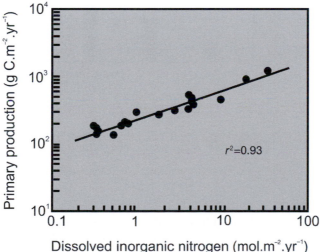

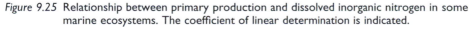

Figure 9.25 Relationship between primary production and dissolved inorganic nitrogen in some marine ecosystems. The coefficient of linear determination is indicated.

Source: Redrawn from Conley (860).

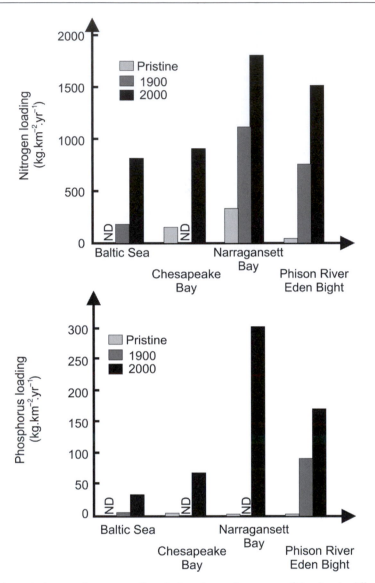

Figure 9.26 Annual nitrogen (upper panel) and phosphorus (lower panel) loading in different estuarine and marine systems of the world under pristine conditions, in 1900 and 2000. ND: no data.

Source: Redrawn from Conley (*860*).

This increase has been observed in the Baltic Sea, Chesapeake Bay, Narragansett Bay and the Phison River/Eden Bight. A 6–50 times increase in nitrogen and 18–180 times increase in phosphate loading have been observed, reducing the ratio N/P (Figure 9.27).

Increase in both frequency and magnitude of eutrophication around the world is the result of population growth and associated food production (e.g. agriculture, chemical fertilisers,

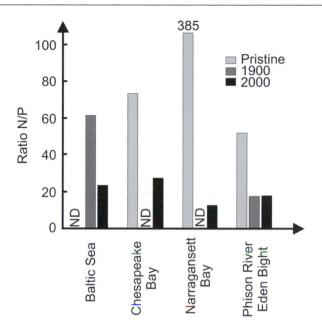

Figure 9.27 Nitrogen/phosphorus ratio loading in different estuarine and marine systems in pristine conditions, 1900 and 2000. ND: no data.

Source: Redrawn from Conley (*860*).

Table 9.8 Percentage of sewage treated in some regions.

Regions	Percentage of sewage treated
Africa	<1
Latin America and Caribbean	14
Asia	35
Europe	66
North America	90

Source: From Martinelli (*863*).

wastewater, urban run-off, aquaculture and atmospheric deposition of nitrogen from burning fossil fuels) (*862*). For example, Table 9.8 shows the percentage of sewage treated in some parts of the globe. This percentage is small in Africa, Latin America, Caribbean and Asia.

9.3.2 The main effects of eutrophication

The main symptom of eutrophication is a substantial increase in chlorophyll a. This, in turn, affects turbidity and triggers the dominance of a few species either in pelagic or benthic habitats. As a consequence, dramatic disruptions of coastal systems have been observed, sometimes leading to potentially toxic or **harmful algal blooms**, depletion in dissolved oxygen and an alteration in biodiversity. Figure 9.28 summarises the cascade of effects caused by

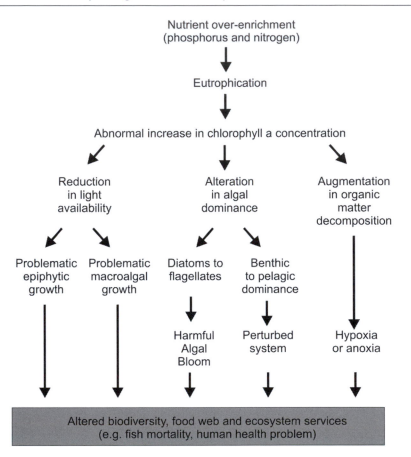

Figure 9.28 Consequences and pathways of eutrophication in estuaries and coastal systems.
Source: Modified from Bricker and co-workers (864).

eutrophication on estuarine and coastal systems. Resulting hypoxia, or anoxia in some extreme cases, often kills most of the local biodiversity.

Excessive nitrogen and phosphorus flows have modified both the structure and the functioning of coastal systems with potential implications for some natural biogeochemical cycles (570). Phosphorus and nitrogen enrichment reduces locally silicate concentration, which causes a shift from diatom-dominated to dinoflagellate or cyanobacteria-dominated systems (860).

HABs have paralleled the increase in nitrogen loading in many regions (865). For example, the relative abundance of the diatom *Pseudonitzschia*, known to produce a neurotoxin called domoic acid and responsible for the human illness termed amnesic shellfish poisoning (ASP), is positively correlated to nitrate loading in the northern part of Mexico (Figure 9.29).

Correlations between paralytic shellfish toxins (PST) produced by dinoflagellates and human population growth over Puget Sound in Washington, DC (USA) have also been documented. In China, blooms of the toxic dinoflagellates *Prorocentrum* spp. and *Karenia mikimotoi* have increased in duration (from weeks to months) and regional extent (from one to tens of km²). This pronounced increase has been attributed to the increase in the

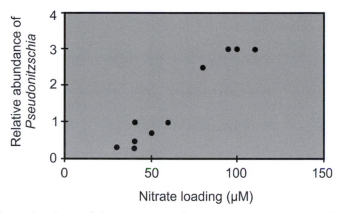
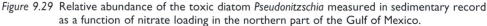

Figure 9.29 Relative abundance of the toxic diatom *Pseudonitzschia* measured in sedimentary record as a function of nitrate loading in the northern part of the Gulf of Mexico.

Source: Redrawn from Heisler and co-workers (865).

concentration of chemical fertilizers in seawaters. The dinoflagellate *Chatonella*, which forms red tide, killed many fish in Florida and PST-producer dinoflagellate *Gonyaulax* killed herring in the Bay of Fundi (Canada). In the Baltic Sea, exceptional blooms of the toxin-producing prymnesiophyte flagellate *Chrysochromulina polyepsis* in May and June 1988 was at the origin of massive mortality (866). Massive fish kills have also been reported in some American estuaries and related to the presence of the dinoflagellate *Pfisteria piscicida*.

In coastal regions of the North Sea, blooms of mucus-forming HAB prymnesiophyte species *Phaeocystis globosa* have been associated with the nitrate content of coastal waters (867). The biomass of such blooms can be impressive (1–5 gC·m^{-3}) and colonies of *Phaeocystis globosa* comprise thousands of cells included in a mucilaginous matrix. Such blooms, which typically appeared at the end of the silica-controlled diatom seasonal maximum in the English Channel and North Sea, are unpalatable and associated with poor zooplankton and fish biomass.

Cyanobacterial species, which fixes atmospheric nitrogen, needs phosphorus. The increase in phosphorus loading is also expected to increase the frequency and the intensity of cyanobacteria blooms such as bright green, yellow-brown and red blooms (868). Severe **eutrophications** have taken place in the Baltic Sea and many large-scale blooms of cyano-bacteria have been attributed to eutrophication (Plate 9.2).

Cyanobacterial blooms modify trophodynamics and biodiversity. Because they are often unpalatable, blooms are subsequently consumed by bacteria, which cause oxygen depletion and lead to hypoxia or anoxia. The reduction in oxygen concentration in turn kills fish and benthic species including shellfish. Summer blooms of *Cyanobacterium nodularia* have been reported in the Harvey Estuary (western part of Australia) and attributed to increasing riverine input from rivers (869). Because cyanobacterial blooms are mainly related to an increase in phosphorus loading, they tend to increase the level of nitrogen and aggravate eutrophi-cation.

Both occurrence frequency and severity of marine diseases have increased since 1980 (870). From field experiments, Bruno and colleagues (870) provided evidence that nutrient enrichment significantly amplified the severity of aspergillosis of the common gorgonian sea fan *Gorgonia ventalina* and yellow band disease of the reef-building corals *Montastraea*

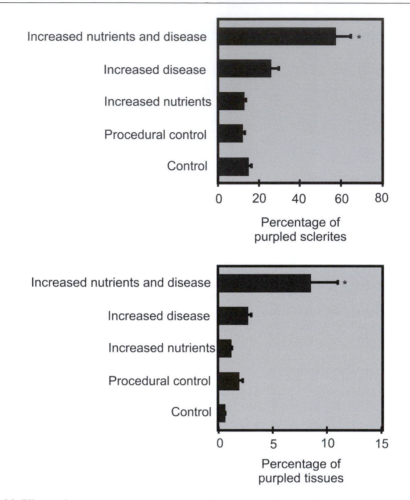

Figure 9.30 Effects of nutrients concentration on the severity of aspergillosis in experimentally infected sea fan colonies. Bars are untransformed mean ±1 standard deviation. Asterisks indicate significant differences at $p < 0.05$. Control is unmanipulated and procedural control includes a bag without nutrients to check for artefacts originating from the inoculations.

Source: Redrawn from Bruno and colleagues (*870*).

annularis and M. *franksii*. For example, the fungus *Aspergillus sydowii* is a pathogen of immune-compromised hosts. Infection is diagnosed by the presence on the coral of darkly pigmented tissue (purpled tissue) located close to necrotic patches that can affect the whole colony. By manipulating nitrogen concentration, Bruno and colleagues showed that the severity of the infection was significantly higher when nutrients concentration was high (Figure 9.30). Marine fungi and bacteria are, in general, limited by nitrogen, and the addition of this limiting factor in the marine environment tends to stimulate their growth and virulence. However, the authors also stressed that climatic factors such as temperature and rainfall may also modulate the severity of epizootics on corals (Chapter 7).

In the same way, nutrients loading increase the replication of marine viruses. Wilson and colleagues (871) showed that the virulence of a cyanophage on the Phycoerythrin-containing *Synechococcus* sp. increased with phosphate concentration.

Coral reefs of Kaneohe Bay (Hawaii, USA) were highly impacted by nutrient enrichment (766). The increase in nitrogen stimulated phytoplankton blooms, which reduced the amount of light that reached the seabed. As a result, sponges in the benthos and green algae overgrew coral species and the reefs regressed substantially in the late 1970s. In the 1980s, the diversion of sewage led to the rapid recovery of the reefs.

9.4 Oxygen depletion

Oxygen is a key variable for most marine species (872). Diaz summarised the importance of this variable by citing the motto of the American Lung Association: 'if you cannot breath, nothing else matters'. Hypoxia also affects biogeochemical cycling of elements and has economic consequences because it removes many exploited species. Oxygen depletion in estuaries and coastal ecosystems are widespread (Plate 9.3). Hypoxia has been reported in many systems along all continents. Although some recoveries have been observed, this remains a major environmental issue. Sensitivity of marine systems to anthropogenic hypoxia is a function of their level of energy (e.g. tidal, wind, currents) and freshwater inputs. Semi-enclosed seas such as the Baltic Sea are particularly vulnerable to hypoxia.

9.4.1 Hypoxia caused by eutrophication

By producing exceptional organic matter, eutrophication is a major cause of hypoxia. Thresholds are difficult to determine, but in general many animals start to be affected when oxygen concentration falls below 2 $mgO_2 \cdot L^{-1}$ (874). Vaquer-Sunyer and Duarte (874) examined 872 published experiments reporting the effect of oxygen depletion on 206 marine species. They found that marine species vulnerability to hypoxia varies from one taxonomic group to another (Figure 9.31). When all organisms were considered, median lethal oxygen concentration (LC_{50}; oxygen concentration at which 50% of organisms die) was 1.60 ± 0.12 $mgO_2 \cdot L^{-1}$ (median and standard deviation) or 2.05 ± 0.09 $mgO_2 \cdot L^{-1}$ (mean and standard deviation). 90% of species had $LC_{50} < 4.59$ $mgO_2 \cdot L^{-1}$. Gastropods are in general less sensitive than bivalves, fish and crustaceans (Figure 9.31A). Sublethal effects (SLC_{50}; i.e. oxygen concentration at which 50% of the organisms show sublethal effects) are, on average, more rapidly observed in fish, crustaceans and molluscs (Figure 9.31B). For example, the Atlantic cod *Gadus morhua* is very sensitive and increases its ventilatory water flow when oxygen concentrations are less than 10.2 $mgO_2 \cdot L^{-1}$. In contrast, the burrowing shrimp *Calocaris macandreae* switches from aerobic to anaerobic metabolism when oxygen concentration falls below 0.085 $mgO_2 \cdot L^{-1}$. Median lethal time LT_{50} (i.e. the time at which 50% of the individuals of a population die after being exposed to hypoxia) ranged from 23 minutes for the flounder *Platichthys flesus* to 32 weeks for the bivalve *Astarte borealis* (Figure 9.31C). Molluscs and cnidarians are highly tolerant to oxygen depletion. The authors also provided evidence that sessile organisms are more tolerant than mobile organisms and detected ontogenic shifts in survival time, earlier stages having shorter survival times.

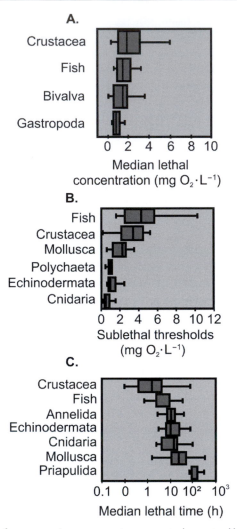

Figure 9.31 Vulnerability of some marine taxonomic groups to hypoxia. (A) Box-plot of LC_{50} for four taxonomic groups. (B) Box-plot of SLC_{50} for six taxonomic groups. (C) Box-plot of LT_{50} for seven taxonomic groups. LC_{50} and SLC_{50}: oxygen concentration at which 50% of the organisms die or show sublethal effects, respectively. LT_{50}: the time at which 50% of the individuals of a population die after being exposed to hypoxia. In this study, the threshold was oxygen $< 2mgO_2 \cdot L^{-1}$.

Source: Simplified from Vaquer-Sunyer and Duarte (874).

Hypoxia is reinforced by some natural processes such as water column stratification, which is ultimately determined by regional hydrodynamics and climate. Hypoxia therefore tends to occur or reinforce in summer in extratropical regions. The sequence of biological events caused by eutrophication-modulated hypoxia is well known (Figure 9.32).

At first, nutrient enrichment stimulates ecosystem production, especially when the system was initially oligotrophic (low nutrients concentration). When the system becomes eutrophic (i.e. high nutrients concentration), seasonal hypoxia starts to be observed and

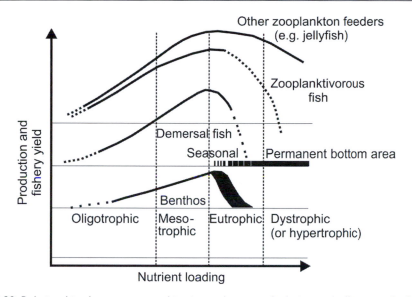

Figure 9.32 Relationships between eutrophication and oxygen depletion and effects on the benthos, fish and other zooplankton feeders.

Source: Redrawn from Diaz (872).

mass mortality episodes occur in summer. The benthos is first impacted, and if nutrient enrichment continues demersal fishes diminish. When eutrophication increases, seasonal hypoxia becomes chronic and zooplanktivorous fish collapses, being replaced by other taxonomic groups such as jellyfish. Dystrophic (or hypertrophic; extreme concentration in nutrients) conditions are characterised by permanent bottom anoxia that are not suitable for benthic invertebrates and demersal fish. Only a few zooplanktivorous fish and jellyfish can live in such conditions.

The Black Sea has been highly impacted by eutrophication (875). This sea contains about one-third (3,700 species) of the Mediterranean Sea species richness (10,000 species) (Table 9.9). Most coastal waters of the Black Sea are eutrophic. The central region is mesotrophic (i.e. moderate concentration in nutrients) and the Sea of Azov and the north-west influenced by nutrients loading from the Danube, Dniester and Dnieper rivers are hypertrophic. As a result, primary production ranges from 570 to 1,200 mg carbon·m^{-2} in hypertrophic areas (876). Eutrophication has aggravated oxygen depletion. Although 87% of this sea is fully anoxic for natural reasons (e.g. geological events, basin shape), both hypoxia and anoxia have extended in time and space, which have affected Black Sea biodiversity. From 1961 to 1994 (aggravation of eutrophication), macrozoobenthic species richness (e.g. crabs) collapsed from 70 to 14 on the Romanian shelf and mass mortality of bottom animals (between 100 and 200 tonnes·km^{-2}) was reported (877). About 60 million tonnes of bottom-living animals died between 1973 and 1990, including 5,000 tonnes of fish (878).

In the Mississippi Delta, the hypoxic zone has extended to 500 km westwards due to nitrogen loading, increasing the total surface of this area from 10,000 to 20,000 km^2 during the last 20 years (879). The hypoxic zone has its seasonal maximum extent in summer where oxygen depletion covers the whole water column. Mass mortality of shrimp and fish has been reported. Biodiversity has been progressively replaced by microbes and jellyfish, and

Table 9.9 Biodiversity of some taxonomic
           groups in the Black Sea.

Taxonomic group	Species richness
Seaweed	>200
Copepoda	36
Polychaeta	192
Mollusca	210
Echinodermata	14
Arthropoda	193
Sipuncula	1
Bryozoa	18
Porifera	28
Fish	171
Bird (Danube delta)	320
Marine mammal	4

Source: From Zaitsev and co-workers (876).

blooms of HABs (e.g. the dinoflagellate *Karenia brevis*, formerly known as *Gymnodinium breve*) have become more frequently recorded.

9.4.2 Natural causes of hypoxia

In addition to **eutrophication**, I shall recall here that hypoxia also occurs naturally in the oceans and seas (880). Permanently hypoxic water masses in the ocean have been termed **oxygen minimum zones (OMZ)**. Helly and Levin estimated that there are 1,148,000km^2 of permanently hypoxic (i.e. water with oxygen concentration <0.5ml·L^{-1}) shelf and bathyal sea floor, of which 59% are located in the northern part of the Indian Ocean (Arabian Sea and Bay of Bengal), 31% occurs in the eastern Pacific Ocean and 10% in the eastern South Atlantic.

Hofmann and colleagues (881) stressed that thresholds, often expressed as mgO$_2$·L^{-1} or mlO$_2$·L^{-1} can be misleading because the same oxygen concentration can create a higher partial pressure of oxygen (P$_{O2}$) at depth and enable animal respiration in the deep-sea ocean. They recalled that partial pressure is a function of oxygen concentration, temperature, salinity and hydrostatic pressure. That is why the authors defined three types of hypoxic waters. Category A hypoxia (P$_{O2}$ < 106 matm) describes waters where species sensitive to hypoxia (e.g. commercial species) exhibit avoidance reactions and migrate in other areas with higher oxygen concentrations (Table 9.10). Category B hypoxia (P$_{O2}$ < 60 matm) describes waters where species sensitive to hypoxia (e.g. commercial species) are absent as a result of low oxygen concentration. Category C hypoxia (P$_{O2}$ < 22 matm) describes waters where unadapted species exhibit mass mortality and where only adapted species live. These authors mapped the spatial distribution of these regions in the global ocean (Plate 9.4). The margins of Mexico, Peru, Chile, Namibia, Pakistan and India are characterised by natural hypoxia. Waters with severe natural hypoxia (category C) are located in the Indian Ocean (200 m), the East Pacific (200 m) and the Black Sea.

Although these OMZs have low biodiversity, their species (e.g. protozoan and metazoan) are adapted to chronic oxygen depletion. Foraminifers and nematodes are prevalent in nearly

Table 9.10 Concentration and P_{O2} for three categories of hypoxia.

Category of hypoxia	$[O_2]$ $mgO_2 \cdot L^{-1}$	$[O_2]$ $ml\ O_2 \cdot L^{-1}$	$[O_2]$ μmol $O_2 \cdot kg^{-1}$	P_{O2} $25°C$	P_{O2} $17°C$	P_{O2} $12°C$
A	3.5	2.45	107	106	93	84
B	2	1.4	61	60	53	48
C	0.71	0.5	22	22	19	17

Note: P_{O2} were calculated for three temperatures, assuming a hydrostatic pressure of $P = 10$ bar and a salinity of 34 PSU (Practical Salinity Unit).

Source: Simplified from Hofmann and colleagues (881).

anoxic water (<0.2 ml·L^{-1}), whereas the macrofauna and the megafauna are uncommon. These areas have been rarely investigated and many species have not been inventoried yet. Chemosynthesis (sulphide oxidation) is probably very common in these areas.

Natural hydro-climatic changes affect OMZ (880). For example, the Eastward Equatorial Current in the Pacific Ocean, which carries oxygenated waters, diminished the vertical extent of the OMZ between 5°N and 5°S in the Eastern Pacific. During the south-west monsoon in the Indian Ocean, the OMZ extends to the shelf and catches of prawn and fish collapse. Any modification in atmospheric circulation, in the path of oceanic currents or in the increase of heat content of the ocean may therefore influence the location, the depth and vertical extent of the OMZs.

Even along the coast, areas of oxygen depletion occur naturally. Upwelling can bring waters with low oxygen concentration to continental-shelf surface. Along the California Current system in summer 2002, substantial changes in the biogeochemistry and the physics of the water took place (882). Inner-shelf hypoxia was identified between 44.65°N and 44.00°N. These changes were the consequences of wind-induced upwelling that advected waters with low oxygen level and high nutrients concentration. We saw above that an extensive OMZ occurs along the continental margin of the north-eastern regions of the Pacific Ocean. Oxygen concentration fell between 0.21 and 1.57 ml·L^{-1}. Such concentration was observed between 2 and 5km from the coasts and hypoxic conditions cover ~820 km^2. This event had substantial consequences on benthic invertebrates and fish (Figure 9.33). The number of schooling and benthic rockfish collapsed in July 2002 (Figure 9.33A). A remotely operated vehicle (ROV) detected many dead invertebrates and fishes (Figure 9.33B–C). Commercial fishery reported a high mortality level ($>75\%$) of crabs (Cancer magister) in pots, whereas this rate is usually nil in normal conditions. Some people found large fish aggregations in shallow but atypical places that reveal that marine species were suffering from oxygen depletion.

9.4.3 Hypoxia caused by anthropogenic climate change

Hypoxia may also be aggravated by anthropogenic climate change, because rising temperature increases ectotherm metabolism (e.g. respiration), reduces seawater oxygen concentration by solubility and exacerbates the surface isolation from deeper water by reinforcing stratification. Observations of an increase in the vertical extent of OMZs have been reported in the tropical oceans (883). Stramma and colleagues showed an expansion of intermediate-depth (300–700 m) low-oxygen zones in the eastern tropical part of the Atlantic Ocean

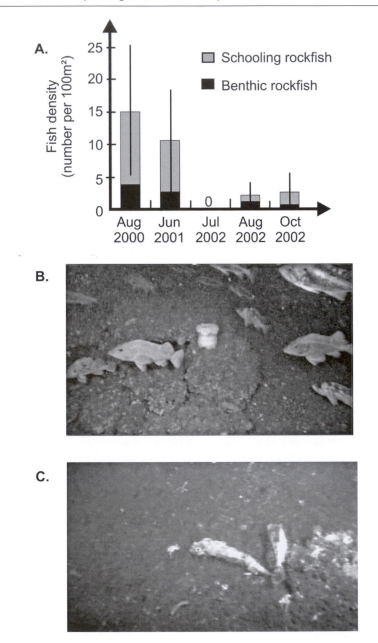

Figure 9.33 Effects of the hypoxic event that took place between July and September 2002 in the California Current system. (A) Densities of schooling and benthic rockfish (*Sebastes* spp.) along the sampling line SH (Strawberry Hill) between August 2000 and October 2002. (B) ROV survey image of rockfish in August 2000. (C) ROV survey image of rockfish in July 2002.

Source: Simplified from Grantham and colleagues (*882*).

and the equatorial part of the Pacific Ocean during the last 50 years. The researchers calculated that the reduction was on average 0.34 µmol $O_2 \cdot kg^{-1} \cdot yr^{-1}$.

Using a low-resolution earth system model, Shaffer and colleagues (884) provided evidence for an increase in OMZ extent with global warming. Expansion of suboxic regions (≤ 10 µmol $O_2 \cdot kg^{-1}$) is expected to take place in the eastern tropical part of the Pacific Ocean and the northern part of the Indian Ocean. These suboxic regions are characterised by an absence of fish and other macroorganisms and by specific biogeochemical cycles that may lead to N_2O (a strong greenhouse gas) production. Using data from the World Ocean Atlas 2005, the authors estimated that maximum suboxic oceanic areas concern 6.3×10^{12} m^2 at 500 m depth, ~1.9% of the world oceanic area. When expressed as a volume, this estimate corresponded to 4.3×10^{15} m^3, ~0.3% of the world ocean volume. Modelled changes show substantial (threefold to sevenfold) expansions in suboxic areas, corresponding to fivefold to 20-fold expansions in suboxic volumes, although these projections depended on warming intensity. OMZ expansion will increase denitrification and may trigger an increase in nitrogen fixers (see also Figure 10.27). Reduction in oxygen solubility in the next centuries to millennia is mainly due to rising SSTs. Oxygen depletion may also be strengthened by reduction in ocean overturning, convection and reduction in atmospheric P_{O2}. In their simulations, atmospheric P_{O2} diminished between 0.2095 atm and 0.2077 atm, depending on warming intensity. Processes contributing to atmospheric O_2 concentration decrease are fossil-fuel combustion, land biomass increase and outgassing from a warmer ocean. Fortunately, this reduction was small (<1%). It is worth here to recall that atmospheric O_2 concentration was considerably reduced during the Permian crisis 251 million years ago (Chapter 5) and was at the origin of atmospheric hypoxia, especially in high-altitude regions (885). The supposed O_2 reduction (~16% versus 21% today) during the late Permian may have compressed the altitudinal ranges of terrestrial species. The combination of global warming and oxygen depletion related to an increase in respiration and CO_2 reinforced marine hypoxia related to rising stratification and lack of oceanic ventilation.

9.5 Introduction and invasion of exotic species

Although species can arrive naturally in a new area, human-caused exotic introduction has become significant. Introduction of non-native species in an environment may be an important cause of ecological alteration (570, 886). In the United States, introduction of 50,000 non-native species has cost ~$137 billion per year (887). So far, marine introductions have been less investigated although their magnitude and frequency may lead to profound ecosystem shifts (888–890). It should be noted that introduction rarely leads to invasion. Only a limited number of species can establish and become invasive, the 'tens rule'. The rule states that only 10% of introduced species can establish in a given area and that only 10% may ultimately become invasive (891). Of course, deviations have been observed and the percentage varies from one system to another, between 5% and 38% (887). Non-indigenous species may own competitive, predatory and defensive strategies for which the autochthonous species have no evolutionary experience (766).

During the twentieth century, 16 exotic phytoplankton species, which belong to the taxonomic groups Bacillariophyceae, Dinophyceae and Raphidophyceae, have been introduced in the North Sea (892). For example, Edwards and colleagues (893) showed that the diatom *Coscinodiscus wailesii* introduced in 1977 in the English Channel now forms an important component of the phytoplankton community structure in spring and autumn in the Channel and the North Sea.

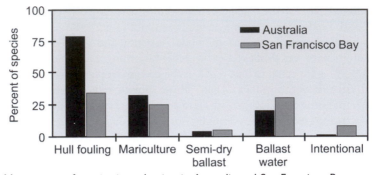

Figure 9.34 Main causes of species introduction in Australia and San Francisco Bay.

Source: Redrawn from Bax and colleagues (*894*).

Species introduction has been accidental or deliberate. Accidental invasions are mainly related to maritime transport (e.g. hull fouling, semi-dry ballast and ballast water) and mariculture. In Australia and San Francisco Bay, the main cause of species introduction is hull fouling, followed by mariculture and ballast water (Figure 9.34). Intentional release is rare in Australia and San Francisco Bay.

9.5.1 Accidental species introductions

There is an increasing concern about ballast water transfer of aquatic organisms by merchant vessels. Ballast waters have been used by ships since 1880 for balance and stability purposes. To do so, boats directly pump seawater from where they are and may transport large quantities of marine species from one port to another distant by thousands of kilometres. These vessels transport 80% of the world trade and 12 billion tonnes of ballast water every year. In Australia, 64 international ports receive about 22,000 commercial ships every year originating from 300 overseas ports (894). The International Maritime Organization (IMO) has recommended exchange of ballast water in the open ocean to prevent exotic species introductions. Carlton and Geller (888) investigated ballast water from 159 cargo ships that came from Japan in Coos Bay (Oregon, USA). They found many different species (~367 different taxa), which indicates that a large spectrum of taxonomic groups can be transported by ballast water (Table 9.11). Intentional release is rare in Australia and San Francisco Bay.

The authors cited some examples of species introduction likely to be caused by ballast water. The copepod *Pseudodiaptomus marinus* from Japan has been introduced in California and has been recently identified in some European seas. The HAB dinoflagellate *Alexandrium minutum* has been recorded along some Australian coasts. It is estimated that ~10,000 species are transported across ecoregions by ballast water (894). Ballast waters can also vehicle viral and bacterial pathogens such as *Vibrio cholerae* and cysts of toxic dinoflagellates.

Sometimes species introductions are effective and the species can establish in a region, affecting strongly native biodiversity. Zebra mussels (*Dreissena polymorpha*), native from western Russia, were first detected in the Great Lakes (Lake Saint Clair) in 1988 possibly from the hulls or the ballast waters of vessels from Europe (e.g. the Black or Caspian Seas). The freshwater species invaded the Great Lakes very quickly and was found in all Great Lakes by December 1993. Before the invasion, these lakes were characterised by high phytoplankton concentration. The species has perturbed many human infrastructures and has been held responsible for some avian botulism outbreaks.

Table 9.11 Number of species per taxonomic group in the ballast waters investigated by Carlton and Geller (*888*).

Taxonomic group	Species number	Taxonomic group	Species number
Crustacea		*Annelida*	
Cirripedia	5	Spionidae	11
Harpacticoida	5	Polynoidae	3
Calanoida and Cyclopoida	25	Other polychaeta	28
Decapoda	14	Hirudinea	1
Euphausiacea	1	*Platyhelminthes*	33
Stomatopoda	1		
Cumacea	3	*Nemertea*	1
Mysidacea	2	*Mollusca*	
Isopoda	4	Bivalvia	9
Caprellidea	1	Gastropoda	10
Gammaridea	8	*Sipuncula*	1
Hyperiidea	1	*Nematoda*	1
Ostracoda	1	*Rotifera*	1
Cladocera	1		
Chelicerata		*Cnidaria*	
Acarina	1	Anthozoa	2
Echinodermata		Scyphozoa	1
Asteroidea	1	Hydrozoa: *Obelia*	1
Echinoidea	2	Other Hydrozoa	21
Ophiuroidea	1	*Radiolaria*	2
Holothuroidea	2	*Foraminifera*	3
Chordata		*Tintinnida*	2
Urochordata	10	*Other Ciliata*	4
Pisces	2	*Dinoflagellata*	4
Hemichordata		*Diatomacea*	128
Enteropneusta	1	*Chlorophyta*	2
Chaetognatha	3	*Rhodophyta*	2
Phoronida	1	*Zosteraceae*	1
Bryozoa	3		

Source: Simplified from Carlton and Geller (*888*).

In the marine realm, the introduction of the green alga *Caulerpa taxifolia* in 1984 by an aquarium in Monaco and its subsequent invasion had severe consequences on native Mediterranean biodiversity. The alga spread rapidly, eliminating native seaweeds and seagrasses as well as associated invertebrates and fish. The alga also releases toxins that repel herbivores such as fish. Although populations introduced into California have been eliminated, C. *taxifolia* cover ~8,842.3 ha along 143.8 km of Mediterranean coasts (*895*). 97% of the available habitat has been covered by C. *taxifolia* between Toulon (France) and Genes (Italy) with strong consequences for native biodiversity. Another species has also invaded the Mediterranean Sea. *Caulerpa racemosa* is now well established in 12 Mediterranean countries (seven for C. *taxifolia*), covering an equivalent surface area of 8,070 ha, corresponding to 163.4 km of coasts.

About 600 introduced species, corresponding to 5% of the known flora and fauna, have been inventoried in the Mediterranean Sea. Such a high biological pollution has been found in San Francisco Bay, the Baltic and the Black Sea. In the Black Sea, the comb jelly *Mnemiopsis leidyi* has been responsible for the collapse of the coastal fisheries with estimated economic loss of many millions of dollars each year. The Asian invasive clam *Potamocorbula amurensis* has reached densities of 10,000 individuals·m^{-2} in San Francisco Bay, affecting coastal fisheries. The New Zealand screwshell *Maoricolpus roseus*, introduced in Tasmania in the 1920s, has invaded the region as far north as Sydney (894). The species has radically modified the habitat and regional biodiversity. Soft sediments have rapidly been covered by the hard shell of this mollusc, allowing other animals to attach on the shell (e.g. the invasive species *Undaria pinnatifida*) and hermit crabs to increase in abundance.

Inter-oceanic invasions may also become more frequent with anthropogenic climate change. In spring 1998, following an unusually large ice-free period in the Arctic, large numbers of the Pacific diatom *Neodenticula seminae* were found in samples collected by the CPR survey in the Labrador Sea and in the North Atlantic. *N. seminae* is an abundant member of the phytoplankton community in the subpolar regions of the northern Pacific Ocean and has a well-known palaeoecological history deduced from deep sea cores. According to modern surface sampling in the North Atlantic since 1948, in addition to palaeoevidence, the occurrence of the diatom in 1998 was the first record of this species in the North Atlantic for at least 800,000 years. It has since spread east and south in the Atlantic Ocean, although a diminution has recently been observed. The invasion of this species is unlikely to originate from ballast water introduction, but may have been caused by the opening up of the Arctic Ocean. Some authors have interpreted this invasion as a sign of the scale and rapidity of changes that are taking place in both Arctic and North Atlantic oceans as a consequence of global warming (896). The appearance of this diatom could be the first evidence of a trans-Arctic biogeographical movement related to Arctic sea-ice extent reduction in contemporary times.

9.5.2 Deliberate species introductions

Although species introductions are most often accidental, some introductions have been intentional. The red king crab *Paralithodes camtschaticus* was introduced in the eastern part of the Barents Sea between 1961 and 1969 by Russian scientists from the western Kamchatka peninsula to improve the regional economy (Figure 9.35). This crab, which can live 20 years, is among the world's largest arthropods, weighing about 10 kg and having a leg span of 1.8 m. They are exploited for their meat, providing a fishery of €20 million in 2011 (897).

About 1.5 million zoea I larvae, 10,000 1–3-year-old juveniles (sex ratio = 0.5) and 2,609 5–15-year-old adults (1,655 females and 954 males) were released in the field (898). Because of large food amount, red king crab population increased rapidly and moved towards Norway. At the present time, population size is estimated to be ~15 million in the Norwegian Sea. Because the crab is an opportunistic omnivore and a scavenger, it is currently strongly affecting native biodiversity of the Barents and Norwegian Seas. For example, the red king crab is in competition with cod for food, which may increase the pressure from overfishing or reduce the potential beneficial effects of global warming on this species in the Norwegian and the Barents Seas.

The Pacific oyster *Crassostrea gigas*, which originates from the Japan/Korea region, has been translocated into a number of countries such as France for aquaculture purposes.

Figure 9.35 Photo of the red king crab *Paralithodes camtschaticus*.

Source: Courtesy of Lis Lindal Jørgensen, Institute of Marine Research, Norway.

In 2006, the species represented an important part of the world shellfish culture estimated to 4.6 million tonnes. Examples are many, and shrimps and fish have been introduced in new regions to supplement fisheries.

Species introductions can have strong local or regional consequences for native biodiversity and severe social and economic consequences. Invasions are likely to continue and to interact with other stressors such as anthropogenic climate change.

9.6 UV-B radiation

The formation of the stratospheric ozone layer (SOL) 2 billion years ago enabled life development on land. SOL protects species against the damaging effects of UV-C (280–100 nm) and a large fraction of UV-B (315–280 nm). Short UV, called vacuum UV (10–100 nm) are filtered by nitrogen. UV-C are filtered at the top of the ozonosphere (35 km of altitude). Ozone is transparent to most UV-A (400–315 nm).

9.6.1 Discovery of the ozone hole

Human activities have increased the production of dangerous substances (e.g. chlorofluoro-methanes) for SOL. In 1985, Farman and colleagues from the British Antarctic Survey (899) discovered the Antarctica ozone hole by showing a drop in ozone concentration after sun reappearance each spring at two research sites (Halley and Faraday). These results confirmed earlier fears that some man-made chemicals, which may remain for a long time in the atmosphere (40–150 years), may lead to the production of a large amount of chlorine in the stratosphere and the destruction of SOL (900). NASA scientists subsequently used satellites to monitor the ozone hole in space and time (Plate 9.5) and some results showed a reduction of 50% of SOL over Antarctica (*901*). The urgency of the situation led to the Montreal Protocol signed by 190 countries in 1987 and subsequent amendments to stop CFC production.

9.6.2 Mechanisms of ozone destruction

Ozone-depleting substances (ODS) are present everywhere in the stratosphere. They enter the stratosphere mainly from the tropical upper troposphere and then travel towards polar regions. The hydroxyl radical (OH^-), the nitric oxide radical (NO^-), atomic chlorine ion (Cl^-) and atomic bromine ion (Br^-) are examples of ODS, acting as free radical catalysts. Although the first two elements are of natural origin, the last two originate from human activities. For instance, the atomic chlorine ion comes from chlorofluorocarbons (CFCs):

$$CFCCl_3 + \text{electromagnetic radiation} \rightarrow CFCl_2 + Cl \qquad (9.2)$$

Once the atomic chlorine is free, it reacts quickly with ozone (O_3) to produce a diatomic oxygen (*902*):

$$Cl + O_3 \rightarrow ClO + O_2 \qquad (9.3)$$

Subsequently, the atomic chlorine ion is reformed, completing the chlorine catalytic cycle:

$$ClO + O \rightarrow Cl + O_2 \qquad (9.4)$$

The atomic oxygen (O) forms from both ozone and oxygen molecules that react by means of solar ultraviolet radiation. These reactions prevail at tropical and middle latitudes where ultraviolet radiation is more intense. The net effect of these reactions is to remove one ozone molecule from the stratosphere:

$$O + O_3 \rightarrow 2O_2 \qquad (9.5)$$

Other reactions prevail over polar regions. The abundance of chlorine monoxide (ClO) is higher in these regions because of the formation of polar stratospheric clouds (PSCs). PSCs appear in the low part of the stratosphere in winter when temperatures are below $-78°C$. In general, stratospheric temperatures are below this threshold in the Arctic between 10 and 60 days (mid-December to mid-February; minimum temperature of $-80°C$) and for the entire winter (mid-May to mid-October; minimum temperature of $-90°C$) in the Antarctic. PSCs mainly form from the condensation of water and nitric acid (HNO_3). Progressively, PSCs move down by gravity, contributing to remove nitric acid from the stratosphere, a process called stratospheric denitrification. Ice particles may also form when the temperatures drop below PSC formation. If these low temperatures persist for several weeks, the ice particles tend to move down and reduce the water content of the stratosphere, a process termed stratospheric dehydration. By reducing the amount of water available for PSC formation, the phenomenon contributes to reduce ozone destruction.

The following series of chemical reactions takes place over polar regions (*902*):

$$(ClO)_2 + \text{visible sunlight} \rightarrow ClO_2 + Cl \qquad (9.6)$$

$$ClO_2 \rightarrow Cl + O_2 \tag{9.7}$$

$$2(Cl + O_3 \rightarrow ClO + O_2) \tag{9.8}$$

The net effect of these chemical reactions is to transform two ozone molecules into three oxygen molecules. When bromine combines with chlorine, this gives a second series of chemical reactions (902):

$$ClO + BrO \rightarrow Cl + Br + O_2 \tag{9.9}$$

Or

$$ClO + BrO \rightarrow ClBr + O_2 + \text{visible sunlight} \rightarrow Cl + Br + O_2 \tag{9.10}$$

$$Cl + O_3 \rightarrow ClO + O_2 \tag{9.11}$$

$$Br + O_3 \rightarrow BrO + O_2 \tag{9.12}$$

These two series of reactions need visible sunlight to sustain the concentration of chlorine monoxide (or bromine monoxide). They do not take place in winter because of continuous darkness, but in late winter and early spring when sunlight comes back. Ozone destruction thereby takes place after midwinter in the polar stratosphere. At this time, ozone formation is weak because the amount of ultraviolet radiation remains low.

Polar systems are isolated from other regions in winter because of strong winds circulating around the poles and forming a polar vortex. This prevents any horizontal mixing that may attenuate the ozone depletion process. Although the strength of the polar vortex over the Arctic Ocean is generally weaker than around Antarctica, ozone depletion reached high levels in 2011 (903). The strong reduction was related to the reinforcement of the stratospheric polar vortex, which led to the reduction of 80% of the ozone between 18 and 20 km altitude (Plate 9.6).

Ozone depletion has also been observed elsewhere on the planet. In 1996, PSCs were temporarily detected over the United Kingdom. In the northern hemisphere, ozone depletion has reached 10–15%. In the southern hemisphere, the reduction is even more pronounced as a result of the ozone hole over Antarctica. In the Southern Ocean, a decline in ozone concentration of 18% from 1979 to 1995 has been observed. Ozone depletion had reached 5–10% over mid-latitudes (904). Over the tropics, the average reduction is estimated at ~4%.

9.6.3 Effects of UV on biodiversity

SOL reduction has led to an increase in UV-B reaching the planet surface. For example, in New Zealand, UV-B radiation increased by 15–20% in 1998–1999 compared to levels measured in 1970. In the terrestrial realm, ecologists conducted some experiments, which revealed the adverse effects of increasing UV-B radiation (e.g. DNA damage) on shrubs in Tierra del Fuego (southern Argentina), an area sometimes affected by the Antarctic ozone hole (905).

In the marine realm, it was first assumed that UV-B radiation increase could not affect biodiversity. However, recent studies have shown that UV-B can alter significantly phytoplankton down to 60 m in the clear waters of the subtropical Atlantic Ocean (906). In the Mediterranean Sea, this depth reaches 26 m. UV-B penetration reaches 12 m in clear coastal waters, although only 10% reach 50 cm in turbid coastal waters. UV-B radiation impairs photosynthesis, growth rates and reproduction. Results from a six-week

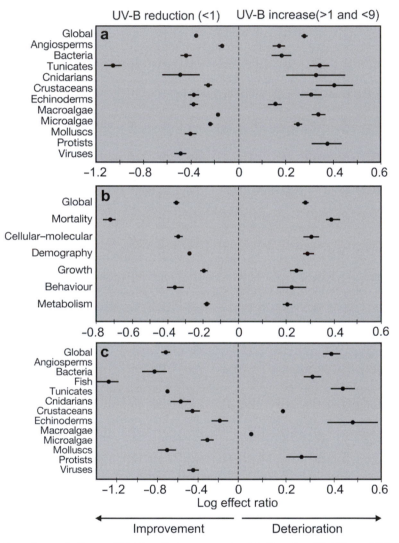

Figure 9.36 Experimental effects of UV-B radiation on marine biodiversity. (A) Effect of UV-B radiation on some marine taxonomic groups including all functions. (B) Effect of UV-B radiation on some functions including all marine taxonomic groups. (C) Effect of UV-B radiation on the mortality of some marine taxonomic groups. The left parts of the figure show an improvement in performance when UV-B radiation is reduced and the right part a deterioration in performance when UV-B radiation is increased.

Source: Redrawn from Llabrés and colleagues (906).

cruise (Icecolors) in the Marginal Ice Zone (MIZ) of the Bellingshausen Sea found UV-B photosynthesis inhibition associated with an increase in the proportion of UV-B reaching the photic zone (*901*). The researchers estimated that ozone depletion was responsible for a reduction in primary production ranging from 6% to 12%.

Llabrès and colleagues performed a meta-analysis on the effects of UV-B radiation on marine biodiversity (*906*). They considered in their analyses 1,784 experiments encompassing a wide range of organisms (Figure 9.36). They found that 91.4% of these experiments detected an adverse effect of UV-B radiation on marine species: (1) a reduction in UV-B improving performance; or (2) an increase in UV-B affecting performance (Figure 9.36A). Processes that were most affected when all taxonomic groups were combined were mortality, followed by cellular and molecular processes, behaviour, demography, growth and metabolism (Figure 9.36B). Mortality of marine species decreases on average by 81% when UV-B declines (Figure 9.36C). Fish and bacteria were the most affected taxonomic groups, followed by molluscs, tunicates and echinoderms.

Llabrès and co-workers considered exclusively species that grow near the oceanic surface or with a life stage occurring in the photic zone (e.g. echinoderms). The analysis provides evidence that UV-B radiation has a strong effect on marine biodiversity. At present, it is not known whether marine species will be able to adapt to rising UV-B radiation by developing protective mechanisms or by photoadaptation. Young stages (e.g. eggs and larvae) will be more affected by rising UV-B radiation.

Rising UV-B radiation may be one of the factors responsible for abundance decline observed in the Southern Ocean between 1970 and 2003 (*471*) (see also temperature effects in Chapters 5–7). This stressor may have also affected coral reefs in addition to warming and may have reduced the abundance of *Prochlorococcus* in oligotrophic oceanic regions. Although the potential influence of UV-B radiation on marine biodiversity is still pending, this stressor may have significant effects and interact with other factors such as eutrophication, acidification and global warming (see Section 9.9.3).

9.7 Tourism

Tourism may have some local effects on marine biodiversity. For example, two-thirds of the degradation on corals in the Red Sea may be due to tourists who take or buy a piece of coral, and to unexperienced divers. Coastal regions are the main areas influenced by tourism. Fragile areas such as mangroves and coral reefs should be protected because these systems are often marine biodiversity hotspots. Ecotourism (i.e. nature-based tourism) has rapidly developed. It has been estimated that 157–236 million people chose ecotourism in 1988, contributing to US$93–233 billion of national incomes (*397*). People are interested in travelling in relatively undisturbed places to enjoy beautiful sceneries or to watch specific animals (e.g. turtles, whales and dolphins) or ecosystems (e.g. coral reefs). For example, 9 million people travelled in 1998 to watch whales and dolphins, spending a total of US$1 billion (*397*). Beaches are also extremely appreciated. However, the tourism industry generally rapidly depreciates a natural site by quickly altering the ecosystems that were at the origin of its development. Although ecotourism may provide specific opportunities for protecting our natural heritage, this activity should be regulated.

9.8 Extinction

Extinction is common at the geological timescale and is normally counteracted by speciation. Consequently, of the 4 billion species that have populated our planet for 3.5 billion years, 99% have gone extinct (907). The five big extinctions that punctuated the planet history were characterised by the disappearance of ~75% of species (Chapter 5). During extinction periods, species became smaller, a phenomenon termed the 'Lilliput effect' (908). The Lilliput effect apparently took place in the aftermath of most major Phanerozoic extinction episodes. The effect has been documented for a variety of animals such as Early Silurian corals, Late Devonian conodonts and Early Danian echinoids (909). The Lilliput effect may currently be observed, as nowadays fishing practices are reducing the size of many commercially exploited species. I also recall that in the Late Quaternary, ~50% of the 167 mammal genera >44 kg went extinct (Holocene megafauna extinction).

9.8.1 Pronounced increase in global extinction rate

Over the past century, extinctions of groups such as bird and mammal have reached levels at least 100 times greater than the average extinction rate observed over 500 million years (the 'background extinction rate') (910) (Figure 9.37). When past major extinction episodes are investigated, a typical conclusion is that it takes several million years before diversity restarts to increase (765). Future projections suggest that the global extinction rate may even continue to rise and becomes 1,000–10,000 times higher than the background extinction rate (Figure 9.37). Our contemporaneous epoch will probably witness the sixth extinction if humans do not find ways to alleviate their impact on the biosphere (911).

9.8.2 Extinction in the oceans and seas

Roberts and Hawkins states that 'Jean Baptiste de Lamarck and Thomas Huxley, two of the foremost thinkers of the eighteenth and nineteenth centuries, believed that humanity could not cause the extinction of marine species' (912). Since the nineteenth century, many studies have provided compelling evidence that human activities in seas and oceans may be at the origin or contribute with other factors to species extinctions. This has already been documented for some organisms. The Steller's sea cow *Hydrodamalis gigas* was a 9-metre-long sirenian feeding on kelp and discovered by the crew of a Russian ship in 1741. Soon after its detection, the species became heavily exploited for its meat and was exterminated in 1768 (766). Four species have been inventoried. The West Indian manatee (*Trichechus manatus*) has been divided into two subspecies: the Florida manatee (*Trichechus manatus latirostris*) and the Antillean manatee (*Trichechus manatus manatus*). The two other species are the West African manatee (*Trichechus senegalensis*) and the dugong (*Dugong dugon*). In addition, the Amazon manatee *Trichechus inunguis* is the only sirenian living exclusively in fresh waters. All sirenians are highly impacted by human activities (e.g. illegal hunting, boat accidents, habitat degradation).

The great auk (*Pinguinus impennis*) was exterminated in the mid-nineteenth century. The species depended on isolated rocky islands for their breeding. The bird, hunted by humans for 100,000 years, was part of the culture of many countries. Scientists attempted to protect the bird but the large demand by private collectors and European museums led to the bird's extinction (the last individual died in 1844). Steller's sea cow and great auk were air-breathing

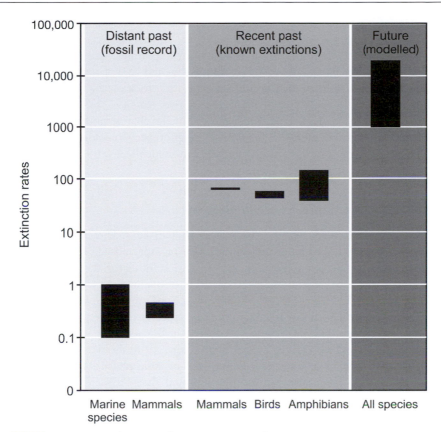

Figure 9.37 Past, current and projected extinction rates of some taxonomic groups.

Source: Redrawn from the Millennium Ecosystem Assessment (*910*).

animals. However, some studies have also documented marine invertebrate extinctions (*912*). For example, the eelgrass limpet *Lottia alveus*, occurring in the north-eastern part of America, went extinct in the 1930s after an epidemic affected its habitat. Others may be on the verge of extinction. The Texas pipefish *Syngnathus affinis* is now very rare in the Gulf of Mexico because of human-caused destruction of its natural habitat. The totoaba (*Totoaba macdonaldi*) is a large fish restricted to the Gulf of California. The species has been reduced dramatically by fishing and habitat degradation.

Although the number of extinctions is hard to estimate, there is little doubt that it is higher than the background extinction rate. It remains unknown if the extinction level is currently greater than extinctions that took place during the Big Five, but extinction is likely to reinforce in the next decades, not only because of the direct effects of human activities, but also because of anthropogenic climate change. The main extinction causes, termed the 'evil quartet of extinction rates', are habitat destruction, overexploitation, introduced species and secondary extinctions (*913*). Compiling a data set of 133 species populations that were extirpated locally, regionally or globally, Dulvy and colleagues (*914*) estimated that 55% of losses in marine biodiversity were related to exploitation, 35% to habitat destruction and the remainder linked to invasive species, pollution, disease and climate change.

Overexploitation has contracted species geographical range and has dramatically reduced population size, making species more vulnerable to anthropogenic climate change and climatic variability (677). Habitat degradation is an important factor. I recall here that between 30% and 60% of mangroves, 50% of saltmarshes and 16% of coral reefs have already been destroyed, and it is likely that unnoticed extinction took place. Trawling affects the equivalent of the world's continental shelves every two years (912). Habitat degradation may prevent species to adapt to anthropogenic climate change and precipitate species extinction. This factor is definitely an important driver of species ecological extinction (i.e. no ecological role because of too small a number of individuals), even if in biological terms some individuals may remain detectable.

9.8.3 Species vulnerability to extinction

Roberts and Hawkins (912) listed some factors that make species more vulnerable to extinction (Table 9.12). Extinction risk is higher for big, specialised or stenograph species or when species have low population density. An important factor of species vulnerability to extinction is restricted spatial range. About 81% of mammalian extinctions that took place during the last 500 years concerned endemic species inhabiting islands (915). In the marine environment, endemism is very common, especially for benthic organisms and fish (916). Roberts and Hawkins cited a study that estimated that 9.2% of a sample of 1,677 coral fish species had a spatial range <50,000 km², a threshold used in the terrestrial domain to classify species as endemic. About 24% of coral reef fish had a geographical range <500,000 km². For other species, they reported that 11.3% of 1,063 hermatypic coral (Anthozoa, Scleractinia), 41% of 169 black corals (Anthozoa, Antipatharia) and 28% of 316 cone shell species (Mollusca, Conidae) had a range <500,000 km². Between 30% and 70% of tropical amphipods are restricted to single islands. Many species such as damselfish are endemic to the Galapagos Islands and are restricted to areas as small as 600 km².

Another factor that modulates species vulnerability to extinction is fecundity, and size and age at maturity (Table 9.12). Rays and skates are highly vulnerable because they have low fecundity and high age and length at maturity (917). The starry ray (*Raja radiate*; length at maturity: 44 cm) is probably more resistant to anthropogenic forcing than the thornback ray (*R. clavata*) and common skate (*R. batis*), which have a length at maturity of 70 and 140 cm, respectively. As many rays, skates and sharks, the barndoor skate (*Dipturus laevis*) only produces 10 eggs per year.

9.9 Interactive effects

Global change has multifaceted effects on marine biodiversity and many anthropogenic stressors interact to adversely affect marine species. Interactive effects, which include antagonisms and synergisms between parameters controlling marine biodiversity, may lead to the mitigation or the amplification of the effects of global change (918). Jackson illustrated this point by taking the example of oysters (879). These organisms have been reduced to a low level by overexploitation. Nowadays, the recovery is made difficult by other anthropogenic influences such as hypoxia that originate from chronic eutrophication of estuaries and coastal ecosystems and the introduction of invasive species that compete

Table 9.12 Factors that make species more vulnerable to extinction.

Factors	Vulnerability	
	High	*Low*
Population turnover		
Longevity	Long	Short
Growth rate	Slow	Fast
Natural mortality rate	Low	High
Production biomass	Low	High
Reproduction		
Reproductive effort	Low	High
Reproductive frequency	Semelparity	Iteroparity
Age or size at sexual maturity	Old or large	Young or small
Sexual dimorphism	Large difference in size between sexes	No difference
Sex change	Protandry	No
Spawning	In aggregations at predictable locations	Not in aggregations
Allee effects at reproduction	Strong	Weak
Capacity for recovery		
Regeneration from fragments	Does not occur	Occurs
Dispersal	Short distance	Long distance
Competitive ability	Poor	Good
Colonising ability	Poor	Good
Adult mobility	Low	High
Recruitment by larval settlement	Irregular and/or low level	Frequent and intense
Allee effects at settlement	Strong	Weak
Range and distribution		
Horizontal distribution	Nearshore	Offshore
Vertical depth range	Narrow	Broad
Geographic range	Small	Large
Patchiness of population within range	High	Low
Habitat specificity	High	Low
Habitat vulnerability to destruction	High	Low
Commonness	Rare	Abundant
Trophic level	High	Low

Source: From Roberts and Hawkins (*912*).

with oysters for food and space. To these, we can add the impact of overfishing, which has released some predators that was previously under control by larger organisms (*919*). I provide here some examples of interactive effects that may affect marine biodiversity. I recognise that this may be quite speculative at the present time, but the need for investigating the potential influence of interactive effects on biodiversity has been stressed by many authors (*906*).

9.9.1 Climate and fishing interaction

Climate probably interacts strongly with fishing. As we saw earlier, some important effects of overfishing include depletion of spawning stock biomass and truncation of the age-size structure of stocks (677). These effects tend to concentrate reproduction in time and space and reduce eggs quantity and quality, which in turn decrease stocks resilience to environmental variability. In extreme cases, fishing makes population dynamics almost exclusively driven by fluctuations in recruitment, which is likely to increase sensitivity of the stocks to climatic variability. Climate and fishing interactions remain difficult to disentangle and quantify in space and time.

Many authors have suggested that fishing increases sensitivity of species or ecosystems to climate (677, 788, 920). However, how many have talked about the increase in species sensitivity to fishing by climate? Only a few, I am afraid. Although this does not make any difference at the end, this reflects the current belief that climate is perhaps not a major driving force in comparison to fishing. This is also exemplified by the Ecosystem Fisheries Based Management (EBFM). EBFM generally considers the impact of fishing on the ecosystem and not the effect of climate-induced changes in the ecosystem state upon its living resources (788). This was probably right for the last decades, although in Chapter 5 we documented clear examples of the power of climatic variability on marine ecosystems (e.g. Pacific regime shift, antagonistic oscillations of anchovies and sardine, Bohuslän periods).

Projections of global temperature change could reach ~4–5°C in worst-case scenarios by the end of this century. Basic biogeographical knowledge teaches us that climate is the main determinant of the arrangement of life in the oceans (Chapters 3–4 and 11). Therefore, global warming is likely to affect the spatial distribution of exploited resources. A major source of uncertainty with regard to global warming lies in the rapidity with which species will respond. As we saw in Chapter 7, recent studies suggest that aquatic ecosystems tend to respond more rapidly than terrestrial ecosystems. Major biodiversity changes have already taken place in the north-eastern part of the North Atlantic Ocean and attributed to ~1°C of temperature change (551, 657, 696). Therefore, in some places, climate warming might become a driving force as great as or greater than fishing (497). Fisheries based on one or two year classes are highly dependent on successful recruitment, and one poor recruitment event may cause stock collapse when fishing effort is high. Managers can control fishing. However, despite controlling greenhouse gas emissions to mitigate climate change, scientists can only document, and perhaps in best cases anticipate, the potential influence of climate on exploited resources (920).

9.9.2 Climate and pollution

Both anthropogenic climate change and ocean acidification might reduce resilience of species and ecosystems to pollution, and some studies showed that rising temperatures and acidification may alter pollutant toxicity (921, 922). With Goberville and colleagues, we showed that climate can also modulate the anthropogenic fertilisation of some coastal ecosystems (770). In this paper, we revealed two interactive effects of climate. The first occurs through the effect of precipitation on river discharges, which amplify the effects of nutrient over-enrichment (923, 924). As reported in the North Sea or in the Mississippi Basin (923, 925), our findings showed that the balance between nutrient storage or their leaching co-vary strongly with direct freshwater inputs. Nutrient delivery was therefore influenced positively by precipitations, themselves related to atmospheric circulation changes (157, 926). Hydrological alterations related to climate change may perturb the water cycle. Some authors suggested that global run-off may increase by 4% if global temperature rises by 1°C (927). The second effect of climate modulation on nutrient over-enrichment occurs when insolation increases. Indeed, insolation reduces nutrient concentration by its positive influence on primary production (928).

9.9.3 Global warming and UV-B

Global warming is expected to reinforce both seasonal and permanent thermoclines, reduce mixing and thereby the quantity of nutrients available for primary production (929). Boyce and co-workers (930) estimated a 1% reduction of global primary production per year from 1979 to 2009. They found a significant decrease in all oceanic regions but the Indian Ocean. High-latitude regions (>60°) showed the greatest proportions of reduction in primary production. These results suggest that global warming may lead to clearer waters, which may allow UV-B radiations to penetrate deeper in the water column and restructure photic zone biodiversity.

We also saw that sea-ice extent, controlled by temperature, may strongly influence krill density in the Southern Ocean (oceanic areas close to the Antarctic Peninsula). Global warming is therefore expected to reduce sea-ice extent and associated algae important for krill growth and survival. Atkinson and co-workers found a pronounced reduction in krill abundance between 1976 and 2003 (471). Krill exploitation may also interact negatively with global warming to precipitate the reduction of *Euphausia superba*. However, at the same time, UV-B radiation has strongly increased over the Southern Ocean as the result of ozone depletion. UV-B radiation has been shown to increase the mortality of juvenile *Euphausia superba* and to affect its activity (931). A 15% rise in UV-B radiation, similar to what has been observed in the Southern Ocean between 1979 and 1995, may diminish marine species performance by 59%. As the species is key for higher tropic levels survival (Chapter 5), krill reduction may have devastating effects on Southern Ocean biodiversity.

Although ozone depletion has led to an increase of 4% of UV-B in tropical oceans, such a small rise may impair coral reef performance by ~30% (906). Lesser and colleagues (932) showed that loss of zooxanthellae was aggravated when the algae were exposed to the synergic influence of both ultraviolet and high temperatures.

9.9.4 Increase in atmospheric CO_2 and temperature

Increasing atmospheric CO_2 is having a double influence on the oceans. First, it is increasing global temperature, which in turn affects global climate. Second, it is also triggering ocean acidification. The synergistic influence of global warming and ocean acidification might strongly affect negatively marine biodiversity in the next decades. Martin and Gattuso (933) investigated the joint effect of P_{CO2} and temperature on the crustose coralline algae *Lithophyllum cabiochae*. This algae has a skeleton composed of high-magnesium calcite and are absent in areas naturally enriched in CO_2 (749). The researchers maintained the algae in four experimental conditions: (1) $[CO_2]$ = 400 ppm and ambient temperature T; (2) $[CO_2]$ = 400 ppm and ambient temperature T + 3°C; (3) $[CO_2]$ = 700 ppm and ambient temperature T; and (4) $[CO_2]$ = 700 ppm and ambient temperature T + 3°C. Forty algae

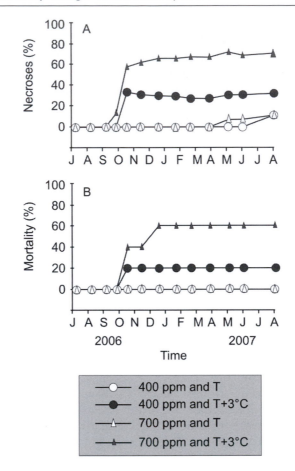

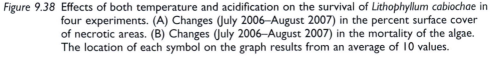

Figure 9.38 Effects of both temperature and acidification on the survival of *Lithophyllum cabiochae* in four experiments. (A) Changes (July 2006–August 2007) in the percent surface cover of necrotic areas. (B) Changes (July 2006–August 2007) in the mortality of the algae. The location of each symbol on the graph results from an average of 10 values.

Source: Simplified from Martin and Gattuso (*933*).

were initially added in each aquarium and the experiments were conducted between 10 July 2006 and 6 August 2007 with Mediterranean water continuously flushed at a rate of $50\%\cdot h^{-1}$. The water was pumped at 10 m depth in Villefranche-sur-Mer. Ambient temperature corresponded to the thermal regime recorded between 13.3°C in March and 22°C in September at 25 m depth in the Bay of Villefranche during the period 1995–2006. The authors examined monthly changes in the percent surface cover of necrotic areas and individual mortality of the coralline algae from July 2006 to August 2007 (Figure 9.38).

When temperatures remain within the natural limits of the site, only restricted necrosis was observed from May at both CO_2 concentrations (400 and 700 ppm). When temperatures were increased by 3°C, necrosis appeared from October at both CO_2 concentrations (Figure 9.38A). The percent surface cover of necrotic area was ~40% at $[CO_2]$ = 400 ppm and reached ~70% at $[CO_2]$ = 700 ppm. In the same way, mortality percentage was nil when temperatures were not artificially increased (Figure 9.38B). At $[CO_2]$ = 400 ppm, the experimental increase of the thermal regime increased the mortality by 20% and at \[CO2] = 700 ppm, the increase reached 60%.

These results show the major effect of temperature on both necrosis and mortality. It also suggests that acidification without global warming would have no effect on the coralline algae. However, acidification clearly aggravates the effects of global warming on the species and this study clearly demonstrates the joint adverse influence of both stressors on the species' individual survival. This experimental study also shows the existence of critical physiological thresholds, above which the species' physiological state starts to worsen. Temperature increase was beneficial on algae calcification in autumn and winter when temperatures were low, whereas temperature rise had an adverse effect on net species calcification in summer (seasonal maximum) when temperatures were >25°C for several weeks. This detrimental effect increased species' sensitivity to disease (immunosuppression), involving large necrosis leading to mortality.

The interplay between temperature and CO_2 concentration was also investigated on an eight-week experiment on Heron Island (Southern Great Barrier Reef) during the 2007 austral summer (February to March) (756). The study compared bleaching, productivity and calcification responses of the crustose coralline algae (*Porolithon onkodes*), the branching (*Acropora intermedia*) and massive (*Porites lobata*) coral species in response to acidification and warming. They increased CO_2 concentration from doubling to fourfold, comparing to today's level. The researchers showed that the effect of rising temperature on bleaching, net productivity and calcification was exacerbated by increasing CO_2 concentration on all species. However, their results were mostly evident for high CO_2 concentration (1,000–1,300 ppm). Reynaud and colleagues (934) also showed experimentally that high P_{CO2} (760 µatm) had no influence on the calcification of the scleractinian coral *Stylophora pistillata* at 25°C, but triggered a pronounced reduction in calcification at 28°C. Experimental studies also showed that the crab *Cancer pagurus* exposed to important P_{CO2} reduced its upper thermal limits of aerobic scope by 5°C (935).

Chapter 10

Marine biodiversity, ecosystem functioning, services and human well-being

Biodiversity has aesthetic, cultural, intellectual and spiritual values that are essential to mankind. It probably has a strong influence on human mentality, what Edward Wilson called biophilia (936). Psychologists have provided compelling evidence that the presence of a good natural environment enables the reduction of human stress. Biodiversity loss may have strong adverse effects on humans.

There is often a trade-off between biodiversity preservation and society (937). Ruijgrok and colleagues (937) describe this trade-off in the following way: 'both society and nature are allowed to change and to inflict change upon each other as long as neither of them suffers serious damage, threatening its existence; it is a matter of mutual benefit'. The dependence of our society on nature is patent (938) but a sick society may not be able to take care of nature (Figure 10.1).

Human-caused alteration of biodiversity through habitat destruction, overexploitation, pollution, invasive species and anthropogenic climate change may have strong effects on ecosystem functioning, goods and services and global biogeochemical cycles (Figure 10.2).

10.1 Biodiversity and ecosystem functioning

Biodiversity is composed of interconnected species within food webs that exchange matter, energy and information and form complex systems. Such complex systems are said to be self-regulating stable systems. The realisation that biodiversity is declining has compelled scientists to investigate what could be the consequences of biodiversity loss for ecosystem functioning (939). In a review, McCann stated that 'we are, in a very real sense, deconstructing the earth under the implicit assumption that ecosystems have evolved the ability to withstand such assault without collapse'. The great variety of life is not only a passive epiphenomenon, but it has an effect on ecosystem functioning and on stocks and fluxes of matter and energy. Here, I define ecosystem functioning as the processes, activities and properties of ecosystems influenced by biodiversity. Biodiversity is sometimes seen to act as 'a biogeochemical catalyst with Michaelis-Menten-like kinetics' to improve ecosystem functions (940). Processes such as photosynthesis influence ecosystem functioning by both oxygen and biogenic carbon production. Activities such as filter-feeding influence the turbidity of the water column. Properties such as mean community body size affect carbon exportation in the pelagic environment. Processes, activities and properties all influence, and are a consequence of, trophic food webs. Biodiversity may affect different ecosystem functions (941):

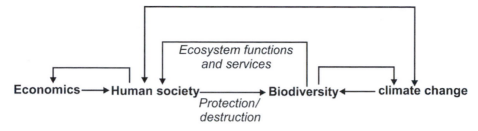

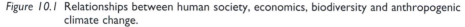

Figure 10.1 Relationships between human society, economics, biodiversity and anthropogenic climate change.

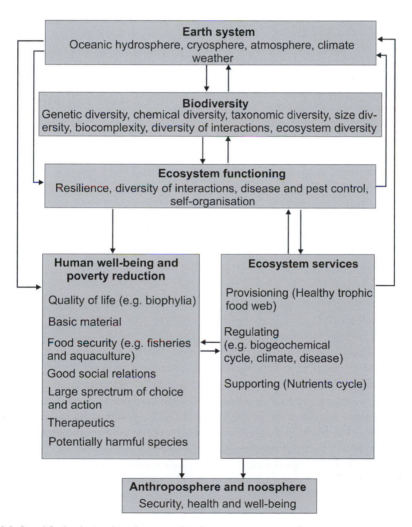

Figure 10.2 Simplified relationships between biodiversity, ecosystem functioning, services and human well-being.

1   Physical, chemical and biological processes that affect self-maintenance of ecosystems (e.g. stability, resistance, resilience, persistence).
2   Interspecific interactions from positive (mutualism) to negative (competition and predation) and trophic food webs.
3   Interaction between species and their environment.
4   Stocks and fluxes of matter and energy (e.g. biomass and production).

Ecosystem functioning should not be confounded with services (see Section 10.2). The latter term is usually exclusively reserved for those ecosystem functions that can be directly and quickly exploited by mankind.

Naeem (940) proposed a three-point framework to explain the potential relationships between biodiversity and ecosystem functioning. The first remarkable point is a theoretical point where no biodiversity remains and where there is therefore no ecosystem function, a *reductio ad absurdum* (Figure 10.3). The second point is where biodiversity remains at a pre-industrial level. The third point is an intermediate point from which any biodiversity loss leads to substantial ecosystem functioning deterioration (inflection point). Possible trajectories between the first and the last point are many, and Naeem suggests three different types. The first is a saturating curve, indicating that any species addition has only a minor influence on biodiversity. The second is a Promethean view, which is based on the assumption that human ingenuity can increase ecosystem function, even if biodiversity diminishes. The third is an Arcadian view, which assumes that nature provides all goods and services to mankind providing it remains in a pristine state. Of course, all types of trajectories between these two extremes are possible, corresponding to different hypotheses (e.g. rivet, redundancy, linear, keystone, idiosyncratic).

How may species richness and ecosystem functioning (productivity, biomass, nutrient cycling and stability) be related? More than 50 hypotheses have been proposed to explain the link between biodiversity and ecosystem functioning. The null hypothesis is an absence of relationship between biodiversity and ecosystem functioning. The hypothesis of linear positive relationships between biodiversity and ecosystem functioning (BEF) assumes that species are mostly singular, each contributing to increase ecosystem function (Figure 10.3A). In contrast, the redundant species hypothesis (943) assumes that only a few species are singular and that most species are redundant (Figure 10.3B). The BEF relationship therefore reaches a horizontal asymptote. Walker (943) qualified important species as 'drivers' and redundant species as 'passengers'. The quantification of the degree of redundancy is, in practice, very difficult. In the keystone hypothesis, a single species is assumed to have a remarkable effect on ecosystem functioning and the removal of such species may have dramatic consequences for ecosystem functioning (Figure 10.3C). Keystone species may be a mechanical engineer, influencing the flow of matter, energy and information, but also species that affect biotic interaction and for whom removal may have long-lasting effects on ecosystem trophodynamics. The removal of such species is difficult to compensate and compensation may take place by new species colonisation (944). The discontinuous hypothesis states that the BEF relationship is complex and that the loss or the gain of a species has an uneven effect on ecosystem functioning (Figure 10.3D). The rivet hypothesis (945) stipulates that the system may tolerate some reduction in species richness up to a point it produces a remarkable effect on ecosystem functioning (Figure 10.3E). The idiosyncratic hypothesis (946) states that the loss or gain of a species has an unpredictable effect on ecosystem

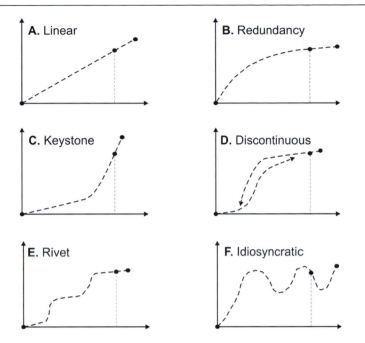

Figure 10.3 Main hypotheses on the relationships between biodiversity and ecosystem functioning. The first point is a point of no biodiversity. The last point is the point of pre-industrial biodiversity. The intermediate point is the expected current biodiversity level.

Source: Modified from Naeem and colleagues (942).

functioning (Figure 10.3F). This hypothesis should not be confounded with the null hypothesis, in which the addition or removal of a species has constantly no effect on ecosystem functioning.

Many hypotheses about BEF relationships can be observed in nature. When a few functions are examined, the redundant species hypothesis may be validated in some circumstances. However, if more ecosystem functions are considered (in practice, a difficult task), the true BEF relationship may be somewhere between a positive linear and an asymptotic relationship. Because not all species have the same weight in the ecosystems, the removal of an engineer species may have strong consequences for ecosystem functions, and therefore the rivet hypothesis may be closer to the reality than any other hypotheses. We can make the parallel with a symphonic orchestra. All musicians have a role. While the removing of one violinist may only be perceived by an expert, if several are removed then the piece starts to be significantly altered. If the contrabass is subsequently excluded, most people may notice the change. Taking away clarinets, drums and piano, then the original symphonic music will definitively be denatured. Although the music of life is more complicated, this example suggests that at a high biodiversity level, it may be difficult to detect the effects of biodiversity erosion. The erosion may only become perceptible when a significant alteration in ecosystem function takes place.

Experimental research has provided evidence for an effect of biodiversity on ecosystem functioning in a variety of terrestrial systems (947). Using the ecotron, Naeem and colleagues

(947) found that reduced biodiversity might alter ecosystem performance such as their capabilities to absorb CO_2. Complementarity effect was hypothesised to be important, allowing a better use of solar radiation by the plants. Since pioneer studies (947–949), BEF research has evolved and subsequently considered the relationships between biodiversity and species interaction, biodiversity and resistance to invasive species and biodiversity and nutrients use efficiency, to just name a few. Theories have also been developed to better comprehend BEF relationships (950).

10.1.1 Biodiversity and stability

10.1.1.1 Different definitions of stability

Among investigated ecosystem functions, stability is a key property resulting from ecosystem **homeostasis**. Stability can be defined as the ability of a system to remain in a constant state (939). This definition assumes the existence of stable equilibrium dynamics. However, stability has been defined in many different ways (951) (Table 10.1).

To better characterise stability, Begon and colleagues (291) used the concepts of **resistance** and **resilience** (Table 10.1), local and global stability and dynamic fragility and robustness (Figure 10.4). Local/global stability occurs when a system can resist to a small/large perturbation, respectively. A community that resists large environmental changes is said to be dynamically robust, whereas a community that is sensitive to small environmental changes is said to be dynamically fragile (Figure 10.4).

In Table 10.1, resilience can also be termed elasticity. Odum (952) defined resistance as the maximum possible alteration in ecosystem function outside the system's normal operating range (Figure 10.5). Resilience was the time needed for the system to come back to its normal operating range.

Resilience can also be defined as the system's capacity to maintain its structure and function through disturbances, without necessarily returning to a particular reference state (953). Holling makes the emphasis on the transient behaviour of systems. A system may

Table 10.1 The different definitions of stability.

Term	Type	Definition
Equilibrium stability	Dynamic stability	A system is stable if it returns to its equilibrium after a small perturbation. When no perturbation takes places, the system does not exhibit any variability.
General stability	Dynamic stability	Stability increases when a population increases its density away from zero.
Variability	Dynamic stability	The degree of change in a population measured as the coefficient of variation, local autocorrelation or multi-scale variance.
Equilibrium resilience	Resilience and resistance stability	A system that returns quickly to its dynamic regime after a perturbation occurs. In this definition, the system has inherent sources of variability.
Resistance	Resilience and resistance stability	The rapidity with which a system changes after a perturbation takes place.

Source: Adapted from McCann (939).

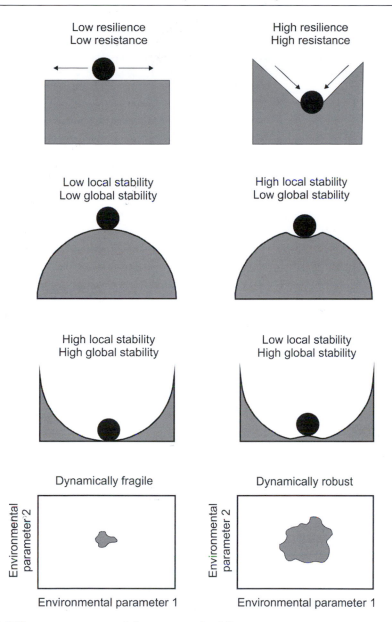

Figure 10.4 Different components of the concept of stability.

Source: Redrawn from Begon and co-workers (*291*).

never reach an equilibrium (it remains in a transient state) because of the existence of perturbations. Within a domain of attraction, the system continues to function in a normal way despite disturbances. Multiple stable states or domains of attraction may exist. To jump from one domain to another, the disturbance must be greater than the resilience associated with a specific domain of attraction. This view implicates the existence of both stabilising and destabilising forces (e.g. disturbances or disruptions).

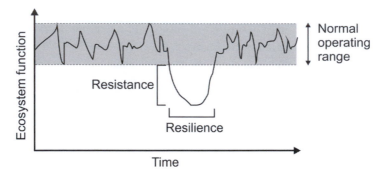

Figure 10.5 Illustration of the concept of resistance and resilience.
Source: Redrawn from Odum (*952*).

Other measures of stability are **persistence** (i.e. the time a species' population persists or the time a community or a system continues to maintain its structure and function complete). Persistence is, in practice, quite difficult to measure, however. Any stability measurement may depend on the scale of observation (*954*). O'Neill (*954*) recalls that no ecosystem is stable and that all ecosystems will be altered completely as time goes by (e.g. geological timescale).

10.1.1.2 Early investigations on the link between biodiversity and stability

As early as 1958, Elton thought that complex communities with many parasites and predators reduce the propensity of populations to grow exponentially and lead to stability (*955*). At that time, this paradigm was also shared by many scientists such as MacArthur (*956*). From simple conceptual models, MacArthur concluded that stability was difficult to obtain when species richness was low because species need to feed on many different trophic levels. Higher species richness tends to trigger species specialisation for certain diets and the number of trophic levels is lower in comparison to species number. By this way, high species richness per trophic level tends to stabilise a community. However, May (*957*) showed by using mathematical models that diversity destabilises population dynamics. He described a food web by three parameters: S the species richness, C the **connectance** (i.e. the number of species interacting directly) and β the average interaction strength. Species interaction strength was fixed randomly. May found that a community was only stable if $\beta(SC)^{1/2} < 1$. The equation means that instability increases when connectance, species richness and/or average species interaction strength increase. Gardner and Ashby illustrated well the potential links between elements (here species), connectance and stability (Figure 10.6).

Whatever the number of species ($n = 4$, 7 or 10), the probability of stability in a community depends strongly and in a complex way upon connectance. When n increases, this dependency becomes even higher and the curve takes the form of a step function and the threshold between stability and instability becomes sharper.

10.1.1.3 The role of species interactions in diversity/stability relationships

Therefore, complexity (and biodiversity) leads to instability. May's results, also found by other authors, were surprising because field observations showed that many ecosystems are

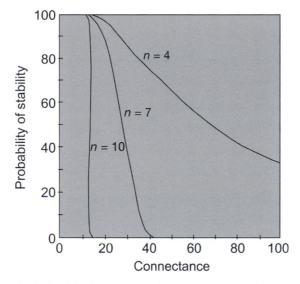

Figure 10.6 Theoretical relationships between species, connectance and community stability. Probability of stability and connectance are expressed as percentage, *n* is the number of species.

Source: Adapted from Gardner and Ashby (*958*).

complex and apparently stable. To solve this paradox, Yodzis (959) examined 40 food webs and found that stability was observed when interactions were close to the ones observed in the field (i.e. weak interactions). Two decades later, McCann and co-workers (960) showed that complex communities are stable when most species exhibit weak interactions with others. Weak interactions between species may stabilise community dynamics by limiting the influence of the strong destabilising resource-consumer interactions (939). Many studies showed that food webs are mainly composed of weakly interacting species with few strong interactions (960). Weak interactions reduce the statistical chance that a population becomes extinct. Biodiversity erosion may increase average species interaction strength and leads to more instability. Bascompte and colleagues (961) used a dynamical ecological model to show that network interaction asymmetry, the low number of strong interactions and the high heterogeneity of species strength enhance long-term coexistence and biodiversity maintenance (Figure 10.7). The figure shows the multitude of species interactions within a food web (Figure 10.7B–E) and the asymmetry of biotic interactions.

10.1.1.4 Predator diversity dampens trophic cascades

Predator biodiversity may also dampen trophic cascades (962). Investigating a salt marsh food web composed of a *Spartina* (primary producers), phloem-feeding *Prokelisia* planthoppers (herbivores) and both mirid bug *Tytthus vagus* and spiders (*Hogna modesta*, *Pardosa littoralis*, *Grammonota trivitatta*), Finke and Denno (962) demonstrated experimentally that arthropod predator biodiversity generates intra-guild interactions. These interactions alleviate the effects of predation on herbivores and dampen cascading effects on primary producers.

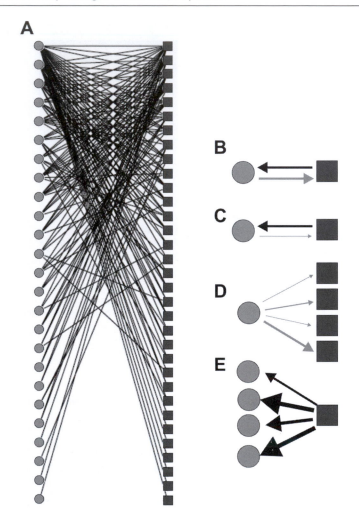

Figure 10.7 Asymmetry of species interactions in a community of plants and their seed dispersers (Cazorla, Spain). (A) Asymmetric interactions between the plants and the animal dispersers. Circles denote plant species and squares their dispersers. (B–E) Examples of types of interaction strength observed in this community. Arrow width denotes interaction strength.

Source: Modified from Bascompte and co-workers (961).

10.1.1.5 High diversity of preys increases fitness

Species feeding on a large variety of preys enhance their fitness. Kleppel (963) found that some copepods (e.g. *Calanus pacificus, Acartia tonsa, Centropages furcatus*) present a high dietary diversity and Kleppel and Burkart (964) showed that dietary diversity increased egg production. A diverse diet allows the species to adjust to changes in available preys and to increase the probability to have a nutritionally complete ration (963). Meta-analyses of controlled experiments provided evidence that a diverse diet increases growth, survival and

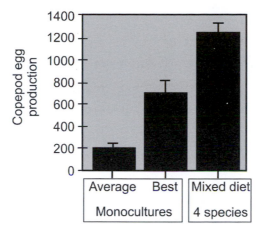

Figure 10.8 Diet diversity increases life-history processes such as copepod egg production.

Source: Redrawn from Worm and co-workers (965).

fecundity of species. Figure 10.8 shows that a mixed diet, composed of four species, enhances copepod egg production rate. Increasing prey diversity enhances consumer nutrition and fitness and reduces top-down control because increased production compensates consumption (966).

10.1.1.6 High biodiversity promotes ecosystem stability

In the context of BEF research, many studies have suggested that higher biodiversity might promote ecosystem stability, a phenomenon known as the **Portfolio effect** (967). The Portfolio effect (968), or the insurance effect (969), is the stabilising effect that biodiversity has on aggregate ecosystem properties. The mechanism proposed by Yachi and Loreau (969) is the differential responses of species to environmental conditions driven by their ecological niche. They therefore think that compensation may occur because the reduction in one species may be alleviated by the increase in another species.

Naeem and Li (970) showed that biodiversity increases **ecosystem reliability**. Using algae (primary producers), bacteria (decomposers) and protists (consumers), they demonstrated that when species richness increased for each trophic group, the predictability of both density and biomass measurements was more elevated (Figure 10.9). Biodiversity can therefore be seen as a form of biological insurance against species loss. The mechanism behind biological insurance may be **compensatory growth**, the increasing growth of a species when another from the same functional group diminishes in abundance, as we saw above. This effect has also been termed the statistical averaging effect.

There is still an active debate on the role of biodiversity in ecosystem stability. Stability has often been invoked to explain why some regions have more species richness (207, 227). Therefore, stability can be a cause or a consequence of high diversity. Some authors also questioned the existence of stable systems in nature (971). Stability may only exist when climate does not change significantly. However, climate fluctuates all the time. Especially today at a time of an unprecedented climate change in both magnitude and speed, stability may not be possible because species will primarily react to climate change individually and

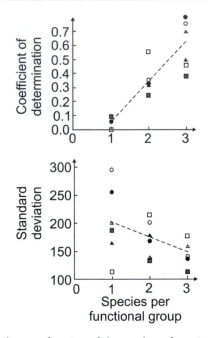

Figure 10.9 Ecosystem reliability as a function of the number of species per functional group. Reliability was estimated from both determination coefficient (r^2; upper panel) and standard deviation (lower panel) of bacterial densities in replicates as a function of the number of functional groups. Determination coefficients originate from a linear regression of the natural logarithm of the autotrophic biomass as a function of the number of functional groups. Open symbols: high light. Filled symbols: low light. Circles: high nutrients. Triangles: intermediate nutrients. Squares: low nutrients.

Source: Redrawn from Naeem and Li (970).

these alterations in species composition will in turn modify communities. As communities will be altered, ecosystem properties will change, which will affect ecosystem functions and services. It is unlikely that high biodiversity promotes more stability with respect to anthropogenic climate change.

Heterogeneity is an important component. Habitat heterogeneity, either in time or space, allows species to coexist and to adjust to a fluctuating environment. O'Neill (954) even stressed that 'a homogeneous system, like an overspecialized species, cannot respond to change and is inherently unstable'. He added that 'the stability of an ecological system cannot be predicted by a theory that ignores heterogeneity'.

10.1.2 Biodiversity and ecosystem productivity

BEF research has also focussed on the relationship between biodiversity and ecosystem production. All life on earth depends upon this ecosystem process. Tilman and colleagues (972) showed that ecosystem production increased with plant biodiversity in 147 grass-lands. Soil nitrogen was more fully exploited by the ecosystem when diversity was higher (Figure 10.10).

Three hypotheses have been proposed to explain the relationship between biodiversity and ecosystem production (939, 973). First, more diverse communities are expected to

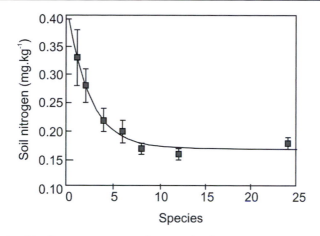

Figure 10.10 Relationships between species richness and soil nitrogen in 147 grassland ecosystems.
Source: Redrawn from Tilman and co-workers (972).

contain more species with dominant traits (e.g. more productive species), a phenomenon called **positive selection effect** or sampling effect (974). The sampling effect is the higher probability that a diverse community contains a species that increases substantially functional traits. This effect may, to some extent, explain a number of positive BEF relationships. Whether this effect is an artefact or a true ecological mechanism remains discussed.

Second, the complementarity effect may explain BEF relationships. Complementarity occurs when the increase in species richness provides new functional traits to the ecosystem, allowing resources to be better exploited. In such a case, the (resource) niches *sensu* MacArthur (975) of the species are non-overlapping. Differences in species architecture may allow the ecosystem to better exploit resources (976). The insurance hypothesis states that high species richness increases traits diversity and the probability that compensation takes place if a species becomes extinct.

Third, the identity effect arises when a few species determine the main pool of functional traits. This is also called the selection effect (i.e. the selection of particular functional traits that affect ecosystem functioning) (977). Some life history traits may strongly influence biomass, productivity or stability. Some authors proposed that ecosystem functions are controlled by the functional characteristics of some species rather than by their species richness (978). Some life-history traits (e.g. higher potential productivity and average competition and thereby high invasion potential) may explain the link between biodiversity and productivity. Loreau (977) proposed that the sampling effect is a particular case of identity/selection effect.

Although higher biodiversity increases biomass production at a regional scale, this is not a guarantee of a greater stability (979). From field experiments (grassland communities from 1 to 32 species), Pfisterer and Schmid (979) observed that species-poor plots (mean size of 8 m × 2 m) were characterised by lower biomass production than species-rich plots. However, species-poor plots were more resistant and resilient to drought perturbations than species-rich plots.

In the marine ecosphere, large-scale relationships between biodiversity and productivity (or, more correctly, production) are negative and diversity tends to be elevated in oligotrophic regions (Chapter 4) (980). When nutrients increase, this often leads to a reduction in species richness, a phenomenon called the paradox of enrichment (280). There is, however, a compatibility between experimental and observational studies when higher diversity promotes higher productivity in low productive environment and when diversity diminishes with productivity in high productive environment. Such relationships are in agreement with the 'hump-backed' diversity curve.

10.1.3 Functional diversity

With the exception of hydrothermal vents and cold seeps where chemosynthesis takes place, deep-sea ecosystems have no primary production and are highly limited by food; between 0.5% and 2% of photic primary production arrives in deep-sea benthic ecosystems. Danovaro (981) showed some positive monotonic relationships between biodiversity and ecosystem functioning (e.g., biomass, ecosystem efficiency, trophic diversity traits) in deep-sea sediments. For example, a rise in species richness is often associated with an increase in functional diversity (Figure 10.11). Whether this relationship indicates causality is not clear, however. Both properties may be influenced by a common abiotic factor (e.g. habitat heterogeneity). Danovaro also documented an exponential relationship between deep-sea biodiversity and ecosystem function and efficiency (i.e. available energy resources that can be channelled into the biomass).

Although interconnectedness and complexity of both pelagic and benthic marine ecosystems make extrapolations from experiments difficult (982), on the Pacific coast, Paine manipulated a rocky shoreline community and found that the removal of the carnivorous starfish *Pisaster ochraceus* substantially altered the food web, and therefore ecosystem functioning (983). A single native species (keystone species) had, therefore, a disproportionate effect on the ecosystem state (984). In BEF vocabulary, this result illustrates an extreme case of identity effect.

Raffaelli and Friedlander also showed that the relationships between biodiversity and ecosystem functioning may not be straightforward. They took the example of the Ythan

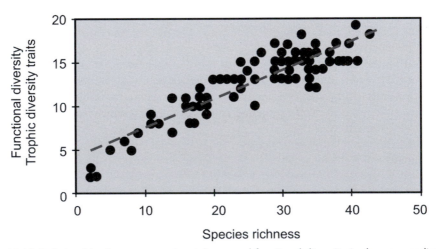

Figure 10.11 Relationships between species richness and functional diversity in deep-sea sediments. Source: From Danovaro (981).

Estuary (Scotland), which undertook dramatic changes from a nearly pristine state in the 1960s to a eutrophic state in the late 1980s. Nutrient enrichment led to biodiversity changes. However, the alteration was seen in species' relative abundance and not in species' loss or gain. The biomasses of the macroalgae *Ulva intestinalis* and *Chaetomorpha* spp. rose by 150% between the 1960s and the late 1980s to 1990s. For example, shelduck *Tadorna tadorna* and redshank *Tringa tetanus* diminished because *Corophium* population, on which they depend, collapsed due to the high macroalgae densities. In contrast, dunlin (*Calidris alpina*), bar-tailed godwit (*Limosa lapponica*) and curlew (*Numenius arquata*) increased in abundance by 50%, 300% and 1,420%, respectively. These compositional changes led to a new operating system state, which increased primary productivity, and perhaps more surprisingly system stability, measured here as resilience (**ascendancy** and **overhead**).

10.1.4 Relationships between marine biodiversity and food webs

Ways by which food web structure (species interactions), body size and abundance interact are important to understand the relationships between biodiversity and ecosystem functioning (Figure 10.12).

Both food web complexity and species interactions increase when biodiversity augments (Chapter 4). We saw above that species interaction strength is quite important in modulating environmental fluctuations (985). We also saw that **connectance** should decrease

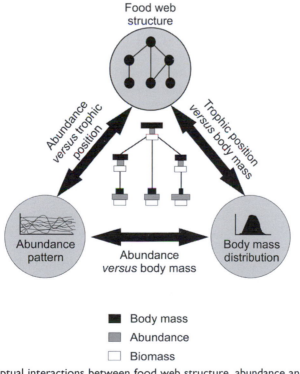

Figure 10.12 Conceptual interactions between food web structure, abundance and body mass.

Source: Redrawn from Emmerson (985).

when species richness increases for the system to remain stable. Connectance is calculated as follows:

$$C = \frac{L}{\left(S^2 - S\right)} \qquad (10.1)$$

With L the number of observed trophic links and S the species richness. Note that sometimes the connectance is calculated as $C = L/S^2$. Connectance is an indication of trophic complexity. Trophic complexity has several components: (1) generality measures the number of preys a species has; (2) vulnerability is the number of predators per species; and (3) connectedness is the number of both prey and predator for each species. Emmerson (985) recalled that these properties depend on species' body mass/body size. The consideration of the relationships between body mass and food web structure is also important because body mass determines the species' position in the food web. This is particularly true in the marine environment. Body size also affects long-term changes in species abundance, which may then ramify through the food web and alter fluxes of matter and energy.

Food web ecology, founded by Elton (986), has often shown that mean size/mass of prey increases with predator body size/mass and higher trophic levels have predator with greater body size/mass (987). In their study of 35 food webs (marine, stream, lake and terrestrial), Riede and colleagues (987) provided evidence for a reduction in predator–prey body mass ratios towards higher trophic levels, contradicting the Eltonian paradigm (i.e. invariance of predator–prey body mass ratios along the food web). According to Emmerson, these findings suggest that big predators may only interact weakly with their prey because they are highly energetic (i.e. big preys). In contrast, small predators are likely to develop stronger interactions with their prey because they may be less energetic and the predator needs to kill more prey per unit of time than big predators. This result suggests that when fishing is elevated, the reduction of top predators (in term of diversity or mean size population structure) may reinforce trophic interactions and destabilise the system more rapidly, making it more sensitive to environmental changes (137).

Emmerson provides a food web framework to understand how biodiversity erosion may affect trophic food webs (Figure 10.13). This example shows how the removal of one or more species may propagate through the food web. This model is, however, simplistic because the species' position within the food web (as well as its ecology) changes as a function of its developmental stage. For example, some adult decapods feed on young Atlantic cod (*Gadus morhua*) but decapods are predated by adult cod (988).

Although the number of pristine ecosystems has become rare, comparison between undisturbed or less affected ecosystems may provide some insights on the effects of biodiversity alteration for ecosystem functioning (982). In the central Pacific, the Hawaiian archipelago can be divided into two areas: (1) the Main Hawaiian Islands (MHI), which have been heavily exploited; and (2) the North-Western Hawaiian Islands (NWHI), which are unaffected (conservation area). Raffaelli and Friedlander (982) compared these two marine ecosystems. Severe biomass depletion occurred in MHI where top predators such as sharks and large jacks were heavily fished. Apex predators represented >54% of the whole standing stock biomass in NWHI, whereas this trophic group accounted for only <3% in MHI. Furthermore, herbivore biomass and low-level carnivores was 30–50% lower in MHI than in NWHI. The biomass pyramid was inverted in NWHI because small trophic levels (short lifespan) were composed of species rapidly channelled by predation to higher trophic levels

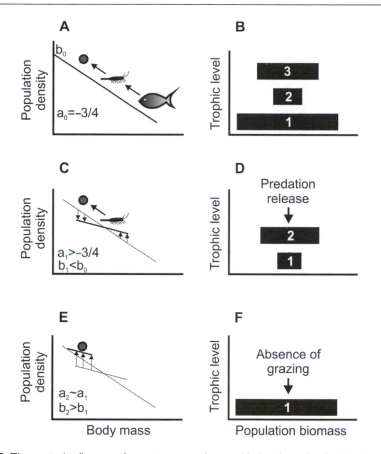

Figure 10.13 Theoretical influence of species removal on an idealised trophic food web composed of a phytoplankton, a zooplankton and a fish. (A) Theoretical relationships between population density and body mass expected from the metabolic theory of ecology (Chapter 4). (B) Expected population biomass of each trophic level. (C) Theoretical relationships between population density and body mass once the predator is removed. (D) Zooplankton population biomass is likely to increase because of predator release. This in turn reduces phytoplankton population biomass. (E) Theoretical relationships between population density and body mass once the predator (fish) and the grazer (herbivore zooplankton) are removed. (F) Phytoplankton population biomass is likely to increase because of grazing release; a is the slope of the relationship and b the intercept. Horizontal black rectangles indicate population biomass of phytoplankton (1), zooplankton (2) and fish (3).

Source: Modified from Emmerson (985).

(longlife span). Apex predation structures the size and the age structure of prey populations. Whereas preys age structure was skewed towards small individuals in NWHI, the opposite situation occurred in MHI where small individuals represented the bulk of prey biomass. The energetics of the system was therefore very different. Diversity of populations and species were small in MHI and both species' complementarity and facilitation make the MHI ecosystems less efficient and probably less resilient to changes in environmental conditions because of lack of redundancy. As we said already, intense fishing modified **life history traits** of the exploited fish and has both direct and indirect consequences on the whole ecosystem. Both lower body size and mean trophic levels alter top-down control. Overexploitation tends to select species that mature, reproduce and grow fast, species that are generalist (e.g. omnivores) and both detritivore and decomposer species.

In the terrestrial realm, animal biodiversity is correlated positively to plant biodiversity because plants structure physically the habitat (989). Because many herbivores are specialised and feed only on a few phytoplankton, macroalgae or plant species, the resource special-isation hypothesis (990) also states that a diverse community may provide a larger variety of resources for herbivore species. The more individuals hypothesis stipulates that high productive ecosystems have more individuals and may therefore hold more species because large populations have less chance to go extinct (267). A high diversity of producers increases productivity (972). This may increase the available resources for consumers and support more individuals, and therefore more species. The more individuals hypothesis is, however, insufficient to explain patterns of observed diversity in some ecosystems such as tree hole ecosystems (991). In these terrestrial ecosystems, more productive holes are associated with more species, but not more individuals.

10.1.5 Natural relationships between diversity and ecosystem functioning at the geological scale

Dornbos and colleagues (992) took advantage of the 600 Ma of marine fossil record to review the BEF relationship at a geological timescale, focusing on the Ediacaran (578–542 Ma) and Cambrian (542–485 Ma) radiations and the End Permian (251 Ma) mass extinction. After the dominance of microorganisms in the first 3 billion years, the Ediacaran period was characterised by a rapid radiation of large, soft-bodied, multicellular eukaryotes, the Ediacara biota. The authors showed that diverse Ediacaran communities were positively correlated to biomass and productivity. The arrival of new functional groups (e.g. first detection of animal bioturbation), possibly driven by niche partitioning, increased ecological complexity and functional redundancy. The radiation continued during the Cambrian, when predation, widespread biomineralisation and deep bioturbation all appeared. Phytoplankton diversification led to an increase in primary production. Animal interaction increased as suggested by the presence of well-developed eyes and both mechano- and chemosensory organs. This radiation also favoured the development of suspension-feeding species. Ecospace occupation enlarged. In contrast, the End Permian extinction, the biggest extinction ever recorded, was associated with a reduction in primary production and in ecosystem stability for ~5 million years after the extinction. These deep-time experiments provide an interesting insight on the relationships between biodiversity and ecosystem functioning. Whether this can be applied to the current level of biodiversity erosion remains to be investigated.

10.1.6 The debate on the role of biodiversity for ecosystem functioning

Criticisms on experimental studies addressing the effects of biodiversity on ecosystem functioning are numerous (993).

A first criticism is that BEF experiments are biased because they cannot account for all processes occurring in nature. Most experiments take place at small spatial and temporal scales and cannot account for demographic processes such as dispersal and immigration. They are therefore difficult (if possible) to extrapolate at the ecosystem scale (994). Many

BEF experiments have been performed at short timescales and on immature communities (995). In the terrestrial realm, this is the case for the Biodiversity and Ecological Processes in Terrestrial Herbaceous Ecosystems (BIODEPTH) experiment (949). Thompson and co-workers (995) compared the relationships between biodiversity and biomass at Bibury (Gloucestershire, UK), an undisturbed site characterised by a mature community and BIODEPTH. Both sites had similar biomass (mean biomass = 553 $g \cdot m^{-2}$ at Bibury versus 656 $g \cdot m^{-2}$ at BIODEPTH). Biodiversity was assessed by measuring species richness, functional group richness and functional biodiversity. Whereas BEF relationships were positive in BIODEPTH (immature community), the ones found at Bibury (mature community) were negative or insignificant. Diversity of traits (functional diversity) was not correlated to biomass at Bibury. The authors propose that immaturity of BIODEPTH communities may explain the positive BEF relationship. Indeed, seeds were planted in spring 1996 and harvested in summer 1997. The authors concluded that there are probably no causal BEF relationships in the BIODEPTH experiments because these immature communities are changing rapidly. These positive relationships may be explained by the higher probabilities of diverse plots to encompass either competitive plants or N-fixers in an environment poor in nitrate (996).

A second criticism, and perhaps the most fundamental issue, is the lack of consideration that biodiversity and stability/biomass/productivity interact with each other. For example, although biodiversity may indeed lead to more stability, biodiversity may also be the product of (environmental) stability itself. It is therefore important to separate the effects of the environmental stability on biodiversity and the effect of biodiversity on community stability (Figure 10.14). We saw that climate controls large-scale to regional and micro-scale environmental stability (Chapters 2–4), which has in turn a strong influence on biodiversity and community stability (207, 225, 659). Although biodiversity should influence locally ecosystem stability, this feedback mechanism investigated in BEF research should become rapidly limited by the climatic regime (Figure 10.14). Failure to tease apart these effects (biodiversity effects on stability and climate-induced stability on biodiversity) has been a serious issue in BEF research, and I think that quantification of the respective influence of the two effects is a key scientific questions. Applying meta-analyses on 32 controlled experiments, Worm and colleagues (965) showed that species and genetic diversity enhance stability, defined as the ability to withstand recurrent perturbations. Although this result is probably true at small scales, the effect of stability for maintaining biodiversity was not mentioned; this is, however, a well-known influence in macroecological research and biogeography (Chapters 3–4) (227).

A third criticism is that many experiments on BEF relationships have been tested by including unrealistic low diversity values (949, 972). Many experiments included only a few species, a scenario far from the reality that may have inflated the relationships between biodiversity and ecosystem functioning (Figure 10.15). More experiments should work on removing species from mature communities to more realistically investigate the effect of biodiversity erosion on ecosystem functioning (997). Such experiments would limit potential flaws of early experiments (974).

Fourth, some authors proposed that the relationships observed between biodiversity and ecosystem functioning may be the result of a spurious sampling effect (see Section 10.1.2 and both first and third criticisms). However, Loreau (977) argued that the sampling effect might be a real ecological phenomenon and a special case of selection or identity effect. The selection effect is the selection of specific traits that influence ecosystem function. Loreau

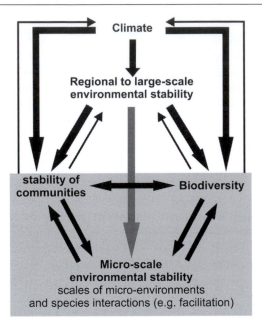

Figure 10.14 Conceptual diagram of the relationships between biodiversity and stability. The size of the arrows is proportional to the expected importance of the relationships or interrelationships between components or processes. BEF research (e.g. biodiversity and stability) has concentrated on the relationship between the measurements of an ecosystem function and biodiversity, occulting other external drivers. Future BEF research should better separate and quantify processes or mechanisms influencing BEF relationships.

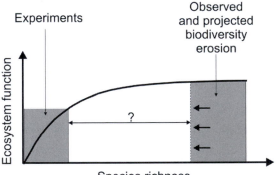

Figure 10.15 Sections of the BEF relationship examined in some experiments (left) and sections of the BEF relationships influenced by human activities (right). The BEF relationship is assumed here to follow the redundant species hypothesis. The grey area symbolises the zone of the BEF relationships concerned in some experiments (left) and by both observed and projected biodiversity erosion (right). The horizontal two-way arrow denotes the hypothetical difference between experiments and both observed and projected biodiversity erosion. The question mark indicates that the magnitude of this difference is unknown and may vary among ecosystems. The three horizontal arrows denote the current direction of the influence of human activities on biodiversity.

said that 'biodiversity matters only for ecosystem functioning to the extent that it provides phenotypic trait variation, or functional diversity, related to the particular ecosystem process considered'. There are many species-specific effects that have been observed in field (998) and laboratory (999) experiments. At Tatoosh Island (United States), a manipulative experiment in a rocky intertidal community showed that the removal of macroscopic grazers (urchins, chitons and limpets) switched the system from less productive perennial species (*Hedophyllum sessile*) to highly productive annual kelps (*Alaria marginata*). In this system, the rise in species richness did not increase annual production. Instead, this is the removal of the consumers that favoured a competitively superior species that increased production (998). Another study that investigated the relationship between biodiversity and productivity in 12 natural mature grassland ecosystems detected competitive effects on production but failed to find any effect of diversity on production (1000). The authors concluded that the local effect of diversity on production in mature ecosystems was weak in comparison to other drivers (e.g. environmental conditions and disturbance history) of biomass production and that the main influences originated from the abiotic environment (Figure 10.16).

A fifth criticism, related to the second one, is that the effects of biodiversity on ecosystem functioning may be insignificant in comparison to the effect of the environment. The environment would influence species composition, and in turn ecosystem functions. The influence of biodiversity on ecosystem functions may be small and rapidly negligible in a world of strong environmental change. This is a serious issue, and much effort is needed on this topic in future BEF research.

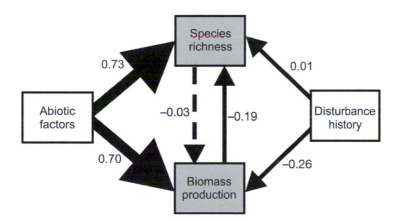

Figure 10.16 Relationships between species richness, biomass production, abiotic conditions and disturbance history from meta-analysis of model paths. The size of the arrows is proportional to the influence of the variables. The influence is estimated from standardised path coefficients. The effects of abiotic conditions, biomass production on richness and disturbance on biomass production were found to be consistently significant in all 12 studies: Utah montane grassland, Tanzanian grasslands, Minnesota prairie, Kansas prairie, Mississippi prairie, Louisiana prairie, Indian tropical savanna, Wisconsin prairie, Finnish meadows, Texas grasslands, Louisiana coastal wetlands and Louisiana riverine marsh. The dashed arrow indicates an absence of a significant relationship.

Source: Redrawn from Grace and co-workers (1000).

10.2 Biodiversity changes and ecosystem goods and services

10.2.1 An economic quantification

Human economy utilises 40% of net primary production (*570*). Assessing the economic values of 17 ecosystem services of 16 biomes (Table 10.2), Costanza and co-workers (*1001*) estimated that services provided by these biomes (terrestrial and marine) were US$16 and 54 trillion (average = 33).

About 63% of the estimated values (US$16–54 trillion) are from marine ecosystem services because of their vast area. These estimates are conservative because some services are difficult to assess. A minimum of ecosystem structure is necessary to ensure functions (i.e. movement of matter, energy and information) that lead to ecosystem goods and services. An ecosystem service may be the product of several ecosystem functions. In contrast, one function may be at the origin of several services. These estimates represented 1.8 times the total GNP (gross national product). This means that if we were to replace all these ecosystem services, we would need 1.8 times the GNP every year. Although this is a *reductio ad absurdum* because all ecosystem services are inimitable and that without ecosystem services there would be no life, this shows the economic importance of ecosystem services.

Marine ecosystems were estimated to produce US$577 per hectare per year, or US$20,949 billion per year in terms of total global flow value (Table 10.3). The larger contributor was coastal systems, which provide US$4,052 per hectare per year.

10.2.2 An overview of the influence of some ecosystems for goods and services

Coastal ecosystems such as seagrasses, macroalgae, mangroves and coral reefs provide a large number of goods and services, embracing the production of plants and animals used for food and medicines, water treatment and protection against flooding and erosion (*189*).

10.2.2.1 Seagrasses

Seagrasses are coastal ecosystems composed of angiosperms (60 species worldwide) belonging to four families: Posidoniaceae, Zosteraceae, Hydrocharitaceae and Cymodoceaceae. These engineer species support biologically rich communities and are critical nursery grounds for fish larvae and juveniles. These ecosystems, which establish in areas as deep as 60 m, are found on tidal mudflats in estuaries from tropical to subarctic systems, in coral reef lagoons, on shallow sandy areas close to the coast and around sand cays (*1002*). Seagrasses are the main diet of dugongs, green turtles and manatees, and provide a habitat for many species such as shellfish, prawns and fish (e.g. finfish and seahorse). Seagrasses deliver key ecological services. They stabilise sediment and absorb nutrients from coastal run-off, helping to keep clear water and act as dampers to wave action. Seagrasses also represent an important source of carbon that can be exported to deep sea. The excess of organic carbon is buried within seagrass sediments, which are hotspots for carbon sequestration (*1003*). With algae beds, seagrass meadows cover 200 million ha and provide important ecosystem services mainly in the form of nutrient cycling estimated to worth $19,004·ha^{-1}·yr^{-1} (*1001*). They also support commercial fisheries of $3,500·ha^{-1}·yr^{-1} (*1004*).

Table 10.2 Ecosystem (terrestrial and marine) services and functions examined in the study of Costanza and co-workers (1001).

Number	Ecosystem goods and services	Ecosystem functions	Examples
1	Gas regulation	Regulation of atmospheric chemistry	CO_2 and O_2 concentration, O_3 formation
2	Climate regulation	Global temperature and precipitation	Greenhouse gases regulation, DMS production
3	Disturbed regulation	Resilience of ecosystems to environmental changes	Storm protection, food control
4	Water regulation	Regulation of hydrological flows	Provisioning of water for agriculture (irrigation)
5	Water supply	Storage and retention of water	Provisioning of water by aquifers, watershed and reservoirs
6	Erosion control and sediment retention	Retention of soil within and ecosystem in lakes and wetlands	Prevention of loss of soil by wind, run-off and storage of stilt
7	Soil formation	Soil formation processes	Weathering of rock and the accumulation of organic material
8	Nutrient cycling	Storage, nutrient cycling and processing and acquisition of nutrients	Nitrogen fixation and other elemental or nutrient cycles
9	Waste treatment	Recovery of mobile nutrients and removal or breakdown of excess of xenic nutrients and compounds	Waste treatment, pollution control, detoxification
10	Pollination	Movement of floral gametes	Provisioning of pollinators for the reproduction of plants
11	Biological control	Trophic and dynamic controls of populations	Control of preys by predators and reduction of the herbivory pressure by predators
12	Refugia	Habitat for resident and transient populations	Nurseries, habitat for migratory species, regional habitats for locally harvested species or overwintering grounds
13	Food production	Part of the gross primary production extractable as food	Production of fish, game, crops, nuts; subsistence farming or fishing
14	Raw materials	Part of the gross primary production extractable as raw materials	Production of lumber, fuel or fodder
15	Genetic resources	Sources of unique biological materials and products	Medicine, materials for science, genes for resistance to plant pathogens and crop pests; ornamental species
16	Recreation	Opportunities for recreational activities	Ecotourism, sport fishing and other outdoor activities
17	Cultural	Opportunities for non-commercial uses	Aesthetic, artistic, educational, spiritual and scientific values of ecosystems

Source: From Costanza and colleagues (1001).

Table 10.3 Average global value of annual marine ecosystem services for 1994.

| Biome | Area (ha × 10⁶) | Ecosystem service number | | | | | | | | | | | Total value per ha ($·ha⁻¹·yr⁻¹) | Total value per ha ($·yr⁻¹ × 10⁹) |
		1	2	3	8	9	11	12	13	14	16	17		
Open ocean	33,200	38	×	×	×	118	5	×	15	0	×	76	252	8,381
Coastal	3,102	×	×	88	3,677	×	38	8	93	4	82	62	4,052	12,568
Estuaries	180	×	×	567	21,100	×	78	131	521	25	381	29	22,832	4,110
Seagrass/algae beds	200	×	×	×	19,002	×	×	×	×	2	×	×	19,004	3,801
Coral reefs	62	×	×	2,750	×	58	5	7	220	27	3,008	1	6,075	375
Shelf	2,660	×	×	×	1,431	×	39	×	68	2	×	70	1,610	4,283
Total	36,302												577	20,949

Note: A cross indicates lack of available information. For the meaning of ecosystem service numbers, see Table 10.2. These estimates are conservative. All values are in $·ha⁻¹·yr⁻¹ except the last column ($·yr⁻¹ × 10⁹).

Source: Modified from Costanza and co-workers (1001).

10.2.2.2 Macroalgae

Macroalgae form widespread underwater forests playing a major role in structuring regional-scale biodiversity. Some seaweeds (e.g. *Laminaria digitata*) are also economically important, being exploited for their alginate and iodine content (*1005*). Seaweeds such as the red macroalga *Porphyra* are consumed in many countries. Macroalgae alter the wave action regime and allow the establishment of species sensitive to waves. They are also nurseries for many fish and are inhabited by species such as urchins and snails. Such ecosystems also favour sediment stabilisation.

10.2.2.3 Mangroves

Globally, ~30% of mangroves have been transformed in the last two decades (*1006*). The degradation of these ecosystems has reduced ecosystem goods and services, including fish/crustacean nurseries, wildlife habitat, flood control, protection from tropical cyclones, storm surges and tsunamis, pollutants filtration, nutrients recycling and sediment trapping. These systems, as seagrass beds mitigate the effect of freshwater discharge, are sinks for both organic and inorganic substances, as well as pollutants. They improve water clarity and reduce nutrients concentration, which may enable the establishment of coral reefs (*1007*).

10.2.2.4 Coral reefs

Alteration of coral reefs can substantially affect ecosystem goods and services. Coral reefs cover ~0.1–0.5% of the oceanic realm, which represents between 617,000 and 1,500,000 km^2 (*1007*). Despite its small surface area, between 9 and 12% of all fish captured annually come from coral reefs, and a third of all described marine fish are found in these systems. More than 10 million people depend on coral reefs for their protein intake and ~100 countries have coral reefs. The types of food provided by coral reef ecosystems range from mussels and sea cucumbers to fish. Giant clams of the genus *Tridacna* and shells such as *Trochus* (mother-of-pearl shells) are not only captured for food, but also as jewellery. Many coral reef species are also sold for marine aquariums, a market of US$25–40 million per year in 1985. Coral reefs protect coasts against erosion, and have aesthetic, cultural and spiritual values. Some corals record climate change. Corals make calcification and constitute both keystone and engineer species that physically structure the habitat. These ecosystems often maintain a high biodiversity level and export invertebrates, fish and materials to adjacent systems (e.g. pelagic food webs). They play a key role in waste assimilation and biological control, and generate an important source of revenue through recreational activities and tourism (Table 10.3).

Collapse in the local biodiversity of the reef can have pronounced adverse effects in terms of resources exploited by local human populations, ecotourism and shoreline protection. In Indonesia, coral destruction has led to coastal erosion, which cost between US$820,000 and 1,000,000. A coral reef barrier strongly protects the coast against erosion. In the Maldives, a 1 km artificial breakwater cost US$12,000,000 (*1007*). From a biogeochemical point of view, coral reefs and their symbiotic cyanobacteria act as nitrogen fixers in a marine environment poor in nitrogen. This nitrogen becomes available for the whole system trophodynamics, which explains why these ecosystems are so productive. This excess also becomes available for adjacent seas. Over geological timescales, these ecosystems act as sinks for CO_2, but this effect may be of minor significance when compared to the global carbon budget (*1008*).

10.2.3 Effects of biodiversity on ecosystem goods and services

10.2.3.1 Meta-analyses

Worm and co-workers (965) investigated the relationships between biodiversity and ecosystem services. Applying meta-analytical techniques to 32 controlled experiments, they suggested that diversity enhances ecosystem processes and services in both primary producers (plants and algae) and consumers (herbivores and predators) (Figure 10.17). The response ratio was positive for all tested variables: resource, production, nutrient cycling and stability.

They then examined a database of biodiversity and ecosystem services for 12 coastal ecosystems. For each ecosystem, they examined between 30 and 80 economically and ecologically important species. They found that species richness reduction was associated with the decrease of a large number of ecosystem services (Figure 10.18). The number of viable fisheries diminished (Figure 10.18A–C). Nursery habitats and the filtering functions reduced (Figure 10.18C). They further associated the increase in beach closure, harmful blooms and oxygen depletion to biodiversity erosion (Figure 10.18D). This study was the first to examine the relationships between marine biodiversity and ecosystem services, investigating both experimental and ecological studies. I should caution, however, that this study did not examine other potentially confounding influences: (1) the effects of climate change on biodiversity and ecosystem services; (2) the consideration of the macroecological relationships between biodiversity, ecosystem services and the climatic regime; and (3) the

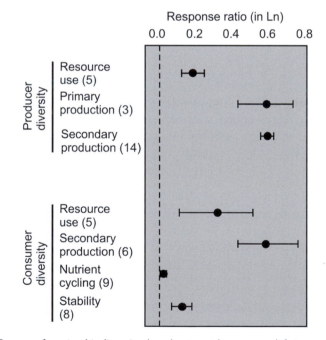

Figure 10.17 Influence of marine biodiversity (producers and consumers) for ecosystem processes and services. The response ratio (in natural logarithm) is shown with a 95% confidence interval. The number of studies on which these results were based is indicated in brackets.

Source: Redrawn from Worm and co-workers (965).

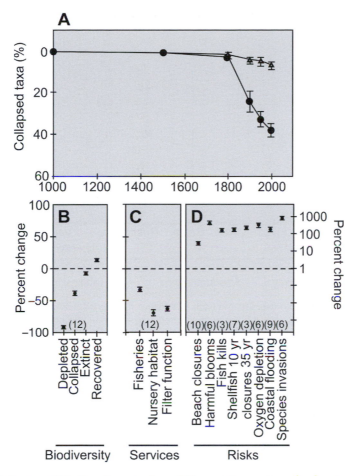

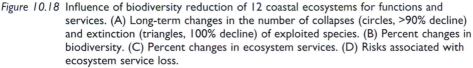

Figure 10.18 Influence of biodiversity reduction of 12 coastal ecosystems for functions and services. (A) Long-term changes in the number of collapses (circles, >90% decline) and extinction (triangles, 100% decline) of exploited species. (B) Percent changes in biodiversity. (C) Percent changes in ecosystem services. (D) Risks associated with ecosystem service loss.

Source: Redrawn from Worm and co-workers (965).

effect of global change (pollution, exploitation, physical alteration), which may affect biodiversity and ecosystem services independently. For example, the increase in harmful blooms may be related to global warming, or to a better monitoring of these phenomena. Oxygen depletion may be related more directly to eutrophication. It is also dangerous to associate the reduction in the biodiversity of these ecosystems to the increase in beach closure, perhaps more directly influenced by water quality. The negative relationships between species richness and collapsed fisheries may be related to the reduction in the fishing pressure between the pole and the equator where species richness increases (their figure 2B).

Some results such as the reduction in many fish stocks and their associated reduction in stability, measured here as the reduction in the resistance and the resilience of the stock,

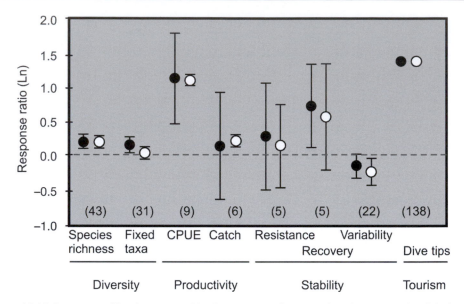

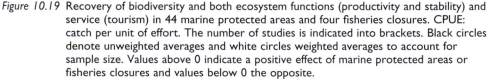

Figure 10.19 Recovery of biodiversity and both ecosystem functions (productivity and stability) and service (tourism) in 44 marine protected areas and four fisheries closures. CPUE: catch per unit of effort. The number of studies is indicated into brackets. Black circles denote unweighted averages and white circles weighted averages to account for sample size. Values above 0 indicate a positive effect of marine protected areas or fisheries closures and values below 0 the opposite.

Source: Redrawn from Worm and co-workers (*965*).

are more robust and substantiated by other studies (*773, 1009*). A very interesting result is the increase in both biodiversity and ecosystem services in marine protected areas. Worm and colleagues also examined data from 44 fully marine protected areas and four large-scale fisheries closures and showed an increase in both target and non-target species (Figure 10.19). In the marine protected areas or after fisheries closures, the rise in biodiversity paralleled an increase in productivity (CPUE and catch) and stability, the latter being measured as resistance, recovery and variability. An increase in tourism revenue was also observed. The last result is quite interesting and demonstrates, to paraphrase the authors, that there is no dichotomy between biodiversity conservation and socio-economic development.

10.2.3.2 Overfishing and biodiversity

As we saw in Chapter 9, overfishing had a strong effect on biodiversity. Eutrophication, pollution and hypoxia have all affected coastal fish. Overfishing has clearly altered food provisioning services. This may have strong effects for human populations that heavily depend on fish as a source of animal protein (Table 10.4).

Fishing has also had an indirect effect on marine biodiversity through trophic cascades and habitat destruction (Chapter 9). Aquaculture will not be a solution, as farmed fish such as salmon and tuna need fishmeal and oil, which derives from the catch of small pelagic fish (Chapter 9).

Table 10.4 Percentage of animal protein originating from fish in 2000.

Region	Percentage of animal protein from fish products (%)
Middle East and North Africa	9.0
Europe	10.6
South America	10.9
North America	11.5
Central America and Caribbean	14.4
Sub-Saharan Africa	23.3
Oceania	24.2
Asia (excluding Middle East)	27.7

Source: From Food and Agriculture Organization of the United Nations.

10.2.3.3 Diversity and species invasion

Diversity plays an indirect role on invasion resistance. Stachowicz and colleagues assessed diversity influence on species invasion in a subtidal marine invertebrate community (1010). In the field, they found a negative correlation between native species richness and species invasion frequency. By manipulative experiments, they found that when native biodiversity was high, the reduced open space protected against the establishment of non-native ascidians (sea squirts). When biodiversity was kept constant experimentally, the increase in the availability of open space augmented invaders settlement (Figure 10.20).

10.2.3.4 Diversity and detoxification services

Marine microbial community exerts important detoxification services. They filter water, reduce eutrophication impacts and decompose toxic hydrocarbons. Eutrophication-caused reduction in oxygen concentration may diminish the microbes' capacity to degrade toxin hydrocarbons.

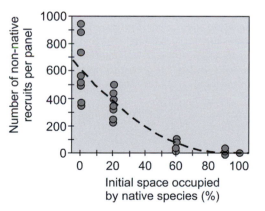

Figure 10.20 Relationships between the initial space occupied by native species and settlement of the non-native ascidian Diplosoma. The determination coefficient was 0.84.

Source: Simplified from Stachowicz and co-workers (1010).

In Chesapeake Bay, the substantial reduction in oyster abundance diminished filtering service and water clarity (776). This study showed that species that provide key functional traits are important, and that beyond a certain limit biodiversity erosion may have strong consequences on ecosystem services.

10.2.4 Biodiversity changes and potentially harmful species

10.2.4.1 Harmful species

Pollution, eutrophication and anthropogenic climate change may have serious consequences for human health. As we saw in Chapter 9, studies have reported an increase in harmful algal blooms (HABs), which has been attributed to eutrophication and global climate change. About 40 marine species produce toxins harmful to humans and other marine species. Species such as *Alexandrium*, *Prorocentrum*, *Nitzschia* and *Dinophysis* are the main species involved (Table 10.5).

Intoxication due to seafood consumption has been reported in many countries around the world (Table 10.6).

10.2.4.2 Ciguatera and climate change

Ciguatera is a foodborne illness that arises when eating fish whose muscle and liver are contaminated with toxins generally released by dinoflagellates (e.g. *Gambierdiscus toxicus*) inhabiting subtropical/tropical waters (e.g. the Pacific Ocean and Caribbean Sea). The toxins (e.g. ciguatoxin, maitotoxin, scaritoxin, palytoxin) are progressively concentrated from low to high trophic levels (biomagnification). Fish such as barracudas, snapper, parrotfishes and amberjacks can contaminate humans and cause gastrointestinal (e.g. abdominal pain, diarrhoea, vomiting) and neurological (e.g. parasthesia) problems. Ciguatera frequency is

Table 10.5 Toxins produced by some marine microalgae and their effects for human health.

Toxins	Microalgae	Carriers	Effects on human health
Diarrhoeic shellfish poisoning (DSP)	Dinoflagellates (*Dinophysis* and *Prorocentrum*)	Filtering shellfish (e.g. oysters, mussels, cockles, clams)	Gastrointestinal symptoms (e.g. vomiting, abdominal pain, diarrhoea)
Paralytic shellfish poisoning (PSP)	Dinoflagellates (*Alexandrium* and *Gymnodinium*)	Filtering shellfish (e.g. oysters, mussels), crustaceans, fish	Muscular paralysis, breathing difficulty, respiratory arrest; fatality in 10% of the people intoxicated
Amnesiac shellfish poisoning (ASP)	Diatoms (*Nitzschia*)	Shellfish (e.g. mussels)	Mental confusion, loss of memory, disorientation, coma; fatality in some elderly people
Neurotoxic shellfish poisoning (NSP)	Dinoflagellates (*Gymnodinium*)	Oysters, clams, crustaceans	Muscular paralysis, shock, sometimes death
Venerupin shellfish poisoning (VSP)	Dinoflagellates (*Prorocentrum*)	Oysters, clams	Gastrointestinal, nervous, haemorrhagic, hepatic symptoms, delirium, hepatic coma

Source: From World Health Organization (1011).

Table 10.6 Intoxication cases due to seafood consumption.

Intoxication	Country	Date	Consequences
PSP	Philippines	1983	300 cases, 21 deaths
	United Kingdom	1968	78 cases
	Spain	1976	63 cases
	France	1976	33 cases
	Italy	1976	38 cases
	Switzerland	1976	23 cases
	Germany	1976	19 cases
DSP	Japan	1976–1982	1,300 cases
	France	1984–1986	4,000 cases
	Scandinavia	1984	300–400 cases
VSP	Japan	1889	81 cases, 51 deaths
	Japan	1941	6 cases, 5 deaths
	Norway	1979	70 cases
ASP	Canada	1987	153 cases, 3 deaths
NSP	Florida	1977(?)	No data

Note: PSP: paralytic shellfish poisoning. DSP: diarrhoeic shellfish poisoning. VSP: venerupin shellfish poisoning. ASP: amnesic shellfish poisoning. NSP: neurotoxic shellfish poisoning. The question mark denotes the absence of accurate data on this event.

Source: From World Health Organization (*1011*).

expected to increase with rising SST and therefore global warming. Projections indicate possible increases in the cases of Ciguatera by 160–430 per thousand for 2050, although this figure depends on the degree of warming (*587*). Sea temperature modulates ciguatoxin production by altering growth rates of toxic organisms, toxin production rates and toxin bioaccumulation. However, the relationship is probably more complex, and caution is needed as too warm waters may also reduce ciguatera frequency (*1012*).

10.2.4.3 Other dangerous species

Documenting and investigating marine biodiversity is important because not all marine species are safe for humans. Anthropogenic climate change will alter the spatial distribution of dangerous species and new (unexperienced) people may enter into contact with these species, leading to serious marine hazards. It is therefore important to know, document and warn against the potential effects of some marine species on humans. Venomous species mainly occur in tropical waters. Some inject a zootoxin through a venom apparatus (e.g. spines, stings, teeth). This is the case for some species of sponges (the genera *Fibulia*, *Neofibularia*, *Tedania*, *Dysideaz*), medusa (*Physalia* spp., *Chironex fleckeri*, *Chiropsalmus quadrigatus*), anthozoans (e.g. *Actinodendron glomeratum*, *Doflenia armata*), worms, molluscs (e.g. conidae such as *Conus geographus* and *C. striatus*), sea urchins (toxopneustidae such as *Toxopneustes* and *Tripneustes*), sea cucumbers (*Bohadschia argus*), starfish (*Acanthaster planci*), octopods (e.g. *Hapalochlaena maculosa*), sea snakes (e.g. *Enhydrina schistosa*) in the Indo-Pacific region and fishes such as stingrays, weever, scorpionfish and stone fish. Some species cannot be eaten because they are poisonous (*1013*). In danger, some sea cucumbers such as *Bohadschia argus* expel viscera with an organ called the Cuvierian organ. In the water, this creates long white sticky threads that adhere to any species. In contact with a human's eye, the released toxin may cause permanent blindness (*1014*). Sharks such as heterodontidae (e.g. *Heterodontus*) and squalidae (e.g. the spiny dogfish *Squalus acanthias*) can be dangerous. The latter species has a spine that can inject a venom causing strong pain lasting several hours. Death has rarely been observed. Some species such as the eyed electric ray (*Torpedo torpedo*) can provide a strong electric shock of up to 200 volts, but they are in general not dangerous. Stingrays (*Myliobatidae*) have venom located in the underside of the animal's spine, which can be fatal to humans, even after its death. Lionfish (*Scorpaenidae*) possess venomous fin rays, the potency of which may represent a serious hazard for divers and fishermen. The lionfish *Pterois venom* causes extreme pain, breathing difficulties and convulsions, and can sometimes lead to the death of young children or elderlies. Other potentially dangerous fish are chimaeridae (chimaeras), trachinidae (e.g. *Trachinus draco*), uranoscopidae (*Uranoscopus scaber*) and acanthuridae (coral reef species).

Ingestion of marine species can also be dangerous. For example, human fatalities from crab ingestion have been documented in Japan, the Philippines, Mauritius and Fiji (*1015*). Xanthid crabs have often been at the origin of these fatalities. These crabs contain paralytic shellfish toxins, tetrodotoxin and a palytoxin-like substance. Human death can occur within several hours after consumption of a poisonous crab. Ingestion of the crab *Zosimus aeneus* is often lethal. 70% of humans die after being bitten by the mollusc *Conus geographicus*. The sea snail feeds on fish and has venom, called conotoxin. Many fishes are also poisonous. The fugu (tetraodontidae, genus *Takifugu*) is a fish highly appreciated by the Japanese. However, the fish is dangerous and ~23 people have died in Japan since 2000 after eating this fish. Intoxication also occurs among cartilaginous fishes (Chondrichthyes). Fatal accidents have also been reported after ingestion of sea turtles (*Chelonia mydas*, *Eretmochelys imbricata*, *Dermochelys coriacea*) in the south-eastern part of Asia and in Madagascar, although the toxins involved are at present unknown. Consumption of the liver of the polar bear (*Ursus maritimus*), rich in vitamin A, may create overvitaminosis.

10.2.4.4 Eutrophication

Eutrophication can also exacerbate the growth of bacteria possibly hazardous for human health. Bacteria such as *Escherichia coli*, *Salmonella* spp. or *Vibrio cholerae* are potentially dangerous for people taking a bath (*1011*). Under ordinary conditions, these bacteria die rapidly in seawater because of the relative nutrient shortage, bacteria exposure to UV radiation (bactericidal effect) and seawater osmolarity. However, when food becomes elevated, UV radiation diminishes and some algae may release chemicals that produce bacteria osmo-protection. These conditions become favourable for bacteria development, which constitutes a serious threat for the human population.

10.2.5 Biodiversity changes and chemicals

10.2.5.1 Biodiversity as a natural biochemistry machinery

Biodiversity can be seen as a natural biochemistry machinery that produces a gigantic quantity of chemical compounds that can be used as food supplement or therapeutics. As stated by William Fenical, 'biodiversity translates to genetic uniqueness, which in turn results in the expression of diverse biochemical processes producing metabolic products' (*1016*). Species diversity also means chemical diversity. Development of drug resistance is pushing the pharmaceutical industry to research new molecules. However, biodiversity loss due to both direct and indirect influences of human activities may prevent the discovery of new drugs and remedies against current but also future illness.

Most drugs used in cancer treatments (60%) and infectious diseases (75%) come from natural sources. Terrestrial biodiversity has been at the basis for the development of many compounds used in medicine. For example, vincristine and vinblastine extracted from the rose periwinkle (*Catharanthus roseus*) endemic to Madagascar provide effective treatments against some forms of leukaemia (e.g. Hodgkin's lymphoma) and other cancers. These molecules are antimitotic, disrupting the formation of cell microtubules. Salicylic acid (aspirin), nowadays chemically synthetised, was originally collected from the leaves and bark of the willow tree. Morphine was formerly extracted from the seedpods of the poppy *Papaver somniferum*.

About half of the potential pharmaceutical drugs currently under investigation come from marine biodiversity (Table 10.7). Competition for space of some benthic organisms (e.g. species associated with coral reefs) has led to the production of chemical defences that might lead to new marine drugs (*1016*). Natural compounds released into the seas by species are rapidly diluted. Therefore, species have developed potent molecules, which have a strong effect at small concentrations (*1018*). Many scientists working in the field of pharmacology believe that marine products are important candidates for treating many forms of illness (*1016, 1019*). Marine compounds may exhibit immunosuppressant, anti-inflammatory, anti-mitotic, anti-fungal, anti-tuberculosis, anthelmintic, anti-protozoal, antibacterial and antiviral activity (Table 10.7). These chemicals often originate from secondary metabolites (i.e. unusual compounds not involved in primary metabolism). Thousands of these metabolites have been found for the last decades (*1020*). Other compounds can also be food supplement. For example, the diatom *Phaeodactylum tricornutum* (Bacillariophyceae) contains long chained polyunsaturated omega-3 and omega-6 fatty acids (*1021*), which prevent cardiovascular disorders, some cancers and contribute to nerve and brain development in foetuses and children. Sea cucumbers produce holothurin, a compound that is used in medicine to regulate cardiac activity and to improve metabolism.

10.2.5.2 Important potential of marine drugs

Hu and colleagues (*1022*) showed that the number of newly discovered marine natural products increased substantially after the invention and development of the high-resolution nuclear magnetic resonance spectrometer in 1984 (Figure 10.21). About 500 new natural compounds were discovered in 2005 (Figure 10.21). Natural products have been divided into seven groups (*1022*): terpenoids, steroids (including saponins), alkaloids, ethers (including ketals), phenols (including quinones), lactones and peptides (Figure 10.22).

Terpenoids are studied for their antibacterial and cytotoxic anti-neoplastic effects (i.e. chemotherapeutic agents used in chemotherapy). Terpenoids are also precursors of steroids.

Steroids, found in plants, animals and fungi (ergosterols), are investigated because they provide treatment against inflammation, allergic conditions, eczema and Crohn's disease.

Alkaloids (e.g. morphine), often toxic to other organisms, also have strong pharmacological effects; vinblastine is an anti-tumour agent, and quinine (from the cinchona tree) is an antipyretic and antimalarial agent. The vincamine (from Apocynaceae plant family) is a vasodilating and an anti-hypertensive agent. The marine alkaloid ecteinascidin-743

Table 10.7 Examples of compounds synthesised from marine biodiversity.

Application	Original source	Status
Pharmaceuticals		
Antiviral drugs (herpes infection, AIDS)	Sponge, *Cryptotethya crypta*	Commercially available
Anti-cancer drug (non-Hodgkin's lymphoma)	Sponge, *Cryptotethya crypta*	Commercially available
Anti-cancer drug	Bryozoan, *Bugula neritina*	Phase II clinical trials
Anti-cancer drug (tumour-cell DNA disruptor)	Sea hare, *Dolabella auriculata*	Phase I clinical trials
Anti-cancer drug	Ascidian, *Ecteinascidia turbinata*	Phase III clinical trials
Anti-cancer drug	Ascidian, *Aplidium albicans*	Advanced preclinical trials
Anti-cancer drug (microtubule stabiliser)	Gastropod, *Elysia rubefescens*	Advanced preclinical trials
Anti-cancer drug	Sponge, *Discodermia dissolute*	Phase I clinical trials
Anti-cancer drug	Sponge, *Lissodendoryx* sp.	Advanced preclinical trials
Anti-cancer drug (G2 checkpoint inhibitor)	Actinomycete, *Micromonospora marina*	Advanced preclinical trials
Anti-cancer drug	Ascidian, *Didemnum granulatum*	In development
Anti-inflammatory agent	Sponge, *Jaspis* sp.	In development
Anti-fungal agent	Marine fungus	In development
Anti-tuberculosis agent	Sponge, *Trachycladus*	In development
Anti-HIV agent	Gorgonian, *Pseudopterogorgia*	In development
Antimalarial agent	Ascidian	In development
Anti-dengue virus agent	Sponge, *Cymbastela*	In development
Anti-cancer drug (tumour-cell DNA disruptor)	Marine crinoid	In development
Molecular probes		
Phosphatase inhibitor	Dinoflagellate	Commercially available
Phospholipase A2 inhibitor	Sponge, *Luffariella variabilis*	Commercially available
Bioluminescent calcium indicator	Bioluminescent jellyfish, *Aequora victoria*	Commercially available
Reporter gene	Bioluminescent jellyfish, *Aequora victoria*	Commercially available
Medical devices		
Orthopaedic and cosmetic surgical implants	Coral, molluscs, echinoderm skeletons	Commercially available
Diagnostics		
Detection of endotoxins (LPS)	Horseshoe crab	Commercially available
Enzymes		
Polymerase Chain-Reaction (PCR) enzyme	Deep-sea hydrothermal vent bacterium	Commercially available
Nutritional supplements		
Polyunsaturated fatty acids used in food additives	Microalgae	Commercially available
Pigments		
Conjugated antibodies used in basic research and diagnostics	Red algae	Commercially available
Cosmetic additives		
Cosmetic (anti-inflammatory)	Gorgonian, *Pseudopterogorgia elisabethae*	Commercially available

Source: Modified from the US Commission on Ocean Policy (*1017*).

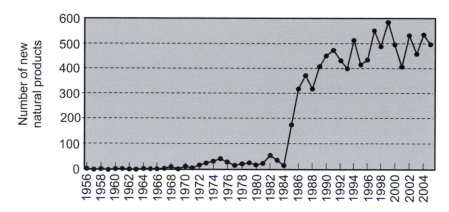

Figure 10.21 Long-term changes in the number of new marine products.

Source: Redrawn from Hu and colleagues (*1022*).

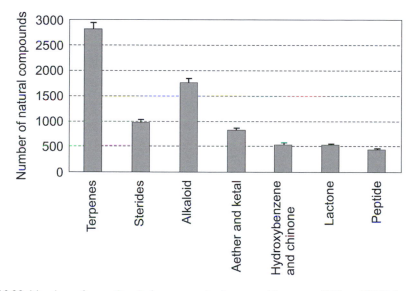

Figure 10.22 Number of new chemical compounds discovered between 1985 and 2008 from marine biodiversity.

Source: Redrawn from Hu and colleagues (*1022*).

isolated from the mangrove ascidian *Ecteinascidia turbinata* is a broad-spectrum anti-tumour agent that may lead to significant improvements in cancer treatment (*1019*). Many alkaloids that show potent inhibitory activity against *Mycobacterium tuberculosis* have been isolated from marine invertebrates. For example, ergorgiaene is a compound produced from a West Indian gorgonian *Pseudopterogorgia elisabethae*, which inhibits the growth of M. *tuberculosis* H37Rv at a concentration of 12.5 g·ml^{-1} (*1019*).

Ethers are mainly produced by marine microorganisms (fungi) as secondary metabolites. Ethers are important in medicine and were first used in 1842 as anaesthetics (e.g. ethyl ether, also known as ether). Codeine (or the methyl ether of morphine) is an effective pain relief. Phenols are used in medicine to treat varicose veins and to shrink haemorrhoids, and some compounds may be used as potential anaesthetics.

Some lactones are used in traditional medicine to treat inflammation and cancer. Bryostatin 1, originating from the bryozoan *Bugula neritina* (0.000001% of the wet animal's weight) is the most effective anti-leukemia agent ever found and may soon be used in the treatment of acute leukaemia. The compound inhibits leukaemia cells in culture.

Peptides are compounds highly used in medicine and pharmacology. Some compounds act as microtubule-interfering agents and may be used in cancer treatment. This is the case for dolastatin-10, which was isolated from the Indian sea slug *Dollabella auricularia*. The chemical originates, however, from its cyanobacterial dietary origin (*1023*). Others are used to induce apoptosis (i.e. programmed cell death) as loss of pro-apoptotic signals may lead to cancer initiation. Because apoptosis does not trigger any immune response, it has become an interesting development in cancer treatment (*1024*). Some marine peptides may also be efficient in inhibiting angiogenesis (i.e. the formation of new blood vessels). For example, Neovastat (AE-941), synthesised from the cartilage of the shark *Squalus acanthias*, directly inhibits tumour cell growth and angiogenesis.

It remains difficult to cultivate invertebrates to extract some compounds in large quantity (*1022*). For example, one tonne (wet weight) of the Caribbean tunicate *Ecteinascidia turbinata* is needed to extract 1 g of yondelis (trabectedin, ET-743), an anti-cancer agent that is currently tested clinically. This means that chemical synthesis techniques are indispensable for the large-scale development of these marine compounds.

10.2.5.3 Microorganisms

Marine microorganisms are also promising groups for the production of new therapeutics (Figure 10.23). For example, prokaryotic marine cyanobacteria are a significant source of structurally bioactive secondary metabolites (*1025*). The antibiotic vancomycin, the immunosuppressive agent cyclosporine and the anti-cancer agent bleomycin originate from these microorganisms. Some potent neurotoxins acting either as blockers (e.g. kalkitoxin) or activators (e.g. antillatoxin) of the eukaryotic voltage-gated sodium channels have also been derived from marine cyanobacteria. The anti-microtubule agents curacin A and

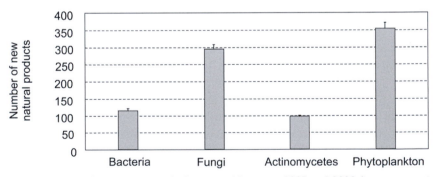

Figure 10.23 Number of new compounds discovered between 1985 and 2008 for some marine microorganisms and phytoplankton.

Source: Redrawn from Hu and colleagues (*1022*).

dolastatin 10 are being clinically tested as anti-cancer drugs. Synthetic analogues (e.g. TZT-1027 or soblidotin) have been found to perform better than existing anticancer molecules such as paclitaxel and vincristine, and are currently undergoing Phase I clinical trials.

Halichondrin B, a compound that shows anti-tumour effects against human ovarian cancers in mice, melanoma and some leukaemia, was isolated from the sponge *Halichondria okadai*. However, the molecule originates from a dinoflagellate called *Prorocentrum lima*, which is concentrated by the sponge during the filter-feeding process.

10.2.5.4 Macroalgae

Marine macroalgae have been exploited for many centuries for the production of agar, carrageenan and alginate. Some species such as *Laminaria digitata* are also exploited for their iodine content. The pharmaceutical industry has started to look for new drugs and compounds isolated from red, brown and green macroalgae (Figure 10.24). The red macroalga *Laurencia* produces many complex terpenoids and acetogenins. The dihydroxytetrahydro-furan, extracted from the marine brown macroalga *Notheia anomala*, exhibits potent and selective nematocidal effect against the nematodes *Trichostrongylus colubriformis* and *Haemonchus contortus*, which represent an important problem in commercial livestock industry and cause human disease (*1019*).

10.2.5.5 Invertebrates

Many compounds have been isolated from marine invertebrates (Figure 10.25). Most products have been derived from Porifera (sponge) and Coelenterates (corals). There are ~8,000 species of sponges living in both freshwater and marine ecosystems. Sponges have no hard outer protective shell, which makes them highly sensitive to predators such as invertebrates, fish and turtles. To protect themselves against these predators, these organisms have developed chemical defences. These chemical compounds also act as a fouling agent against pathogenic bacteria, algae and fungi. Sponges and their associated microorganisms have provided ~5,300 natural compounds so far, and new ones are identified on an annual

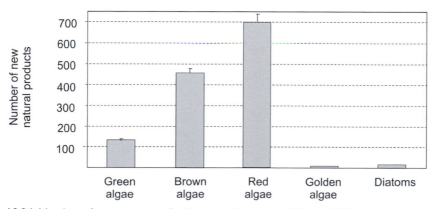

Figure 10.24 Number of new compounds discovered between 1985 and 2008 for some groups of marine macroalgae and diatoms (phytoplankton).

Source: Redrawn from Hu and colleagues (*1022*).

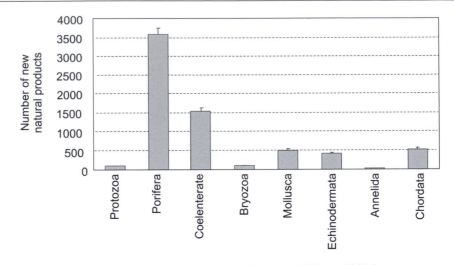

Figure 10.25 Number of new compounds discovered between 1985 and 2008 for some marine protozoa and invertebrates.

Source: Redrawn from Hu and colleagues (*1022*).

basis (*1026*). Toxic defensive chemicals produced by sponges have been used to treat people with cancer, AIDS, tuberculosis, bacterial infections and cystic fibrosis. For example, the two nucleosides spongothymidine and spongouridine were isolated in 1950 from the Caribbean sponge *Cryptotethya crypta* (*1026*). These molecules have enabled the synthesis of Ara-C, an anti-cancer agent, and Ara-A, an antiviral drug (Figure 10.26).

Vidarabine or Ara-A, a synthetic analogue of spongouridine, inhibits viral DNA synthesis of herpes, vaccinia and varicella zoster viruses (*1026*). This discovery has been used to synthetise the AZT (azidothymidine), whose generic name is zidovudine, a compound currently part of the tritherapy used to slow down the evolution of acquired immune deficiency syndrome (AIDS). Other compounds such as alkaloids, sterols, terpenes and peroxides have also been isolated from sponges. A total of 40 compounds are being tested in the treatment of viral ailments and for their anti-HIV activity. Papuamides (A, B, C,

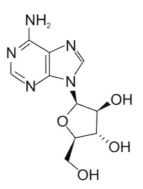

Figure 10.26 Structure of Ara-A.

Source: From Sagar and colleagues (*1026*).

D), anti-HIV and cytotoxic cyclic depsipeptides, were isolated from the sponges *Theonella mirabilis* and *Theonella swinhoei* along the north coast of Papua New Guinea. Avarol, a sesquiterpenoid hydroquinone, has been isolated from the marine sponge *Disidea avara*. The molecule has a dose-dependent inhibitory effect on the replication of the etiologic agent of AIDS.

Horseshoe crabs are estimated to be at least 300 million years old, being the closest living relative of trilobites. Today, there are four species of horseshoe crabs. The arthropod *Limulus polyphemus* (subclass Xiphosura) is found along the eastern coast of the United States of America south of the Maine and in the Yucatan Peninsula. Other species inhabit the Indo-Pacific region. Hemocyanin (copper-based blood) of horseshoe crabs also contains amebocytes, which defend the organism against **Gram**-negative pathogenic and non-pathogenic bacteria (the most abundant bacteria found in both freshwater and marine environments). In the 1960s, Dr Frederik Bang from the Biological Laboratory (Woods Hole) discovered that injection of marine bacteria into the hemocyanin of *Limulus polyphemus* triggered massive clotting. A few years later, the researcher helped by Dr Jack Levin, established that the clotting was related to an **endotoxin** and was triggered by the amebocytes of the horseshoe crab. The resulting cell-free reagent was termed *Limulus* Amebocyte Lysate (LAL) and is currently used to test the purity of medicines. LAL enables the detection of bacterial endotoxins and living bacteria, allowing the pharmaceutical manufacturers to ensure that drugs, vaccines and medical devices do not contain living microorganisms and endotoxin. Even non-pathogenic bacteria such as *Escherichia coli* when included in human blood may cause sepsis and death. Endotoxin injection in the human blood is pyrogenic. The production of LAL is done by collecting 30% of the blood of the horseshoe crabs. The chitin from horseshoe crabs is utilised in the manufacturing of chitin-coated filament for suturing and chitin-coated wound dressing for burn victims because they diminish healing time by 35–50%.

10.2.5.6 Vertebrates

Vertebrates are also a source of new chemical compounds. For example, squalamine is an aminosterol extracted from the dogfish shark *Squalus acanthias*. Squalamine has a potent antimicrobial activity against *Staphylococcus aureus* and exhibits antiangiogenic and antitumor activity.

Marine biodiversity has also been used as a model to better understand the function of some human organs. For example, the study of electrical impulses from the optic nerve of horseshoe crab eyes led to some important principles about the function of human eyes.

10.2.5.7 Conclusions

It is therefore essential to preserve marine biodiversity because it is a source of therapeutics. Marine biodiversity loss will affect species interactions, which have been the engine of the multiplicity of chemical defences invented by species to protect themselves against their consumers and predators. Biodiversity erosion will severely impair our capability to develop new therapeutics in the future.

10.3 Potential effects of changes in marine biodiversity for global biogeochemistry

Marine biodiversity moves a large quantity of elements and compounds, and it is likely that current alteration in marine biodiversity due to both direct and indirect effects of human activities will affect global biogeochemistry.

10.3.1 Marine biodiversity and the global carbon cycle

Climate change (natural and/or anthropogenic) associated with an alteration in the composition and structure of the pelagic biocoenosis might have strong consequences for the biological carbon pump (i.e. the sequestration of carbon from the atmosphere to the deep sea due to the biological activity in the oceans). Estimates indicate that the oceans absorb approximately one-third to one-half of anthropogenic CO_2 emitted from fossil fuels and industrial processes (Chapter 8) (717). Kump and colleagues (54) stressed that if the biological carbon pump was fully efficient (100%) (i.e. consumed all nutrients present in the water column), it would decrease atmospheric CO_2 concentration to 165 ppm. In contrast, if the biological carbon pump were stopped completely, atmospheric CO_2 concentration would increase to 720 ppm. This hypothetical example illustrates well the importance of the biological carbon pump for the regulation of atmospheric CO_2 concentration. Results from ocean-atmosphere general circulation models suggest that the carbon pump might be reduced if the current rise in global temperature continues (717). I only reviewed some biological mechanisms that might be of interest to explain the potential decrease in ocean carbon exportation, but a lot remains to be done in this research area.

10.3.1.1 A case study: North Atlantic copepods

The extratropical North Atlantic Ocean is an important region for carbon export (929), and it is thought that the biological pump will be less efficient in a warmer world because of changes in phytoplankton types (floristic shifts) and reduced upward mixing of nutrients due to increased ocean stratification (1027). Deepening of the nutricline, as a result of increased stratification, may shift phytoplankton community from diatoms (major carbon exporters) to coccolithophores (1028). Beaugrand and colleagues (207) showed that a biodiversity increase in marine calanoid copepods paralleled a reduction in the mean size of the copepod community (Plate 10.1). This general reduction in copepod size was associated with eastern North Atlantic warming (Plate 10.1E). We investigated the potential consequences of global warming for ecosystem functioning and carbon cycles (Plate 10.1B–C and E–F). Information on size structure of copepods was converted into the minimum turnover time of carbon incorporated in these organisms and the mean residence time above 50 m of sinking copepod particles (i.e. faecal pellets) by using allometric equations (562). In general, the minimum turnover time of carbon incorporated in organisms is directly related to organism size, the mean residence time of sinking particles produced by organisms in surface waters is inversely proportional to size and the downward export of sinking particles is inversely related to their mean residence time at surface, hence downward export is directly related to the size of organisms that produce sinking particles.

Because of the strong inverse relationships between biodiversity and size of copepods (Plate 10.1), there were highly significant correlations between copepod biodiversity and both the minimum turnover time of carbon in copepods (negative correlation) and the mean residence time of sinking copepod particles (positive correlation). The negative correlation between biodiversity and the minimum turnover time of carbon in copepods indicates that northward biodiversity increase was accompanied by a reduction in the minimum turnover time of carbon in these organisms (i.e. quicker circulation of the carbon incorporated in smaller organisms that had shorter life cycles). This quicker circulation of carbon would be one component of the increase in ecosystem metabolism that is likely to accompany the rise in temperature in the Atlantic Ocean.

The positive correlation between biodiversity and mean residence time of sinking copepod particles indicated that biodiversity increase was also accompanied by an increase in mean residence time of sinking copepod faecal pellets above 50 m. Although it is debated whether or not copepod faecal pellets contribute significantly to the downward biological carbon pump (1029), if the increase in the mean residence time of biogenic carbon in surface waters affected the whole plankton ecosystem, we can reasonably speculate that food web controlled carbon export could be weaker in a warmer world and thus contribute to a positive feedback (i.e. amplification) to climate change.

These results indicate that the biological carbon pump may diminish not only because of lower nutrient inputs into the euphotic zone (1030), but also because organic carbon would reside longer in surface waters where it would be processed through smaller-sized zooplankton and thus dissipated through more complex food webs, and additionally because total copepod biomass may decrease (262). A diminution in community size in relation to more stable and warmer environments was also found for Atlantic phytoplankton (265) and North Sea fish (1031). It therefore appears that increasing temperature leads to smaller-sized community assemblages across multiple pelagic trophic levels that may adversely affect the biological carbon pump.

10.3.1.2 Calcifying organisms

The biological carbon pump is also controlled by inorganic carbon export in the form of $CaCO_3$ particles and shells built by many marine-calcifying organisms. Marine molluscs produced between 50 and 1,000 g of $CaCO_3 m^{-2} \cdot yr^{-1}$ (1032). The rate of calcification for coral reefs is ~10 kg$CaCO_3 m^{-2} \cdot yr^{-1}$ (1033). Planktonic foraminifers produce calcite flux rate estimated at 1.3–3.2 Gt·yr^{-1}, equivalent to 23–56% of the total open marine $CaCO_3$ flux (1034). They also play an important role as organic carbon transport into the ocean interior (1035). **Coccolithophores** are also key players of the regulation of the planetary carbon cycle and other biogeochemical compounds. At high latitudes (40–60°), blooms of Emiliania huxleyi are estimated to produce each year ~0.4–1.3 million tonnes of $CaCO_3$ (731). This estimate is, however, low in comparison to the global estimates, probably because satellites do not identify all coccolithophore blooms, missing those occurring below the surface. Milliman assessed that $CaCO_3$ sedimentation was ~0.46 GtC·yr^{-1} (730).

As we saw in Chapter 8, Moy and colleagues (758) showed a reduction of 30–35% between pre-industrial and current shell weights of the foraminifera Globigerina bulloides in the Southern Ocean. If these results can be generalised to the whole foraminifera group, size reduction of these organisms might diminish $CaCO_3$ flux from the surface to the bottom and decrease carbon export. It is, however, difficult from current knowledge to provide an estimate on this positive feedback. Shell weight may decline, but their number could increase. In addition, foraminifera blooms might become more widespread and frequent over cold regions in the future because of the melting of summer Arctic sea-ice. Recent studies also suggest that at least other species (e.g. crustaceans) might benefit from CO_2 fertilisation and increase their exoskeleton weight (negative feedback) (726). Which of the two effects will be prominent in terms of carbon or calcium carbonate export remains an open question. This is another example of large potential uncertainty associated with long-term climatic projections.

10.3.2 Marine biodiversity and the nitrogen cycle

In the open ocean, the biological production is often constrained by new limiting nutrients such as nitrogen and phosphorus. Both global warming and ocean acidification will affect marine biodiversity, which will in turn alter the cycle of these macronutrients. Because global change implications are currently less known for the phosphate than the nitrogen cycle, I only review briefly the implications of CO_2 fertilisation on the nitrogen cycle.

Nitrogen is provided from deep-water sources or from the activity of nitrogen-fixing prokaryotes called diazotrophs (Figure 10.27). An example of a diazotroph is the tropical cyanobacteria *Crocosphaera watsonii* or the genus *Trichodesmium*. This genus is responsible for ~50% of total marine nitrogen fixation (1036).

Nitrogen loss takes place through two processes. The first process, called denitrification, reduces oxidised nitrogen (especially NO_3^-). The process can be summarised as follows:

$$NO_3^- \rightarrow NO_2^- \rightarrow NO + N_2O \rightarrow N_2 \qquad\qquad (10.1)$$

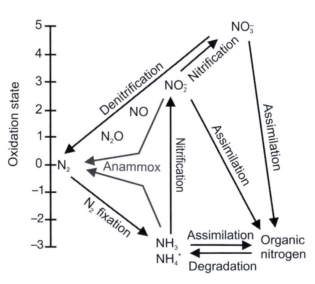

Figure 10.27 Main types of transformation of the nitrogen in the ocean. Each chemical species is plotted as a function of its oxidation state.

Source: Redrawn from Hutchins and co-workers (1037).

The full process is a redox reaction:

$$2NO_3^- + 10e^- + 12H^+ \rightarrow N_2 + 6H_2O \tag{10.2}$$

The second process, identified in 1999, is termed 'ANAMMOX', for ANaerobic AMMonium (NH_4^+) OXidation, and is performed by some bacteria of the group *Candidatus Scalindua*. The reaction is:

$$NH_4^+ + NO_2^- \rightarrow N_2 + 2H_2O \tag{10.3}$$

This reaction is responsible for ~50% of the nitrogen loss in the oceans.

Nitrogen transformation into the oceans occurs through the food web (assimilation) or by degradation (Figure 10.27). Aerobic nitrifying bacteria (*Nitrospina* and *Nitrococcus*) and archaea (*Nitrosopumilus maritimus*) can oxidise ammonium (NH_4^+) to produce nitrite (NO_2^-) and then nitrate (NO_3^-). This process is called nitrification.

Dissolved inorganic carbon enrichment is expected to increase nitrogen activity of the cyanobacteria of the genus *Crocosphaera* and *Trichodesmium* (*1037*). Some biogeochemical models provided evidence that CO_2 enrichment is likely to stimulate both nitrogen fixation and photosynthetic rates of these organisms, which might have an influence on the total nitrogen pool in surface with potential ramifications for the whole system of trophodynamics (*1037*). Increase in nitrogen fixation ranges from 35% to 65% (sometimes more) when CO_2 rose from 375 to 750 ppm. Because nitrogen is expected to decrease in the surface layer with global warming, competition between NO_3 users and N_2 fixers modulated by the balance between the energetic cost of fixing N_2 and the advantage of acquiring nitrogen when NO_3 concentration is small, may favour nitrogen-fixers abundance (*1038*).

10.3.3 Marine biodiversity and the global silicon cycle

The global rate of biosilification has been estimated to range between 200 and 280 $TmolSi \cdot yr^{-1}$ (*1039*). **Diatoms** are the world's largest species contributing to biosilification, other species such as radiolaria and silicoflagellates (Figure 10.28) being of marginal importance nowadays (*1040*).

Diatoms account for ~40% of the total marine primary production (*1039*) and dominate extratropical spring blooms, equatorial divergence zones, upwellings, macrotidal coastal ecosystems and oceanic systems after wind-mixing events. When diatoms die, they sink rapidly to the bottom of the oceans as marine snows. Although their organic compounds deteriorate rapidly, their **frustule** (Figure 10.29), composed of silicified cell walls, transform progressively to give biogenic opal.

Diatoms have therefore a key role in the current ocean's silicon cycle. Diatoms uptake ~265 $TmolSi \cdot yr^{-1}$ (Figure 10.30). Of this, half is remineralised in the surface layer and half go to deeper layers. About 125 $TmolSi \cdot yr^{-1}$ are dissolved there and 6.62 $TmolSi \cdot yr^{-1}$ sediment on the seafloor. In the United States, these important deposits are used commercially to provide fine abrasive (in silver polish and toothpaste), filters, insulating materials and anti-caking agents. Because the marine silica cycle is nowadays controlled by diatoms that are large contributors to the oceanic carbon cycle through the biological pump, opal formation can be used as a proxy for carbon sequestration. Biogenic opal forms in all

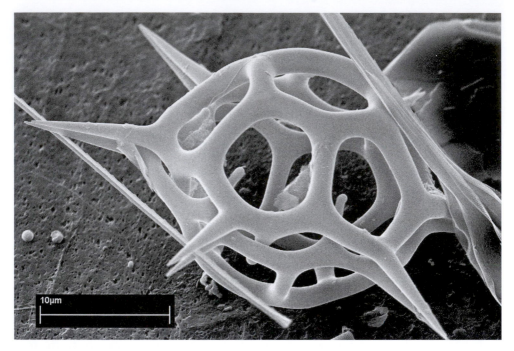

Figure 10.28 Scanning electron microscopy image of a silicoflagellate showing the skeleton made in silica.

Source: Courtesy of Dr Courcot, Laboratoire d'Océanologie et de Géosciences.

Figure 10.29 Scanning electron microscopy image showing the frustule of the diatom *Thalassiosira rotula*.

Source: Courtesy of Dr Courcot, Laboratoire d'Océanologie et de Géosciences.

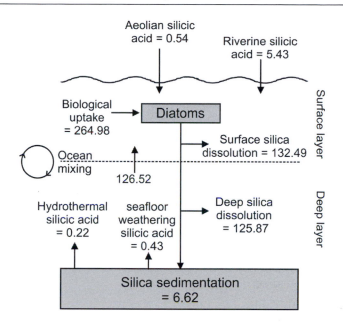

Figure 10.30 Modelled influence of the diatoms on the ocean's silicon cycle. All fluxes are in TmolSi·yr^{-1}. One Tmol (teramole) is 1,012 moles. Silicic acid is also termed silicate.

Source: Redrawn from Yool and Tyrrell (*1041*).

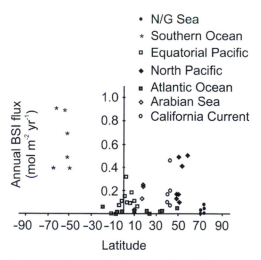

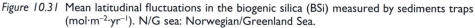

Figure 10.31 Mean latitudinal fluctuations in the biogenic silica (BSi) measured by sediments traps (mol·m^{-2}·yr^{-1}). N/G sea: Norwegian/Greenland Sea.

Source: Modified from Ragueneau and co-workers (*1040*).

latitudes and depths from the Southern Ocean to ecoregions such as the equatorial Pacific, the Atlantic Subpolar Gyre (south of Greenland) and the South African coastal upwelling system (Plate 10.2). Opal fluxes spatially fluctuate by more than a factor 100 (*1040*). The highest fluxes take place in the Southern Ocean where values from trap sediments range from 0.6 to 0.9 molSi·m^{-2}·yr^{-1} (Figure 10.31). Other elevated values are found in the North Pacific at about 50°N, and to a lesser extent in the equatorial Pacific and in the California Current.

When opal sinks and accumulates in deep-sea sediments, it forms siliceous ooze. Diatom ooze is mainly located towards high-latitude regions and along some continental margins where upwelling takes place, whereas radiolarian ooze is concentrated in equatorial areas. Siliceous ooze is subsequently transformed during burial and forms bedded cherts.

10.4 Potential feedbacks

In this section, I provide two examples on how changes in marine biodiversity may alter the earth system. Many other mechanisms detailed in the previous and present chapters and sections are not mentioned here.

10.4.1 Marine biodiversity and tropical cyclones

Oceanic global primary production has been estimated to be ~48.5 petagrams of carbon (1 Pg = 1GtC) per year (*35*). There has been a reduction in phytoplankton biomass since the beginning of the twentieth century (*930*). This reduction, which remains controversial (*1042*), may have strong consequences for tropical cyclone formation. We know that phytoplankton concentration in the water column influences water albedo (Chapter 2). When water has few phytoplankton, solar radiations penetrate deeper in the water column. When phytoplankton increases at surface, the albedo increases and light is absorbed, which increases sea surface temperature and cools deeper waters. By increasing stratification, and therefore reducing nutrient concentration in the photic zone, global climate change could reduce phytoplankton abundance and diminish the albedo. Such changes may affect the intensity, the frequency and the path of tropical cyclones (*1043*). Gnanadesikan and co-workers (*1043*) showed by modelling that if the North Pacific had no phytoplankton, this hypothetical situation would lead to a reduction in cyclone activity by 67% and would concentrate cyclone trajectory along the equator.

10.4.2 Marine biodiversity and cloud formation

Dimethylsulphide (DMS) is a volatile sulphur compound produced by marine phytoplankton (*1044*). DMS, contributing to the global sulphur cycle, is oxidised in the atmosphere to produce sulphate aerosols involved in the formation of cloud condensation nuclei. Significant DMS production is restricted to a few phytoplanktonic taxonomic groups, which are the Dinophyceae (dinoflagellates) and the Prymnesiophyceae, including coccolithophores and *Phaeocystis*. These organisms produce intracellular dimethylsulphoniopropionate (DMSP), a precursor of DMS. For example, at high latitudes (40–60°), blooms of *Emiliania huxleyi* are estimated to produce each year ~10,000 tonnes of DMS sulphur (*731*). DMS is then transferred from the ocean to the atmosphere via photochemical oxidation, consumption by bacteria and sea-air exchange where it turns into a source of acidic aerosols (*83*).

Charlson and colleagues proposed a feedback interaction between marine biota and climate (*1045*). The authors stated:

> Because the reflectance (albedo) of clouds (and thus the earth's radiation budget) is sensitive to cloud-condensation nuclei density, biological regulation of the climate is possible through the effects of temperature and sunlight on phytoplankton population and dimethylsulphide production. To counteract the warming due to doubling of atmospheric CO_2, an approximate doubling of cloud-condensation nuclei would be needed.

However, although a rise in oceanic DMS emission would increase the number of cloud condensation nuclei, it remains unknown in which ways global warming could affect DMS phytoplankton producers (*1046*).

Part 3

Theorising and scenarising biodiversity

Chapter 11

Theorising and scenarising biodiversity

11.1 Introduction

As we saw in Chapter 7, global climate is expected to change at a magnitude rarely reached in recent history, threatening the organisation and the distribution of ecosystems and their biodiversity. Ecologists and bioclimatologists are urged to understand and anticipate both biological and ecological changes because they may have strong consequences for supporting, provisioning and regulating services (Chapters 7 and 10). How will species redistribute and what are the likely effects for ecosystem properties? How will diversity and other functional attributes change? What will be the consequences for ecosystem functioning, services and biogeochemical cycles? What will be the new ecosystems of some currently productive seas? These questions have lately gained enormous importance, especially since climate-induced changes in ecosystems have already been observed (Chapter 7).

During a first phase, a phase I believe not yet fully completed, investigations on the effects of climate change on marine species and ecosystems have focused on single patterns (e.g. phenological or biogeographical shifts), single functional attributes (e.g. diversity, size) or ecoregions. As we saw in Chapters 5 and 7, the principal research aim was to determine whether global climate change may affect biological and ecological systems and whether the effect of warming was already perceptible in natural systems (427, 663). This remains a difficult task because climatic variability (Chapters 1–5) is hard to disentangle from anthropogenic climate change (Chapters 5 and 7), ocean acidification (Chapter 8) and sometimes more direct anthropogenic effects such as fishing (Chapter 9). Nevertheless, to date, potential effects of climate change on individual behaviour (1047) and physiology (1048), population abundances (694), species phenology (657, 658) and biogeography (551, 685), communities (1049), ecosystems (540) and biomes (929) have been documented (Chapters 5–7). At the community and ecosystem levels, long-term decadal shifts, including abrupt ecosystem shifts, have been reported (59, 483, 494, 1050). Even though some hypotheses have been proposed (60), the underlying processes behind those shifts remain unclear (60). Climate-induced trophic cascades have also been identified (140, 695) and hypotheses and theories proposed to explain how the climatic signal might propagate through the food web (495, 559) (Chapter 7). Ecosystems being interlinked, the effect of climate change on benthic-pelagic coupling (533) and ocean/land coupling (702) have been identified. Other researchers have focused on key properties of marine systems such as diversity and size (see also Chapter 10). Diversity is expected to reorganise as climate warms (207, 1031, 1051) and the mean size of individual species and communities is expected to diminish (207), a phenomenon sometimes called dwarfism (1052) (Chapters 6–7 and 10).

When diversity increases, size diminishes and significant negative relationships have been revealed at a macroecological scale (*60, 207*). All these patterns and processes have been investigated independently, however. Are they connected? By which processes and mechanisms?

Results from these empirical studies have greatly improved our understanding of the influence of climate on marine biodiversity and their responses to climate change. Therefore, in a second phase, general theories could be developed and tested against the growing number of global biological datasets to predict how species and ecosystems are organised (Chapters 2–4) and how they might respond to climate change (Chapters 1–6 and 7). The need for the elaboration of theories in ecology has been stressed by many authors. Ramon Margalef (*293*) stated that:

> Ecologists have been reluctant to place their observations and their findings in the frame of a general theory. Present day ecology is extremely poor in unifying and ordering principles. A certain effort should be made in constructing a general frame of reference, even though some of the speculation may be dangerous or misleading.

Recognising the multiplication of theories and hypotheses at small or intermediate spatial scales, Michel Loreau wrote more recently that 'what is lacking is a synthetic theoretical framework to organize these results into a coherent set of alternative or complementary hypotheses that yield testable predictions'. Eileen O'Brien and colleagues stated that the development of an ecological theory might have been delayed by the difficulties of identifying both fundamental laws and principles that govern biodiversity distribution (*274*). John Lawton stressed, however, that rules or laws proposed so far have been contingent, meaning that these laws are true for a specific taxonomic group, a given region or a specified condition (e.g. Bergmann's and Rapoport's patterns; Chapter 3).

Lonnie Aarsen (*1053*) wrote: 'Ecologists appear to be endlessly uncertain and critical of the scope and methodology of their discipline'. John Lawton (*1054*) even stressed that ecologists have had 'an almost suicidal tendency . . . to celebrate complexity and detail at the expense of bold first order phenomena'. I concur, and propose here a theory that I have recently developed with my colleagues and called the 'MacroEcological Theory on the Arrangement of Life', or the METAL theory. This theory proposes that the arrangement of life and biodiversity responses to climate and environmental changes is primarily the result of the interaction between the species ecological niche (*sensu* Hutchinson) and both the climate and the environment. As the theory is based on the concept of the ecological niche, I first make a brief overview of the concept in the next section.

11.2 The concept of the ecological niche

11.2.1 Towards the notion of ecological niche

Thomas Malthus showed in an essay in 1798 that all species have the possibility to increase in number indefinitely. This was summarised in Gause's book *The Struggle for Existence* by the following mathematical equation (*489*):

$$N_t = 2^{bt} \tag{11.1}$$

Where N_t is the number of individuals of a given species at time t and b the exponent of the geometric progression. From Equation 11.1, it is clear that a species could increase indefinitely if there were no external mechanisms to control its progression (Figure 11.1).

All organisms may populate the whole planet at a different speed depending on b in Equation 11.1. The rate of reproduction was already known to be a function of organism size (489). Therefore, plants or animals with high dispersal capabilities could have a pandemic distribution if they were not limited by the constraints of the environment affecting their physiology. This is summarised by Becking/Beijerinck's law, which states that 'everything is everywhere, but the environment selects' (10, 1055). This law is especially true for plankton and probably explains why these organisms adjust their spatial distribution very quickly to changing climatic and environmental conditions. Species with high dispersal capabilities are rapidly constrained by the effect of the environment on their physiology. It is likely that when the stress imposed by the environment becomes high, the effect of interspecific relationships such as competition and predation (or resistance to parasite or disease) become more patent, reinforcing the direct effect of changing climatic and environmental conditions on the individuals (Chapter 4).

Therefore, the environment exerts a control on each individual species. This has been well recognised for decades. For example, the law of minimum of Justus von Liebig (1840) stipulates that production fluxes of an organism are an exclusive function of a limiting factor (830). This law has received large empirical support (1056). The law of tolerance (Shelford, 1913) extended the law of Justus von Liebig by stating that a species is limited by its range of tolerance for environmental factors (1057). However, many refinements can be added. The tolerance range varies during the life cycle of the organisms, younger organisms being generally more sensitive than older stages (65) (Chapter 6). The tolerance range varies among ecological factors. A species can be 'euryecious' to many factors while being 'stenoecious' to another. The range of species tolerance to a given factor may also be modified by another (interactive effects).

11.2.2 The ecological niche

Both the law of minimum and the law of tolerance led to the concept of the ecological niche. The first idea of the niche was envisioned by Joseph Grinnell (1058) as the habitat

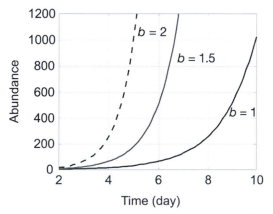

Figure 11.1 Population growth predicted by the exponential model for different values of b.

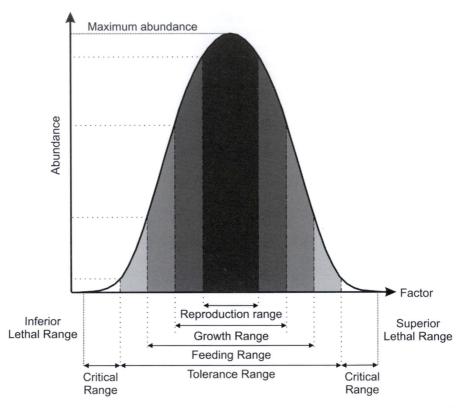
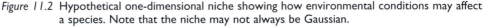

Figure 11.2 Hypothetical one-dimensional niche showing how environmental conditions may affect a species. Note that the niche may not always be Gaussian.

Source: Modified from Helaouët and Beaugrand (*538*).

requirement of a species. In his paper, Grinnell related the small spatial distribution of the California thrasher (*Toxostoma redivivum*) to the adaptation of this bird through physiological and psychological processes to a narrow range of environmental conditions (e.g. temperature and atmospheric humidity). Charles Elton envisioned differently the niche concept (*986*). The zoologist defined the niche in terms of functional attributes. The niche is the place a species occupies in the food web and the influence it exerts on its environment (*1059*).

The concept of the niche of Hutchinson (*236*) is probably the definition of the niche that has been the most widely applied by ecologists. The Hutchinsonian niche represents the combination of environmental tolerances and resources required by an organism. Hutchinson (1957) conceptualised this notion with the 'n-dimensional hypervolume', in which n ideally corresponds to all environmental (biotic and abiotic) factors. This way to define the niche was considered to be revolutionary because it was operational, enabling a straightforward quantification of the species niche. Figure 11.2 shows a simplified representation of the niche of a hypothetical species characterised by only one dimension (one environmental factor).

Only optimal conditions generate high abundances and allow for successful reproduction (*538*). When the environment becomes less favourable, this affects consecutively offsprings production, growth and feeding (Figure 11.2; see also Chapter 6). Extreme conditions become critical and may eventually affect survival (*161*). Towards the niche extremities, the energy taken from the environment becomes more assigned to maintenance and to ensure homeostasis. The energy is therefore dissipated as heat. Towards the niche centre, energy is allocated to productivity and stored in organic matter as biochemical (endo-somatic) energy, becoming accessible to other organisms (*266*).

The Hutchinsonian niche can also include biological factors, although the operational concept rarely incorporates biological interactions. The Hutchinsonian niche is close to the niche of Grinnell, and some authors consider that the Hutchinsonian niche is simply a better (and mathematical) conceptualisation of the niche of Grinnell (*1060*).

11.2.3 Improvements and enlargements of the niche concept

Many authors have worked on the niche concept (1061, 1062). Chase and Leibold (1059) proposed a revised definition of the niche as 'the joint description of the environmental conditions that allow a species to satisfy its minimum requirements so that the birth rate of a local population is equal or greater than its death rate along with the set of per capita effects of that species on these environmental conditions'. They then pursued in a second definition by stating that the niche is 'the joint description of the zero net growth isocline (ZNGI) of an organism along with the impact vectors on that ZNGI in the multivariate space defined by the set of environmental factors that are present'. This definition is close to the Huchinsonian niche but perhaps more difficult to implement.

11.2.3.1 Fundamental and realised niches

Several other amendments of the Hutchinsonian niche have subsequently been proposed. I discuss here only those that I think are the most important for the discussion that will follow. Ecologists early noted that the niche tended to be larger in the absence of biotic interactions. This led Hutchinson to add two new concepts: the **fundamental** and the **realised niches**. The fundamental niche represents the response of all species' physiological processes to the synergistic effects of environmental factors with the absence of negative biotic interactions (e.g. competition, parasitism or predation). The realised niche is the fundamental niche modified by factors such as dispersal that increase niche breadth or factors such as competition that tighten it. Although Hutchinson stated that the fundamental niche was larger than the realised niche, Bruno and colleagues (*1063*) drew attention to the fact that positive species interactions, called facilitation (Chapter 4), might increase the size of the realised niche when compared to the fundamental one (Figure 11.3). The authors showed that facilitation might counteract the influence of negative species interaction. They took the example of intertidal seaweed that decreases both thermal and desiccation stresses, allowing some species to live higher on the shore than they normally do (*1064*). In the Gulf of Maine, Bertness and colleagues showed that *Ascophyllum nodosum* canopies significantly improved recruitment, growth and survival of some benthic organisms such as crab and mussel. The authors posited that such facilitation might be pervasive in harsh environments.

Pulliam (*1065*) proposed a model (the source-sink dynamics model) to explain and assess differences between fundamental and realised niches. Pulliam showed that the realised niche was smaller when factors reducing survival such as competition predominated. However,

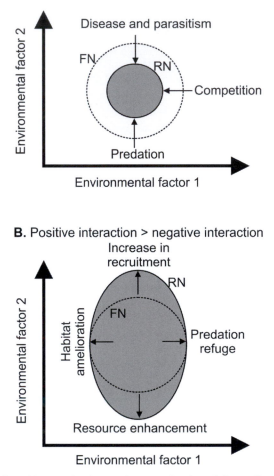

A. Negative interaction > positive interaction

B. Positive interaction > negative interaction

Figure 11.3 Influence of positive species interactions on the size of the realised niche. (A) Negative > positive interactions. (B) Positive > negative interactions. Dispersal was not considered in this model. FN: fundamental niche. RN: realised niche.

Source: Redrawn from Bruno and co-workers (*1063*).

he also provided evidence that the realised niche could be greater than the fundamental niche when dispersal was high (Figure 11.4).

11.2.3.2 Gause's principle

Gause's principle of competitive exclusion stipulates that in a community at equilibrium, two species with a similar ecology (i.e. similar niche) cannot live in the same place (*489, 1066*). Or in a few words, 'complete competitors cannot coexist' or 'ecological differentiation is the necessary condition for coexistence' (*1067*). This principle is said to be a 'corollary of the process of evolution by natural selection' (*1067*). The contention of the Russian microbiologist (*489*) has been at the origin of many works and reviews to explain how so

A. Dispersal < negative interaction

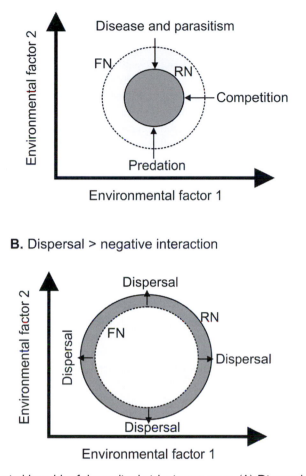

B. Dispersal > negative interaction

Figure 11.4 Hypothetical breadth of the realised niche in two cases. (A) Dispersal < negative interaction. (B) Dispersal > negative interaction. Positive interactions were not considered here. FN: fundamental niche. RN: realised niche.

Source: Redrawn from Pulliam (*1065*).

many species coexist in the same place (*1066*). This principle, often used when species assembly rules are examined, is applied in the models developed as part of the METAL theory.

11.2.3.3 Variation of the niche shape throughout the life cycle

The niche does not generally remain constant throughout the whole species life cycle. The environmental tolerance varies during life, younger stages often being characterised by a narrower niche than older stages (Chapter 6). Reygondeau and I (*65*) provided evidence for an increase in niche breadth of the marine copepod *C. finmarchicus* from young copepodite stages to stage C5. Interestingly, the adult stage was characterised by a niche narrower than stage C5, probably because the species must lay their eggs in an optimal environment. This result is in agreement with Figure 6.17. Grubb (*1067*) stressed that it is likely that modifications in competitive ability and both biotic (e.g. predation) and abiotic (sensitivity to a particular environmental factor) sensitivities during the different developmental phases explain diversity maintenance:

> It becomes apparent that species-richness may be maintained by a heterogeneous environment acting on all a series of stages in the reproduction of the plants present. The variable environment determines whether there is or is not at a particular place and time for a particular species an abundance of flower-formation, pollinisation, good seed-set, dispersal, germination, establishment of the juvenile plant and passage of the juvenile to the adult. The idea that all these stages are important in the maintenance of species-richness has been all too often ignored.

11.2.3.4 The regeneration niche

The recognition of difference in requirements and sensitivity during the life cycle of a species led Grubb (*1067*) to propose the term of **regeneration niche**, which was defined as the 'expression of the requirements for a high chance of success in the replacement of one mature individual by a new mature individual of the next generation'. Simply stated, the regeneration niche is the part of the fundamental niche that enables successful reproduction (*1060*). The botanist gave examples of important processes such as flowering, pollination, seed settling and dispersal germination, and establishment and characters such as size and shape that may have different environmental tolerances. This concept has been used to examine the relationships between the optimal part of the fundamental niche (the regeneration niche) and the realised niche of the marine copepod species *Calanus finmarchicus* (Figure 11.5).

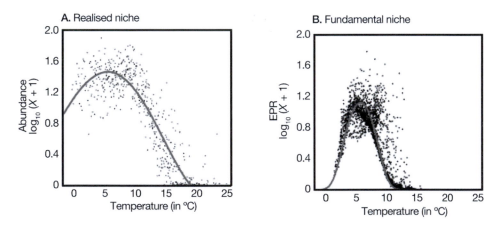

Figure 11.5 The realised (A) and regeneration (B) niches of the copepod *Calanus finmarchicus* assessed using sea surface temperature. The realised niche was computed by averaging all abundance data per category of temperature in the North Atlantic Ocean. In (A), the curve was fitted by non-linear regression to the data. The regeneration niche was assessed here by the calculation of potential Egg Production Rate (EPR). In (B), the curve was the result of the application of a macrophysiological model.

Source: Modified from Helaouët and Beaugrand (*538*).

We found that the breadth of the regeneration niche was narrower than the realised niche. We also provided evidence that the copepod occurred overall in places where it can reproduce (i.e. high abundance level was associated with places where egg production was maximal). The link between egg production and spatial distribution may be explained by the fact that an expatriated population is unlikely to persist for long, and therefore to be detected at the scale of our study. Pulliam (1065) stressed that a population may persist as long as the immigration rate from source regions nearby is sufficient. Dispersion from source habitats (i.e. region where local reproduction exceeds mortality) (1068) seems to be rapidly counteracted by mortality related to physiological stress particularly pronounced, as Figure 11.5 shows, during the reproductive period. The physiological stress experienced during the sensitive reproductive period is likely to be exacerbated by interspecific competition in sink habitats (regions where mortality exceeds local reproduction).

In the example above, the regeneration niche occurred in the optimal part of the fundamental niche but this is perhaps not always the case. Jackson and colleagues imagined a scenario where the regeneration niche did not coincide with the optimal part of the fundamental niche (1060). They drew attention to the fact that different developmental stages may have different niches. Using the concept of **regeneration niche** developed by Grubb, they compared situations where only a global ecological niche model was considered (they called it the unitary niche model) with situations where distinction was made between adult and regeneration niches under different theoretical scenarios of environmental fluctuations (Figure 11.6).

When the unitary niche model was considered, alternation between colonisation and extinction can be forecasted corresponding to situations when the environment reached a threshold (Figure 11.6A and C). When the adult and regeneration niches were distinguished, it was possible to separate ephemeral recruitment from the establishment of a more persistent population (Figure 11.6B and D). When the first threshold (T1; the adult threshold) is reached, adults can migrate to the area but cannot reproduce. Despite the fact that the species can be detected, the region represents a sink area (*sensu* Pulliam). When the second threshold (T2; the regeneration threshold) is attained, the species population can start to regenerate and the region can eventually become a source area. The density of the population increases and it becomes more and more resilient to environmental fluctuations. This is the result of the **Allee effect** (1069). During years with adverse environmental fluctuations, the population may not reproduce, but can still persist if the negative forcing does not continue too long. This model explains the sudden arrival of species and its disappearance the following year.

11.2.3.5 The species' genetic background sets up its physiology and ecological niche

Species are set up genetically to experience a specific environmental or a climatic regime. When random or seasonal changes in environmental conditions occur, it has to be adjusted by a corrective response. The **law of requisite variety** formulated by Ashby (1070) stipulates that a system that can respond in a great variety of ways can better compensate for a variety of perturbations. Therefore, a species can only occur where it has the physiological potential to resist to climatological or environmental variability.

Huntley and colleagues (1071) hypothesised from first principles that species may respond to climate change in two different ways. First, the response can be through natural selection

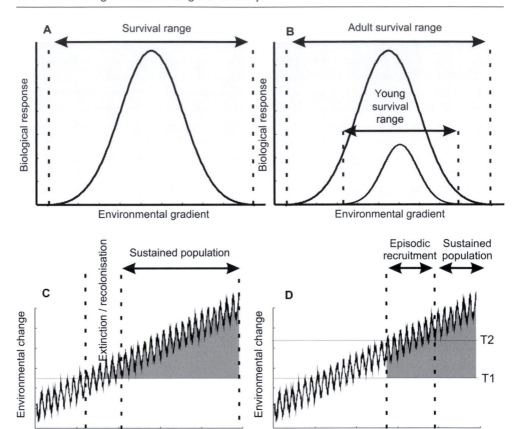

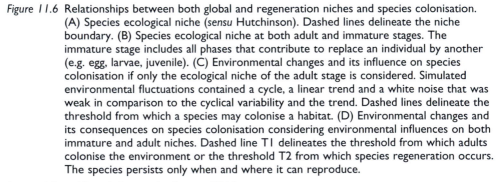

Figure 11.6 Relationships between both global and regeneration niches and species colonisation. (A) Species ecological niche (*sensu* Hutchinson). Dashed lines delineate the niche boundary. (B) Species ecological niche at both adult and immature stages. The immature stage includes all phases that contribute to replace an individual by another (e.g. egg, larvae, juvenile). (C) Environmental changes and its influence on species colonisation if only the ecological niche of the adult stage is considered. Simulated environmental fluctuations contained a cycle, a linear trend and a white noise that was weak in comparison to the cyclical variability and the trend. Dashed lines delineate the threshold from which a species may colonise a habitat. (D) Environmental changes and its consequences on species colonisation considering environmental influences on both immature and adult niches. Dashed line T1 delineates the threshold from which adults colonise the environment or the threshold T2 from which species regeneration occurs. The species persists only when and where it can reproduce.

Source: Adapted from Jackson and colleagues (*1060*).

of genotypes with selection of individuals best adapted to the new climatic or environmental conditions (Chapter 7). Second, species might exhibit a temporal (phenological) or spatial (biogeographic) response, allowing species to match their bioclimatic envelope (Chapter 7). How can genetic difference alter the ecological niche? We have seen in Chapter 5 that long-term climate changes from the tectonic frequency to the historical band can engender genetic changes. Genetic changes at the species level more often observed at the tectonic scale cause undoubted changes in the limits of the ecological niche. Towards higher frequencies from the orbital frequency band, genetic changes manifest themselves at the population level. However, these changes are unlikely to affect the species ecological niche. If climate change is too rapid, it might be above the species' capability to adapt (364). How changes in genotypes and phenotypes will alter the limits of the species' bioclimatic envelope remains an open question (Chapter 7).

11.3 Rationale of the METAL theory

In previous chapters (Chapters 2–7), we saw that the responses of marine ecosystems and their biodiversity to climate and the environment involve many organisational levels from the the gene to the ecosphere and across many scientific disciplines (Figure 11.7).

Does this complexity mean, however, that we need to consider all processes to anticipate the effect of climate change on marine species and ecosystems? Genetics set up the limits of the physiological and the biological responses of an individual to climate change (Chapter 6). Processes such as mutation rate, selection, gene flow and inbreeding depression are classically examined because they influence either phenotypic plasticity or adaptation and affect species' survival on long-term timescales (1072) (Figures 11.7 and 11.8A). To understand the effects of climate change at the individual level, physiological responses are particularly important (1048) (Figures 11.7 and 11.8B), not least processes contributing to thermal tolerance (e.g. oxygen limitation, ventilation, circulation processes) are particularly scrutinised (158) (Chapter 6). At this level, behaviour is also significant, and some species may reduce the effects of climate change on their physiology using a specific behaviour (e.g. the search for microrefugia or microniches) (1047) (Chapters 6–7). At the population level, processes such as reproduction, growth, dispersal and mortality rate all become relevant (1073) (Figures 11.7 and 11.8C) and individual-based models (IBMs), also known as process-based models, have been used to investigate the effect of climate change at both temporal and spatial scales.

When the whole species range is investigated, views diverge on how to tackle the problem (541, 1074). A common school of thought considers that nearly all the above mechanisms are important and applies process-based models (the reductionist approach). Mechanistic models have therefore been developed providing a reasonable fit to actual data and improving our understanding of the spatial distribution of some marine species (1073). A reductionist approach is not possible for many marine species, however, because the biological responses of marine organisms to climate-induced changes in their environment are rarely known (541). Furthermore, the integration of all biological processes into a model (sometimes based on the sum or the product of hundreds of equations) may lead to information overload and increase the uncertainties on the estimates, counterproductively (see Section 11.6.2).

In contrast to this reductionist method to study ecological phenomena, another school of thought takes the macroecological approach that proposes to develop macroscopes to reveal emergent patterns and processes (1075, 1076). The macroscope is a theoretical concept, coined by de Rosnay (1077), which looks for properties of a system at a macroscopic level. In the same way a microscope is needed to observe the infinite small or a telescope is needed to observe the infinite big, the macroscope is a theoretical concept to investigate complex systems (1077). The concept of emergence in ecology stresses that an ecological entity or unit cannot be predicted from the study of its elements. This is also called non-reducible property (952). Emergent patterns and processes are therefore not simple

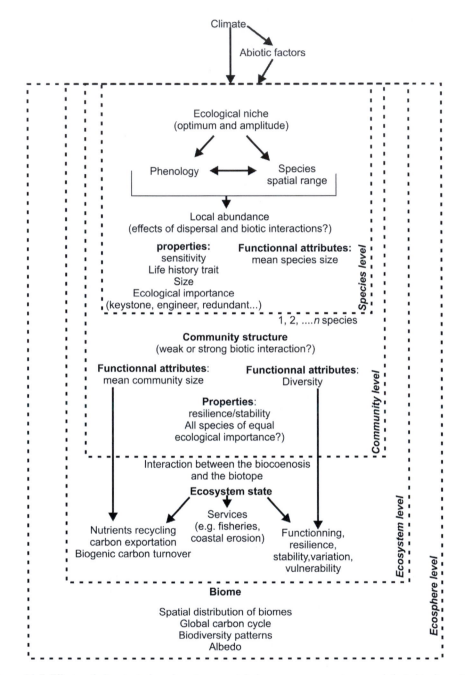

Figure 11.7 Effects of climate-induced environmental changes on ecosystems and their biodiversity at different organisational levels.

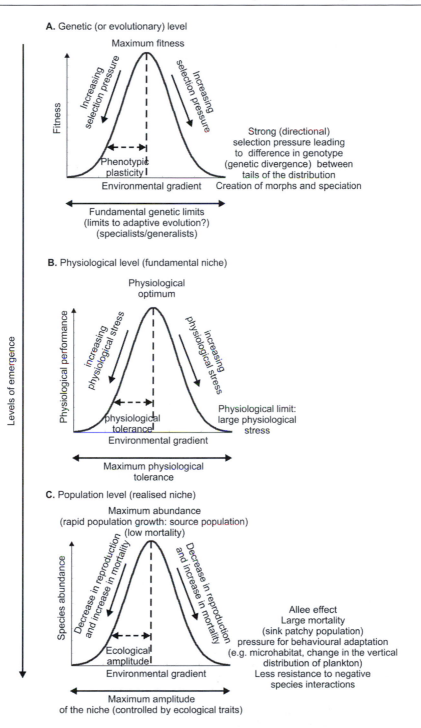

A. Genetic (or evolutionary) level

Maximum fitness

Fitness

Increasing selection pressure

Increasing selection pressure

Phenotypic plasticity

Environmental gradient

Strong (directional) selection pressure leading to difference in genotype (genetic divergence) between tails of the distribution

Creation of morphs and speciation

Fundamental genetic limits (limits to adaptive evolution?) (specialists/generalists)

B. Physiological level (fundamental niche)

Physiological optimum

Physiological performance

increasing physiological stress

increasing physiological stress

physiological tolerance

Environmental gradient

Physiological limit: large physiological stress

Maximum physiological tolerance

C. Population level (realised niche)

Maximum abundance
(rapid population growth: source population)
(low mortality)

Species abundance

Decrease in reproduction and increase in mortality

Decrease in reproduction and increase in mortality

Ecological amplitude

Environmental gradient

Allee effect
Large mortality
(sink patchy population)
pressure for behavioural adaptation
(e.g. microhabitat, change in the vertical distribution of plankton)
Less resistance to negative species interactions

Maximum amplitude of the niche (controlled by ecological traits)

Levels of emergence

Figure 11.8 Interpretation of the effects of environmental changes along a hypothetical one-dimensionional gradient at three biological levels: (A) genetic, (B) physiological and (C) population. Note that the shape of the niche may not always be Gaussian.

èxtrapolations from microscopic studies to larger scales. In a complex adaptive system (CAS), even when the components and the assembling rules are known, it is rarely possible to predict the details of the resulting system because of the emergence of new properties at a greater level (*1078*). At the macroscopic level, prediction of the structure and behaviour of CASs may be possible by focusing on emergent patterns, some of which being statistical (*1075, 1076*). For example, the concept of the ecological niche *sensu* Hutchinson (*1079*) can be considered as an elementary emergent macroscopic mechanism because it integrates the sum of all physiological and demographic processes occurring at both the individual and population levels, which is not possible to realistically implement in process-based models for most species (*541*).

11.4 The METAL theory

The MacroEcological Theory on the Arrangement of Life (METAL theory) states that biodiversity is strongly, and in a quasi-deterministic way, influenced by climate and the environment. This influence mainly occurs through the interactions between the species ecological niche and both climatic and environmental changes. The use of the ecological niche concept allows the consideration of underlying processes (genetic and physiological) that are difficult to identify and parametrise for a large number of species. Interactions between the niche and climatic and environmental changes should propagate through different organisational levels from the species to the community and the ecosystem. These interactions should also be perceptible from the smallest ecosystems to the whole ecosphere (Figure 11.7). This theory offers a way to make testable ecological and biogeographical predictions to understand how life is organised and how it will respond to global environmental changes. Hereafter, I provide some examples of the use of the METAL theory and show how the theory may unify spatial and temporal patterns and processes in ecology. At the time of writing, the theory is being developed, and I can only provide a short overview here.

11.4.1 Understanding of large-scale biodiversity gradients

As we saw in Chapter 4, the latitudinal biodiversity gradient reflects the tendency of many taxonomic groups to show an increase in biodiversity from the poles to the equator. In the marine domain, groups such as foraminifers, pteropods, euphausiids and fish exhibit this pattern. However, its cause has been vigorously debated for decades and more than 25 hypotheses have been proposed. Using the METAL theory, we showed how temperature (mean and variability), modulated by climate, can create gradients in species richness by interacting with the species thermal niche (*217*). In this work, we only used a one-dimensional thermal niche. This simplification was acceptable for the marine domain. In the terrestrial realm, a measure of water availability to plants or precipitation would have been as necessary as temperature to recreate the ecogeographical pattern (Figure 3.3).

We created a simple bioclimatic model, called the Species Niche and Climate Interaction (SNCI) model, which generated a range of all possible pseudo-species characterised by a specific thermal tolerance (or thermal niche). The thermal niches were rectangular to relax the constraints due to the niche shape. Within the model, each pseudo-species, from strict stenotherms to universal eurytherms, was allowed to inhabit a given oceanic region so long as it could endure weekly SST changes. To potentially colonise a vacant geographical cell, a pseudo-species had to remain inside its thermal niche. Some species could temporarily occur outside their thermal tolerance. For example, some pelagic species such as copepods exhibit a dormancy stage in their life cycle and are able to survive far below their thermal optimum for short time periods (see Section 6.4.3). Finally, in order for the pseudo-species to survive after colonisation, a minimum of geographical cells must be occupied by the species to conform with the size of minimum viable population (*1080*). Niche non-saturation often occurs (*241*), and to consider this fact some of the potential niches could also be removed either randomly or by increasing the rate of random elimination polewards to consider the hypothesis that repeated glaciations of the Pleistocene might have removed a larger proportion of species polewards (*241, 244*).

The theory reconstructed well the latitudinal gradient in species richness and explained more than 90% of latitudinal biodiversity patterns of foraminifers and copepods (Plate 11.1). Interestingly, the latitudinal variation was not a constant cline, but displayed a hump-shaped relationship, a pattern often observed for foraminifers and copepods (*69, 225*). These theoretical results also suggest that the latitudinal gradient can set up rapidly, as was observed recently among invasive species (*243*). Sax (*243*) showed that the establishment of the latitudinal gradient in species diversity occurred in a few decades for European and North American exotic species. Our findings suggest that hypotheses based on past glaciation events or speciation rate do not represent the primary cause of the biodiversity pattern, although our results were improved when direct effects for: (1) higher niche vacancies polewards due to the repeated Pleistocene glaciations (*242*); and (2) indirect effects of higher speciation rates in the tropics (*227*) were included in the model.

A strong Mid-Domain Effect (MDE) (Chapter 4) arose from the generation of all thermal niches in the Euclidean space but climate ultimately selected pseudo-species that could establish in a section of the thermal gradient. This effect was probably not negligeable, but it is important to note that it did not take place in the geographical space.

Theoretical spatial biodiversity patterns were examined according to the degree of pseudo-species stenothermy/eurythermy. The dissection of the patterns explained discrepancies observed in the biodiversity patterns of some taxonomic groups such as pinnipeds (Plate 4.14D). Expected species richness of strict stenotherms was high in polar regions, and to a lesser extent over the tropics where both seasonal and year-to-year fluctuations in weekly SST are lower (Plate 11.2). The pattern observed in Plate 11.2A has been found for some marine fish (*1081*), and it is well known that many polar animals occurring in the Arctic Ocean are stenothermic (*1082*). When the degree of eurythermy increased, expected biodiversity was higher in subpolar regions and over the tropics (Plate 11.2B). Biodiversity peaked in all regions but polar biomes with a maximum diversity detected over mid-latitude oceanic regions (Plate 11.2C). Modelled richness of eurytherms exhibits a maximum in temperate regions (Plate 11.2D), and as expected for high eurytherms a maximum was detected towards polar regions (Plate 11.2E). This dissection of the global biodiversity pattern shows that according to the degree of stenothermy/eurythermy, related to the phylogenetic origin and to some extent to the degree of evolution or speciation of a group, taxa can exhibit different diversity patterns. Therefore, the METAL theory explains why some taxonomic groups should exhibit different latitudinal biodiversity patterns.

Expectedly, we observed that the breadth of the modelled thermal niches was positively correlated to the latitudinal range amplitude of pseudo-species and with the number of geographical cells inhabited by pseudo-species. As discussed in Chapter 4, Rapoport's pattern states that the mean latitudinal range of a species increases with latitude (215), which led Stevens to conjecture that the ecogeographical biodiversity pattern may be the result of this rule (Chapter 4). Although Rapoport's pattern was not formally tested (1083), as niche breadth positively correlated to both the latitudinal range amplitude and the number of geographical cells occupied by a species, results from the METAL theory suggest that Rapoport's pattern cannot be general. If this was the case, a progressive increase in pseudo-species richness would be detected polewards when the breadth of the thermal niche increases (and so the latitudinal range amplitude). This was not observed for strict stenotherms, which exhibited an inverse pattern (Figure 11.2A) and for stenotherms that showed a bimodal pattern (Figure 11.2B). However, from medium to high eurytherms, Rapoport's effect was detected, although intermediate maxima were observed (Figure 11.2C–D). The METAL theory therefore enables us to better understand the lack of generality of Rapoport's effect (1083). As with exception of the latitudinal gradient in species richness, inconsistencies (other than methodological) should be related to the breadth of the thermal niches.

The METAL theory applied to global biodiversity patterns shows that a large part of these patterns results from the interaction of species' thermal tolerance with both seasonal and year-to-year fluctuations in climate. Some aspects of the theory were earlier hypothesised by authors such as Stevens, although not formally described (the climatic variability hypothesis) (Chapter 4). The METAL theory explains the primary cause of the latitudinal biodiversity gradient, although it is clear that other factors are also influential. For example, genetic differentiation and speciation are central to establish new niches. The theory holds true at a global scale but it is also patent from field observations that interspecific relationships (e.g. competition and mutualism) locally inflates biodiversity (Chapter 4). All these factors have synergistically contributed to shape the present biodiversity and have a strong influence at more local scales. Note that this study could be redone with more niche dimensions to better identify regional biogeographic regions.

11.4.2 Understanding the responses of ecosystems and their biodiversity to climate change

We saw in Chapter 7 that global warming is influencing communities and their biodiversity (427, 657). One mechanism by which a species may respond is to track habitat changes, either in time, through a phenological shift, or in space, by a biogeographical shift (427, 657, 663). At the community level, long-term community alterations that occur during abrupt ecosystem shifts (AESs) have been reported and occasionally attributed to climate change (497, 1084). Although climatic and/or environmental changes are generally hypothesised to play a key role in these responses, the underlying pathways by which the environment may produce phenological, biogeographical and community shifts remain unanswered. We showed that the METAL theory allows the connection of phenological and biogeographical shifts at the species level and long-term shifts at the community level by invoking a single mechanism. Again, this mechanism is the interaction between the species ecological niche and environmental changes that propagate from the individual to the species and community level.

11.4.2.1 Theoretical considerations

Phenology (i.e. the study of periodic biological phenomena) is expected to be altered as climate changes. Processes behind phenological shifts are complex but temperature is often advocated in both the terrestrial and the marine realms (656) (Chapters 6–7). For example, phenological shifts in many planktonic species (e.g. dinoflagellates and copepods) have been reported in the North Sea (657) and attributed to regional warming. Mackas and colleagues (658), in a review of phenological shifts in different oceanic systems, also proposed temperature as the main driver, although it was apparent that other factors (e.g. day length, mixed-layer depth) may play a significant role (657). Biogeographical movements, also termed species tracking (647), are often seen as latitudinal species shifts in response to regional warming (427, 551, 662, 663). In the marine realm, biogeographical shifts have taken place, consistent with latitudinal changes expected under climate warming and hydro-climatic variability. For example, in the eastern North Atlantic, a poleward movement of warm-water calanoid copepods has been associated with concomitant northward retraction of cold-water species (427) and similar biogeographical movements have been observed in both exploited and unexploited fish (551, 685). Phenological and biogeographical changes were presented in detail in Chapter 7. How might both phenological and biogeographical shifts be connected, and by which mechanisms?

Let us first work on a simple case, where the niche is one-dimensional and only represented by SST, to establish at the species level the theoretical foundations between species niche, species spatial distribution and phenology. The response curve of the abundance E of a pseudo-species s in a given site i and time j to change in SSTs x can be modelled as follows (1085):

$$E_{i,j,s} = c_s \, \exp \, -\left(\frac{(x_{i,j} - u_s)}{2t_s^2} \right) \tag{11.2}$$

With $E_{i,j,s}$ the expected abundance of a pseudo-species s at location i and time j; c_s the maximum value of abundance for species s fixed to one; $x_{i,j}$ the value of temperature at location i and time j; u_s the thermal optimum and t_s the thermal amplitude for species s. The thermal tolerance is an estimate of the breadth (or thermal amplitude) of the species thermal niche (1085). A one-dimensional (thermal) niche with a Gaussian shape has a few remarkable points (Figure 11.9). When the species is at the centre of its thermal niche (between T_{S1} and T_{S2}; S for stable), the species is resistant to climate-induced temperature changes, the resistance being dependent on niche shape and the breadth of the optimal zone (464, 545). When the species is present in an environment with a thermal regime that corresponds to the edge of its thermal niche (T_{HV}; HV for high variability), it is highly sensitive to climate-induced changes in temperatures. Towards the limits of the niche, the variability levels off (T_D; D for detection) before reaching lethal temperature T_L. Beyond T_D, the climatic influence becomes more difficult to detect and depends on the noise/signal ratio (464). This framework enables predictions of the species' responses to climate-induced temperature changes in space and time. This framework was applied to plankton (*Calanus finmarchicus* and *Acartia tonsa*), fish (Atlantic cod) and seabirds (little auk) (543, 545, 1086, 1087). A thermal niche can be modelled and expected abundance can be determined in space (spatial or latitudinal distribution) or time (monthly scale) from the knowledge of sea temperature. For example, we modelled the niche, spatial distribution and phenology

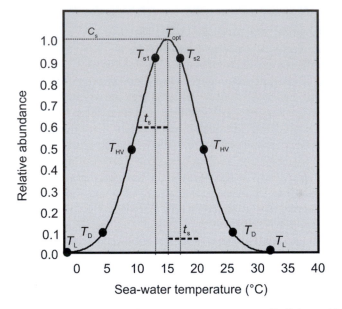

Figure 11.9 A hypothetical thermal niche. T_{opt}: optimum temperature. T_S (S for stable): temperature
between which no strong alteration in species abundance is expected. T_{HV} (HV for high
variability): point at which large variation in abundance is expected as a function of
temperature. T_D (D for detection): point from which the relationship between
temperature and species abundance becomes more difficult to detect. T_L (L for lethal):
temperature at which the animal dies.

Source: Redrawn from Beaugrand (464).

of a eurytherm psychrophile (u_s = 15°C and t_s = 5°C; Plate 11.3A–C). Using only the thermal
dimension of the niche as an example, we modelled the expected mean annual distributional
range of a hypothetical temperate species with a broad thermal niche (Plate 11.3A; thermal
optimum us = 15°C and thermal tolerance t_s = 5°C; Equation 11.1). For this pseudo-species
(Plate 11.3A), the METAL theory predicts a mean annual extratropical range with a poleward
limit to the south of the Polar Biome and an equatorward limit north of the Atlantic Trade
Wind Biome (*152*) (Plate 11.3B). The calculation of the expected species' abundance as a
function of latitude and month leads to three predictions relating latitudinal range and
phenology (Plate 11.3C). First, in the southern part of its distributional range (zone 1; Plate
11.3C), the organism has a seasonal maximum in winter or spring, the latter period is more
likely when ecological factors such as photosynthetically active radiation (PAR) affect the
species either directly through its influence on photosynthesis (e.g. phytoplankton), or
indirectly through trophodynamics (e.g. herbivorous zooplankton). PAR influence on
primary production is prominent polewards (*164*). At its southern range, such a species is
unlikely to adjust its phenology in response to an increase in sea temperature, resulting in
a local reduction of its annual mean and a northward biogeographical shift. Both the
resistance and the resilience of this cold-water species to warming are expected to be small,
but the opposite is expected in the case of a cooling. Second, at the centre of its range (zone
2; Plate 11.3C), the species will exhibit its maximum seasonal extent, the duration being
modulated by the breadth of its thermal niche *ts* (here, the species can occur all months of

the year, so long as other niche dimensions such as PAR or length of day (LOD) do not exert a controlling influence). Here, sea warming is expected to trigger a shift towards an earlier phenology. All else being equal or held constant, erosion of the seasonal occurrence period in late summer should be compensated at an annual scale by higher abundance towards spring or early summer, and consequently no substantial alteration in annual species abundance is expected; species resistance and resilience to warming are greatest in this area. Third, at the northern edge of its distributional range (zone 3; Plate 11.3C), the species is likely to peak in summer or late summer. In this case, if temperatures warm, the cold-water species can extend its occurrence in early summer and spring, resulting in an increase of its annual mean abundance. If SST warms north of its northern boundary, a northward range shift will occur, although first occurrences are likely to be detected in late summer when sea temperatures are highest.

In the case of a theoretical stenotherm (thermal optimum u_s = 15°C and thermal tolerance t_s = 1°C), the thermal sensitivity increases and the overall species' abundance declines (Plate 11.3D–E versus Plate 11.3A–C). Although the same phenological pattern is expected in its poleward and equatorward distributional range, this species does not occur throughout the year at its range centre (Plate 11.3F). In this case, zone 2 (Plate 11.3F) will be narrow and the species will only persist for a few months. It follows that a stenotherm will react to a substantial increase in temperature even at the centre of its range. The stenotherm is therefore expected to be more sensitive to a climate-mediated shift in temperature. Predictions can be established for other species (e.g. tropical species).

Comparing the two theoretical species, local density of the eurytherm is much higher than the stenotherm (Plate 11.3). This result is in agreement with Brown's theory (649) that relates the local density of a species to the size of its spatial distribution and its ecological niche; this implies that a stenotherm should have both a lower local density and a smaller spatial range, as predicted by the METAL theory. Although Brown's theory, which holds true for many marine organisms (1089), was exclusively proposed to explain the relationship between niche breadth and distributional range, our results indicate that the species ecological niche can also explain how species and communities will respond to changes in temperature in time and space.

11.4.2.2 Phenological and biogeographical shifts

We tested the METAL theory against real data choosing the marine copepod *Calanus finmarchicus* (1088). The niche was modelled using the Non-Parametric Probabilistic Ecological Niche (NPPEN; see Section 11.7.1) model (180), which calculated the expected abundance of *C. finmarchicus* as a function of monthly SST, monthly PAR, monthly chlorophyll a concentration and bathymetry. These four parameters form the most important niche dimensions for *C. finmarchicus* (538, 1090). Seasonal changes in PAR are highly correlated positively with length of day, which has been assumed to be an important controlling factor of the initiation and termination of *C. finmarchicus* diapause (1091). As NPPEN is non-parametric, the niche was not Gaussian, in contrast to the example developed previously.

The four-dimensional (not Gaussian) ecological niche was then utilised to forecast the spatial distribution of the calanoid (1958–2009; Plate 11.4A). High theoretical abundances were found north of the Oceanic Polar Front (OPF) (1092), and to a lesser extent south of Newfoundland and in the northern part of the North Sea (Plate 11.4A). This corresponds well to the observed spatial distribution (1958–2009) inferred from the Continuous Plankton Recorder (Plate 11.4B).

When the expected abundance of C. *finmarchicus* is represented as a function of latitude and month (1960–1979, a relatively cold period (*497*); eastern North Atlantic between 30°W and 10°W), the model predicts that: (1) the species should have a seasonal maximum in spring at the southern edge; and (2) between spring and summer towards the centre of its spatial distribution (Plate 11.4C). Observed abundance of C. *finmarchicus* as a function of latitudes and months provides strong support for both predictions (Plate 11.4D). The spatio-temporal pattern in expected abundance is, however, more concentrated than observed abundance (Plate 11.4C–D). At the end of the species' seasonal occurrence (in summer towards higher latitudes), the abundance level remains high, whereas expected abundance decreases. This lag may be explained by the fact that when the environment becomes less suitable, the species may remain a certain amount of time before decreasing in abundance (diapause initiation and source/sink dynamics) (*1068, 1093*). At the beginning of the seasonal occurrence, the lag between expected and observed abundance is much less pronounced and can be explained by the time needed for the species to increase its abundance level (reproduction and individual growth) (*1093*).

The abundance of C. *finmarchicus* was subsequently predicted as a function of both latitude and month for the warm period 1990–2009 (*1094*) (Plate 11.4E–F). Expectations were: (1) a reduction in the abundance level in spring resulting in an erosion of the species spatial distribution at its southern margin; and (2) a reduction in the species abundance in late summer to the north. Good support was found for both predictions, and 70.56% of the variance of combined phenological and biogeographical changes was explained by the model (Plate 11.4C–F). We observed a biogeographical shift of C. *finmarchicus* northwards in the eastern North Atlantic (see the equatorward range limit in Plate 11.4D for 1960–1979 versus Plate 11.4F for 1990–2009). As expected, and because the species overwinters in deep water and PAR is positively correlated to phytoplankton production in these regions (*164*), the calanoid cannot compensate for the warming observed between March and September in the southern part of its current distribution by adjusting its phenology in winter. In the central part of its range (~60°N), rising temperature had a negative effect on the species abundance in summer. A warming is expected to generate a poleward shift in the species spatial distribution. The METAL theory predicts that individuals might be first detected in late summer, a prediction that is confirmed by observations of the first occurrence of southern zooplankton species (e.g. *Centropages typicus*, C. *violaceus*, *Temora stylifera*) along European coasts (*532, 1095*).

These results suggest that phenological shifts are not exclusively the result of evolutionary adaptation or phenotypic plasticity (*1096*) (Chapter 7). Rather, our results suggest that phenological shifts can simply reflect the species adjustment with respect to its thermal niche (*60*). Investigating the long-term phenological shift of the North Sea plankton, Edwards and Richardson (*657*) found no phenological shift for spring diatoms. They proposed that the phenology of these diatoms did not shift because the reproduction of the taxonomic group was mainly controlled by the photoperiod. The METAL theory allowed us to complete this hypothesis. Spring North Sea diatoms cannot shift their phenology in winter because most species are located to their southern boundary and cannot compensate for an increase in sea temperature in winter because reduced winter PAR limits their proliferation. Spring North Sea diatoms should therefore decline in the future if the warming continues. In contrast, Edwards and Richardson (657) found that summer diatoms and dinoflagellates exhibited an earlier seasonal occurrence (their Figure 1), a result predicted by the METAL theory.

11.4.2.3 Long-term community shifts

How climate change may trigger long-term community shifts remains unresolved in many systems. In many ecoregions such as the North Sea although temperature is the most cited potential factor (483, 497, 1084), the exact mechanism by which the parameter may affect community remains poorly understood. The METAL theory proposes that the interaction between the species ecological niche and the environment (including climate) should propagate from the species to the community level. To show how this should occur, we generated a number of hypothetical thermal niches for both eurytherms and stenotherms (Plate 11.5; see Equation 11.2) (1088). We estimated hypothetical changes in the abundance of each species according to their niches.

If the METAL theory applies at the community level, community shifts may be detected when substantial temperature shifts occur. For example, a 1°C increase in temperature may trigger substantial changes in the abundance of some eurytherms, especially those for which the initial thermal regime is close to T_{HV} (Plate 11.5A–B and Figure 11.9). In the fictive example, note that no change is observed for one pseudo-species (blue sky; Plate 11.5A–B) because the initial thermal regime corresponds to its thermal optimum $Topt$, and the region of the niche between the two points Ts is stable (Figure 11.9). The magnitude of the change is expected to be larger for some stenotherms (Plate 11.5D–E) (1097). Note that some stenotherms for which the initial thermal regime was close to the edge of their thermal niche (outside TD) and for which the initial thermal regime was close to their thermal optimum ($Topt$; Figure 11.9) exhibit small changes in abundance (Plate 11.5D–E). The same expectations emerge when pseudo-species abundance (eurytherms and stenotherms) are assessed from a time series of temperature (mean average = 20°C) having the same variability as long-term changes in temperatures observed in the North Sea for the period 1958–2010 (Plate 11.5C and F). When the 12 pseudo-species (six eurytherms and six stenotherms) are combined and their long-term changes analysed by standardised principal component analysis (PCA), the first principal component (80.05% of the total variance) reveals major changes in this pseudocommunity, including a stepwise and rapid shift in the late 1980s (Plate 11.5G). Note that this corresponds to the time of a major shift observed in North Sea pelagic ecosystems (Chapters 7).

This theoretical example illustrates how the non-linear interaction between the thermal niche of each species in a community and temperature may lead to long-term changes and abrupt community shifts (ACSs). Rapid changes in regional temperatures may originate from a rapid change in atmospheric circulation (59) or climate change (497), and are more frequent at thermal frontal zones, often located at the boundaries between systems (e.g. transitional regions between the temperate and the polar biomes) (497). In such ecosystems, the METAL theory predicts that the response of the community is directly a function of the magnitude of the thermal shift (Plate 11.5C and F).

The application of the METAL theory at the community level leads to several predictions: (1) A stepwise shift in temperature leads to an ACS. (2) Long-term shift in temperature will be amplified at the community scale by the non-linear interaction between the species thermal niche and the thermal regime, this being more prominent when the number of stenotherms is higher in the community or when the thermal regime of the region is close to point T_{HV} of eurytherms (1097). This prediction may explain the amplification of small climate-induced environmental changes by the community (1098). (3) Some species do not exhibit a stepwise response to change in temperature during an ACS (e.g. species close to their optimum and species at the edge of their thermal niche beyond TD). This prediction may explicate why a large number of species do not show any substantial shifts during an ACS (499, 1099).

We tested these predictions using data on copepods in the North Sea where substantial community changes took place in the 1980s (497). A total of 90 pseudo-species each characterised by different thermal niches from stenotherms to eurytherms (Figure 11.10A–B) were created and their expected abundance (as annual mean) was estimated as a function of monthly SSTs.

The niche was modelled exclusively as a function of monthly SSTs because: (1) bathymetry does not change on a year-to-year basis; and (2) both PAR and chlorophyll a concentration were mostly important to reconstruct the seasonal and the distributional range of C. finmarchicus. None of the species had the same thermal niche following the principle of competitive exclusion of Gause (489). The goal was to show how, by creating a pool of species with niches differing by their optimum and amplitude, the sum of the temporal changes occurring for each species could create long-term community shifts similar to those observed in the North Sea. As the number of pseudo-species generated for the model exceeded the actual number of copepods occurring in the North Sea, 27 pseudo-species were sampled at random to correspond to the number of observed copepods (27 species or taxa). We performed a standardised PCA on the table years (1958–2009) × annual expected abundance (27 pseudo-species), and repeated this procedure 10,000 times, examining the correlations between long-term (first principal components) expected and observed changes for copepods (Figure 11.10C). These components were compared to the first principal component originating from a standardised PCA applied on a table years (1958–2009) × annual observed abundance (27 species or taxa).

Significant positive relationships ($p < 0.05$) between expected and observed changes in the North Sea copepods were found. The correlations between expected and observed long-term changes were in general (88.90% of the 10,000 simulations) greater than the correlation calculated between annual SSTs and observed changes. The METAL theory therefore explains 56.25% of long-term changes in copepods in the North Sea and provides a mechanism to understand how climate-caused changes in temperatures may influence long-term community shifts.

11.4.2.4 Another explanation for regime shifts

Abrupt community shifts are often explained by referring to the theory of alternative stable states (494, 953, 1100). The theory states that the shift from one alternative stable state (or attractor) to another depends upon the size of the attraction basin and the strength of both positive and negative feedbacks. The transition is difficult to anticipate, and return to initial environmental conditions is insufficient for the system to switch back in the case of hysteresis (1101) (see also Section 5.1.4.3). Although some rapid ecosystem shifts may well originate from the presence of attractors, their existence remains hard to establish (1102). Some processes and alternative stable states leading to ACSs have been recognised in lakes

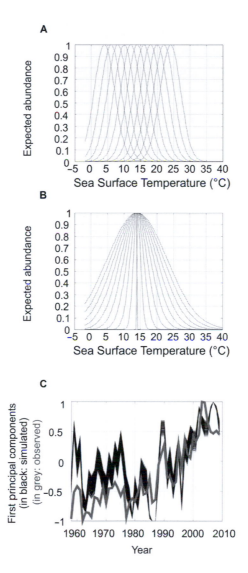

Figure 11.10 The community shift in the North Sea (4°W–10°E; 51°N–60°N) reconstructed from the application of the METAL theory. Examples of some simulated niches based on (A) different thermal optimums u_s and a constant thermal tolerance t_s and (B) a constant average u_s and different thermal tolerances t_s. Only pseudo-species that could establish in the North Sea were used in the analyses. (C) First principal components (10,000 first principal components; in black) from standardised PCAs applied on each simulated table 52 years × 27 pseudo-species and the first principal component (in grey) from a standardised PCA performed on the table 52 years × 27 copepods.

Source: Modified from Beaugrand and colleagues (1088).

and in some marine ecosystems (*494, 1101, 1103*). For example, algal overgrowth in Caribbean coral reefs is mainly attributed to: (1) nutrient loading, which stimulates algal growth; and (2) overfishing, which reduces the number of herbivorous fish that control algal proliferation (*1103*). Similar mechanisms have also been invoked to explain regime shifts in kelp forest ecosystems.

In contrast to these local benthic ecosystems, processes and mechanisms leading to ACSs in pelagic ecosystems remain difficult both to identify and understand (*774, 1104*) and scientists have mainly progressed on pattern recognition (*59, 483, 695, 710*). While climate and temperature have repeatedly been hypothesised to play an important role in the origin of some regime shifts such as the Pacific Ocean (*59*), in the North Sea (*483, 1084*) and in the eastern North Atlantic (*695, 710*) mechanisms by which the climate signal triggers these ecosystem responses (*551, 658*) remain elusive. As we saw previously, the METAL theory offers an alternative mechanism, which has also been enunciated under the form of two theorems (*1097*).

The first theorem states that 'providing that the niche has a Gaussian shape and that there is no species interaction, a substantial increase (e.g. a 1°C of mean annual temperature) in the thermal regime of an ecosystem triggers an abrupt community shift. The magnitude of the community shift depends on the extent of the temperature change and the degree of eurythermy (stenothermy) of the species composing a community. A higher degree of stenothermy increases the sensitivity of the community to temperature change'. Therefore, at the community level, a substantial change in temperature triggers rapid community shifts. In practice, the first theorem is robust from departure to the condition on the niche shape (*1085*). The only condition is that the niche must be unimodal. The magnitude of the observed shift is likely to strongly depend on the noise/signal ratio, which is influenced by data quality (e.g. sampling and species identification). The noise/signal ratio influences the location of the point T_D along the thermal niche (Figure 11.9). This phenomenon may also be at the origin of substantial changes in the timing of a shift. This problem is probably observed when changes in the thermal regime are relatively small. I will talk about the potential influence of species interaction in the next sections.

The second theorem states that 'during a climate-driven abrupt community shift, the response of species is individualistic, depending upon the characteristics of their thermal niche, the initial thermal regime and the magnitude of the thermal shift. It follows that not all species are expected to show a shift (e.g. species located around T_{opt} or outside T_D) and that some may react earlier (stenotherms for an initial thermal regime close to T_{HV}) than others (eurytherms or species with an initial thermal regime close to T_{opt} or outside T_D)' (Figure 11.9). This second theorem has strong practical implications. According to species or taxonomic groups, the magnitude and the timing of the shift may vary, independently of the type of statistical procedures that also influences the timing (*499*). In the North Sea, it has been suggested that the timing of the shift varies according to the selection of species, taxonomic group and species assemblage (*483, 497, 1050, 1084*). It is also possible that some studies focusing on the same ecosystem find a shift, whereas others do not, independently of other issues related to statistical technics or sampling programmes.

Since temperatures are projected to rise rapidly with anthropogenic climate change, ACSs are likely to increase in both frequency and intensity. The term regime shift has often been used in the past to describe stepwise changes in the ecosystem/community state. Because time series were relatively short (i.e. a few decades), two full dynamic regimes (apparently stable states) and a shift were frequently observed (*59, 483, 1050*). Since then, time series have increased in length and have started to reveal more complex temporal patterns (*497*). The METAL theory explains why such complex patterns form. It may soon become apparent that we are observing ecosystem/community state vacillations, where ACSs alternate with periods of more relative stability. These periods of relative stability may not be confounded with stable states. The METAL theory offers a way to predict future climate-induced ACSs that may be at the origin of trophic cascades and amplifications.

11.5 Strength and assumptions of the METAL theory

11.5.1 Strength of the theory

The METAL theory is a unifying theory that aims to explain the arrangement of life in the oceans and seas, and to understand and anticipate how marine biodiversity may be altered by natural climatic/environmental variabilities and anthropogenic climate change. The theory posits that both biological and ecological systems tend to be relatively quickly in equilibrium with their climatic and environmental regimes of variability, the rapidity depending on the taxonomic groups under study. The theory also enables us to understand the relative success of ecological niche models or species distribution models to estimate the past, present and future species spatial distribution (*1105*) and suggests that these works can be extended from the species to the community and ecosystem levels in space and time (*1088, 1097*). I have only provided a few examples of application of the METAL theory. It may soon become apparent that the theory explains many ecogeographical patterns found in nature (Chapter 3) and most natural responses observed in the context of anthropogenic climate change.

In the past, useful empirical patterns on how species or communities can be affected by environmental changes (Chapters 2–7) have been detected (see Chapters 5 and 7), but these patterns were not tested against a solid theoretical basis. The METAL theory allows the establishment of predictions that can then be tested using observed data. In this way, the theory could radically change the way we investigate the effects of both climate and environmental changes on marine biodiversity. The theory may also unify the whole fields of ecology devoted to the understanding of life distribution in both the terrestrial and the marine ecospheres.

The METAL theory is primarily deterministic, but a stochastic component could be implemented in the different models used so far. Using time series investigated to characterise the 1976/77 Pacific regime shift, Hsieh and colleagues (*1106*) showed whereas time series of physical variables (e.g. Pacific Decadal Oscillation Index, Southern Oscillation Index) were linear stochastic, biological time series (e.g. CALCOFI copepods, sockeye salmon, chum salmon) exhibited a non-linear signature. These results found as part of the METAL theory supported the idea of a non-linear amplification of stochastic hydro-climatic forcing.

Even though stochastic effects due to complex abiotic and biotic interactions throughout a species' life cycle, and from demographic effects that control vital processes (e.g. fecundity, survival), make it difficult to forecast the response of species and ecosystems to climate change (*1107, 1108*), results from the METAL theory suggest that a significant proportion of biogeographical, phenological and long-term community shifts are deterministic and predictable at some observed emergent spatio-temporal scales. A fixed (non-linear) ecological niche offers a way to understand how communities and their species may respond to environmental variability and global climate change. Although the ecological niche is already applied to anticipate the response of a species distributional range to environmental changes by means of ecological niche models (1105), the concept of the ecological niche has never been used to link phenological, biogeographical and community shifts, which are among the main documented responses to climate change so far (427, 497, 657, 663).

The METAL theory provides an explanation for why climate-induced long-term environmental changes – especially temperature changes – and changes in species and communities are so often tightly correlated (559). We saw in the first seven chapters of the book the often strong relationships between marine biodiversity and both the climate and the environment. Both the arrangement of biodiversity and its spatial and temporal changes are, in part, the product of the interaction between the species ecological niche and both climatic and environmental changes in time and space. At the community scale, a large part of climate-caused long-term community shifts is the result of climate-modulated environmental changes on individual species niche, which explains why many species remain stable during an abrupt ecosystem shift, and why some may react earlier than others (499).

11.5.2 Current assumptions of the theory

11.5.2.1 A complex ecological space

Ecosystems and their biodiversity are influenced by many environmental parameters; we could even say an infinite number of dimensions (Chapter 3). It is impossible to use all niche dimensions, and so it is important to select a few that can control a large part of the species spatial distribution. The METAL theory outlined here used a limited number of ecological dimensions. Simplification is needed, but on the other hand oversimplification could become hazardous. In Chapters 3–4, we saw that climate is among the most important parameters controlling species spatial distribution (10). Pearson and Dawson wrote 'It is a central premise of biogeography that climate exerts a dominant control over the natural distribution of species' (1074). The climatic variability hypothesis states that the latitudinal range of species is primarily determined by their thermal tolerance (215). Temperature is indeed a key variable in the marine realm because it is the result of many hydro-climatic processes (497) and because the factor exerts an effect on many fundamental biological and ecological processes (1089) (Chapters 2–7). This assumption especially holds true for thermal-range conformers such as marine species (including invertebrates and fish) whose physiological thermal amplitude closely determines their latitudinal range at large spatial and temporal scales (1089). Phenological, biogeographical and long-term community shifts have often been correlated to temperature (427, 657, 658, 695). Other environmental parameters (photosynthetically active radiation, bathymetry, chlorophyll a) have also been considered in models testing the METAL theory. These environmental parameters were particularly important in reconstructing seasonal patterns of abundance. When the spatial and temporal scales diminish, the number of ecological dimensions should increase, and therefore an application of the METAL theory at smaller spatial scales may necessitate the consideration of more ecological dimensions. Further environmental dimensions will be implemented in future refinements of the theory.

11.5.2.2 Species interaction

The METAL theory does not consider species interaction yet. In our simplest models, we assume that the spatial and temporal abundance of a species is mainly determined by its fundamental niche (i.e. the niche without the effect of dispersal or species interactions such as competition, predation and facilitation) (*1079*). Species interactions are, however, often involved in the positive feedbacks that shift a system from one state to another (*494*). Because of these species interactions, climate-induced species shifts may propagate through the food web (*559, 702*). Bottom-up and top-down controls, trophic cascades and amplifications have been observed (Chapter 7). Trophic interactions can be implemented in the METAL theory and will probably lead to an increased sensitivity of the system to the environment and climate.

11.6 Limits to predictions in the context of global change

It would be presumptuous to think that one day we will be able to know exactly how species and ecosystems vary in space and time. Even meteorologists, which have the best networks of observations all around the planet and state-of-the-art models and computers have some difficulties to forecast weather for the next day; and yet the number of variables they consider is much smaller than the one we should consider for understanding and anticipating spatial and temporal changes in biological and ecological systems. Although it is difficult to elaborate general predictions, an increasing body of knowledge, including the METAL theory, suggests it is possible. What level of accuracy we can possibly achieve is an important issue. It is crucial that scientists communicate on the level of precision they can reach when they establish projections. This should not only be attributed to the reliability of a science, but also to the awareness that absolute limits may exist. I illustrate my view with some concrete examples. Can a medical doctor say when you will develop a cancer if you continue to smoke? He or she can only say that your risk of developing a cancer is increased if you smoke. Can a glaciologist say when exactly an avalanche will happen on some parts of a mountain if snow continues to fall? He or she can only provide a level of risk. Other unanswerable questions could be: Can we determine the exact weather in the next decades with climate change? When will the increasing level of greenhouse gases initiate dangerous climate change?

11.6.1 Chaos

Boreo and co-workers (*971*) questioned whether biological predictions are possible. At the community level, they reminded that prediction of system behaviour composed of three species has already become too complex and that simple mathematical models may exhibit complex dynamics (*1109*). Boreo and co-authors made the parallel with physicists that spent tremendous time to try to discover perpetual motion until thermodynamics demonstrated it was impossible. Determinism does not guaranty predictability. Phenomena are non-linear and are sensitive to initial conditions, the 'butterfly effect' (*1110*). The meteorologist Edward Lorenz from the Massachusetts Institute of Technology (MIT) summarised this point by the following sentence: 'the present determines the future, but the approximate present does not approximately determine the future'. To demonstrate his point, the meteorologist simplified the atmospheric convection using three (non-linear) ordinary differential equations:

$$\frac{dx}{dt} = -Px + Py \qquad (11.3)$$

$$\frac{dy}{dt} = rx - y - xz \tag{11.4}$$

$$\frac{dz}{dt} = xy - bz \tag{11.5}$$

With x, y, and z three atmospheric parameters and P, r and b three constants. Integrating these three equations first revealed the existence of a strange attractor, also called Lorenz's attractor or the butterfly's attractor (Figure 11.11). Such a strange attractor appeared in the phase spaces of chaotic systems. Using slightly different initial conditions, Lorenz showed that the trajectories rapidly differed (Figure 11.11; top panel). I reanalysed these equations. The upper panel shows that the trajectory of the system is highly sensitive to initial conditions (the grey and black trajectories are highly different despite having close initial conditions).

The chaos theory teaches us that a system is very sensitive to initial conditions. Each state depending upon the previous one, the sensitivity of the predictions to initial conditions

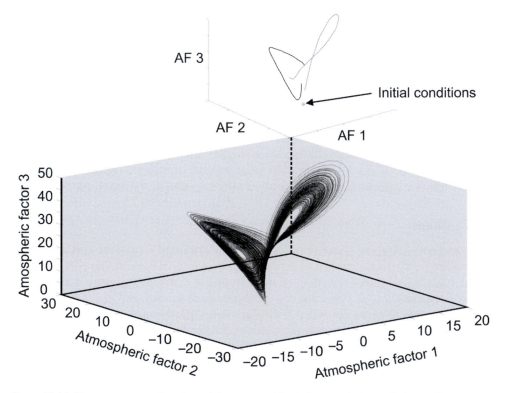

Figure 11.11 Phase space as a function of three atmospheric factors that reveals Lorenz's attractor. The figure was performed using three ordinary differential equations with $P = 10$, $r = 28$ and $b = 8/3$, and 1,000 iterations. The top figure shows the sensitivity of the dynamic system to initial conditions. Two simulations were conducted with very few differences for x, y and z at t_0. In simulation 1, $x_0 = 0.50$, $y_0 = 0.50$ and $z_0 = 0.50$ (black circle). In simulation 2, $x_0 = 0.55$, $y_0 = 0.55$ and $z_0 = 0.55$ (grey point). This led rapidly to two different trajectories (in black and in grey) in the phase space. AF1–AF3: atmospheric factor 1, 2 and 3.

is such that even the best organised observation networks of the meteorologists are not sufficient to know accurately the initial system state. Uncertainties are such that the system, through its own dynamics, rapidly becomes unpredictable. Boreo and co-workers (*971*) concluded that:

> If the weather (and climate) cannot be predicted with precision over the medium and long-term, and if the weather is only one of the variables affecting the performance of ecosystems, most with unpredictable behaviour, it is obvious that we cannot predict ecological phenomena (with precision).

Boreo stressed: 'We can predict that frost will kill plants, but this is a physiological prediction, and it does not help much in ecology, since we cannot predict the occurrence of frost in proper advance and precision'. However, the authors forgot that Lorenz also showed that attractors do exist (Figure 11.11; bottom panel). We can see that trajectories remain around two attractors in Figure 11.11. The challenge is therefore to be able to identify such attractors to anticipate future 'average' conditions, even if exact trajectories cannot be detected. The identification of attractors is an important goal in the context of the METAL theory.

11.6.2 Complexity

Understanding the responses of marine ecosystems remains challenging because they are complex adaptive systems (CAS), meaning that macroscopic dynamics emerge from numerous non-linear interactions at smaller hierarchical (spatio-temporal) scales (*1111*) and that multiple responses can occur when pressures are exerted on the system. Ecosystems are characterised by critical thresholds, which lead to abrupt changes between alternative modes of operation. Often, the state of the dynamic system under pressure initially shows little obvious changes until a **critical threshold** or **tipping point** is crossed, moving the system to another dynamic regime (*494*). Climate and environmental changes will influence natural systems through multiple effects that will propagate into the systems in complex ways (*1112*).

Levin and Lubchenco (2008) distinguished a number of properties that should be monitored: diversity and heterogeneity, which reflect the adaptive capacity of a system, redundancy and degeneracy, and space/time modularity, as well as modularity in organisational structures and the characteristics of feedback loops. Non-linearity could thereby lead to surprising outcomes, difficult to forecast and of potential strong consequences for biological and socio-economic systems.

May showed that even simple models might have complicated dynamics (*1109*). May worked on the logistic equation:

$$X_{t+1} = rX_t(1 - X_t) \tag{11.6}$$

With X the abundance of a species population at time t or $t + 1$ and r the population growth rate. When a species population is small, X tends to increase substantially. In contrast, when the population is high, it tends to decrease. May showed that even this simple deterministic equation might exhibit a complex behaviour, alternating between phases of stable points and chaos (Figure 11.12).

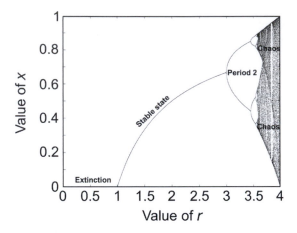

Figure 11.12 The bifurcation diagram for the logistic equation. Calculations were performed for r varying between 0 and 4 by increments of 0.001. Results are shown for t between 450 and 500. There are zones on the bifurcation diagram where stable points (1 < r < 3) exists and others where cycles (e.g. r = 3.2 for a cycle of period 2 and r = 3.5 for a cycle of period 4) can be observed. Above 3.75, a chaotic behaviour is detected.

The bifurcation diagram shows that a small uncertainty on the growth parameter r in the logistic equation can lead to a very different behaviour (here estimated between $t = 450$ and $t = 500$). When $r < 1$, extinction of the population takes place and uncertainty is likely to have a limited influence at equilibrium. When $1 < r < 3$, uncertainty on r has only limited effects. After 3, a cycle of period 2 is observed on the diagram. Above 3.5, the behaviour starts to be more complicated and above 3.75, it becomes chaotic (Figure 11.12); a small uncertainty on r can lead to a very different outcome. This result suggests that individual based models (IBMs), based on many parameters, are likely to be sensitive to parameters estimate. As equations pool in the models, uncertainties become large and model outcomes more uncertain.

A dynamic system cannot remain in a steady state, and changes continuously. Disturbance occurs regularly. Once a change happens, it is rarely reversible. The system cannot return to quasi-identical conditions. Historicity/contingency is therefore important. In the case of the North Sea, if temperature conditions return to the cold-dynamic equilibrium of the 1960s–1970s, a backward dynamic could occur, implying similar but not identical development, evolution and succession.

Although it is not possible to predict exactly what could appear in the next century, we can work on trying to determine future regimes, states or attractors. We have to communicate on the level of accuracy we can currently achieve because of: (1) the chaotic nature of the system; and (2) the number of components and interactions far more elevated than in meteorology.

11.6.3 Sustainability

Managers, policymakers and conservationists all search sustainability. Sustainability is at the confluence of economics, society and the environment (Figure 11.13A).

A. Normal environmental changes

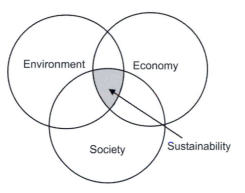

B. Strong environmental changes

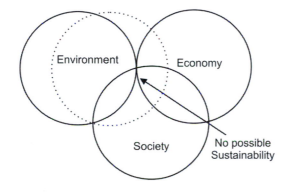

Figure 11.13 Venn diagram showing that sustainability is at the intersection of the environmental, societal and economic spheres. (A) Normal environmental variability. Sustainability (grey area) depends on society and economy. (B) Pronounced environmental and climate changes. Sustainability may well become independent on economy and society. Note that positive environmental changes (not shown) may well be observed in the context of anthropogenic climate change.

A recent study indicates that between 10% and 48% of the climatic regimes observed currently on the planet may disappear by the end of this century. In parallel, between 12% and 39% of new climates may appear (*1113*). In this context, biodiversity will be altered and will reorganise and ecosystem states will evolve (Chapter 7). In the context of global change (Figure 11.13B), sustainability for a given resource may well be an unachievable task because the prerequisite for sustainability is a constant environmental regime, which is not expected in the context of global change, especially with anthropogenic climate change. For a stock to remain sustainable in the context of adverse environmental change, the only possibility is to alleviate the socio-economic pressure, a hard task since global demography is increasing and more and more people are living close to the sea.

11.6.4 Other assumptions

11.6.4.1 Uniformitarianism

When ecological niche models estimate the spatial distribution of a species based on past spatial distribution, they implicitly assume uniformitarianism. The principle of uniformitarianism is applied by biogeographers to account for present and past species distribution. According to Lomolino and colleagues (10), uniformitarianism is based on the 'assumption that the basic physical and biological processes now operating on the earth have remained unchanged throughout time because they are manifestations of universal scientific laws'. They pursued by stating that it is perhaps better to use the term actualism that means methodological uniformitarianism suggesting that although the law does not change, intensity of the processes changes.

11.6.4.2 Niche conservatism

Niche conservatism is the tendency of a species to preserve ancestral ecological traits and environmental distribution (644). Crisp and colleagues examined the hypothesis of phylogenetic biome conservatism for 11,000 plant species at timescales of tens of millions years and at large spatial scales covering all continents in the southern hemisphere. They found strong support for the hypothesis, showing that a species rarely changed biome. Evolutionary divergence was only associated with biome shift in 396 events (3.6%) on a total of 11,000. Interestingly, the events of evolutionary divergence only occurred between biomes with higher ecological similarities. The authors also made the parallel with introduced species, which generally tend to invade regions that have similar ecological characteristics (niches, *sensu* Hutchinson). Crisp and colleagues concluded by stating that species evolutionary success does not rely in adapting to a new biome, but rather in tracking biomes with similar environmental conditions. Overpeck and colleagues (322) also stressed that only small macroevolution was apparent during the Quaternary (Chapter 5). They explained this observation by the rapidity of climate and environmental changes during this period. However, morphological fluctuations were observed in some groups such as fossil molluscs (the gastropod *Pupilla muscorum*) during the Quaternary, although even extreme ecotypes were located within the range of contemporaneous species (1114).

Evolutionary conservatism in climatic tolerances is assumed by some methods (e.g. the **nearest living relatives**), which are applied to reconstruct palaeoclimates. However, the nearest living relatives approach has been criticised because the assumption of conservatism does not hold over millions or tens of millions of years. Over such timescales, the technique can only reflect broad variations in climate. However, over shorter timescales, the use of the approach reinforces the hypothesis of contemporary conservatism in ecological tolerances. I define here contemporary conservatism in ecological tolerances as the species tendency to maintain constant its fundamental niche at timescales from year-to-year to multi-centennial.

11.6.4.3 Acclimatisation

The possibility of species **acclimatisation** is real. This means that the thermal envelope of a species can be altered by climate change. However, all species will not acclimatise to a change in the thermal regime in the same way. Stillman (1115) investigated four species

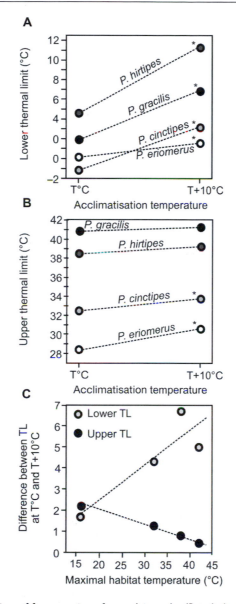

Figure 11.14 Acclimatisation of four species of porcelain crabs (*Petrolisthes*) to temperature. (A) Lower thermal limits of cardiac function to temperature. The reference temperature was 8°C for *P. cinctipes* and *P. eriomerus* and 15°C for *P. gracilis* and *P. hirtipes*. (B) Upper thermal limits of cardiac function to temperature. An asterisk denotes a significant change between the acclimatation temperature and the reference temperature. (C) Differences between acclimatation temperature (*T* + 10°C) and the reference temperature (*T*°C) for both lower and upper thermal limits as a function of the maximum temperature observed in the habitat of the four species. TL: thermal limit.

Source: Redrawn from Stillman (*1115*).

of porcelain crabs of the genus *Petrolisthes*. Two warm-water species, *P. gracilis* and *P. hirtipes*, are endemic to the northern part of the Gulf of California, whereas the two colder-water species, *P. cinctipes* and *P. eriomerus*, are located in the cold-temperate zone of the north-eastern Pacific Ocean. The author determined experimentally the lower and upper thermal limits of the cardiac function of the four crabs (Figure 11.14). Although all four species significantly changed their lower thermal limits of cardiac function (heart rate) when temperature increased by 10°C, only cold-water species were able to increase their upper thermal limits significantly (Figure 11.14A–B). The researcher thereby showed that thermophilic species acclimatise less than their more psychrophile congeneric species (Figure 11.14C). The upper thermal limits were more affected in cold-water than warm-water species. In contrast, although not significant, the lower thermal limits were more altered in warm-water than colder-water species.

This study suggests that species are likely to play on their lower thermal limit more easily in case of a warming than their upper thermal limit. The contrary is probably true in the case of a cooling. Such results are in phase with the Hutchinsonian niche. Indeed, the lower thermal limit can be more easily adjusted in case of a warming because the alteration is directed towards the centre of the ecological niche, whereas the upper thermal limit can be adjusted with more difficulties because it is directed towards the edge of the niche.

11.6.4.4 Dispersal

Ackerly alleged that the capacity of species to track or adapt to climate change depends on their dispersal capacity and the heterogeneity of their **realised environment** (609). Species characterised by high dispersal potential are likely to migrate rapidly with little adaptive change (Chapter 7). In contrast, species with poor dispersal ability are likely to either adapt or go extinct. Characterising the habitat heterogeneity is essential, as climate change is likely to have more impact on species in heterogeneous than contiguous habitats.

11.7 Scenarising biodiversity

As we saw in Chapters 5 and 7, many works have reported on the influence of climatic variability and anthropogenic climate change on organisms (141, 696, 1116). It is very likely that current and future climate change will destabilise biomes and provinces and will greatly alter biological and ecological marine systems (1117). The regional consequences will be considerable. At the species level, abundance will change, which in turn will modify communities (659). Biotic interactions will be altered in sign and strength (559).

Ecological niche models (ENMs) and species distribution models are frequently applied to examine the potential trajectories of biological and ecological systems in the past and future. ENMs are sometimes based on the ecological niche concept, especially the realised niche. The determination of the realised niche offers great potential for evaluating the likely influence of climate change on species distribution. ENMs can also model the fundamental niche when enough information on the species' physiology is known (541). ENMs have been used in conservation to manage endangered species (1118), to predict the species' responses to climate change (1119), to forecast past distribution (361) and to estimate the potential invasion of a non-native species (1120). When quantitative data are available, regression techniques such as generalised linear models (GLMs) (1121) or generalised additive models (GAMs) (*1122*), ordination or neural networks have been frequently applied (*1123, 1124*). When only binary (presence-absence) data are available, there are far fewer techniques that can be applied, although regression techniques such as GAMs can still be utilised. Traditional models such as BIOCLIM based on multilevel rectilinear envelope and DOMAIN based on point-to-point similarity metric (*1125*) tend to be relatively simple. More sophisticated models have been developed recently such as the ecological niche factor analysis (ENFA) (*1126*) based upon principal component analysis (PCA), MAXENT (*1127*) based on the principle of maximum entropy and the NPPEN (non-parametric probabilistic ecological niche) model (*180, 1117*). Below, I describe briefly the NPPEN model that can be used in conjunction with the METAL theory because it primarily estimates the species ecological niche.

11.7.1 The non-parametric probabilistic ecological niche model

The NPPEN model is an ENM that only requires presence data (*180*). The first step consists in constructing a reference matrix ($Z_{m,p}$) of the environmental data corresponding to the presence records. In a second step, the Mahalanobis generalised distance is calculated between the observations and the homogenized reference matrix as follows:

$$D_{x,z}^2 = \left(x - \overline{Z}\right)' R^{-1} \left(x - \overline{Z}\right) \tag{11.7}$$

With x the vector of length p, representing the values of the environmental data to be tested, $R_{p,p}$ the correlation matrix of reference matrix $Z_{m,p}$ and $\overline{Z}$ the average environmental condition inferred from $Z_{m,p}$. The use of the Mahalanobis distance instead of a classical Euclidian distance presents a double advantage: it enables the correlation between variables to be taken into account and is independent of the scales of the descriptors (Figure 11.15). Comparison between the chord and the Mahalanobis generalised distance shows that when correlations between two parameters are significantly different to 0, the model performs better when it is based on the Mahalanobis generalised distance (Figure 11.15C–D). When environmental parameters are not correlated, the two distances give the same patterns of probability (Figure 11.15A–B).

In the third step, the model calculates the probability of each grid point to belong to the reference matrix by using a simplified version of the multiple response permutation procedure (MRPP). This probability (v) is the number of times the simulated distance was found greater or equal than the observed average distance:

$$v = \frac{q_{\varepsilon_s \geq \varepsilon_0}}{n} \tag{11.8}$$

With ε_0 the average observed distance, ε_s the recalculated distance after permutation and n the maximum number of permutations. If the probability is ~1, the environmental value of the tested point are at the centre of the ecological niche. A probability of ~0 indicates that the environmental conditions of the point are outside the ecological niche. Finally, the last step consists in mapping the probability of species occurrence. This method is applied to estimate the species ecological niche (*sensu* Hutchinson), to model its spatial and future distribution.

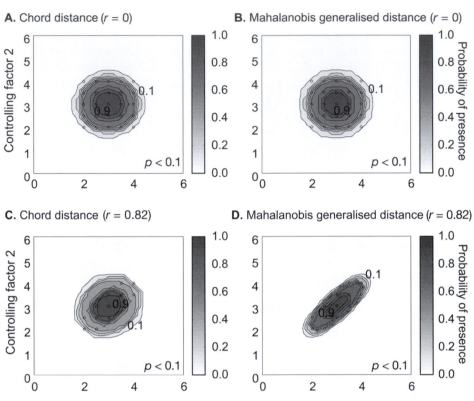

Figure 11.15 Fictive examples that show that the Mahalanobis generalised distance performs better than the chord distance when ecological variables are correlated. First, the reference matrix is composed of 25 observations with two non-correlated controlling factors (A and B). (A) Probabilities based on the chord distance ($r = 0$). (B) Probabilities based on the Mahalanobis generalised distance ($r = 0$). Second, the reference matrix is composed of 13 observations with two parameters. The correlation between the two controlling factors is high ($r = 0.82$; C and D). (C) Probabilities based on the chord distance ($r = 0.82$). (D) Probabilities based on the Mahalanobis generalised distance ($r = 0.82$). Black circles denote the reference observations (reference matrix).

Source: From Beaugrand and co-workers (*180*).

11.7.2 Examples of application of the niche approach

I present here three examples of application of the NPPEN model. As I said earlier, other ENMs have been proposed, and it is recommended to use several ENMs to better understand uncertainties on the estimations of probability of occurrence (or estimated abundance).

11.7.2.1 Projection of long-term changes in Atlantic cod

The NPPEN model was used to model the future spatial distribution of Atlantic cod (*Gadus morhua*) under different climatic scenarios. The ecological niche of cod was first estimated (Figure 11.16).

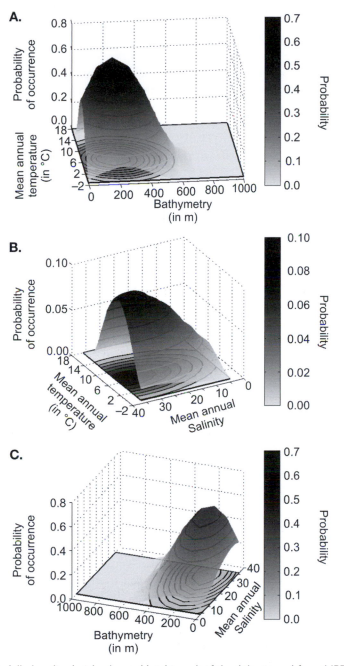

Figure 11.16 Modelled realised niche (*sensu* Hutchinson) of the Atlantic cod from NPPEN.
(A) Probability of cod occurrence as a function of bathymetry and mean annual SST.
(B) Probability of cod occurrence as a function of mean annual sea surface salinity
and SST. (C) Probability of cod occurrence as a function of mean annual sea surface
salinity and bathymetry.

Source: From Beaugrand and colleagues (*180*).

The modelled niche exhibited high probabilities of occurrence at bathymetry ranging from 0 to 500 m (mode from 100 m to 300 m), at annual SST from –1°C to 14°C (mode from 4°C to 8°C) and at annual sea surface salinity ranging from 0 to 36 (mode from 25 to 34).

Maps of probability of presence were then assessed from the knowledge of the niche. NPPEN revealed the pronounced effect of current climate change on cod spatial distribution and projections for the end of this century showed that the fish may eventually disappear as a commercial species from regions (e.g. North Sea) where a sustained decrease or collapse has already been documented (Plate 11.6).

Other studies suggested similar effects of climate on the same species (*1128–1130*). Indeed, projections indicate that in the case of moderate to strong warming, cod might strongly diminish in both abundance and distribution, and reach the level of commercial extinction in the North Sea (Figure 11.6C–E). In contrast, the abundance of cod is likely to increase in the Barents Sea. The rebuilding of cod stocks in the North Sea might be difficult. Instead, our effort should perhaps focus on what resource is likely to become available over the next decades to enable fishermen to anticipate changes in the resources should this ecoregion continue to warm.

IPCC projections indicate a likely warming of the earth of >2°C by the end of this century, and possibly even 4–5°C, which would represent the temperature difference between the LGM and the Holocene (*83*). Even warming to half this difference will have a major impact on the spatial distribution of many species. In the case of cod, it is unlikely that this species will be able to adapt to such warming on a short timescale in the North Sea. Controlling fishing effort thereby does not guarantee the sustainability of this resource in the North Sea. However, the absence of control of fishing effort is likely to precipitate the collapse of the stock (Chapter 9).

11.7.2.2 Projected changes in Laminaria digitata

Kelp ecosystems form widespread underwater forests playing a major role in structuring regional biodiversity (*1131, 1132*). Seaweeds such as *Laminaria digitata* are also economically important, being exploited for their alginate and iodine content. Although some studies showed that kelp ecosystems are regressing and that multiple causes are likely to be at the origin of the disappearance of certain populations, the extent to which global climate change may play a role remains elusive (*1133*). Many populations of *L. digitata* along European coasts are on the verge of local extinction due to a climate-caused increase in sea temperature. By modelling the spatial distribution of the seaweed, we evaluated the possible implications of global climate change for the geographical patterns of the macroalga using temperature data from the Coupled Model Intercomparison Project phase 5 (Plate 11.7). Projections of the future range of *L. digitata* throughout the twenty-first century showed that large shifts in the suitable habitat of the kelp are likely. In particular, a northward retreat of the southern limit of its current geographic distribution of *L. digitata* from France to Danish coasts and the southern regions of the United Kingdom is forecasted in most scenarios. Obviously, these projections depend on the intensity of warming.

A medium to high warming is expected to lead to the species' extirpation along French coasts as early as the first half of the twenty-first century and there is high confidence that regional extinction will spread northwards. These changes are likely to cause the decline of species whose life cycle is closely dependent upon *L. digitata* and lead to the establishment of new ecosystems with lower ecological and economic values.

11.7.2.3 Projected changes of an introduced species

The jackknife clam *Ensis directus* is a marine bivalve native to the American coasts and introduced in Europe at the end of the 1970s because of ballast waters. The bivalve is a successful invader and colonised the English Channel and the North Sea, from French to Norwegian coasts including the United Kingdom. Although many studies focused on its biology, ecology and colonisation, the extent to which *E. directus* may invade European and Nordic seas remains poorly known. Raybaud and colleagues (*1005*) used the model NPPEN, calibrated on the native area of the mollusc, to evaluate its potential distributional range over European seas at equilibrium (Plate 11.8).

These results showed that *E. directus* is likely to continue to progress its invasion towards the southern coasts of France (Arcachon Bay). We used the latest generation of climate models (five climate models and the most recent set of IPCC scenarios) to assess possible changes in the species geographical range for the end of the century. Compared with the current period, projections revealed that the probability of occurrence should increase from Denmark to France. In contrast to what is generally observed, an extension southwards of the range of *E. directus* is more likely, even if climate continues to warm.

11.7.3 Limitations of ecological niche models

The limitations of ENMs are similar to some limitations enounced previously in the context of the METAL theory. One of the main drawbacks of ENMs is the necessity of determining precisely the limits of the niche. Those limits can be assessed providing that the number of samples is sufficient, this number increasing with the number of niche dimensions. The niche should be determined on the basis of the whole spatial distribution of species. If the niche is truncated, even local projections may be strongly biased (*541*).

11.7.3.1 The realised environment

Another issue that complicates the identification of the niche is related to the concept of realised environment (*1134*). As we saw earlier, the realised environment is the combination of the environmental factors (i.e. the fundamental environment) that are realised in a specific landscape. Indeed, not all combinations of environmental factors are realised in a given region. For example, Figure 11.17 shows that there are some combinations of environmental variables (bathymetry and temperature) that are not observed in the North Atlantic Ocean.

The concept of realised environment is also related to habitat heterogeneity. If the realised environment is too different from the **fundamental environment** (i.e. all potential combinations of environmental factors within limits imposed by a given landscape), the habitat becomes highly heterogeneous and is likely to contain some '**environmental islands**' (a discontinuous habitat), a concept formalised by Ackerly (*609*). In a habitat that contains many environmental islands, adaptive evolution may be more frequent, making projections of distributional ranges from ENMs more difficult.

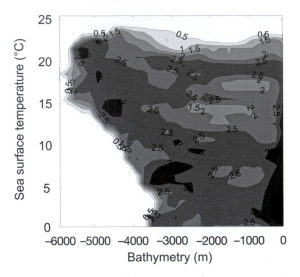

Figure 11.17 Number of geographical cells (1° longitude × 1° latitude) that belongs to a category of bathymetry and sea surface temperature in the North Atlantic Ocean (80.5°W–70.5°E and 35.5°N–80.5°N). The number of geographical cells is expressed as $\log_{10}(x+1)$.

Jackson and Overpeck (*1134*) proposed the concept of **potential niche**, which was defined by Ackerly as located at the junction of the realised niche with the realised environment.

11.7.3.2 Spatial autocorrelation

Spatial autocorrelation may inflate the probabilities inferred from ENMs (*1135*). However, ENMs working on presence-only data are much less subject to this problem than other types of ENMs (e.g. GLMs). The problem is that only a few studies have considered local autocorrelation functions (*224, 1136*). Most corrections are based on global autocorrelation functions, which assume isotropy (e.g. Moran's Index, global semi-variograms), a condition rarely met in the field (*224*). Spatial autocorrelation of environmental data is higher in a continuous environment where forecasting by ENMs is likely to be more accurate, whereas spatial autocorrelation is expected to decrease in an environment containing many environmental islands where projections by ENMs are likely to be less reliable due to high pressures on organisms for adaptive changes.

11.7.3.3 Transient disequilibrium

Another assumption underlying the application of ENMs is that species distribution must be in equilibrium with climate (*541*). Transient disequilibrium may severely alter projection of species spatial distribution. In the context of global warming, transient disequilibria may lead to an underestimation of the impact of warming on species distribution. While equilibrium is likely to be rapid with plankton and nekton, this might not be the case for intertidal, mangrove and coral systems. No-analogue communities may also modify the realised niche and affect projections provided by ENMs.

11.7.3.4 Habitat tracking process

Another point is related to habitat tracking processes, which tend to promote stabilising selection (Chapter 7). Species that can track environmental changes are likely to be less subject to adaptive changes because they tend to remain inside their ecological niche. For such species (mainly motile species such as plankton), ENMs should provide interesting projections of change in distributional range. It is unknown if apparently less motile species can carry out habitat tracking processes. For benthic organisms, habitat tracking can be better achieved if they have larval developmental stages in the plankton. Knowing Thorson's rule (Chapter 3), we can expect that benthic species inhabiting ecosystems towards the equator will be able to more easily track favourable habitat in space and time because more species have larvae that disperse in the plankton. This might be more difficult for species inhabiting systems polewards because benthic organisms tend to exhibit direct development.

11.7.3.5 Ecological projections depend upon the reliability of general circulation models

A large source of uncertainties in ecological projections originates from atmosphere-ocean general circulation models (1137). Goberville and colleagues (1138) modelled the spatial distribution of the sweet chestnut (*Castanea sativa*) for the next decades in Europe using environmental data originating from seven general circulation models (GCMs) and four IPCC (Intergovernmental Panel on Climate Change) RCP (Representation Concentration Pathways) scenarios (Plate 11.9).

They found that uncertainties on projections caused by the different GCMs (different colours in Plate 11.9) were of the same magnitude as uncertainties related to the intensity of the radiative forcing (between 2.6 and 8.5 $W \cdot m^{-2}$; different line styles in Plate 11.9). At the end of this century, uncertainties related to GCMs were even more elevated when radiative forcing (RCP scenarios) was high. This study suggests that if nothing is done to alleviate our greenhouse gases emissions, the resulting extreme global warming is likely to make biological responses to anthropogenic climate change even more uncertain.

Chapter 12

Conclusions

I conclude this book by highlighting three key issues we should resolve rapidly to better understand and anticipate how biodiversity will be altered in the context of global change. These issues are related to: (1) the development of a macroscopic approach; (2) the elaboration of joint global monitoring programmes; and (3) the foundation of a unifying ecological theory.

12.1 A macroscopic approach

Global change necessitates radically altering the view both biologists and ecologists have on monitoring the marine ecosphere. We should rethink the way we currently monitor and investigate the marine ecosphere by placing the ocean under the 'macroscope' (*1139*). As we saw, the macroscope is a theoretical concept, coined by Joel de Rosnay (*1140*), which looks for properties of a system at a macroscopic level. In the same way as a microscope is needed to observe the extremely small or a telescope is needed to observe the extremely big, the macroscope is a theoretical concept to investigate complex systems (*1140*). This concept is often implicitly applied in the new emerging science of macroecology (*1075, 1141*).

The macroscopic view took time to emerge in the scientific community, especially among biologists, who have a long history of reductionist approach. Meteorologists know well the complementary relationships between both microscopic and macroscopic approaches. Indeed, it is impossible to describe the behaviour of each atom or molecule in the atmosphere to forecast the dynamics of weather. Therefore, meteorologists adopt a macroscopic view by describing the average microscopic behaviour that they consider as the elementary macroscopic brick. When a shoal of sardines moves, what is important to anticipate is not the behaviour of the individuals, but the behaviour of the shoal. The macroscopic approach leads to the use of macroscopic indicators or parameters. In meteorology, the macroscopic approach of the state of the atmosphere leads to the use of the pressure and temperature as the mean state of the atmosphere (*53*).

12.2 Global monitoring

At a time when a lot of effort is devoted to better understand how ecosystems work at a microscale (high-frequency but localised monitoring), it is as crucial to build global data sets and to design new joint (global-scale) monitoring strategies to tackle the issue of not only anthropogenic climate change, but global change. It is obvious that neither a single sampling programme nor a single laboratory can do that alone. This effort should be done as part of a consortium in which sampling strategies and protocols are designed to be used in every oceanic ecome and province (*152, 1142*), and new numerical techniques elaborated. We might wish to create a new observatory based on modern techniques, a global observatory of the marine ecosphere that might help us to better investigate issues related to global change (*541*).

A global effort would not only offer data of better quality for calibrating all types of models, but this may also represent an alternative to modelled data. We rely too much on ecosystem models to determine parameters that inform on the good ecological status of marine biodiversity. Global observation networks have long been recognised as essential by meteorologists and physical oceanographers. Especially in meteorology, global-scale monitoring is done on a routine basis and has proved its relevance to: (1) determining weather; (2) the climatic state; and (3) serving as a baseline for both meteorological and atmosphere-ocean general circulation models. Although local studies still tend to dominate, global change compels us to plan and implement broader joint scientific schemes.

The funding needed to monitor the marine environment is often dependent on private support gained as part of local studies on the impact of human activities. Many samples are being collected during a few months, or at best a few years, and the result of such ephemeral sampling is often lost. Even in the terrestrial realm, the continuation of long-term monitoring programmes has been elusive and the most successful time series had to allow for the vicissitudes of funding (*1143*). To the funding problem, it is possible to add the time needed to identify and enumerate biodiversity. These difficulties may explain why biologists have often been reluctant to share their data.

As a result of the lack of available biological data, basic information such as species spatial distribution remains difficult to acquire. Although information on the spatial distribution of some commercially exploited species exists (*361*), for most fish species (e.g. pipefish, sprat, sandeel) it remains complicated to obtain maps of spatial distribution because of the lack of occurrence/abundance data. In these conditions, it is challenging to have a clear idea on the spatial and temporal fluctuations in marine biodiversity and track the effects of global change. Information on species occurrence has only recently started to be gathered in large data sets such as the Ocean Biogeographic Information System (OBIS; www.iobis.org), an initiative developed as part of the Census of Marine Life programme. Such initiatives should clearly be encouraged. Monitoring should become global, and sharing of biological data is an absolute prerequisite to better understand how ecosystems and their biodiversity will respond to global environmental change.

12.3 Towards a unifying ecological theory

It is high time to develop a unifying ecological theory to understand and predict how biodiversity is organised and how it responds to climate change. A unifying theory is outlined in this book and based on Hutchinson's concept of the ecological niche, which integrates genetic, molecular and physiological processes that are difficult to parametrise in classical approaches because of the large number of parameters that need to be estimated and issues associated with oversimplification and assumptions. The theory should make predictions from the species to the biome level and from small to large spatial/temporal scales.

Theories will enable predictions to be carried out, guide future research by defining priorities and identify both weaknesses and strengths in our current scientific knowledge. As we saw in Chapter 11, a large part of the arrangement of biodiversity and its responses to environmental and climate changes is quasi-deterministic, and can therefore be anticipated.

References

1 AL Wegener, *The origin of continents and oceans*. (Dover Publications, New York, 1966).

2 JE Lovelock, L Margulis, *Tellus*.**26**,2(1974).

3 P Westbroek, *Terre! Des menaces globales à l'espoir planétaire*. (Seuil, Paris, 2009).

4 P Ward, *The Medea hypothesis. Is life on Earth ultimately self-destructive?* (Princeton University Press, Princeton, NJ, 2009).

5 PJ Crutzen, in *Earth system science in the Anthropocene: emerging issues and problems*, E Ehlers, T Krafft, eds (Springer, Heidelberg, 2006), pp. 13–18.

6 SL Pimm *et al.*, *Science*.**269**,347(1995).

7 RM May, *Philosophical Transactions of the Royal Society of London B*.**343**,105(1994).

8 C Mora *et al.*, *PLoS Biology*.**9**,e1001127(2011).

9 R May, *PLoS Biology*.**9**,e1001130(2011).

10 MV Lomolino *et al.*, *Biogeography*. (Sinauer Associates, Sunderland, 3rd edn, 2006), pp. 845.

11 S Frontier, D Pichot-Viale, *Ecosystèmes. Structure, fonctionnement, évolution*. (Masson, Paris, 1993).

12 P Herring, *The biology of the deep ocean*. (Oxford University Press, Oxford, 2002), pp. 314.

13 MJ Benton, *Geological Journal*.**36**,211(2001).

14 S Widdicombe, PJ Somerfield, in *Marine biodiversity and ecosystem functioning: frameworks, methodologies and integration*, M Solan *et al.*, eds (Oxford University Press, Oxford, 2012), pp. 1–15.

15 C Linnæus, *Systema naturæ, sive regna tria naturæ systematice proposita per classes, ordines, genera, & species*. (Lugduni Batavorum, Haak, 1735).

16 RH Whittaker, *Science*.**163**,150(1969).

17 CR Woese, GE Fox, *Proceedings of the National Academy of Sciences of the United States of America*.**74**,5088(1977).

18 T Cavalier-Smith, *Biological Reviews*.**73**,203(1998).

19 T Cavalier-Smith, *Proceedings of the Royal Society B: Biological Sciences*.**271**,1251(2004).

20 B Groombridge, MD Jenkins, *Global biodiversity: Earth's living resources in the 21st century*. (World Conservation Press, Cambridge, 2000).

21 ML Reaka-Kudla, in *Biodiversity II*, ML Reaka-Kudla *et al.*, eds (Joseph Henry Press, Washington, DC, 1997), pp. 83–108.

22 P Bouchet, in *The exploration of marine biodiversity: scientific and technological challenges*, CM Duarte, ed. (Fundacion BBVA, Paris, 2006), pp. 1–64.

23 JA Koslow *et al.*, *Marine Ecology Progress Series*.**213**,111(2001).

24 Census of Marine Life, 'Scientific results to support the sustainable use and conservation of marine life: a summary of the census of marine life for decision makers'. (Census of Marine Life, Washington, DC, 2011).

25 CA Suttle, *Nature*.**437**,356(2005).

26 M Breitbart, *Annual Review of Marine Science*.**4**,425(2012).

27 DJ Scanlan, NJ West, *FEMS Microbiology and Ecology*.**40**,1(2006).

28 CS Ting *et al.*, *Trends in Microbiology*.**10**,134(2002).

29 RE Blankenship *et al.*, in *Evolution of primary producecrs in the sea*, PG Falkowski, AH Knoll, eds (Elsevier Academic Press, London, 2007), pp. 21–35.

30 RR Kirby, *Ocean drifters: a secret world beneath the waves*. (Studio Cactus, Singapore, 2010).

31 JM Sieburth *et al.*, *Limnology and Oceanography*.**23**,1256(1978).

32 SY Moon-van der Stray *et al.*, *Limnology and Oceanography*.**45**,98(2000).

33 S Gasparini, J Castel, *Journal of Plankton Research*.**19**,877(1997).

34 RW Sanders *et al.*, *Marine Ecology Progress Series*.**86**,1(1992).

35 CB Field *et al.*, *Science*.**281**,237(1998).

36 A Sournia, *Progress in Oceanography*.**34**,109(1994).

37 AC Pierrot-Bults, in *Marine biodiversity: patterns and processes*, RFG Ormond *et al.*, eds (Cambridge University Press, Cambridge, 1997), pp. 69–93.

38 A Bucklin *et al.*, in *Life in the world's oceans*, AD McIntyre, ed. (Blackwell, Chichester, 2010), pp. 247–265.

39 MV Angel, *Conservation Biology*.**7**,760(1993).

40 C Mora *et al.*, *Proceedings of the Royal Society B: Biological Sciences*.**275**,149(2008).

41 CN Bianchi, C Morri, *Marine Pollution Bulletin*.**40**,367(2000).

42 PVR Snelgrove, *Bioscience*.**49**,129(1999).

43 A Wulff *et al.*, *Estuaries*.**20**,547(1997).

44 A Pascual *et al.*, *Journal of Marine Systems*.**72**,35(2008).

45 NC Hulings, JS Gray, 'A manual for the study of Meiofauna'. (Smithsonian Contributions to Zoology, Washington, DC, 1971).

46 KR Carman, MA Todaro, *Journal of Experimental Marine Biology and Ecology*.**198**,37(1996).

47 DG Raffaelli, CF Mason, *Marine Pollution Bulletin*.**12**,158(1981).

48 J-C Dauvin, *Marine Pollution Bulletin*.**36**,669(1998).

49 RE Grizzle, *Marine Ecology Progress Series*.**18**,191(1984).

50 E Macpherson *et al.*, *Zoosystema*.**27**,709(2005).

51 JS Levinton, *Marine biology: function, biodiversity, ecology*. (Oxford University Press, Oxford, 2009).

52 A Hufty, *Introduction à la climatologie*. (de Boeck, Québec, 2005).

53 S Malardel, *Fondamentaux de météorologie*. (Cépaduès-Editions, Toulouse, 2005).

54 LR Kump *et al.*, *The earth system*. (Pearson Education, Upper Saddle River, NJ, 2004).

55 WGI Intergovernmental Panel on Climate Change, *Climate change 2007: the physical science basis*. (Cambridge University Press, Cambridge, 2007).

56 J-P Vigneau, *Géoclimatologie*. (Ellipses, Paris, 2000).

57 JE Lovelock, *Gaia: medicine for an ailing planet*. (Gaia Books, London, 2005).

58 I Rombouts *et al.*, *Limnology and Oceanography*.**55**,2219(2010).

59 SR Hare, NJ Mantua, *Progress in Oceanography*.**47**,103(2000).

60 G Beaugrand, *Deep-Sea Research II*.**56**,656(2009).

61 PJ Robinson, A Henderson-Sellers, *Contemporary climatology*. (Pearson Education, Edinburgh, 1999), pp. 315.

62 F Rubel, M Kottek, *Meteorologische Zeitschrift*.**19**,135(2010).

63 M Kottek *et al.*, *Meteorologische Zeitschrift*.**15**,259(2006).

64 RR Dickson *et al.*, *Progress in Oceanography*.**20**,103(1988).

65 G Reygondeau, G Beaugrand, *Journal of Plankton Research*.**33**,119(2010).

66 M Tomczak, JS Godfrey, *Regional oceanography: an introduction*. (Daya, Delhi, 2003), vol. 1.

67 GL Pickard, WJ Emery, *Descriptive physical oceanography: an introduction*. (Pergamon, New York, 1990).

68 J Merle, *Océans et climat*. (IRD Editions, Paris, 2006).

69 I Rombouts *et al.*, *Proceedings of the Royal Society B.***276**,3053(2009).

70 C Wunsch, *Science.***298**,1179(2002).

71 S Rahmstorf, in *Encyclopedia of quaternary sciences*, SA Elias, ed. (Elsevier, Amsterdam, 2006), pp. 1–10.

72 A Ganachaud, C Wunsch, *Nature.***408**,453(2000).

73 J Blindheim, in *The Norwegian Sea ecosystem*, HR Skjoldal, ed. (Tapir Academic Press, Trondheim, 2004), pp. 65–96.

74 JR Toggweiler, B Samuels, *Deep-Sea Research.***42**,477(1995).

75 R Seager *et al.*, *Quarterly Journal of the Royal Meteorological Society.***128**,2563(2002).

76 J-F Deconinck, *Paléoclimats. L'enregistrement des variations climatiques.* (Vuibert, Villefranche-de-Rouergue, 2006).

77 DV Hoyt, KH Schatten, *Solar Physics.***181**,491(1998).

78 TJ Crowley, K-Y Kim, *Geophysical Research Letters.***23**,359(1996).

79 GC Reid, *EOS, Transactions of the American Geophysical Union.***74**,23(1993).

80 PAP Moran, *Journal of Animal Ecology.***18**,115(1949).

81 T Oguz *et al.*, *Journal of Marine Systems.***60**,235(2006).

82 CA Guisande *et al.*, *Marine Ecology Progress Series.***269**,297(2004).

83 TM Cronin, *Principles of paleoclimatology.* (Columbia University Press, New York, 1999).

84 KY Vinnikov *et al.*, *Science.***286**,1934(1999).

85 R Bradley, *Science.***288**,1353(2000).

86 TP Barnett *et al.*, *Science.***292**,270(2001).

87 S Levitus *et al.*, *Science.***292**,267(2001).

88 JW Hurrell *et al.*, *Science.***291**,603(2001).

89 RR Dickson, WR Turrell, in *The ocean life of Atlantic salmon. Environmental and biological factors influencing survival*, D Mills, ed. (Fishing News Books, Bodmin, 2000), pp. 92–115.

90 R Quadrelli, JM Wallace, *Journal of Climate.***17**,3728(2004).

91 R Dickson *et al.*, *Progress in Oceanography.***38**,241(1996).

92 JW Hurrell, *Science.***269**,676(1995).

93 MHP Ambaum, BJ Hoskins, *Journal of Climate.***15**,1969(2002).

94 J Cherry *et al.*, *Water Resources Management.***19**,673(2005).

95 DWJ Thompson *et al.*, *Journal of Climate.***13**,1018(2000).

96 JW Hurrell, RR Dickson, in *Marine ecosystems and climate variation*, NC Stenseth *et al.*, eds (Oxford University Press, Oxford, 2004), pp. 15–32.

97 J Cohen, M Barlow, *Journal of Climate.***18**,4498(2005).

98 H Hatun *et al.*, *Science.***309**,1841(2005).

99 SA Josey, R Marsh, *Journal of Geophysical Research.***110**,C05008(2005).

100 JW Hurrell, CK Folland, *Exchanges.***25**,1(2002).

101 J Bjerknes, *Monthly Weather Review.***97**,163(1969).

102 GT Walker, *Memoirs of the Indian Meteorological Department.***24**,75(1923).

103 RW Katz, *Statistical Science.* **17**,97(2002).

104 SGH Philander, *Nature.***302**,295(1983).

105 RS Kovats, *Bulletin of the World Health Organization.***78**,1127(2000).

106 C Wang, PC Fiedler, *Progress in Oceanography.***69**,239(2006).

107 KE Trenberth *et al.*, *Journal of Geophysical Research.***107**,4065(2002).

108 JL Lean, DH Rind, *Geophysical Research Letters.***35**(2008).

109 SGH Philander *et al.*, *El Niño, La Niña and the southern oscillation.* (Academic Press, San Diego, CA, 1989).

110 SI An, *Theoretical and Applied Climatology.***97**,29(2009).

111 K Wyrtki, *Journal of Geophysical Research.***90**,7129(1985).

112 K Ashok *et al.*, *Journal of Geophysical Research.***112**,C11007(2007).

113 S-W Yeh *et al.*, *Nature.***461**,511(2009).

114 Y Kosaka, S-P Xie, *Nature*. **501**,403(2013).

115 SG Philander, in *Encyclopaedia of sciences*, J Steele, ed. (Academic Press, Oxford, 2001), pp. 827–832.

116 SG Philander, A Federov, *Annual Review of Earth and Planetary Sciences*.**31**,579(2003).

117 M Suarez, PS Schopf, *Journal of Atmospheric Sciences*.**45**,3283(1988).

118 NJ Mantua et al., *Bulletin of the American Meteorological Society*.**78**,1069(1997).

119 NJ Mantua, SR Hare, *Journal of Oceanography*.**58**,35(2002).

120 CK Folland et al., *Geophysical Research Letters*.**29**(2002).

121 N Schneider, BD Cornuelle, *Journal of Climate*.**18**,4355(2005).

122 KE Trenberth, JW Hurrell, *Climate Dynamics*.**9**,303(1994).

123 JM Wallace, DS Gutzler, *Monthly Weather Review*.**109**,784(1981).

124 C Zhang, *Reviews of Geophysics*.**43**,RG2003(2005).

125 J Vialard et al., *Geophysical Research Letters*.**35**,L19608(2008).

126 AJ Matthews, *Quarterly Journal of the Royal Meteorological Society*.**134**,439(2008).

127 BS Barrett, LM Leslie, *Monthly Weather Review*.**137**,727(2009).

128 DE Waliser et al., *Bulletin of the Aerican Meteorological Society*.**January**,33(2003).

129 KE Trenberth et al., *Journal of Climate*.**18**,2812(2005).

130 HA Rashid, I Simmonds, *Journal of the Atmospheric Sciences*. **62**,1947(2005).

131 DB Enfield et al., *Geophysical Research Letters*.**28**,2077(2001).

132 ST Gray et al., *Geophysical Research Letters*.**31**,L12205(2004).

133 JR Knight et al., *Geophysical Research Letters*.**32**,L20708(2005).

134 NS Keenlyside et al., *Nature*.**453**,84(2008).

135 KE Trenberth, DJ Shea, *Geophysical Research Letters*.**33**,L12704(2006).

136 M Edwards et al., *PLOS ONE*.**8**,e57212(2013).

137 K Brander, *ICES Journal of Marine Science*.**62**,339(2005).

138 JM Fromentin, B Planque, *Marine Ecology Progress Series*.**134**,111(1996).

139 FP Chavez et al., *Science*.**299**,217(2003).

140 H Hatun et al., *Progress in Oceanography*.**80**,149(2009).

141 JE Cloern et al., *Geophysical Research Letters*.**37**,L21602(2010).

142 FE Clements, VE Shelford, *Bio-ecology*. (John Wiley & Sons, Ann Arbor, MI, 1939).

143 A Longhurst, *Ecological geography of the sea*. (Academic Press, London, 1998).

144 J Schultz, *The ecozones of the world. The ecological divisions of the geosphere*. (Springer, Leipzig, 2005).

145 Bailey, *Ecosystem geography*. (Springer-Verlag, New York, 1996).

146 L Watling et al., *Progress in Oceanography*.**111**,91(2013).

147 MD Spalding et al., *Bioscience*.**57**,573(2007).

148 H Walter, *Walter's vegetation of the earth. The ecological systems of the geo-biosphere*. (Springer, Berlin, 2002).

149 RH Whittaker, *Communities and ecosystems*. (Macmillan, New York, 2nd edn, 1975).

150 WP Dinter, *Biogeography of the OSPAR maritime area*. (Federal Agency for Nature Conservation, Bonn, 2001).

151 BJ Rothschild, TR Osborn, *Journal of Plankton Research*.**10**,465(1988).

152 A Longhurst, *Ecological geography of the sea*. (Elsevier, Amsterdam, 2007).

153 MT Jolly et al., *Limnology and Oceanography*.**54**,2089(2009).

154 G Asrar et al., in *Theory and application of optical remote sensing*, G Asrar, ed. (Wiley, New York, 1989), pp. 252–295.

155 E Goberville et al., *Marine Pollution Bulletin*.**62**,1751(2011).

156 C de Boyer Montégut et al., *Journal of Geophysical Research*.**109**,C12003(2004).

157 E Goberville et al., *Marine Ecology Progress Series*.**408**,129:147(2010).

158 HO Pörtner, *Naturwissenschaften*.**88**,137(2001).

159   HU Sverdrup, *Journal du Conseil Permanent International pour l'Exploitation de la Mer*.**18**, 287(1953).

160   P Helaouët, G Beaugrand, *Marine Ecology Progress Series*.**345**,147(2007).

161   K Schmidt-Nielsen, *Animal physiology: adaptation and environment*. (Cambridge University Press, New York, 4th edn, 1990).

162   JC Orr *et al.*, *Nature*.**437**,681(2005).

163   KJ Kroeker *et al.*, *Ecology Letters*.**13**,1419(2010).

164   MJ Behrenfeld, *Ecology*.**91**,977(2010).

165   JH Martin, *Oceanography*.**4**,52(1991).

166   BW Frost, *Marine Ecology Progress Series*.**39**,49(1987).

167   JH Martin *et al.*, *Nature*.**371**,123(1994).

168   MJ Dunbar, in *Zoogeography and diversity of plankton*, S van der Spoel, AC Pierrot-Bults, eds (Arnold, London/Bunge, Utrecht, 1979), pp. 112–125.

169   G Reygondeau *et al.*, *Journal of Biogeography*.**39**,114(2012).

170   WH Adey, RS Steneck, *Journal of Phycology*.**37**,677(2001).

171   JC Briggs, *Marine zoogeography*. (McGraw-Hill, New York, 1974).

172   JC Briggs, BW Bowen, *Journal of Biogeography*.**39**,12(2012).

173   K Sherman, AM Duda, *Marine Ecology Progress Series*.**190**,271(1999).

174   S van der Spoel, *Progress in Oceanography*.**34**,121(1994).

175   CW Beklemishev, *Okeanologia*.**5**,1059(1961).

176   F Ramade, *Eléments d'écologie. Ecologie fondamentale*. (Ediscience International, Paris, ed. 2, 1994).

177   S Frontier *et al.*, *Ecosystèmes. Structure, fonctionnement et évolution*. (Dunod, Paris, 3rd edn, 2004).

178   G Beaugrand *et al.*, *Marine Ecology Progress Series*.**232**,179(2002).

179   G Reygondeau *et al.*, *Global Biogeochemical Cycles*.**27**,1046(2013).

180   G Beaugrand *et al.*, *Marine Ecology Progress Series*.**424**,175(2011).

181   G Beaugrand *et al.*, *Progress in Oceanography*.**58**,235(2003).

182   G Beaugrand *et al.*, *Marine Ecology Progress Series*.**219**,205(2001).

183   A Merico *et al.*, *Geophysical Research Letters*.**30**,1337(2003).

184   TK Westberry, DA Siegel, *Global Biogeochemical Cycles*.**20**,GB4016(2006).

185   ZI Johnson *et al.*, *Science*.**311**,1737(2006).

186   UNEP, 'Ecosystems and biodiversity in deep waters and high seas'. (UNEP/IUCN, Switzerland, 2006).

187   JC Briggs, *Global biogeography*. (Elsevier, Amsterdam, 1995), vol. 14.

188   NG Vinogradova, *Advances in Marine Biology*.**32**,325(1997).

189   JH Primavera, *Ocean and Coastal Management* **49**,531(2006).

190   KL Smith *et al.*, *Proceedings of the National Academy of Sciences of the United States of America*.**106**,19211(2009).

191   CR Smith *et al.*, *Trends in Ecology and Evolution*.**23**,518(2008).

192   CF Grassle, NJ Maciolek, *The American Naturalist*.**139**,313(1992).

193   CF Grassle *et al.*, in *The Earth in transition*, GM Woodwell, ed. (Cambridge University Press, Cambridge, 1990), pp. 385–393.

194   PJD Lambshead, *Oceanis*.**19**,5(1993).

195   C Bachraty *et al.*, *Deep Sea Research Part I: Oceanographic Research Papers*.**56**,1371(2009).

196   C Yesson *et al.*, *Deep Sea Research Part I: Oceanographic Research Papers*.**58**,442(2011).

197   MJ Kaiser *et al.*, *Marine ecology: processes, systems and impacts*. (Oxford University Press, Oxford, 2005).

198   S Meiri, *Journal of Biogeography*.**30**,331(2003).

199   C Watt *et al.*, *Oikos*.**119**,89(2010).

200 KG Ashton, *Diversity and Distributions*.**7**,289(2001).

201 KG Ashton, CR Feldman, *Evolution*.**57**,1151(2003).

202 TA Mousseau, *Evolution*.**51**,630(1997).

203 SK Berke *et al.*, *Global Ecology and Biogeography*.**22**,173(2013).

204 S Meiri, *Global and Planetary Change*.**20**,203(2011).

205 MA Olalla-Tarraga, *Oikos*.**120**,1441(2011).

206 TM Blackburn *et al.*, *Diversity and Distributions*.**5**,165(1999).

207 G Beaugrand *et al.*, *Proceedings of the National Academy of Sciences of the USA*.**107**,10120(2010).

208 D Atkinson, *Advances in Ecological Research*.**3**,1(1994).

209 J Forster, AG Hirst. *Functional Ecology*.**26**,483(2012).

210 D Atkinson, *Journal of Thermal Biology*.**20**,61(1995).

211 DH Steele, *Internationale Revue der gesamten Hydrobiologie und Hydrographie*.**73**,235(1988).

212 E Edeline *et al.*, *Global Change Biology*.**19**,3062(2013).

213 JS Alho *et al.*, *Journal of Evolutionary Biology*.**24**,59(2010).

214 CJ Bidau, DA Marti, *Ecology, Behavior and Bionomics*.**37**,370(2008).

215 GS Stevens, *The American Naturalist*.**133**,240(1989).

216 E Macpherson *et al.*, in *Marine macroecology*, JD Witman, K Roy, eds (University of Chicago Press, Chicago, IL, 2009), pp. 122–152.

217 G Beaugrand *et al.*, *Global Ecology and Biogeography*.**22**,440(2013).

218 G Thorson, *Biological Reviews of the Cambridge Philosophical Society*.**25**,1(1950).

219 JB Foster, *Nature*.**202**,234(1964).

220 V Millien, *Evolution*.(2011).

221 JRG Turner, BA Hawkins, in *Frontiers of biogeography: new directions in the geography of nature*, MV Lomolino, LR Heaney, eds (Sinauer Associates, Sunderland, MA, 2004), pp. 171–190.

222 FG Stehli *et al.*, *Science*.**164**,947(1969).

223 JL Reid *et al.*, in *Advances in oceanography*, H Charnock, G Deacon, eds (Plenum Press, New York, 1976), pp. 65–130.

224 G Beaugrand, F Ibañez, *Marine Ecology Progress Series*.**232**,197(2002).

225 S Rutherford *et al.*, *Nature*.**400**,749(1999).

226 W Prell *et al.*, *The Brown University Foraminiferal Data Base. IGBP PAGES/World Data Center-A for Paleoclimatology Data Contribution Series # 1999-027.* (NOAA/NGDC Paleoclimatology Program, Boulder, CO, 1999).

227 K Rohde, *Oikos*.**65**,514(1992).

228 K Rohde, *Oikos*.**79**,169(1997).

229 ML Rosenzweig, EA Sandlin, *Oikos*.**80**,172(1997).

230 NJ Gotelli, BJ McGill, *Ecography*.**29**,793(2006).

231 SP Hubbell, *The unified neutral theory of biodiversity and biogeography*. (Princeton University Press, Princeton, NJ, 2001).

232 KJ Gaston, SL Chown, *Functional Ecology*.**19**,1(2005).

233 RK Colwell, GC Hurtt, *The American Naturalist*.**144**,570(1994).

234 RK Colwell, DC Lees, *Trends in Ecology and Evolution*.**15**,70(2000).

235 F Bokma, M Mönkkönen, *Trends in Ecology and Evolution*.**15**,287(2000).

236 GE Hutchinson, *Cold Spring Harbor Symposium Quantitative Biology*.**22**,415(1957).

237 PJ Darlington, *Zoogeography: the geographical distribution of animals*. (Wiley, New York, 1957).

238 RH MacArthur, EO Wilson, *The theory of island biogeography*. (Princeton University Press, Princeton, NJ, 1967).

239 J Terborgh, *The American Naturalist*.**107**,481(1973).

240 ML Rosenzweig, *Species diversity in space and time*. (Cambridge University Press, Cambridge, 1995), pp. 436.

241 K Rohde, *Nonequilibrium ecology*. (Cambridge University Press, Cambridge, 2005).

242 JA Crame, in *Fontiers of biogeography I: new directions in the geography of nature*, MV Lomolino, LR Heaney, eds (Sinauer Associates, Sunderland, MA, 2004), pp. 272–292.

243 DF Sax, *Journal of Biogeography*.**28**,139(2001).

244 AG Fisher, *Evolution*.**14**,64(1960).

245 TD Walker, JW Valentine, *The American Naturalist*.**124**,887(1984).

246 GB West *et al.*, *Science*.**276**,122(1997).

247 JH Brown *et al.*, *Ecology*.**85**,1771(2004).

248 M Kleiber, *Hilgardia*.**6**,315(1932).

249 RH Peters, *The ecological implications of body size*. (Cambridge University Press, Cambridge, 1983).

250 GB West *et al.*, *Science*.**284**,1677(1999).

251 JF Gillooly *et al.*, *Science*.**293**,2248(2001).

252 JF Gillooly, AP Allen, *Ecology*.**88**,1890(2007).

253 PA Allen *et al.*, *Science*.**297**,1545(2002).

254 I Rombouts *et al.*, *Oecologia*.**166**,349(2011).

255 AP Allen *et al.*, in *Scaling biodiversity*, D Storch *et al.*, eds (Cambridge University Press, New York, 2007), pp. 283–299.

256 AP Allen *et al.*, *Proceedings of the National Academy of Sciences of the United States of America*.**103**,9130(2006).

257 AP Allen *et al.*, *Science*.**297**,1545(2002).

258 MP O'Connor *et al.*, *Oikos*.**116**,1058(2007).

259 BA Hawkins *et al.*, *Ecology*.**88**,1898(2007).

260 NR Record *et al.*, *Oecologia*.**170**,289(2012).

261 JH Brown *et al.*, *Science*.**299**,512(2003).

262 B-L Li *et al.*, *Ecology*.**85**,1811(2004).

263 WG Sprules, M Munawar, *Canadian Journal of Fisheries and Aquatic Sciences*.**43**,1789(1986).

264 B Biddanda *et al.*, *Limnology and Oceanography*.**46**,730(2001).

265 WKW Li, *Nature*.**419**,154(2002).

266 JH Connell, E Orias, *The American Naturalist*.**98**,399(1964).

267 RE Ricklefs, GL Miller, *Ecologie*. (De Boeck, Brussels, 1st edn, 2005).

268 BA Hawkins *et al.*, *Ecology*.**84**,3105(2003).

269 DH Wright, *Oikos*.**41**,496(1983).

270 DJ Currie, *The American Naturalist*.**137**,27(1991).

271 DM Gates, *Energy and ecology*. (Sinauer Associates, Sunderland, MA, 1985).

272 J-P Pérez *et al.*, *Physique: une introduction*. (de Boeck, Brussells, 1st edn, 2008).

273 SE Jorgensen, in *Thermodynamics and ecological modelling*, SE Jorgensen, ed. (Lewis, Washington, DC, 2001), pp. 305–347.

274 EM O'Brien *et al.*, *Oikos*.**89**,588(2000).

275 DO Hessen *et al.*, *Ecography*.**30**,749(2007).

276 RJ Huggett, *Fundamentals of biogeography*, J Gerrard, ed., Routledge fundamentals of physical geography series. (Routledge, London, 2nd edn, 2004).

277 WK Purves *et al.*, *Le monde du vivant*. (Flammarion Médecine-Sciences, Paris, 2nd edn, 2000).

278 MJ Behrenfeld, PG Falkowski, *Limnology and Oceanography*.**42**,1(1997).

279 BA Hawkins, EE Porter, *The American Naturalist*.**161**,40(2003).

280 M Rosenzweig, *Science*.**171**,385(1971).

281 S Roy, J Chattopadhyay, *Journal of Biosciences*.**32**,421(2007).

282 MC Rex *et al.*, *The American Naturalist*.**165**,163(2005).

283 G Bonan, *Ecological climatology*. (Cambridge University Press, Cambridge, 2nd edn, 2008).

284 CJ Lolis *et al.*, *International Journal of Climatology*.**24**,1803(2004).

285 L Legendre, S Demers, *Naturaliste Canadien*.**112**,5(1985).

286 G Beaugrand et al., Marine Ecology Progress Series.**200**,93(2000).

287 J-N Druon et al., Marine Ecology Progress Series.**439**,223(2011).

288 G Beaugrand et al., Marine Ecology Progress Series.**204**,299(2000).

289 CR McMahon, G Hays, Global Change Biology.**12**,1330(2006).

290 R Margalef, Oceanologica Acta.**1**,493(1978).

291 M Begon et al., Ecology. From individuals to ecosystems. (Blackwell, Oxford, 4th edn, 2006).

292 DT Tittensor et al., Nature.**466**,1098(2010).

293 R Margalef, The American Naturalist.**97**,357(1963).

294 C Little, JA Kitching, The biology of rocky shores, C Little et al., eds, Biology of habitats. (Oxford Universiy Press, Oxford, 1996).

295 MW Schonbeck, TA Norton, Journal of Experimental Marine Biology and Ecology.**31**,303(1978).

296 KE Conlan et al., Marine Ecology Progress Series.**166**,1(1998).

297 E Reimnitz et al., Geology.**5**,405(1977).

298 RT Paine, The American Naturalist.**100**,65(1966).

299 WF Ruddiman, Science.**164**,1164(1969).

300 GR Bigg, in The ocean life of Atlantic salmon. Environmental and biological factors influencing survival, D Mills, ed. (Fishing News Books, Bodmin, 2000), pp. 137–152.

301 B Hansen, S Osterhus, Progress in Oceanography.**45**,109(2000).

302 LN Gillman, SD Wright, Ecology.**87**,1234(2006).

303 J Gillooly et al., Proceedings of the National Academy of Sciences of the United States of America.**102**,140(2005).

304 SR Palumbi, Journal of Experimental Marine Biology and Ecology.**203**,75(1996).

305 AJ Webster et al., Science.**301**,478(2003).

306 TA Norton, in The ecology of rocky coasts, PG Moore, R Seed, eds (Hodder & Stoughton, London, 1985), pp. 7–21.

307 GC Hays, Nature.**376**,650(1994).

308 GC Hays, Limnology and Oceanography.**40**,1461(1995).

309 GC Hays, Deep-Sea Research I.**43**,1601(1996).

310 JW Dippner et al., Journal of Marine Systems.**25**,23(2000).

311 AJ Underwood, Journal of Experimental Marine Biology and Ecology.**9**,239(1972).

312 JH Costello et al., Proceedings of the National Academy of Sciences of the USA.**87**,1648(1990).

313 R Margalef, Scientia Marina.**61(supplement 1)**,109(1997).

314 C Marrassé et al., Proceedings of the National Academy of Sciences of the USA.**87**,1653(1990).

315 E Saiz, M Alcaraz, Marine Ecology Progress Series.**80**,229(1992).

316 NA Campbell et al., Biology. (Pearson Education, San Francisco, CA, 8th edn, 2008).

317 W Kühnelt, Ecologie générale. (Masson et Compagnie, Paris, 1969).

318 World Conservation Monitoring Centre, Global biodiversity: status of the earth's living ressources. (Chapman & Hall, London, 1992).

319 AS Grutter, Nature.**398**,672(1999).

320 AS Grutter, R Bshary, Animal Behaviour.**68**,583(2004).

321 R Bshary, Journal of Animal Ecology.**72**,169(2003).

322 J Overpeck et al., in Paleoclimate, global change and the future, KD Alverson et al., eds (Springer Verlag, Heidelberg, 2003), pp. 81–111.

323 CH Lineweaver, D Schwartzman, in Origins, J Seckbach, ed. (Kluwer Academic Press, Dordrecht, 2004), pp. 233–248.

324 DL Royer et al., GSA Today.**14**,4(2004).

325 JR Petit et al., Nature.**399**,429(1999).

326 J Jouzel et al., Science **317**,793(2007).

327 A Moberg et al., Nature.**433**,613(2005).

328 ME Mann et al., Science.**326**,1256(2009).

329 JE Overland *et al.*, *Journal of Marine Systems.***79**,305(2010).

330 C Sagan, C Chyba, *Science.***276**,1217(1997).

331 JL Kirschvink, in *The proterozoic biosphere*, JW Schopf, C Klein, eds (Cambridge University Press, Cambridge, 1992), pp. 51–52.

332 PF Hoffman, DP Schrag, *Terra Nova.***14**,129(2002).

333 D Schwartzman, *Life, temperature, and the earth*. (Columbia University Press, New York, 1999).

334 JE Lovelock, *The ages of Gaia: a biography of our living earth*. (W.W. Norton & Company, New York, 1988), pp. 255.

335 D Schwartzman, CH Lineweaver, in *Non-equilibrium thermodynamics and the production of entropy: life, earth and beyond*, A Kleidon, R Lorenz, eds (Springer, Berlin, 2005), pp. 207–217.

336 J Veizer *et al.*, *Nature.***408**,698(2000).

337 DL Royer, *Geochimica and Cosmochimica.***2006**,5665(2006).

338 AL Wegener, *Geologische Rundschau.***3**,276(1912).

339 JT Wilson, *Nature* **211**,676(1966).

340 C Lyell, *Principles in geology, being an attempt to explain the former changes of the earth's surface, by reference to causes now in operation* (John Murray, London, 1830).

341 HD Scher, EE Martin, *Science.***312**,428(2006).

342 A Hallam, in *Evolution from molecules to men*, DS Bendall, ed. (Cambridge University Press, Cambridge, 1983), pp. 367–386.

343 CR Marshall, *Science.***329**,1156(2010).

344 J Alroy, *Science.***329**,1191(2010).

345 MJ Benton, *Science.***268**,52(1995).

346 PR Bown *et al.*, in *Coccolithophores: from molecular processes to global impact*, HR Thierstein, JR Young, eds (Springer Verlag, Heidelberg, 2004), pp. 481–514.

347 PR Bown, *Micropaleontology.***51**,299(2005).

348 M Pagani *et al.*, *Science.***308**,600(2005).

349 P Hallock *et al.*, *Paleogeography, Paleoclimatology, Paleoecology.***83**,49(1991).

350 PJ Mayhew *et al.*, *Proceedings of the Royal Society B.***275**,47(2008).

351 P Wignall, *Elements.***1**,293(2005).

352 GH Denton *et al.*, *Science.***328**,1652(2010).

353 S Rahmstorf, *Nature.***419**,207(2002).

354 J-C Duplessy, *Comptes rendus. Géosciences.***337**,881(2005).

355 A Paul, C Schäfer-Neth, *Paleoceanography.***18**,1058(2003).

356 TM Smith *et al.*, *Journal of Climate.***21**,2283(2008).

357 A Paytan *et al.*, *Science.***274**,1355(1996).

358 JC Herguera, WH Berger. *Geology.***22**,629(1994).

359 JH Martin, *Paleoceanography.***5**,1(1990).

360 PG Falkowski, *Scientific American.***287**,54(2002).

361 GR Bigg *et al.*, *Proceedings of the Royal Society London B.***275**,163(2008).

362 HP Comes, JW Kadereit, *Trends in Plant Science.***3**,432–438(1998).

363 JT Weir, D Schluter, *Proceedings of the Royal Society B.***271**,1881(2004).

364 AD Barnosky, BP Kraatz, *BioScience.***57**,523(2007).

365 AD Barnosky, *Journal of Mammalian Evolution.***12**,247(2005).

366 KJ Willis, KJ Niklas, *Philosophical Transactions of the Royal Society B.***359**,159(2004).

367 GR Coope, *Philosophical Transactions of the Royal Society B.***359**,209(2004).

368 JC Avise *et al.*, *Proceedings of the Royal Society B.***265**,1707(1998).

369 NK Johnson, C Cicero, *Evolution.***58**,1122(2004).

370 JP Severinghaus, *Nature.***457**,1093(2009).

371 J-C Duplessy *et al.* Comptes rendus. Géosciences.**337**,888(2005).

372 RB Alley, *Quaternary Science Reviews.***19**,213(2000).

373 W Dansgaard *et al.*, *Nature*.**339**,532(1989).

374 W Dansgaard, H Oeschger, in *The environmental records in glaciers and ice sheets*, H Oeschger, CC Langway Jr, eds (John Wiley & Sons, Chichester, 1989), pp. 287–318.

375 J-C Duplessy, *Quand l'océan se fâche. Histoire naturelle du climat.* (Editions Odile Jacob, Paris, 1996), pp. 277.

376 M Eliot *et al.*, *Quaternary Science Reviews*.**21**,1153(2002).

377 D Paillard, *Nature*.**409**,147(2001).

378 JCH Chiang, CM Bitz, *Climate Dynamics*.**25**,477(2005).

379 WS Broecker *et al.*, *Paleoceanography*.**3**,1(1988).

380 RJ Stouffer *et al.*, *Journal of Climate*.**19**,1365(2006).

381 I Eisenman *et al.*, *Paleoceanography*.**24**,PA4209(2009).

382 TC Jennerjahn *et al.*, *Science*.**306**,2236(2004).

383 J-A Flores *et al.*, *Marine Micropaleontology*.**76**,53(2010).

384 BM Vinther *et al.*, *Nature*.**461**,385(2009).

385 MW Kerwin *et al.*, *Paleoceanography*.**14**,200(1999).

386 N Koç *et al.*, *Quaternary Science Reviews*.**12**,115(1993).

387 Q Schiermeier, *Nature*.**433**,562(2005).

388 V Trouet *et al.*, *Science*.**324**,78(2009).

389 E Witon *et al.*, *Palaeogeography, Palaeoclimatology, Palaeoecology*.**239**,487(2006).

390 EA Hadly, *Palaeogeography, Palaeoclimatology, Palaeoecology*.**149**,389(1999).

391 EA Hadly *et al.*, *PLOS Biology*.**2**,1600(2004).

392 A Incarbonara *et al.*, *Climate of the Past*.**6**,795(2010).

393 S Silenzi *et al.*, *Global and Planetary Change*.**40**,105(2004).

394 DM Raup, *Proceedings of the National Academy of Sciences of the United States of America*.**91**,6758(1994).

395 RM May *et al.*, in *Extinction rates*, JH Lawton, RM May, eds (Oxford University Press, Oxford, 1995), pp. 1–24.

396 SM Stanley, *Microevolution.* (W.H. Freeman & Company, San Francisco, CA, 1979).

397 KJ Gaston, JI Spicer, *Biodiversity: an introduction.* (Blackwell, Hong Kong, 2004).

398 D Jablonski, *Science*.**253**,754(1991).

399 SA Bowring *et al.*, *Science*.**280**,1039(1998).

400 PJ Brenchley *et al.*, *Geological Journal*.**36**,329(2001).

401 S Finnegan *et al.*, *Science*.**331**,903(2011).

402 CMO Rasmussen, DAT Harper, *Palaeogeography, Palaeoclimatology, Palaeoecology*.**311**,48(2011).

403 DJ McLaren, in *Geological implications of large asteroids and comets on the Earth*, LT Silver, PH Schultz, eds (Geological Society of America Special Paper, New York, 1982), pp. 477–484.

404 P Copper, *Acta Palaeontologica Polonica*.**43**,137(1998).

405 SI Kaiser *et al.*, *Palaeogeography, Palaeoclimatology, Palaeoecology*.**240**,146(2006).

406 CW Pitrat, *Palaeogeography, Palaeoclimatology, Palaeoecology*.**8**,49(1970).

407 LH Tanner *et al.*, *Earth-Science Reviews*.**2004**,103(2004).

408 MJ Benton, RJ Twitchett, *Trends in Ecology and Evolution*.**18**,358(2003).

409 DM Raup, *Science*.**206**,217(1979).

410 A Saunders, M Reichow, *Chinese Science Bulletin*.**54**,20(2009).

411 MJ Benton, *When life nearly died.* (Thames & Hudson, London, 2003).

412 PB Wignall, *Earth-Science Reviews*.**53**,1(2001).

413 PB Wignall, *Geobiology*.**5**,303(2007).

414 AP Jones *et al.*, *Earth and Planetary Science Letters*.**202**,551(2002).

415 G Ryskin, *Geology*.**31**,741(2003).

416 ES Krull *et al.*, *New Zealand Journal of Geology and Geophysics*.**43**,21(2000).

417 JT Kiehl, CA Shields, *Geology*.**33**,757(2005).

418 PB Wignall, RJ Twitchett, *Science*.**272**,1155(1996).

419 RM Hotinski *et al.*, *Geology*.**29**,7(2001).

420 JEN Veron, *Corals in space and time: the biogeography and evolution of the Scleractinia*. (Cornell University Press, Ithaca, NY, 1995).

421 LW Alvarez *et al.*, *Science*.**208**,1095(1980).

422 MJ Jeffries, *Biodiversity and conservation*. (Routledge, London, 1997).

423 FS Russel, *Journal du Conseil. Conseil International pour l'Exploration de la Mer*.**14**,171(1939).

424 AC Hardy, *Hull Bulletins of Marine Ecology*.**1**,1(1939).

425 JM Colebrook, *Journal du Conseil. Conseil International pour l'Exploration de la Mer*.**42**,179(1985).

426 P Reid, M Edwards, in *Encyclopaedia of sciences*, J Steele, ed. (Academic Press, Oxford, 2001), pp. 2194–2200.

427 G Beaugrand *et al.*, *Science*.**296**,1692(2002).

428 CRC Sheppard *et al.*, *The biology of coral reefs*. (Oxford University Press, Oxford, 2009), pp. 339.

429 TA Gardner *et al.*, *Ecology*.**86**,174(2005).

430 DR Bellwood *et al.*, *Nature*.**429**,827(2004).

431 D Sarmiento *et al.*, *LTER Network News*.**22**,11(2009).

432 RH Emson *et al.*, *Marine Biology*.**140**,723(2002).

433 SB Goldenberg *et al.*, *Science*.**293**,474(2001).

434 JR Knight *et al.*, *Geophysical Research Letters*.**33**,1(2006).

435 C Little *et al.*, *The biology of rocky shores*. (Oxford University Press, Oxford, 2009).

436 BF Zaitchik *et al.*, *International Journal of Climatology*.**26**,743(2006).

437 M Houdart, 'Impacts des conditions climatiques de l'été 2003 sur la faune et la flore marines'. (IFREMER, Paris, 2004).

438 F Gomez, S Souissi, *Journal of Sea Research*.**58**,283(2007).

439 KH Mann, JRN Lazier, *Dynamics of marine ecosystems: biological–physical interactions in the oceans*. (Blackwell Science, Oxford, 2nd edn, 1996).

440 HH Gran, *Rapports et Procès-Verbaux des réunions du Conseil International pour l'Exploration de la Mer*.**75**,37(1931).

441 HW Ducklow, RP Harris, *Deep-Sea Research II*.**40**,1(1993).

442 K Lochte *et al.*, *Deep-Sea Research II*.**40**,91(1993).

443 I Joint *et al.*, *Deep-Sea Research II*.**40**,423(1993).

444 CS Yentsch, *Journal of Plankton Research*.**12**,717(1990).

445 HJ Isemer, L Hasse, *The bunker climate atlas of the North Atlantic Ocean. Volume 2: air-sea interaction*. (Springer Verlag, Berlin, 1987).

446 J Backhaus *et al.*, *Marine Ecology Progress Series*.**189**,77(1999).

447 T Dale *et al.*, *Sarsia*.**84**,1(1999).

448 HW Harvey, *The chemistry and fertility of sea waters*. (Cambridge University Press, Cambridge, 1955).

449 TJ Hart, in *Discovery reports*. (Cambridge University Press, Cambridge, 1942), vol. 21, pp. 263–348.

450 VH Strass, JD Woods, in *Toward a theory on biological-physical ineractions in the world ocean*, BJ Rothschild, ed. (Kluwer Academic Press, Dordrecht, 1988), pp. 113–136.

451 MJ Behrenfeld *et al.*, in *State of the climate in 2008*, TC Peterson, MO Baringer, eds (Bulletin of the American Meteorological Society, 2009), pp. 568–573.

452 EM Hurlburt, *Journal of Plankton Research*.**12**,1(1990).

453 HJ Nanninga, T Tyrrell, *Marine Ecology Progress Series*.**136**,195(1996).

454 WM Balch *et al.*, *Continental Shelf Research*.**12**,1353(1992).

455 NC Stenseth *et al.*, *Science*.**297**,1292(2002).

456 E Post, NC Stenseth, *Ecology*.**80**,1322(1999).

457   RR Dickson et al., Journal of Plankton Research.**10**,151(1988).

458   PC Reid, B Planque, in The ocean life of Atlantic salmon. Environmental and biological factors influencing survival, D Mills, ed. (Fishing News Books, Bodmin, 2000), pp. 153–169.

459   J Alheit, E Hagen, Fisheries Oceanography.**6**,130(1997).

460   J Marshall et al., International Journal of Climatology.**21**,1863(2001).

461   G Beaugrand et al., Biofuturs.**270**,26(2006).

462   JA Stephens et al., Journal of Plankton Research.**20**,943(1998).

463   MR Heath et al., Fisheries Oceanography.**8(supplement 1)**,163(1999).

464   G Beaugrand, Marine Ecology Progress Series.**445**,293(2012).

465   B Planque, PC Reid, Journal of the Marine Biological Association of the UK.**78**,1015(1998).

466   FJ Mueter et al., Progress in Oceanography.**81**,93(2009).

467   K Drinkwater et al., in Marine ecosystems and global change, M Barange et al., eds (Oxford University Press, Oxford, 2010), pp. 11–39.

468   GL Hunt et al., Deep-Sea Research II.**49**,5821(2002).

469   FJ Mueter et al., Canadian Journal of Fisheries and Aquatic Sciences.**64**,911(2007).

470   E Marris, Nature.**432**,4(2004).

471   A Atkinson et al., Nature.**432**,100(2004).

472   C Wilkinson, Status of the coral reefs of the world 2004. (Australian Institute of Marine Sciences, Townsville, 2004).

473   P Lehodey et al., Nature.**389**,715(1997).

474   C Barbraud, H Weimerskirch, Nature.**411**,183(2001).

475   WB White, RG Peterson, Nature.**380**,699(1996).

476   M Edwards et al., ICES Journal of Marine Science.**58**,39(2001).

477   M Edwards et al., Marine Ecology Progress Series.**239**,1(2002).

478   GA Becker, M Pauly, ICES Journal of Marine Science.**53**,887(1996).

479   G Beaugrand, Marine Ecology Progress Series.**269**,69(2004).

480   WR Turrell et al., ICES Journal of Marine Science.**53**,899(1996).

481   G Becker, H Dooley, Ocean Challenge.**6**,52(1995).

482   MR Heath et al., Nature.**352**,116(1991).

483   PC Reid et al. Fisheries Research.**50**,163(2001).

484   JA Lindley et al., Journal of the Marine Biological Association of the UK.**70**,679(1990).

485   JA Lindley et al., Journal of Experimental Marine Biology and Ecology.**173**,47(1993).

486   A Lindquist, FAO Fisheries Report.**291**,813(1983).

487   A Bakun, SJ Weeks, Fish and Fisheries.**7**,316(2006).

488   A Bakun, P Cury, Ecology Letters.**2**,349(1999).

489   GF Gause, The struggle for coexistence. (Williams & Wilkins, Baltimore, MD, 1934).

490   FS Russell et al., Nature.**234**,468(1971).

491   AJ Southward et al., Advances in Marine Biology.**47**,1(2005).

492   AJ Southward, Nature.**249**,180(1974).

493   M Scheffer et al., Nature.**413**,591(2001).

494   M Scheffer, Critical transitions in nature and society. (Princeton University Press, Princeton, NJ, 2009).

495   P Cury et al., in Responsible fisheries in the marine ecosystem, M Sinclair, G Valdimarsson, eds (FAO/CAB International, Rome, 2003), pp. 103–123.

496   B deYoung et al., Trends in Ecology and Evolution.**23**,402(2008).

497   G Beaugrand et al., Ecology Letters.**11**,1157(2008).

498   CH Greene, AJ Pershing, Science.**315**,1084(2007).

499   G Beaugrand, Progress in Oceanography.**60**,245(2004).

500   JJ Polovina et al., Deep-Sea Research.**42**,1701(1995).

501   LW Botsford et al., Science.**277**,509(1997).

502   EL Venrick et al., Science.**238**,70(1987).

503 PJ Anderson, JF Piatt, *Marine Ecology Progress Series*.**189**,117(1999).

504 B deYoung *et al.*, *Progress in Oceanography*.**60**,143(2004).

505 SR Hare *et al.*, *Fisheries*.**21**,6(1999).

506 C de Duve, *Vital dust: life as a cosmic imperative*. (Basic Books, New York, 1995).

507 G Wächtershäuser, in *Thermophiles: the keys to molecular evolution and the origin of life?*, J Wiegel, M Adams, eds (Taylor & Francis, London, 1998), pp. 47–57.

508 D Schwartzman, CH Lineweaver, in *Bioastronomy 2002: life among the stars ASP conf series*, R Norris, F Stootman, eds (2003), vol. 213, pp. 355–358.

509 D Schwartzman *et al.*, *BioScience* **43**,390(1993).

510 JC Michael, N Mrosovsky, *Canadian Journal of Zoology*.**82**,1302(2004).

511 FH Pough, *The American Naturalist*.**115**,92(1980).

512 Y Turquier, A Toulmond, *L'organisme en équilibre avec son milieu*. (Doin éditeurs, Paris, 1994).

513 CL Scott *et al.*, *Polar Biology*.**23**,510(2000).

514 K Schmidt-Nielsen, *Scaling: why is animal size so important?* (Cambridge University Press, Cambridge 1995).

515 MS Blumberg, *Body heat: temperature and life on earth*. (Harvard University Press, Cambridge, MA, 2002).

516 JM Napp *et al.*, *Journal of Plankton Research*.**21**,1633(1999).

517 JG Kingsolver, RB Huey, *Evolutionary Ecology Research*.**10**,251(2008).

518 D Atkinson, RM Sibly, *Trends in Ecology and Evolution*.**12**,235(1997).

519 MJ Angilletta *et al.*, *Journal of Thermal Biology*.**27**,249(2002).

520 VE Shelford, *Ecology*.**12**,455(1931).

521 M Frederich, HO Pörtner, *American Journal of Physiology*.**279**,51531(2000).

522 JL Johansen, GP Jones, *Global Change Biology*.**17**,2971(2011).

523 DA Ratkowsky *et al.*, *Journal of Bacteriology*.**149**,1(1982).

524 LS Peck *et al.*, *Functional Ecology*.**18**,625(2004).

525 DA Ratkowsky *et al.*, *Journal of Bacteriology*.**154**,1222(1983).

526 ME Huntley, MDG Lopez, *The American Naturalist*.**140**,201(1992).

527 JR Hazel, *Annual Reviews of Physiology*.**57**,19(1995).

528 M Sinensky, *Proceedings of the National Academy of Sciences of the USA*.**71**,522(1974).

529 JR Hazel, SR Landrey, *American Journal of Physiology*.**255**,622(1988).

530 HO Pörtner, R Knust, *Science*.**315**,95(2007).

531 H Chamayou, *Elements de bioclimatologie*. (Agence de coopération culturelle et technique, Paris, 1994).

532 G Beaugrand *et al.*, *Progress in Oceanography*.**72**,259(2007).

533 RR Kirby *et al.*, *Marine Ecology Progress Series*.**330**,31(2007).

534 J Chevalier *et al.*, *Annales des Sciences Naturelles*.**20**,147(1999).

535 DO Conover, BE Kynard, *Science*.**213**,577(1981).

536 N Ospina-Alvarez, F Piferrer, *PLOS ONE*.**3**,1(2008).

537 HO Pörtner, AP Farrell, *Science*.**322**,690(2008).

538 P Helaouët, G Beaugrand, *Ecosystems*.**12**,1235(2009).

539 RC Newell, *Adaptation to environment*. (Butterworth, London, 1976).

540 R Kirby *et al.*, *Limnology and Oceanography*.**53**,1805(2008).

541 G Beaugrand *et al.*, *Progress in Oceanography*.**111**,75(2013).

542 N Karnovsky *et al.*, *Marine Ecology Progress Series*.**415**,283(2010).

543 J Fort *et al.*, *PLOS ONE*.**7**,e41194(2012).

544 GN Somero *et al.*, in *Animals and temperature: phenotypic and evolutionary adaptation*, IA Johnston, AF Bennett, eds (Cambridge University Press, Cambridge 2010), pp. 53–78.

545 G Beaugrand, RR Kirby, *Climate Research*.**41**,15(2010).

546 I Mori, Y Ohshima, *Nature*.**376**,344(1995).

547 SH Jónasdóttir, M Koski, *Journal of Plankton Research*.**33**,85(2011).

548 V Di Santo, WA Bennett, *Journal of Fish Biology*.**78**,195(2011).

549 HL Wallman, WA Bennett, *Environmental Biology of Fishes*.**75**,259(2006).

550 DW Sims *et al.*, *Journal of Animal Ecology*.**75**,176(2006).

551 AI Perry *et al.*, *Science*.**308**,1912(2005).

552 Y Cherel, GL Kooyman, *Marine Biology*.**130**,335(1998).

553 C Gilbert *et al.*, *The Journal of Experimental Biology*.**211**,1(2008).

554 Y Le Maho *et al.*, *American Journal of Physiology*.**231**,913(1976).

555 JP Robin *et al.*, *American Journal of Physiology*.**254**,61(1988).

556 C Gilbert *et al.*, *Physiology and Behavior*.**88**,479(2006).

557 A Ancel *et al.*, *Nature*.**385**,304(1997).

558 NJ Aebischer *et al.*, *Nature*.**347**,753(1990).

559 RR Kirby, G Beaugrand, *Proceedings of the Royal Society London B*.**276**,3053(2009).

560 M Daufresne *et al.*, *Proceedings of the National Academy of Sciences of the USA*.**106**,12788(2009).

561 G Woodward *et al.*, *Trends in Ecology and Evolution*.**20**,402(2005).

562 L Legendre, J Michaud, *Marine Ecology Progress Series*.**164**,1(1998).

563 G Beaugrand, Ph.D., 'North Atlantic pelagic biodiversity and hyro-meteorological variability', Université Pierre et Marie Curie (2001).

564 G Hardin, *Science*.**162**,1243(1968).

565 BS Halpern *et al.*, *Science*.**319**,948(2008).

566 W Mauser, in *Earth system science in the Anthropocene: emerging issues and problems*, E Ehlers, T Krafft, eds (Springer, Heidelberg, 2006), pp. 3–4.

567 R Mackay, *The atlas of endangered species*. (Earthscan, London, 2005).

568 PR Ehrlich, *The population bomb*. (Ballantine Books, New York, 1968), pp. 201.

569 PR Ehrlich, JP Holdren, *Science*.**171**,1212(1971).

570 PM Vitousek *et al.*, *Science*.**277**,494(1997).

571 W Steffen *et al.*, *Global change and the earth system: a planet under pressure*. (Springer Verlag, Heidelberg, 2004).

572 JR McNeill, *Something under the sun*. (WH Norton & Company, New York, 2000).

573 S Arrhenius, *Philosophical Magazine and Journal of Science*.**5**,237(1896).

574 VM Kattsov *et al.*, in *Arctic climate impact assessment*. (Cambridge University Press, Cambridge, 2005), pp. 99–182.

575 R Revelle, HE Suess, *Tellus*.**9**,18(1957).

576 D Lüthi *et al.*, *Nature*.**453**,379(2008).

577 M Pagani *et al.*, *Nature Geoscience*.**3**,27(2009).

578 MR Allen *et al.*, *Surveys in Geophysics*.**27**,491(2006).

579 A Berger, MF Loutre, *Science*.**297**,1287(2002).

580 S Rahmstorf *et al.*, *Science*.**316**,709(2007).

581 ME Mann *et al.*, *Nature*.**392**,779(1998).

582 TP Barnett *et al.*, *Science*.**309**,284(2005).

583 S Levitus *et al.*, *Geophysical Research Letters*.**32**,LO2604(2005).

584 JI Antonov *et al.*, *Geophysical Research Letters*.**32**,L12602(2005).

585 S Levitus *et al.*, *Science*.**287**,2225(2000).

586 H Beltrami *et al.*, *Geophysical Research Letters*.**29**,10.1029/2001GL014310(2002).

587 Intergovernmental Panel on Climate Change, *Climate change 2007: impacts, adaptation and vulnerability*. (Cambridge University Press, Cambridge, 2007).

588 RE Zeebe *et al.*, *Nature Geoscience*.**2**,576(2009).

589 JC Zachos *et al.*, *Science*.**302**,1551(2003).

590 RH Moss *et al.*, *Nature*.**463**,747(2010).

591 N Ray, JM Adams, *Internet Archaeology*.**11**,1(2001).

592 GN Somero, *The Journal of Experimental Biology*.**213**,912(2010).

593 CJ Crossland *et al.*, *Coral Reefs*.**10**,55(1991).

594 RD Gates *et al.*, *The Biological Bulletin*.**182**,324(1992).

595 BG Hatcher, *Trends in Ecology and Evolution*.**3**,106(1988).

596 O Hoegh-Guldberg *et al.*, *Science*.**318**,1737(2007).

597 SD Donner *et al.*, *Global Change Biology*.**11**,2251(2005).

598 E Banin *et al.*, *Water, Air and Soil Pollution*.**123**,337(2000).

599 O Hoegh-Guldberg, *Marine Freshwater Research*.**50**,839(1999).

600 TJ Goreau *et al.*, *Conservation Biology*.**14**,1(2000).

601 PW Glynn, *Annual Review of Ecology and Systematics*.**19**,309(1988).

602 TP Hughes *et al.*, *Science*.**301**,929(2003).

603 AC Baker *et al.*, *Nature*.**430**,741(2004).

604 R Rowan, *Nature*.**430**,742(2004).

605 E Rosenberg, Y Ben-Haim, *Environmental Microbiology*.**4**,318(2002).

606 A Kushmaro *et al.*, *Nature*.**380**,396(1996).

607 Y Ben-Haim, E Rosenberg, *Marine Biology*.**141**,47(2002).

608 JK Reaser *et al.*, *Conservation Biology*.**14**,1500(2000).

609 DD Ackerly, *International Journal of Plant Sciences*.**164**,S165(2003).

610 S Via *et al.*, *Trends in Ecology and Evolution*.**10**,212(1995).

611 CK Ghalambor *et al.*, *Functional Ecology*.**21**,394(2007).

612 S Bearshop *et al.*, *Science*.**310**,502(2005).

613 DH Nussey *et al.*, *Science*.**310**,304(2005).

614 WE Bradshaw, CM Holzapfel, *Science*.**312**,1477(2006).

615 AD Bradshaw *et al.*, *Evolution*.**58**,1748(2004).

616 AD Bradshaw, CM Holzapfel, *The Annual Review of Physiology*.**72**,147(2010).

617 E Mayr, in *Evolution as a process*, J Huxleya, ed. (Allen & Unwin, London, 1954), vol. 167, pp. 157–180.

618 FI Woodward, *Philosophical Transactions of the Royal Society B*.**326**,585(1990).

619 PA Umina *et al.*, *Science*.**308**,691(2005).

620 SJ Franks *et al.*, *Proceedings of the National Academy of Sciences of the USA*.**104**,1278(2007).

621 B Huntley, *Heredity*.**98**,247(2007).

622 SR Palumbi, *Trends in Ecology and Evolution*.**7**,114(1992).

623 SP Holmes *et al.*, *Marine Ecology Progress Series*.**268**,131(2004).

624 L Levin, T Bridges, in *Ecology of marine invertebrate larvae*, L McEdward, ed. (CRC Press, Boca Raton, FL, 1995), pp. 1–48.

625 BP Kinlan, SD Gaines, *Ecology*.**84**,2007(2003).

626 SD Gaines *et al.*, *Oceanography*.**20**,90(2007).

627 SD Ayata, Ph.D., 'Relative importance of hydro-climatic factors and life history traits on the larval dispersal and connectivity at different spatial scales (English Channel, Bay of Biscay)', Université Pierre et Marie Curie (2010).

628 SR Palumbi, AC Wilson, *Evolution*.**44**,403(1990).

629 D Jablonski, *Bulletin of Marine Science*.**39**,565(1986).

630 MN Dawson, WM Hamner, *Proceedings of the National Academy of Sciences of the USA*.**102**, 9235(2005).

631 J Mauchline, in *The biology of calanoid copepods*, JHS Blaxter *et al.*, eds (Academic Press, San Diego, CA, 1998).

632 A Bucklin *et al.*, *Journal of Plankton Research*.**22**,1237(2000).

633 CE Lee, *Evolution*.**53**,1423(1999).

634 J Randall, *Zoological Studies*.**37**,227(1998).

635 SD Ayata *et al.*, *Continental Shelf Research*.**29**,1605(2009).

636 SE Swearer *et al.*, *Nature*.**402**,799(1999).

637 GP Jones *et al.*, *Nature*.**402**,802(1999).

638 HA Lessios *et al.*, *Evolution*.**55**,955(2001).

639 MS Taylor, ME Hellberg, *Science*.**299**,107(2003).

640 SR Palumbi, RR Warner, *Science*.**299**,51(2003).

641 LA Rocha *et al.*, *Proceedings of the Royal Society B*.**272**,573(2005).

642 AD Bradshaw, T McNeilly, *Annals of Botany*.**67**,5(1991).

643 TN Kristensen *et al.*, *Proceedings of the National Academy of Sciences of the USA*.**105**,216(2008).

644 MD Crisp *et al.*, *Nature*.**458**,754(2009).

645 JR Bridle, TH Vines, *Trends in Ecology and Evolution*.**22**,140(2007).

646 AA Hoffmann, PA Parsons, *Extreme environmental change and evolution*. (Cambridge University Press, Cambridge, 1997).

647 AT Peterson *et al.*, in *Climate change and biodiversity*, TE Lovejoy, L Hannah, eds (Yale University Press, London, 2005), pp. 211–228.

648 V Kellermann *et al.*, *Science*.**325**,1244(2009).

649 JH Brown, *The American Naturalist*.**124**,255(1984).

650 C Chapperon, L Seuront, *Functional Ecology*.**25**,1040(2011).

651 HQP Crick *et al.*, *Nature*.**388**,526(1997).

652 RB Myneni *et al.*, *Nature*.**386**,698(1997).

653 PC Reid *et al.*, *Nature*.**391**,546(1998).

654 A Menzel, P Fabian, *Nature*.**397**,659(1999).

655 AH Fitter, RSR Fitter. *Science*.**296**,1689(2002).

656 TL Root, L Hughes, in *Climate change and biodiversity*, TE Lovejoy, L Hannah, eds (Yale University Press, New Haven, CT, 2005), pp. 61–74.

657 M Edwards, AJ Richardson, *Nature*.**430**,881(2004).

658 DL Mackas *et al.*, *Progress in Oceanography*.**97**,31(2012).

659 C Parmesan, J Matthews, in *Principles of concervation biology*, MJ Groom *et al.*, eds (Sinauer Associates, Sunderland, MA, 2006), pp. 333–360.

660 JM Colebrook, GA Robinson, *Bulletin of Marine Ecology*.**6**,123(1965).

661 JM Colebrook, *Marine Biology*.**51**,23(1979).

662 CD Thomas, JJ Lennon, *Nature*.**399**,213(1999).

663 C Parmesan, G Yohe, *Nature*.**421**,37(2003).

664 CD Thomas, *Diversity and Distributions*.**16**,488(2010).

665 R Hickling *et al.*, *Global Change Biology*.**12**,450(2006).

666 A Merico *et al.*, *Deep Sea Research I*.**51**,1803(2004).

667 G Beaugrand *et al.*, *Nature Climate Change*.**3**,263(2013).

668 DB Botkin *et al.*, *BioScience*.**57**,227(2007).

669 CD Thomas *et al.*, *Nature*.**427**,145(2004).

670 P Chevaldonné, C Lejeusne, *Ecology Letters*.**6**,371(2003).

671 TP Dawson *et al.*, *Science*.**332**,53(2011).

672 LS Peck *et al.*, *Polar Biology*.**32**,399(2009).

673 LS Peck, *Antarctic Science*.**17**,497(2005).

674 KDT Nguyen *et al.*, *PLOS ONE*.**6**,e29340(2011).

675 TJ Goreau, RL Hayes, *Ambio*.**23**,176(1994).

676 G Beaugrand *et al.*, *Nature*.**426**,661(2003).

677 C-H Hsieh *et al.*, *Nature*.**443**,859(2006).

678 LA Gosselin, PY Qian, *Marine Ecology Progress Series*.**146**,265(1997).

679 J Hjort, *Rapports et procés-verbaux des réunions – Conseil International pour l'Exploration de la Mer*.**20**,1(1914).

680 T Platt *et al.*, *Nature*.**423**,398(2003).

681 ED Houde, *American Fisheries Society Symposium*.**2**,17(1987).

682 JI Spicer, KJ Gaston, *Physiological diversity and its ecological implications*. (Blackwell Science, Oxford, 1999).

683 JW Horwood, RS Millner, *Journal of the Marine Biological Association of the United Kingdom.* **78**,345(1998).

684 G Beaugrand *et al.*, *Global Change Biology.***15**,1790(2009).

685 K Brander *et al.*, *ICES Marine Science Symposium.***219**,261(2003).

686 J-C Quero *et al.*, *Oceanologica Acta.***21**,345(1998).

687 W Thuiller, *Nature.***448**,550(2007).

688 B Huntley, *Annals of Botany.***67**,15(1991).

689 RG Pearson, *Trends in Ecology and Evolution.***21**,111(2006).

690 C Parmesan *et al.*, *Nature.***399**,579(1999).

691 C Parmesan, in *Climate change and biodiversity*, TE Lovejoy, L Hannah, eds (Yale University Press, New Haven, CT, 2005), pp. 41–55.

692 PC Reid *et al.*, *Marine Ecology Progress Series.***215**,283(2001).

693 M McGlone, JS Clark, in *Climate change and biodiversity*, TE Lovejoy, L Hannah, eds (Yale University Press, New Haven, CT, 2005), pp. 157–159.

694 G Beaugrand, RR Kirby, *Global Change Biology.***16**,1268(2010).

695 C Luczak *et al.*, *Biology Letters.***7**,702(2011).

696 AJ Richardson, DS Schoeman, *Science.***305**,1609(2004).

697 DH Cushing, *Advances in Marine Biology.***26**,249(1990).

698 JM Durant *et al.*, *Ecological Letters.***8**,952(2005).

699 JM Durant *et al.*, *Marine Ecology Progress Series.***474**,43(2013).

700 MR Heath, RG Lough, *Fisheries Oceanography.***16**,169(2007).

701 KT Frank *et al.*, *Science.***308**,1621(2005).

702 C Luczak *et al.*, *Biology Letters.***8**,821(2012).

703 H Thomas *et al.*, *Science.***304**,1005(2004).

704 G Beaugrand *et al.*, *Marine Ecology Progress Series.***502**,85(2014).

705 PC Reid, M Edwards, *Senckenbergiana Maritima.***32**,107(2001).

706 I Kroncke *et al.*, *Marine Ecology Progress Series.***167**,25(1998).

707 RM Warwick *et al.*, *Marine Ecology Progress Series.***234**,1(2002).

708 G Beaugrand, *Fisheries Oceanography.***12**,270(2003).

709 PC Reid *et al.*, *Fisheries Oceanography.***12**,260(2003).

710 G Beaugrand *et al.*, *Nature Climate Change.***3**,263(2012).

711 N Scafetta, BJ West, *Physics Today.***3**,50(2008).

712 G Beaugrand, PC Reid, *Global Change Biology.***9**,801(2003).

713 G Ottersen *et al.*, *Oecologia.***128**,1(2001).

714 JR Raven *et al.*, 'Ocean acidification due to increasing atmospheric carbon dioxide'. (The Royal Society, London, 2005).

715 C Pelejero *et al.*, *Trends in Ecology and Evolution.***25**,332(2010).

716 RA Feely *et al.*, *Oceanography.***22**,36(2009).

717 CL Sabine *et al.*, *Science.***305**,367(2004).

718 K Caldeira, *Oceanography.***20**,188(2007).

719 JT Wootton *et al.*, *Proceedings of the National Academy of Sciences.***105**,18848(2008).

720 RA Kerr, *Science.***328**,1500(2010).

721 K Caldeira, ME Wickett, *Nature.***425**,365(2003).

722 RE Zeebe, *Annual Review of Earth and Planetary Sciences.***40**,141(2012).

723 M Steinacher *et al.*, *Biogeosciences.***6**,515(2009).

724 VJ Fabry *et al.*, *ICES Journal of Marine Science.***65**,414(2008).

725 IE Hendriks *et al.*, *Estuarine, Coastal and Shelf Science.***86**,157(2010).

726 JB Ries *et al.*, *Geology.***37**,1131(2009).

727 AF Vézina, O Hoegh-Gulberg, *Marine Ecology Progress Series.***373**,199(2008).

728 J-P Gattuso *et al.*, *Global and Planetary Change.***18**,37(1998).

729 U Riebesell *et al.*, *Nature.***407**,364(2000).

730 JD Milliman, *Global Biogeochemical Cycles*.**7**,927(1993).

731 CW Brown, *Oceanography*.**8**,59(1995).

732 HO Pörtner, *Marine Ecology Progress Series*.**373**,203(2008).

733 J-P Gattuso *et al.*, *Oceanography*.**22**,190(2009).

734 SC Doney *et al.*, *Oceanography*.**22**,16(2009).

735 GE Hofmann *et al.*, *The Annual Review of Ecology, Evolution, and Systematics*.**41**,127(2010).

736 J Silverman *et al.*, *Geophysical Research Letters*.**36**,L05606(2009).

737 C Pelejero *et al.*, *Science*.**309**,2204(2005).

738 RA Feely *et al.*, *Science*.**320**,1490(2008).

739 S Comeau *et al.*, *PLOS ONE*.**5**,e11362(2010).

740 D Clark *et al.*, *Marine Biology*.**156**,1125(2009).

741 M Byrne, *Oceanography and Marine Biology: An Annual Review*.**49**,1(2011).

742 H Kurihara, *Marine Ecology Progress Series*.**373**,275(2008).

743 A Beckerman *et al.*, *Trends in Ecology and Evolution*.**17**,263(2002).

744 S Dupont *et al.*, *Marine Ecology Progress Series*.**373**,285(2008).

745 HL Wood *et al.*, *Proceedings of the Royal Society B*.**275**,1767(2008).

746 RA Gooding *et al.*, *Proceedings of the National Academy of Sciences of the United States of America*.**106**,9316(2009).

747 R Rodolfo-Metalpa *et al.*, *Nature Climate Change*.**1**,308(2011).

748 JB Ries, *Nature Climate Change*.**1**,294(2011).

749 JM Hall-Spencer *et al.*, *Nature*.**454**,96(2008).

750 CJF ter Braak, IC Prentice, *Advances in Ecological Research*.**18**,271(1988).

751 AU Form, U Riebesell, *Global Change Biology*.**18**,843(2012).

752 BE Casareto *et al.*, *Aquatic Biology*.**7**,59(2009).

753 PW Boyd, *Nature Geoscience*.**4**,273(2011).

754 TJ Crowley, *Global Biogeochemical Cycles*.**9**,377(1995).

755 HM Putman *et al.*, *Biological Bulletin (Woods Hole)*.**215**,135(2008).

756 KRN Anthony *et al.*, *Proceedings of the National Academy of Sciences of the United States of America*.**105**,17442(2008).

757 G De'ath *et al.*, *Science*.**323**,116(2009).

758 AD Moy *et al.*, *Nature Geoscience*.**2**,276(2009).

759 JM Guinotte, VJ Fabry, *Annals of the New York Academy of Sciences*.**1134**,320(2008).

760 PC Reid, G Beaugrand, *Journal of the Marine Biological Association of the United Kingdom*.**92**,1435(2012).

761 J Ahn, EJ Brook, *Science*.**322**,83(2008).

762 A Indermühle *et al.*, *Geophysical Research Letters*.**27**,735(2000).

763 ST Gille, *Science*.**295**,1275(2002).

764 JC Fyfe, *Geophysical Research Letters*.**33**,L19701(2006).

765 AD Barnosky *et al.*, *Nature*.**486**,52(2012).

766 EAE Norse, *Global marine biological diversity. A strategy for building conservation into decision making.* (Center for Marine Conservation, Washington, DC, 1993), pp. 383.

767 JE Cloern, *Marine Ecology Progress Series*.**210**,223(2001).

768 JC Dauvin, *Aquatic Invasions*.**4**,467(2009).

769 D Pauly, *Global Change NewsLetter*.**55**,21(2003).

770 E Goberville *et al.*, *Ecological Indicators*.**11**,1290(2011).

771 JA Hutchings, JD Reynolds, *Bioscience*.**54**,297(2004).

772 D Pauly, V Christensen, *Nature*.**374**,255(1995).

773 D Pauly *et al.*, *Science*.**279**,860(1998).

774 M Casini *et al.*, *Proceedings of the National Academy of Sciences of the USA*.**106**,197(2009).

775 D Pauly *et al.*, *Nature*.**418**,689(2002).

776 JBC Jackson *et al.*, *Science*.**293**,629(2001).

777   RA Myers, B Worm, *Nature*.**423**,280(2003).

778   FAO Fisheries and Aquaculture Department, 'The state of world fisheries and aquaculture 2010'. (Food and Agriculture Organization of the United Nations, Rome, 2010).

779   JA Hutchings, *Nature*.**406**,882(2000).

780   JK Baum *et al.*, *Science*.**299**,389(2003).

781   SA Berkeley *et al.*, *Ecology*.**85**,1258(2004).

782   ICES, 'Report of the worshop on gadoid stocks in the North Sea during the 1960s and the 1970s'. The fourth ICES/GLOBEC backward-facing workshop. MR Heath, KM Brander, eds (ICES, Copenhagen, 2001), vol. 244, pp. 55.

783   EM Olsen *et al.*, *Nature*.**428**,932(2004).

784   S Jennings *et al.*, *Marine fisheries ecology*. (Blackwell Science, Oxford, 2001).

785   G Marteinsdottir, GA Begg, *Marine Ecology Progress Series*.**235**,235(2002).

786   BS Green, *Advances in Marine Biology*.**54**,1(2008).

787   A Corten, *Reviews in Fish Biology and Fisheries*.**11**,339(2002).

788   B Planque *et al.*, *Journal of Marine Systems*.**79**,403(2010).

789   C Wilkinson, 'Status of coral reefs of the world: 2008'. (Global Coral Reef Monitoring Network Reef and Rainforest Research Centre, Townsville, 2008).

790   Y Sadovy *et al.*, *Reviews in Fish Biology and Fisheries*.**13**,327(2003).

791   MR Clark, DM Tracey, *Fishery Bulletin*.**92**,236(1994).

792   C Minto, CP Nolan, *Earth and Environmental Science*.**77**,39(2006).

793   DE McAllister, *Galaxea*.**7**,161(1988).

794   HE Fox, RL Caldwell. *Ecological Applications*.**16**,1631(2006).

795   S Jennings, MJ Kaiser, *Advances in Marine Biology*.**34**,201(1998).

796   KE Carpenter, AC Alcala, *Philippine Journal of Fisheries*.**15**,217(1977).

797   RL Lewison *et al.*, *Trends in Ecology and Evolution*.**19**,598(2004).

798   ORJ Anderson *et al.*, *Endangered Species Research*.**14**,91(2011).

799   RL Lewison *et al.*, *Ecology Letters*.**7**,221(2004).

800   AJ Read *et al.*, *Conservation Biology*.**1**,163(2006).

801   R Callaway *et al.*, *ICES Journal of Marine Science*.**59**,1199(2002).

802   SF Thrush, PK Dayton, *Annual Review of Ecology and Systematics*.**33**,449(2002).

803   M Cryer *et al.*, *Ecological Applications*.**12**,1824(2002).

804   RL Naylor *et al.*, *Nature*.**405**,1017(2000).

805   M Holmer *et al.*, *Marine Pollution Bulletin*.**44**,685(2002).

806   K Haya *et al.*, *ICES Journal of Marine Science*.**58**,492(2001).

807   T Aoki *et al.*, *Bulletin of the Japan Society for Scientific Fisheries*.**53**,1821(1987).

808   CK Lin, *World Aquaculture*.**20**,19(1989).

809   P Hansen *et al.*, *Aquaculture Fisheries Management*.**24**,777(1993).

810   MR Gross, *Canadian Journal of Fisheries and Aquatic Sciences*.**55(supplement 1)**,131(1998).

811   C Orr, *North American Journal of Fisheries Management*.**27**,187(2007).

812   K Whelan, 'A review of the impacts of the salmon louse, *Lepeophtheirus salmonis* (Kroyer, 1837) on wild salmonids'. (Atlantic Salmon Trust, Perth, 2010).

813   PA Heuch, TA Mo, *Diseases of Aquatic Organisms*.**45**,145(2001).

814   EO Wilson, *The creation*. (W.W. Norton & Company, New York, 2006).

815   RB Clark *et al.*, *Marine pollution*. (Oxford University Press, Oxford, 4th edn, 1997).

816   G Wurpel *et al.*, 'Plastics do not belong in the ocean. Towards a roadmap for a clean North Sea'. (IMSA Amsterdam, Amsterdam, 2011).

817   JGB Derraik, *Marine Pollution Bulletin*.**44**,842(2002).

818   M Edwards *et al.*, 'Ecological status report: results from the CPR survey 2009/2010'. (Sir Alister Hardy Foundation for Ocean Science, Plymouth, 2011).

819   KL Law *et al.*, *Science*.**329**,1185(2010).

820   H Kanehiro *et al.*, *Fisheries Engineering*.**31**,195(1995).

821   ML Moser, DS Lee, *Colonial Waterbirds*.**15**,83(1992).

822   WRP Bourne, MJ Imber, *Marine Pollution Bulletin*.**13**,20(1982).

823   CM Boerger *et al.*, *Marine Pollution Bulletin*.**60**,2275(2010).

824   MS Islam, M Tanaka, *Marine Pollution Bulletin*.**48**,624(2004).

825   JF Piatt *et al.*, *The Auk*.**107**,387(1990).

826   JA Wiens *et al.*, *Ecological Applications*.**6**,828(1996).

827   CH Peterson *et al.*, *Science*.**302**,2082(2003).

828   ITOPF, 'Effects of oil pollution on the marine environment'. (The International Tanker Owners Pollution Federation, Canterbury, 2011).

829   CS Albers *et al.*, *Marine Chemistry*.**55**,347(1996).

830   HJW Baar, *Progress in Oceanography*.**33**,347(1994).

831   Y-J Huang *et al.*, *Environmental Monitoring and Assessment*.**176**,517(2011).

832   CP Titley-O'Neal *et al.*, *Water Quality Research Journal of Canada*.**46**,74(2011).

833   H de Wolf *et al.*, *Marine Pollution Bulletin*.**48**,587(2004).

834   G de Metrio *et al.*, *Marine Pollution Bulletin*.**46**,358(2003).

835   P Kaladharan *et al.*, *Indian Journal of Fisheries*.**37**,51(1990).

836   MM Storelli *et al.*, *Rapport de la Commission Internationale de la Mer Méditerranée*.**35**,288(1998).

837   M Metian *et al.*, *Science of the Total Environment*.**407**,3503(2009).

838   M Metian *et al.*, *Environmental Pollution*.**152**,543(2008).

839   M Metian *et al.*, *Journal of Experimental Marine Biology and Ecology*.**353**,58(2007).

840   HA Jastania, AR Abbasi, *Pakistan Journal of Zoology*.**36**,247(2004).

841   A Kakuschke, A Prange, *International Journal of Comparative Psychology*.**20**,179(2007).

842   A Kakuschke *et al.*, *Archives of Environmental Contamination and Toxicology*.**55**,129(2008).

843   K Ronald *et al.*, *The Science of the Total Environment*.**8**,1(1977).

844   A Shlosberg *et al.*, *Journal of Wildlife Diseases*.**33**,135(1997).

845   KA Sloman, RW Wilson, in *Behaviour and physiology of fish*, KA Sloman *et al.*, eds (Elsevier Academic Press, London, 2006), vol. 24, pp. 413–468.

846   T Zhou, JS Weis, *Journal of Fish Biology*.**54**,44(1999).

847   I Dahllöf, JH Andersen, 'Hazardous and radioactive substances in Danish marine waters. Status and temporal trends'. (National Environmental Research Institute, Aarhus University, Aarhus, 2009).

848   National Academy of Sciences, 'Radioactivity in the marine environment (RIME)'. (National Academy of Sciences, Washington, DC, 1971).

849   X Zhao *et al.*, *Marine Ecology Progress Series*.**222**,227(2001).

850   K Buesseler *et al.*, *Environmental Science and Technology*.**45**,9931(2011).

851   KO Buesseler *et al.*, *Proceedings of the National Academy of Sciences of the United States of America*.**109**,5984(2012).

852   DJ Madigan *et al.*, *Proceedings of the National Academy of Sciences of the United States of America*.**109**,9483(2012).

853   A Fernandez *et al.*, *Veterinary Pathology*.**42**,446(2005).

854   Z-B Jiang *et al.*, *Journal of Experimental Marine Biology and Ecology*.**368**,196(2009).

855   MS Hoffmeyer *et al.*, *Iheringia. Série Zoologia*.**95**,311(2005).

856   V Krishnakumar, *Encology*.**9**,6(1994).

857   C Lardicci *et al.*, *Marine Pollution Bulletin*.**38**,296(1999).

858   P Karas, *Marine Pollution Bulletin*.**24**,27(1992).

859   J von Liebig, in *Cycles of essential elements. Benchmark papers in ecology*, LR Pomeroy, ed. (Benchmark Papers in Ecology, UK, 1855), pp. 11–28.

860   DJ Conley, *Hydrobiologia*.**410**,87(2000).

861   RJ Diaz, R Rosenberg, *Science*.**321**,926(2008).

862   PM Glibert *et al.*, *Oceanography*.**18**,198(2005).

863 LA Martinelli, in *Global change and human impacts*, JM Melillo *et al.*, eds (Island Press, Washington, DC, 2003), vol. 61, pp. 193–210.

864 SB Bricker *et al.*, 'National estuarine eutrophication assessment: effects of nutrient enrichment in the nation's estuaries'. (National Oceanic and Atmospheric Administration, National Ocean Service, Special Projects Office and the National Centers for Coastal Ocean Science, Silver Spring, 1999).

865 J Heisler *et al.*, *Harmful Algae*.**8**,3(2008).

866 VH Smith *et al.*, *Environmental Pollution*.**10**,179(1999).

867 C Lancelot, *Science of the Total Environment*.**165**,83(1995).

868 VH Smith, DW Schindler, *Trends in Ecology and Evolution*.**24**,201(2009).

869 RJ Lukatelich, AJ McComb, *Journal of Plankton Research*.**8**,597(1986).

870 JF Bruno *et al.*, *Ecology Letters*.**6**,1056(2003).

871 WH Wilson *et al.*, *Journal of Phycology*.**32**,506(1996).

872 RJ Diaz, *Journal of Environmental Quality*.**30**,275(2001).

873 RJ Diaz *et al.*, in *Eutrophication and hypoxia: nutrient pollution in coastal waters*. (World Resources Institute, Washington, DC, 2010).

874 R Vaquer-Sunyer, CM Duarte, *Proceedings of the National Academy of Sciences of the United States of America*.**105**,15452(2008).

875 O Borysova *et al.*, 'Eutrophication in the Black Sea region: impact assessment and causal chain analysis'. (University of Kalmar, Kalmar, 2005).

876 YP Zaitsev *et al.*, 'Europe's biodiversity: the Black Sea, an oxygen-poor sea'. (European Environment Agency, Ispra, 2001).

877 Y Zaitsev, VO Mamaev, *Biological diversity in the Black Sea: a study of change and decline*. BSE Series, ed., (United Nations Publishing, New York, 1997), vol. 3, pp. 208.

878 YP Zaitsev, *Fisheries in Oceanography*.**1**,180(1992).

879 JBC Jackson, *Proceedings of the National Academy of Sciences of the USA*.**105**,11458(2008).

880 JJ Helly, LA Levin, *Deep-Sea Research I*.**51**,1159(2004).

881 AF Hofmann *et al.*, *Deep Sea Research I*.**58**,1212(2011).

882 BA Grantham *et al.*, *Nature*.**429**,749(2004).

883 L Stramma *et al.*, *Science*.**320**,655(2008).

884 G Shaffer *et al.*, *Nature Geoscience*.**2**,105(2009).

885 RB Huey, PD Ward, *Science*.**308**,398(2005).

886 IAW Macdonald, in *Biodiversity and global change*, OT Solbrig *et al.*, eds (CAB International, Wallingford, 1994), pp. 199–209.

887 SA Ludsin, AD Wolfe, *Bioscience*.**51**,780(2001).

888 JT Carlton, JB Geller, *Science*.**261**,78(1993).

889 A Ricciardi, HJ MacIsaac, *Trends in Ecology and Evolution*.**15**,62(2000).

890 O Jousson *et al.*, *Nature*.**408**,157(2000).

891 M Williamson, *Biological Invasions*. (Chapman & Hall, New York, 1996).

892 S Nehring, *Arch. Fish. Mar. Res*.**46**,181(1998).

893 M Edwards *et al.*, *Journal of the Marine Biological Association of the UK*.**81**,207(2001).

894 N Bax *et al.*, *Marine Policy*.**27**,313(2003).

895 J Klein, M Verlaque, *Marine Pollution Bulletin*.**56**,205(2008).

896 PC Reid *et al.*, *Global Change Biology*.**13**,1910(2007).

897 AM Hjelset, Ph.D., 'Female life-history parameters in the introduced red king crab (*Paralithodes camtschaticus*, Tilesius 1815) in the Barents Sea: a study of temporal and spatial variation in three Norwegian fjords In Department of Arctic and Marine Biology', University of Tromso, Faculty of biosciences, fisheries and economics (2012).

898 LL Jorgensen, *Biological Invasions*.**7**,949(2005).

899 JC Farman *et al.*, *Nature*.**315**,207(1985).

900 MJ Molina, FS Rowland, *Nature*.**249**,810(1974).

901 RC Smith *et al.*, *Science*.**255**,952(1992).

902 JG Anderson *et al.*, *Science*.**251**,39(1991).

903 GL Manney *et al.*, *Nature*.**478**,469(2011).

904 S Solomon, *Reviews of Geophysics*.**37**,275(1999).

905 MC Rousseaux *et al.*, *Global Change Biology*.**7**,467(2001).

906 M Llabrès *et al.*, *Global Ecology and Biogeography*.**22**,131(2012).

907 AD Barnosky *et al.*, *Nature*.**471**,51(2011).

908 A Urbanek, *Historical Biology*.**7**,29(1993).

909 RJ Twitchett, *Palaeogeography, Palaeoclimatology, Palaeoecology*.**252**,132(2007).

910 Millennium Ecosystem Assessment, 'Ecosystems and human well-being: biodiversity synthesis'. (World Resources Institute, Washington, DC, 2005).

911 R Leaky, R Lewin, *The sixth extinction. Biodiversity and its survival.* (Doubleday, New York, 1995).

912 CM Roberts, JP Hawkins, *Trends in Ecology and Evolution*.**14**,241(1999).

913 JM Diamond, in *Conservation for the twenty-first century*, D Western, MC Pearl, eds (Oxford University Press, Oxford, 1989), pp. 37–41.

914 NK Dulvy *et al.*, *Fish and Fisheries*.**4**,25(2003).

915 JL Gittleman, ME Gompper, *Science*.**291**,997(2001).

916 CG Jones *et al.*, *Oikos*.**69**,373(1994).

917 PA Walker, HJL Heesen, *ICES Journal of Marine Science*.**53**,1085(1996).

918 PW Boyd *et al.*, *Limnology and Oceanography*.**55**,1353(2010).

919 RA Myers *et al.*, *Science*.**315**,1846(2007).

920 RI Perry *et al.*, *Journal of Marine Systems*.**79**,427(2010).

921 FJ Millero *et al.*, *Oceanography*.**22**,72(2009).

922 RH Byrne *et al.*, *Marine Chemistry*.**25**,163(1988).

923 JE Vermaat *et al.*, *Estuarine Coastal and Shelf Science*.**80**,53(2008).

924 W Ludwig *et al.*, *Progress in Oceanography*.**80**,199(2009).

925 SD Donner *et al.*, *Global Biogeochemical Cycles*.**16**,1043(2002).

926 CDG Harley *et al.*, *Ecology Letters*.**9**,228(2006).

927 D Labat *et al.*, *Advances in Water Resources*.**27**,631(2004).

928 I Valiela, *Marine ecological processes.* (Springer Verlag, New York, 1995).

929 JL Sarmiento *et al.*, *Global Biogeochemical Cycles*.**18**,1(2004).

930 DG Boyce *et al.*, *Nature*.**466**,591(2010).

931 SJ Newman *et al.*, *Polar Biology*.**22**,50(1999).

932 MP Lesser *et al.*, *Coral Reefs*.**8**,225(1990).

933 S Martin, J-P Gattuso, *Global Change Biology*.**15**,2089(2009).

934 S Reynaud *et al.*, *Global Change Biology*.**9**,1660(2003).

935 R Metzger *et al.*, *Journal of Thermal Biology*.**32**,144(2007).

936 EO Wilson, *Biophilia.* (Harvard University Press, Cambridge, MA, 1984).

937 E Ruijgrok *et al.*, *Ecological Economics*.**28**,347(1999).

938 J Diamond, *Collapse: how societies choose to fail or succeed.* (Penguin Books, New York, 2005).

939 KS McCann, *Nature*.**405**,228(2000).

940 S Naeem, in *Marine biodiversity and ecosystem functioning: frameworks, methodologies and integration*, M Solan *et al.*, eds (Oxford University Press, Oxford, 2012), pp. 34–51.

941 DM Paterson *et al.*, in *Marine biodiversity and ecosystem functioning: frameworks, methodologies and integration*, M Solan *et al.*, eds (Oxford University Press, Oxford, 2012), pp. 24–33.

942 S Naeem *et al.*, in *Biodiversity and ecosystem functioning – synthesis and perspectives*, M Loreau *et al.*, eds (Oxford University Press, Oxford, 2002), pp. 3–11.

943 BH Walker, *Conservation Biology*.**6**,18(1992).

944 SKM Ernest, JH Brown, *Science*.**292**,101(2001).

945 PR Ehrlich, AH Ehrlich, *Extinction. The causes and consequences of the disappearance of species.* (Random House, New York, 1981).

946 JH Lawton, *Oikos*.**71**,367(1994).

947 S Naeem *et al.*, *Nature*.**368**,734(1994).

948 D Tilman, *Science*.**286**,1099(1999).

949 A Hector *et al.*, *Science*.**286**,1123(1999).

950 D Tilman, *Ecology*.**80**,1455(1999).

951 AR Ives, SR Carpenter, *Science*.**317**,58(2007).

952 EP Odum, *Fundamentals of ecology*. (Saunders, Philadelphia, PA, 1971).

953 CS Holling, *Annual Review of Ecology and Systematics*.**4**,1(1973).

954 RV O'Neill, *Ecology*.**82**,3275(2001).

955 C Elton, *Ecology of invasions by animals and plants*. (Chapman & Hall, London, 1958).

956 RH Mac Arthur, *Ecology*.**36**,533(1955).

957 RM May, *Stability and complexity in model ecosystems*. (Princeton University Press, Princeton, NJ, 1973), pp. 268.

958 MR Gardner, WR Ashby, *Nature*.**228**,784(1970).

959 P Yodsis, *Nature*.**289**,674(1981).

960 KS McCann *et al.*, *Nature*.**395**,794(1998).

961 J Bascompte *et al.*, *Science*.**312**,431(2006).

962 DL Finke, RF Denno, *Nature*.**429**,407(2004).

963 GS Kleppel, *Marine Ecology Progress Series*.**99**,183(1993).

964 GS Kleppel, CA Burkart, *ICES Journal of Marine Sciences*.**52**,297(1995).

965 B Worm *et al.*, *Science*.**314**,787(2006).

966 JE Duffy *et al.*, in *Marine biodiversity and ecosystem functioning: frameworks, methodologies and integration*, M Solan *et al.*, eds (Oxford University Press, Oxford, 2012), pp. 164–184.

967 F Figge, *Biodiversity and Conservation*.**13**,827(2004).

968 DF Doak *et al.*, *The American Naturalist*.**151**,264(1998).

969 S Yachi, M Loreau, *Proceedings of the National Academy of Sciences of the United States of America*.**96**,1463(1999).

970 S Naeem, S Li, *Nature*.**390**,507(1997).

971 F Boero *et al.*, *Ecological Complexity*.**1**,101(2004).

972 D Tilman, *Nature*.**379**,718(1996).

973 OL Petchey, *The American Naturalist*.**155**,696(2000).

974 LW Aarssen, *Oikos*.**80**,183(1997).

975 RH Mac Arthur, *Proceedings of the National Academy of Sciences of the United States of America*.**43**,293(1957).

976 A Hector, *Oikos*.**82**,597(1998).

977 M Loreau, *Philosophical Transactions of the Royal Society B*.**365**,49(2010).

978 J Leps *et al.*, *Vegetatio*.**50**,53(1982).

979 AB Pfisterer, B Schmid, *Nature*.**416**,84(2002).

980 MA Huston, *Biological diversity. The coexistence of species on changing landscapes*. (Cambridge University Press, Cambridge, 1994).

981 R Danovaro, in *Marine biodiversity and ecosystem functioning: frameworks, methodologies and integration*, M Solan *et al.*, eds (Oxford University Press, Oxford, 2012), pp. 115–126.

982 D Raffaelli, AM Friedlander, in *Marine biodiversity and ecosystem functioning: frameworks, methodologies and integration*, M Solan *et al.*, eds (Oxford University Press, Oxford, 2012), pp. 149–163.

983 RT Paine, *The American Naturalist*.**100**,65(1966).

984 HEW Cottee-Jones, RJ Whittaker, *Frontiers of Biogeography*.**4**,117(2012).

985 MC Emmerson, in *Marine biodiversity and ecosystem functioning: frameworks, methodologies and integration*, M Solan *et al.*, eds (Oxford University Press, Oxford, 2012), pp. 85–100.

986 C Elton, *Animal ecology*. (Sidgwick & Jackson, London, 1927).

987 JO Riede *et al.*, *Ecology Letters*.**14**,169(2011).

988 RR Kirby et al., Ecosystems.**12**,548(2009).

989 NM Haddad et al., Ecology Letters.**12**,1029(2009).

990 GE Hutchinson, The American Naturalist.**93**,145(1959).

991 DS Srivastava, JH Lawton, The American Naturalist.**152**,510(1998).

992 SQ Dornbos et al., in Marine biodiversity and ecosystem functioning: frameworks, methodologies and integration, M Solan et al., eds (Oxford University Press, Oxford, 2012), pp. 52–72.

993 JE Duffy, Frontiers in Ecology and the Environment.**7**,437(2009).

994 PS Giller et al., Oikos.**104**,423(2004).

995 K Thompson et al., Functional Ecology.**19**,355(2005).

996 MA Huston, AC McBride, in Biodiversity and ecosystem functioning, M Loreau et al., eds (Oxford University Press, Oxford, 2002), pp. 47–60.

997 DA Wardle, M Jonsson, Frontiers in Ecology and the Environment.**8**,10(2009).

998 RT Paine, Science.**296**,736(2002).

999 ON Bjornstad et al., Nature.**409**,1001(2001).

1000 JB Grace et al., Ecology Letters.**10**,680(2007).

1001 R Costanza et al., Nature.**387**,253(1997).

1002 FT Short, S Wyllie-Echeverria, Environmental Conservation.**23**,17(1996).

1003 M Waycott et al., Proceedings of the National Academy of Sciences of the United States of America.**106**,12377(2009).

1004 RA Watson et al., Australian Journal of Marine and Freshwater Research.**44**,211(1993).

1005 V Raybaud et al., PLOS ONE.**8**,e66044(2013).

1006 I Valiela et al., Bioscience.**51**,807(2001).

1007 F Moberg, C Folke, Ecological Economics.**29**,215(1999).

1008 J-P Gattuso et al., Science.**271**,1298(1996).

1009 D Pauly et al., Philosophical Transactions of the Royal Society B.**360**,5(2005).

1010 JJ Stachowicz et al., Ecology.**83**,2575(2002).

1011 World Health Organization, 'Eutrophication and health'. (Office for Official Publications of the European Communities, Luxemburg, 2002).

1012 LE Llewellyn, Toxicon.**56**,691(2010).

1013 P Geistdoerfer, M Goyffon, EMC – Toxicologie Pathologie.**1**,35(2004).

1014 DB James, Fishing Chimes.**30**,39(2010).

1015 LE Llewellyn et al., Toxicon.**40**,1463(2002).

1016 W Fenical, Oceanography.**9**,23(1996).

1017 US Commission on Ocean Policy, 'An ocean blueprint for the 21st century'. (U.S. Commission on Ocean Policy, Washington, DC, 2004).

1018 B Haefner, Drug Discovery Today.**8**,536(2003).

1019 M Donia, MT Hamann, The Lancet. Infectious Diseases.**3**,338(2003).

1020 ME Hay, W Fenical, Oceanography.**9**,10(1996).

1021 LA Zaslavskaia et al., Journal of Phycology.**36**,379(2000).

1022 G-P Hu et al., Marine Drugs.**9**,514(2011).

1023 H Luesch et al., Journal of Natural Products.**64**,907(2001).

1024 L-H Zheng et al., Marine Drugs.**9**,1840(2011).

1025 LT Tan, Phytochemistry.**68**,954(2007).

1026 S Sagar et al., Marine Drugs.**8**,2619(2010).

1027 L Bopp et al., Geophysical Research Letters.**32**,L19606(2005).

1028 P Cermeno et al., Proceedings of the National Academy of Sciences of the USA.**105**,20344(2008).

1029 JT Turner, Aquatic Microbial Ecology.**27**,57(2002).

1030 T Hashioka, Y Yamanaka, Ecological Modelling.**202**,95(2007).

1031 JG Hiddink, R ter Hofstede, Global Change Biology.**14**,453(2008).

1032 JL Gutiérrez et al., Oikos.**101**,79(2003).

1033 KE Chave et al., Marine Geology.**12**,123(1975).

1034 R Schiebel, *Global Biogeochemical Cycles.***16**,1065(2002).

1035 U Passow, *Geochemistry, Geophysics, Geosystems.***5**,Q04002(2004).

1036 C Mahaffey *et al.*, *American Journal of Science.***305**,546(2005).

1037 DA Hutchins *et al.*, *Oceanography.***22**,128(2009).

1038 T Tyrrell, *Nature.***400**,525(1999).

1039 DM Nelson *et al.*, *Global Biogeochemical Cycles.***9**,359(1995).

1040 O Ragueneau *et al.*, *Global and Planetary Change.***26**,317(2000).

1041 A Yool, T Tyrrell, *Global Biogeochemical Cycles.***17**,1103(2003).

1042 A McQuatters-Gollop *et al.*, *Nature.***472**,E6(2011).

1043 A Gnanadesikan *et al.*, *Geophysical Research Letters.***37**,L18802(2010).

1044 DW Townsend, MD Keller, *Marine Ecology Progress Series.***137**,229(1996).

1045 RJ Charlson *et al.*, *Nature.***326**,655(1987).

1046 K Aranami *et al.*, *Journal of Oceanography.***57**,315(2001).

1047 C Chapperon, L Seuront, *Global Change Biology.***17**,1740(2011).

1048 HO Pörtner *et al.*, *Continental Shelf Research.***21**,1975(2001).

1049 MJ Attrill, M Power, *Nature.***417**,275(2002).

1050 G Beaugrand, F Ibanez, *Marine Ecology Progress Series.***284**,35(2004).

1051 JA Sheridan, D Bickford, *Nature Climate Change.***1**,401(2011).

1052 JJ Smith *et al.*, *Proceedings of the National Academy of Sciences of the United States of America.***106**,17655(2009).

1053 LW Aarssen, *Oikos.***80**,177(1997).

1054 JH Lawton, *Oikos.***84**,177(1999).

1055 R De Wit, T Bouvier, *Environmental Microbiology.***8**,755(2006).

1056 SALM Kooijman, *Philosophical Transactions of the Royal Society B.***356**,331(2001).

1057 M Lynch, W Gabriel, *The American Naturalist.***129**,283(1987).

1058 J Grinnell, *Auk.***34**,427(1917).

1059 JM Chase, MA Leibold, *Ecological niches: linking classical and contemporary approches.* (University of Chicago Press, Chicago, IL, 2003).

1060 ST Jackson *et al.*, *Proceedings of the National Academy of Sciences of the USA.***106**,19685(2009).

1061 RH MacArthur, *Geographical ecology.* (Princeton University Press, Princeton, NJ, 1972).

1062 D Tilman, *Resource competition and community structure.* (Princeton University Press, Princeton, NJ, 1982).

1063 JF Bruno *et al.*, *Trends in Ecology and Evolution.***18**,119(2003).

1064 MD Bertness *et al.*, *Ecology.***80**,2711(1999).

1065 R Pulliam, *Ecology Letters.***3**,349(2000).

1066 G Hardin, *Science.***131**,1292(1960).

1067 PJ Grubb, *Biology Review.***52**,107(1977).

1068 HR Pulliam, *The American Naturalist.***132**,652(1988).

1069 WC Allee *et al.*, *Principles of animal ecology.* (Saunders, Philadelphia, PA, 1949).

1070 WR Ashby, *An introduction to cybernetics.* (John Wiley & Sons, New York, 1958).

1071 B Huntley *et al.*, *Ibis.***148**,8(2006).

1072 GM Hewitt, *Philosophical Transactions of the Royal Society B.***359**,183(2004).

1073 DC Speirs *et al.*, *Fisheries Oceanography.***14**,333(2005).

1074 RG Pearson, TP Dawson, *Global Ecology and Biogeography.***12**,361(2003).

1075 JH Brown, *Macroecology.* (University of Chicago Press, Chicago, IL, 1995).

1076 KJ Gaston, TM Blackburn, *Pattern and process in macroecology.* (Blackwell, Oxford, 2000).

1077 J de Rosnay, *The macroscope: a new world scientific system.* (Harper & Row, New York, 1979).

1078 SA Levin, *Ecosystems.***1**,431(1998).

1079 GE Hutchinson, *An introduction to population ecology.* (Yale University Press, New Haven, CT, 1978).

1080 ML Shaffer, *Bioscience.***31**,131(1981).

1081 JR Brett, in *Marine ecology: a comprehensive integrated treatise on life in the oceans and coastal waters*, O Kinne, ed. (John Wiley & Sons, New York, 1970), vol. 1, pp. 515–573.

1082 RJ Menzies *et al.*, *Abyssal environment and ecology of the world oceans*. (John Wiley & Sons, New York, 1973).

1083 KJ Gaston *et al.*, *Trends in Ecology and Evolution*.**13**,70(1998).

1084 M Weijerman *et al.*, *Marine Ecology Progress Series*.**298**,21(2005).

1085 CJF Ter Braak, *Unimodal models to relate species to environment*. (DLO-Agricultural Mathematics Group, Wageningen, 1996), pp. 266.

1086 P Helaouët, *et al.*, PLOS ONE.**8**,10(2012).

1087 A Chalaali *et al.*, *PLOS ONE*.**8**,1(2013).

1088 G Beaugrand *et al.*, *Proceedings of the Royal Society B*.**281**(2014).

1089 JM Sunday *et al.*, *Nature Climate Change*.**1**(2012).

1090 G Reygondeau, G Beaugrand, *Global Change Biology*.**17**,756(2011).

1091 O Fiksen, *ICES Journal of Marine Science*.**57**,1825(2000).

1092 G Dietrich, *Research Geophysic*.**2**,291(1964).

1093 P Helaouët *et al.*, *Progress in Oceanography*.**91**,217(2011).

1094 M Zhao, SW Running, *Science*.**329**,940(2010).

1095 JA Lindley, S Daykin, *ICES Journal of Marine Science*.**62**,869(2005).

1096 R Przybylo *et al.*, *Journal of Animal Ecology*.**69**,395(2000).

1097 G Beaugrand, *Philosophical Tansactions of the Royal Society B*.**In press**(2014).

1098 AH Taylor *et al.*, *Nature*.**416**,629(2002).

1099 G Beaugrand *et al.*, *Marine Ecology Progress Series*.**502**,85.(2014).

1100 M Scheffer, SR Carpenter, *Trends in Ecology and Evolution*.**18**,648(2003).

1101 M Scheffer, EH van Nes, *Progress in Oceanography*.**60**,303(2004).

1102 DA Seekell *et al.*, *Theoretical Ecology*.**6**,385(2013).

1103 M Nystrom *et al.*, *Trends in Ecology and Evolution*.**15**,413(2000).

1104 B de Young *et al.*, *Trends in Ecology and Evolution*.**23**,402(2008).

1105 MB Araujo, A Guisan, *Journal of Biogeography*.**33**,1677(2006).

1106 C-H Hsieh *et al.*, *Nature*.**435**,336(2005).

1107 MS Boyce *et al.*, *Trends in Ecology and Evolution*.**21**,141(2006).

1108 DA Keith *et al.*, *Biology Letters*.**4**,560(2008).

1109 RM May, *Nature*.**261**,459(1976).

1110 EN Lorenz, *Journal of the Atmospheric Sciences*.**20**,130(1963).

1111 SA Levin, J Lubchenco, *Bioscience*.**58**,27(2008).

1112 E Ehlers, T Krafft, in *Earth system science in the Anthropocene: emerging issues and problems*, E Ehlers, T Krafft, eds (Springer, Heidelberg, 2006), pp. 5–12.

1113 JW Williams *et al.*, *Proceedings of the National Academy of Sciences of the United States of America*.**104**,5738(2007).

1114 DD Rousseau, in *Past and future rapid environmental changes: the spatial and evolutionary responses of terrestrial biota*, B Huntley *et al.*, eds (Springer Verlag, Berlin, 1997), pp. 303–318.

1115 JH Stillman, *Science*.**301**,65(2003).

1116 DL Mackas *et al.*, *Progress in Oceanography*.**75**,223(2007).

1117 S Lenoir *et al.*, *Global Change Biology*.**17**,115(2011).

1118 V Sanchez-Cordero *et al.*, *Biodiversity Informatics*.**2**,11(2005).

1119 PM Berry *et al.*, *Global Ecology and Biogeography*.**11**,453(2002).

1120 AT Peterson, DA Vieglais, *Bioscience*.**51**,363(2001).

1121 P McCullagh, JA Nelder, *Generalized linear models*. (Chapman & Hall, London, 1983).

1122 TJ Hastie, RJ Tibshirani, *Generalized additive models*. (Chapman & Hall, London, 1990).

1123 A Guisan, NE Zimmermann. *Ecological Modelling*.**135**,147(2000).

1124 A Guisan, W Thuiller, *Ecology Letters*.**8**,993(2005).

1125 G Carpenter *et al.*, *Biodiversity and conservation*.**2**,667(1993).

1126 AH Hirzel *et al.*, *Ecology*.**83**,2027(2002).

1127 SJ Philips *et al.*, *Ecological Modelling* **190**,231(2006).

1128 RA Clarck *et al.*, *Global Change Biology*.**9**,1669(2003).

1129 K Drinkwater, *ICES Journal of Marine Science*.**62**,1327(2005).

1130 WWL Cheung *et al.*, 'Modelling present and climate-shifted distribution of marine fishes and invertebrates'. (Fisheries Centre, University of British Columbia, Vancouver, 2008).

1131 RS Steneck *et al.*, *Environmental Conservation*.**29**,436(2002).

1132 PK Dayton, *Annual Review of Ecology and Systematics*.**16**,215(1985).

1133 D Birkett *et al.*, *Vol. VII. An Overview of Dynamic and Sensitivity Characteristics for Conservation Management of Marine SACs*.174(1998).

1134 ST Jackson, JT Overpeck, *Paleobiology*.**26(Supplement)**,194(2000).

1135 V Bahn, BJ McGill, *Global Ecology and Biogeography*.**16**,733(2007).

1136 CF Dormann *et al.*, *Ecography*.**30**,609(2007).

1137 E Goberville *et al.*, *Ecology and Evolution*.(Submitted).

1138 E Goberville *et al.*, *Diversity and Distributions*.(Submitted).

1139 A Belgrano, JH Brown, *Nature*.**419**,128(2002).

1140 J de Rosnay, *The macroscope: a new world scientific system*. (Harper & Row, New York, 1979).

1141 TM Blackburn, KJ Gaston, in *Macroecology: concepts and consequences*, TM Blackburn, KJ Gaston, eds (Blackwell, Malden, MA, 2003), pp. 1–14.

1142 K Sherman, *Ocean and Coastal Management*.**29**,165(1995).

1143 T Clutton-Brock, BC Sheldon, *Nature*.**327**,1207(2010).

Index

abrupt ecosystem shifts 173, 252, 410
absolute vorticity 39
abyssal plain 74
abyssobenthic zone 74
accidental introduction 330
accidental pollution 303
acclimation 241
acclimatisation 426–427
acetogenins 381
acidification 261–263
actaeplanic larvae 231
active competition 113
actualism 426
adaptation 119, 188–189, 193–194, 203, 227–234
adaptive capacity 240, 423
adaptive evolution 227–234, 433
adaptive phenotypic plasticity 227
aerobic scope 189, 191, 345
albedo 23–24, 33, 105–107, 121, 131, 212–214, 390–391
Aleutian Low 51–54, 105, 173
alkaloids 377
Allee effect 103, 234, 341, 403
alleles 227
allelopathy 113, 117
Allen's pattern 81–82
allopatric speciation 113, 125,234
ambient energy theory 99, 101
amensalism 117
ammonoids 144, 146
amnesic shellfish poisoning 320, 375
Amoco Cadiz 304–305
anchiplanic larvae 230
ancillary energy 106
anoxia 145–146, 149, 298, 320–321, 325
anoxygenic photosynthesis 8
antagonisms 340

Antarctic Bottom Water 40, 137
Antarctic Circumpolar Current 41, 89, 126
Antarctic Circumpolar Wave 167
Antarctic Oscillation 56
Antarctic ozone hole 212, 335
anthelmintic 377
Anthropocene 1, 209
anthropogenic climate change 1, 15, 209
antibacterial 377
anticyclone 26, 33
antifouling agent 298, 308
antimitotic 377
antituberculosis 377
antiviral activity 377
aphotic 65
aquaculture 296–302
aragonite 260–282
archaea 8, 123, 148, 174–175
Arctic Oscillation 45–48
area hypothesis 87–88
articulate brachiopods 145
aspergillosis 321–322
asphyxiation 304–305
assortative mating 228
Atlantic Multidecadal Oscillation 58–59, 153, 171, 257, 278–279
atmosphere 19–21
atmospheric albedo 22–24, 107
atmospheric circulatory system 25
atmospheric convection 50, 421
atmospheric pressure 19, 34, 48, 54, 100, 105
auroras 21
autoimmunity 312

background extinction rate 338–339
bacteria 4, 5–6, 8, 79, 96, 117, 174, 299, 304, 312, 322, 330, 337, 355, 376, 381–383, 387, 390

barocline 100
barotrope 100
bathymetry 66, 68, 73, 108, 245, 304, 413, 416, 420, 431–432
bathyobenthic zone 74
Becking/Beijerinck's law 397
bedded cherts 390
behavioural thermoregulation 200
Benthos 3, 11–14, 75, 163–164, 251, 323, 325
Bergmann's pattern 79–81
Big Bang 174
Big Freeze 138
bioaccumulation 305, 309, 375
bioclimatic envelope 64, 98, 198–199, 200, 203, 234, 404, 406
biocomplexity 293
bioconcentration 306
biogenic opal 387, 390
biogeochemical partitions 74
biogeographic realms 61–62
biogeography 61–83
biological carbon pump 8–9, 384–385
biological insurance 355
bioluminescence 78–79
biomagnification 306, 308–309, 313, 374
biome 8, 62, 64, 66, 68, 72–73, 80, 87, 98, 121, 134, 140, 171, 228, 234, 246, 255, 366, 368, 395, 409, 412, 415, 426, 428, 437
biosilification 387
biosphere 1, 14, 31, 62–63, 68, 72, 131, 174, 259–260, 283, 338
biotopes 68, 117, 233
Black Band Disease 227
blast fishing 291
Bohuslän herring periods 169
Bølling-Allerød warm period 138
Boltzmann's factor 90–92
Brown's theory 413
Brunt-Vaisala frequency 72
Buffon's law 68
buoyancy 26–27, 40, 104
butterfly effect 421
butterfly's attractor 422
bycatch 292–293

calcareous ooze 129
calcite 133, 260, 263, 265–266, 275, 343, 385
Cambrian 124, 272, 362
cancer treatments 377
capacity adaptation 243
carbon isotope excursion 147–148

carbon sink 259
Cenozoic 120, 126, 128
chaotic systems 422
chemicals 25, 259, 291, 298–299, 302, 306–307, 317, 333, 376–377, 382
chemosynthesis 11, 65, 74–75, 327, 358
Chicxlub crater 149
Chromalveolata 9
chronic pollution 305
ciguatera 374–375
circumpolar vortex 29–30
cirrostratus 105
cirrus 24, 105
clathrate gun hypothesis 149
climate 9, 14–15, 19–60, 120–173, 174, 177, 188–9, 195, 201, 206, 209–258, 278–279, 324, 327, 332–333, 339–340, 342–343, 346, 355–356, 363, 369–374, 384–385, 390, 395–436
climatic instabilities 135–138
climatic variability 19–60, 109, 115, 120–173, 215, 230, 237, 252, 255–257, 340, 342, 395, 410–411, 420, 428
climatogram 31
climograph 64–65
cloudiness 48–49, 66, 101, 104–105, 227
coccolithophores 3, 10, 67, 73, 128, 160, 265, 268, 271, 278–279, 384–385
coccoliths 141, 265
coccosphere 265
cold seeps 65, 74, 358
cold spots 76
commensalism 116–117
compensation 243, 266, 348, 355, 357
compensation depth 155
compensatory growth 355
competition 81, 113, 143, 162, 170, 199, 220, 332, 348, 357, 377, 387, 397, 399, 403, 410, 421
competition hypothesis 81
competitive exclusion 111, 170, 249, 400, 416
complementarity effect 350, 357
complex adaptive system 408, 423
congruence 102, 107–108, 127–128
connectance 352–353, 359–360
conservatism in migration 288
conservative pollutants 307, 309
continental flood basalt eruptions 148
contingency 424
Continuous Plankton Recorder 94, 107, 114, 159, 235, 255, 279, 414

controlled freezing 194
convection 23, 26, 34, 37, 40–42, 47, 49–50,
    55, 59, 104, 106, 155, 196, 261, 329, 421
convergence 27, 29–30, 37, 39, 41, 49–50, 54,
    65, 101, 119, 138, 165
coral bleaching 121, 220–223
coral bleaching hotspot 223
coral reef 56, 74, 111, 116, 121, 127, 150, 152,
    165, 182, 184, 220–227, 233, 241–243, 260,
    267–268, 290–291, 323, 337, 340, 343, 366,
    368–369, 376–377, 385, 418
coral 14, 54, 67, 78, 113, 117, 119, 142,
    144–146, 149, 176, 220–227, 260, 264, 265,
    268, 270–271, 275, 277, 279, 295, 309, 312,
    321–322, 337–338, 340, 343–345, 369, 378,
    381, 434
Coriolis effect 37
Coriolis force 26, 28, 37–38, 40–41, 106
cosmic ray flux 123
countercurrent circulation 197
Cretaceous/Tertiary 128, 149
critical threshold 183, 193, 255–256, 423
cryophiles 199
cryoprotective substance 193–194
cumulus 26
cyanide fishing 291
cyanobacteria 3, 8, 123, 199, 268, 320–321,
    369, 380, 386–387
cyclomorphosis 180, 229
cyclone 14, 26, 33, 55–56, 111, 140, 150,
    152–154, 225, 369, 390

Dalton minimum 42
dangerous species 375
deep convection 47, 50, 55, 59, 155
deep-sea sediment habitats 74
defensive mutualism 116
degree-day model 191
delayed oscillator hypothesis 52
demes 231
denitrification 329, 334, 386
density dependence 103
desert 23, 32, 64–65
Devonian 124, 129, 143–144, 338
diatom ooze 390
diatoms 4, 11, 67, 107–108, 128–129, 133, 140,
    154, 158, 171, 238, 317, 374, 381, 384, 387,
    389, 414–415
diazotrophs 386
diel vertical migration 114, 201
Dilution-Recoupling Hypothesis 158–159

Dipole Mode Index 57
disinfectants 298
dispersal mutualism 116
dissolution 260, 266, 268, 274–275, 277, 282
dissolved inorganic compounds 259
dissolved oxygen 66, 149, 319
disturbance 11, 111–112, 122, 150, 152, 293,
    350–351, 365, 424
diversification 3–4, 109, 119, 127, 134, 362
Drake Passage 41, 56, 89, 125–126
drive netting 292
Dryas octopetala 138
dwarfism 395
dynamic fragility 350
dysoxia 149
dystrophic 325

eccentricity 129, 131–132, 149
ecogeographical pattern in biodiversity 84–119
ecological complexity 228, 362
ecological drift 86
ecological extinction 340
ecological niche 15, 72, 82, 86–87, 90, 97, 112,
    114, 119, 134, 162, 170, 234–235, 237,
    239–240, 249, 275, 355, 396–405, 408, 410,
    413, 415, 419, 420, 426, 428–435
ecome 62, 64–65, 68, 87, 98, 121, 134, 255,
    437
ecoregions 62–63, 68–69, 71, 222, 240, 265,
    315, 330, 390, 395, 415
ecosphere 1, 14, 19, 61–63, 65–66, 68–69,
    72–73, 77, 85, 102, 122–123, 174, 209, 212,
    220, 244, 260, 286, 357, 405, 408, 419,
    436–437
ecosystem engineer 117
Ecosystem Fisheries Based Management 342
ecosystem functioning 15, 239, 251, 346–365,
    395
ecosystem reliability 355
ecotourism 337
ectoparasite 94, 117, 300
ectotherm 79–80, 92–94, 175–177, 180–181,
    185, 188–189, 193, 195, 199, 206, 327
Ediacara biota 362
Eemian 131–132
Ekman spiral 37–38
Ekman transport 37–38, 173
El Niño 48–52, 57, 59, 165, 176, 215, 217,
    221–222, 224, 257, 289
El Niño Southern Oscillation 48–52, 165
elasticity 350

emergence 405, 408
emergence time 123, 174
empirical classifications of climates 31–32
end-Cretaceous mass extinction 149–150
endoparasite 117
endosomatic energy 65, 75, 101, 107, 161, 251, 297, 399
endotherm 79–80, 82, 92, 175, 195–196, 198, 203
end-Permian extinction 120, 145–149, 362
energetic equivalence rule 92
entanglement 302
entropy 100
environmental stability hypothesis 109–110
epeiric seas 126
epifauna 11, 295
equatorial fronts 32
eruptions 20, 111, 121, 129, 147–148, 214
ethers 377, 379
eukaryotes 4–5, 8–9, 123, 187, 362
euphotic zone 65, 75, 78, 106, 155, 158, 200, 385
euryecious 235, 397
euryecy 243
eurygraph 73
eustatic 124, 126, 144, 149, 173
euxinia 149
evolution 14–15, 61, 66, 84, 94, 113, 116, 174, 209, 219, 227–228, 230, 232, 234, 400, 433
exergy 100
exobase 20–21
exosomatic energy 101, 103, 106–107
exposure 240–242
extinction 14–15, 85, 94, 98, 120–121, 126–127, 129–130, 134, 142–150, 211–212, 219, 232, 235, 239–240, 293, 338–341, 362, 371, 403, 424, 432
extremophile 8
Exxon Valdez 303–304

facilitation 220, 361, 399, 421
faint young sun paradox 121
Ferrel cell 27, 29–30, 65, 84
fertilisation 242, 269–270, 342, 386
fishing 48, 165, 170–171, 246, 284–295, 302, 338–339, 341–342, 360–361, 367, 371–372, 395, 432
fitness 116, 170, 220, 227–228, 234, 303, 354–355
food supplement 376–377
food web ecology 360

foraminifers 3, 11, 67, 76, 84, 138, 260, 268, 278–280, 326, 385, 408–409
Foster's pattern 83
founder effect 229
frustule 317, 387–388
fugacity 263
Fukushima Daiichi nuclear accident 313–315
fundamental niche 399–403, 421, 426, 428

Gaia theory 174
gas and fat embolic syndrome 316
Gause's principle of competitive exclusion 400
Geist's pattern 83
genetic classification of climates 32–33
genetic drift 85
genetic polymorphism 230
geographical speciation 234
ghost fishing 302
giant conveyor belt 40
Gibbs free energy 101
gigantism 79
Gleissberg sunspot envelope-modulation cycle 43
global oceanic circulation 34, 36
Gondwana 124, 144,
Gran effect 155
grazing control hypothesis 67
greenhouse climate 126, 129
greenhouse effect 15, 22–24, 105, 145, 148–149, 212
gross photosynthesis 155
growth rates 184–186, 291, 375
Gulf Stream 34–35, 46, 73, 104

habitat 1, 11, 61–62, 66–67, 69, 74, 78, 80, 106, 108, 111–112, 116–117, 119, 128–129, 134, 165, 182, 194, 199, 203, 220, 230, 232, 235, 240, 245, 283, 293, 297–298, 319, 331–332, 338–341, 346, 356, 358, 362, 366–367, 369–370, 372, 397, 403–404, 410, 427–428, 432–433, 435
habitat heterogeneity 112, 356, 358, 428, 433
Hadley cell 27, 29–31, 45, 50, 65
Hadley circulation 49
hadobenthic zone 68, 74
Hale cycle 43
half-life 312
halocline 66
halogenated hydrocarbons 306–309
halophiles 8
haplotype 233

harmful algal blooms 283, 319, 374
heat conservation hypothesis 80
heat shock protein 187, 193
heat storage capacity 33
heat transportation 35
heatwaves 14, 121, 154, 219, 237
heavy metal 309–310, 312
Heinrich events 136
hemispheric see-saw effect 135
Henry's law 259
heterotherms 175
heterothermy 175
High Nutrient Low Chlorophyll 67, 73
high-latitude ecosystems 162, 251, 267, 269
high-magnesium calcite 260, 275, 343
historicity 424
history 13–14, 61–62, 67, 69, 89–90, 95, 120,
   123, 134, 141–142, 174, 186, 227, 233, 235,
   240, 242–243, 289–290, 332, 338, 355, 357,
   365, 395
Holocene megafauna extinction 338
holothurin 377
homeostasis 25, 109–110, 175, 197, 211, 350,
   399
homeotherm 79–80, 176
homeoviscous adaptation 188
homozoic zones 72
hot conditions 154
hot spots 76, 255
huddling 203–204
hull fouling 330
Huronian glaciations 123
hydrogen bonds 187–188
hydrophobic effects 187
hydrothermal vent ecosystems 76
hydrothermal vents 13, 65, 74–75, 358
hypercapnia 266
hyperthermic stress 187, 195, 219, 221
hyperthermophile 8, 123, 174
hypertrophic 325
hypothermia 303
hypothermic stress 193, 196, 203
hypoxia 13, 15, 266, 283, 298, 320–321,
   323–327, 329, 340, 372
hysteresis 136–137, 284, 416

Ideal Gas Law 26, 100
identity effect 357–358, 363
idiosyncratic hypothesis 348–349
immunocompetence 300
immunoenhancement 312

immunosuppressant 377
immunosuppression 312, 345
imposex 308
incoming shortwave radiation 23
incoming solar energy 25, 99, 101
incoming solar radiation 21–24, 33, 42, 103,
   131, 155
Indian Ocean Dipole 57
inertial homeothermy 175, 180, 194
infauna 11, 146, 295
inquilinism 117
insecticides 307
instantaneous growth rates 185
insurance effect 355
interactive effects 15, 209, 340–342, 397
Inter-Decadal Pacific Oscillation 52–54
inter-oceanic invasions 332
intersex 308
interspecific competition 81, 113, 170, 403
Intertropical Convergence Zone 29–30, 41, 50,
   54, 65, 101, 138
intraspecific competition 81, 113
invasion 15, 283, 300, 329–333, 357, 373, 428,
   433
ionic interactions 187
ionosphere 20–21
IPCC Assessment Report 5 219
iron hypothesis 67
irradiance 44, 72, 103, 142, 155, 158–159, 257
isobars 100
isopycns 100
isotherms 72, 100, 199

James's rule 80
Jordan's pattern 82

K/T boundary 128, 149
Kelvin waves 52
keystone hypothesis 348–349
Kleiber's Law 90
kleptothermy 203
knolls 74, 77
krill 106, 114, 165–167, 203, 343
Kuroshio 34–35, 38, 54, 104, 313

La Niña 49, 51–52, 57, 140, 217
Labrador Sea 40, 46–47, 59, 136, 332
lactones 377, 380
Late Devonian mass extinction 144
Late Ordovician mass extinction 144
Late Permian mass extinction 145–149

Late Triassic mass extinction 145
latent heat 23, 34–35, 104–107, 155, 157, 176
Laurasia 124, 144
Law of Conduction 176–177
Law of Fick 176, 180
law of minimum 397
Law of Q10 180
law of requisite variety 403
Law of Tolerance 182, 397
Law of Van t'Hoff 176, 180
Lecithotrophic larvae 82
lethal temperature 181, 196, 198–199, 411–12
Liebig's law 109
Lilliput effect 338
local/global stability 350
long wave radiation 23, 50, 103–104, 106–107,
    157
Lorenz's attractor 422
luciferase 78
luciferin 78
Lyell's theory 125

macroalgae 78, 119, 359, 362, 366, 369, 381
macroscope 405, 436
Madden-Julian oscillation 54–55, 155
mangroves 74, 121, 152, 297, 304, 313, 337,
    340, 366, 369, 379, 434
marine snows 387
mass extinction 126–127, 129, 142–149, 362
match/mismatch hypothesis 249, 288
maternal effect 288
maximum aerobic metabolism 184
Meridional Overturning Circulation 25, 40, 47,
    56, 59, 138
meroplankton 66, 82, 230–231, 316
mesopause 20–21
mesophile 8, 174
mesosphere 20–21
mesotrophic 325
Mesozoic 120, 124, 126, 149
metabiosis 117
metabolic theory of ecology 90–96, 185, 361
metals 307, 309–312
metapopulations 289
meteo-oceanic indices 59
meteors 21
microclimates 66, 134
micronutrients 67, 102, 173, 317
Mid-Domain Effect 85–87, 409
migration ability hypothesis 81
Milankovitch 121, 131–132, 136

Milankovitch Theory 129
minimum of Maunder 42, 44
minimum of Sporer 43
minimum viable population 409
mitochondrial DNA 134, 231, 233
Mixed Layer Depth 66, 72, 108, 155, 238, 411
monsoon 30–32, 55, 72, 138, 154, 327
Montreal Protocol 212, 334
Moran's theorem 162
more individuals hypothesis 362
mutualism 98, 116–117, 348, 410

nearest living relatives 426
Neogene 123
neritic regions 68, 74
neuston 303
neutral theory of diversity and biogeography
    85–86
neutrosphere 20
niche conservatism 235, 426
niche-assembly theory 97–98
nimbostratus 105
nitrifying bacteria 387
nitrogen cycle 386
noise pollution 316
North Atlantic Deep Water 40, 47, 136
North Atlantic Oscillation 45–47, 140,
    161–162, 170, 235, 257
North Atlantic spring bloom 155–160, 162
North Pacific decadal variability 52
North Pacific index 53–54
Northern Annular Mode 45
nutricline 384
nutrient enrichment 102, 317, 321, 323–325,
    359

obliquity 129, 131, 149
oceanic hydrosphere 14, 33–41, 65, 217
oil pollution 303–305
oligotrophic 9, 79, 158, 161, 324, 337, 357
Ordovician 123, 129, 143–144
orogeny 124
Orton's pattern 161
outcoming long wave radiation 23
overexploitation 227, 283–284, 291, 339–340,
    346, 362
oxygen depletion 321, 323, 325–327, 329, 371
oxygen minimum zones 326–327
oxygenic photosynthesis 8, 123, 327
ozone layer 20, 212, 333
ozone-depleting substances 334

Pacific Decadal Oscillation 53–54, 59, 419
Pacific/North American Index 53–54
Palaeocene-Eocene Thermal Maximum 148, 218
Palaeozoic corals 145
paleobiosphere 120
paleoecosphere 120
Paleotethys Ocean 147
paleothermometry 144
Pangaea 124–125, 145, 147
Pangaea Ultima 125
panniculus 196
Panthalassic Ocean 124, 147
paradox of enrichment 358
paradox of nutrient enrichment 102
paralytic shellfish toxins 320, 376
parapatric speciation 233
parasite 94, 116–118, 298, 300, 352, 397
parasitism 116–117, 119, 159, 189, 199, 220, 243, 399
parasitoidism 117
Particulate Organic Carbon 73–76, 161
passive competition 113
peptides 377, 380
periostracum 274
Permian 120, 124, 127, 129, 143–145, 147–149, 329, 362
Permian crisis 145–149, 329
persistence 219, 304–305, 348, 352
persistent organic pollutants 305–309
phagocytosis 312
Phanerozoic 120, 123–124, 127, 130, 143, 145, 338
phenology 161, 191, 228, 237, 249, 252, 395, 411–414
phenols 377, 380
phenotypic plasticity 227–230, 237, 240, 405, 414
phoresy 117, 237
photosphere 23, 42
photosynthetically active radiation 66, 102, 105, 107, 155, 173, 412, 420
phylogenetic biome conservatism 426
physiological compensation 243
physiological heterochromy 243
physiological integration 182, 187
physiological races 193
phytoplankton respiration 155–156
picoeukaryotes 9
Pinatubo 20, 148, 257
planetary circulation 23

planetary vorticity 38–39
plankton 8–10, 14, 39, 59, 66, 69, 81, 102, 107–108, 116, 128, 138, 165, 171, 173, 199, 205, 230–232, 238, 240, 242, 244–247, 249, 257, 265, 278–280, 302, 304, 309, 316, 385, 397, 411, 434–435
planktotrophic larvae 82, 232
plantotrophic 102
plastic debris 302–303, 307
Pleistocene 90, 120, 133–134, 139–140, 234, 263, 280, 409
Pliocene 213, 218
pneumatophores 304
poikilotherms 175–177, 188
poikilothermy 175
Polar cell 27, 29–30, 45, 65
Polar Front 29, 32–33, 46, 80, 89, 107, 245, 413
Polar Frontal Zone 89
polar jet stream 29, 65, 138
polar stratospheric clouds 334
pollution 1, 11–12, 15, 221, 227, 283, 302–305, 309, 312, 314, 316, 332, 339, 342, 346, 367, 371–372, 374
polycyclic aromatic hydrocarbons 306
Portfolio effect 355
positive feedbacks 131, 149, 421
positive selection effect 357
precession of equinoxes 129, 131
precipitation 20, 23, 27, 30–32, 41, 44, 46–51, 54–56, 64, 66, 84, 99, 101, 105, 118, 138, 140, 161, 226, 244–245, 260, 265, 342–343, 367, 408
predation 81, 114–115, 117, 119, 150, 170, 189, 199, 220, 243, 312, 316, 348, 353, 360–362, 397, 399, 402, 421
predator pit loop 170
predator specialization 114
predator-prey interaction 79, 114, 206, 249, 251
prey-to-predator loop 170
principle of competitive exclusion 170, 249, 400, 416
principle of planetary energy balance 23
production theory 86, 101–102, 109
protective organic layer 274
protein denaturation 181
Proterozoic 120, 123–124
province 8, 37, 62–63, 65, 67–69, 72–74, 76, 129, 148, 155, 428, 437
pseudohermaphroditism 308
pteropod 10, 67, 171, 260, 268–269, 279, 408
pycnocline 66, 155, 158

quaternary 120, 129, 131, 147, 338, 426

radioactive substances 312–315
radiolarian ooze 390
Rapoport's pattern 82, 86, 396, 410
recharge oscillator hypothesis 50
reductionist approach 405, 436
redundant species hypothesis 348–349, 364
regeneration niche 402–404
Reid's paradox 246
relative vorticity 39
reproduction 82, 103, 109, 119, 161, 180, 182,
    189, 191, 193, 219, 221, 227–228, 242, 266,
    269–271, 282, 293, 305, 316, 336, 341–342,
    367, 397–398, 402–403, 405, 414
resilience 109, 287, 342, 348, 350–352, 359,
    367, 371, 412–413
resistance 81, 109, 193, 195–196, 221, 230,
    235, 243, 282, 299, 348, 350, 352, 367,
    371–373, 376, 397, 411–413
resistance adaptation 243
resonance 154, 190–191, 227, 255–256, 377
resource availability hypothesis 81
resting metabolic rate 175, 184
rete mirabile 175, 194
rivet hypothesis 348–349
robustness 350
rocky shores 66, 111–115, 152–154, 237, 275,
    304, 358
Rodinia 124
Rossby 29, 45, 52, 72
Rossby waves 45, 52
runaway climate change 149
Russell's cycle 169, 171

salinity 34, 40–41, 47, 67, 72–74, 101, 108,
    115, 136, 138, 165, 167, 195, 221, 261,
    326–327, 431–432
salp 165, 271
sampling effect 357, 363
saturation horizon 260–261, 268
saturation state 124, 259–261, 263, 269
school trap mechanism 170
Schwabe's cycle 42
seagrass 74, 331, 366, 368–369
seamounts 5, 69, 77–78, 268, 295
seasonal acclimatization 192–194
seed dispersal mutualism 116
sensible heat 22–23, 35, 104–106, 155
sensitivity 41, 124, 136, 161, 181, 187–188,
    191, 193–194, 199, 218, 221, 226, 238,

240–243, 245, 263, 266, 274, 288–289,
    304–305, 312, 323, 342, 345, 402, 413, 418,
    421–422
serial depletion 289
shark finning 292
shear vorticity 39
shelf-shading 102, 113
siliceous ooze 129, 390
silicon cycle 387–390
sixth extinction 338
skeletogenesis 269–270, 282
Slope-Abyss Source-Sink hypothesis 102
snowball earth hypothesis 121
soft-sediment ecosystems 75–76, 102, 161
solar cycle 42–43
source-sink dynamics model 399
Southern Annular Mode 56
Southern Oscillation 48–54, 165, 419
speciation 84–86, 89, 94, 98, 113–114, 121,
    125, 134, 141–142, 232–234, 338, 409–410
speciation rates 113, 134, 409
species assembly models 85
species introduction 2, 329–333
species packing 98
species-assembly theory 98
species-energy theory 99, 101
specific heat capacity 33, 176
spring bloom 67, 72, 106, 155, 159–162, 387
squalamine 383
starvation resistance hypothesis 81
statistical averaging effect 355
Stefan-Boltzmann's law 104, 212
stenoecious 160, 235, 397
stenoecy 243
stenograph 73, 340
stenolaemate bryozoans 145
stenotherm 33, 76, 199, 219, 240–241,
    244–245, 408–410, 413, 415–416, 418
steroids 149, 377
stochastic resonance 227, 255–256
storage effect 287
stratiform cloud 105
stratopause 21
stratosphere 20, 23, 148, 333–335
stratospheric dehydration 334
stratospheric denitrification 334
stratospheric ozone concentration 148, 212
stratospheric ozone layer 212, 333
stratus 24, 105, 154
subcalicoblastic space 268
sulphur cycle 390

sunspots 42–44
supercontinent cycle 124
supersaturation 260, 264–265
surface energy balance equation 106
surface currents 34, 37–38, 66, 73
Surface Law 91, 176–177
surface oceanic circulation 34, 37
Sverdrup's critical depth 155, 157–158
Sverdrup's model 155
symbiosis 116, 145, 220–221
sympatric speciation 113, 234
synergisms 340

taxonomic biogeography 72
tectonic frequency band 124
tectonics 25, 121, 124–125, 173
tegumental circulation 198
tegumental vasoconstriction 197
teleplanic larvae 231
temperature-dependent sex determination 191
temperature-size rule 79–80, 180, 185, 206, 251,
    282
tens rule 329
terminations 131, 263
terpenoids 377, 381
Tethys Ocean 146–147
Tetracycline 299
the equilibrium/nonequilibrium hypothesis 90
theory of alternative stable states 416
theory of continental drift 1, 124
theory of ergoclines 106
theory of Island biogeography 85, 87, 229
thermal conductivity 176, 196
thermal denaturation 187
thermal independence hypothesis 80
thermal niche 199, 219, 229, 232, 235–236,
    242, 275, 316, 408–416, 418
thermal pollution 316
thermocline 34, 49–52, 56, 59–60, 66, 343
thermogenesis 175, 196, 203
thermohaline ocean circulation 40
thermophysiological method 191
thermosphere 20–21
thermostasis 175
thermotaxis 199–201, 203
Thorson's pattern 82
three-cell model 26, 29
Toarcian 148
tourism 15, 337, 369, 372
toxic metals 309–310, 312
trade winds 27, 46, 48–50

transgenerational effects 312
translocation experiments 229
trench 65, 69
Triassic 124, 128–129, 143, 145–146, 149, 261
tributyltin 306, 308
trilobites 145, 383,
trophic amplification 248
trophic cascades 247, 255, 284, 353, 372, 395,
    419, 421
trophic complexity 360
trophic mutualism 117
tropical rainforest 32, 64
tropopause 20, 26–27, 45–46
troposphere 19–20, 23, 34, 54, 103–104, 153, 334
tropospheric vertical shear 153
tundra 24, 32, 64–65, 138

undersaturation 260, 263, 265, 269
uniformitarianism 426
uropygiols 196

van der Waals' force 187–188
Van't Hoff-Arrhenius relation 90
vasoconstriction 197
vasodilatation 197
venomous species 375
vernal stratification 155
vicariance 90
virus 5, 7–9, 159, 298, 323, 378, 382
Vital Dust hypothesis 174
volcanism 31, 144–146
vorticity 38–39
vulnerability 15, 240, 316, 323–324, 340–341,
    360

Walker's circulation 48–49, 52
water accommodated fraction 305
water vapour 19, 24, 27, 195
weather regime 60
White Band Disease 227
White Pox 227
White Syndromes 227
Wien's displacement law 22
Wilson's cycle 121

xenobiotic substances 307
xenophyophores 78–79

yellow band disease 321
Younger Dryas 138–140, 257

zooxanthellae 117, 145, 220, 309, 343